Protein Folding

MOLECULAR BIOLOGY

An International Series of Monographs and Textbooks

Editors: BERNARD HORECKER, NATHAN O. KAPLAN, JULIUS MARMUR, AND HAROLD A. SCHERAGA

A complete list of titles in this series appears at the end of this volume.

Protein Folding

CHARIS GHÉLIS
JEANNINE YON

Laboratoire d'Enzymologie Physiocochimique et Moleculaire
Centre National de la Recherche Scientifique
Université de Paris-Sud
Orsay, France

1982

ACADEMIC PRESS

A Subsidiary of Harcourt Brace Jovanovich, Publishers

New York London
Paris San Diego San Francisco São Paulo Sydney Tokyo Toronto

ACADEMIC PRESS, INC.
111 Fifth Avenue, New York, New York 10003

United Kingdom Edition published by
ACADEMIC PRESS, INC. (LONDON) LTD.
24/28 Oval Road, London NW1 7DX

Library of Congress Cataloging in Publication Data

Ghélis, Charis.
Protein folding.

(Molecular biology series)
Bibliography: p.
Includes index.
1. Proteins. 2. Biochemorphology. 3. Molecular structure. I. Yon, Jeannine, Date. II. Title. III. Series.
QP551.G48 547.7'5 82-6830
ISBN 0-12-281520-3 AACR2

PRINTED IN THE UNITED STATES OF AMERICA

82 83 84 85 9 8 7 6 5 4 3 2 1

Contents

I CONSIDERATIONS OF PROTEIN FOLDING DEDUCED FROM CHARACTERISTICS OF FOLDED PROTEINS

2 Structural Characteristics of Folded Proteins

3 Energetics of Protein Conformation: Conditions Restricting the Allowed Conformations

4 Theoretical Approach to Protein Folding

II EXPERIMENTAL APPROACHES

5 Simulation of Protein Folding: Studies of *in Vitro* Denaturation–Renaturation

6 Physicochemical Studies of the Unfolding–Folding Equilibrium

7 Kinetic Studies of Unfolding and Folding Processes

8 Detection and Characterization of Intermediates

III CONCLUDING COMMENTS

12 Mechanisms of Protein Folding

Preface

The objective of this book is to reassemble the most important information in the field of protein folding and to understand, from the amount of experimental data now available, the main principles that govern formation of the three-dimensional structure of a protein from a nascent polypeptide chain, and how the functional properties appear.

The interest in this field has increased considerably over the years and information has reached a level that may allow certain generalizations. A corollary difficulty is encountered: New pieces of information are appearing continually, giving us the uncomfortable impression of progressing on shifting sand, but also the exciting feeling of moving in an important area of interest.

The problem of protein folding has been questioned for a long time. Before World War II Anson and Mirsky (1934, 1935), having shown the reversibility of the denaturation process in the case of hemoglobin, elucidated some of its characteristics. These first studies were followed by many works on protein denaturation. Progressively, the increased knowledge on protein structure allowed more precise and more significant work. One of the most important and stimulating contributions of modern biochemistry is undoubtedly the elucidation of the three-dimensional structure of biological macromolecules. Certainly, the increasing number of protein structures obtained by high resolution X-ray crystallography represents important information to understand the mechanisms of protein folding and the genesis of a biological function.

The book is divided into three parts, which follow an introductory chapter where the main problems of protein folding are presented and discussed at the cellular level. The conformation of native globular proteins is described in Part I. Definitions and rules of nomenclature are given in Chapter 2, including characteristics of structural organization of globular proteins deduced from X-ray crystallographic data. Folding mechanisms were tentatively deduced from the observation of invariants in the architecture of folded proteins. Energetics of protein structure indicating principles of thermodynamic stability of the native structure are found in Chapter 3. In Chapter 4, theoretical computation studies of protein folding, structure prediction, and folding simulation are rapidly reviewed.

Part II is a presentation of various experimental approaches. Reversibility of the unfolding–folding process is discussed in Chapter 5. Chapters 6 and 7 contain equilibrium studies and kinetic studies, respectively. The different ways used to detect and characterize intermediates in protein folding are reported in Chapter 8, omitting immunochemical approaches which are analyzed in a separate chapter (Chapter 9). Folding and assembly of smaller units into a protein are examined in Chapter 10, and Chapter 11 treats problems specific to oligomeric proteins. Some generalizations are made in the last chapter which is certainly not an exhaustive review, but contains information that seemed to us to be most significant.

This book was written for research scientists to contain in a unique volume information scattered throughout the literature. It was planned as an advanced survey on protein folding, not only for specialists but for biochemists in general. It was also written for students of biochemistry and biology, to present more advanced knowledge in protein structure and folding and to propose some basis for discussion.

And last, as said in the Tao Te King, "The man who speaks does not know, the man who knows does not speak"; we can add, "The man who writes, tries to understand." We welcome comments from the readers to know if this goal has been reached.

Charis Ghélis

Jeannine Yon

Acknowledgments

It is a pleasure to acknowledge Dr. George Némethy for his help and advice during preparation of this book and for carefully reading and criticizing the complete manuscript. We sincerely thank Drs. S. Bernhard, E. Bricas, J-R Garel, M. Goldberg, G. Hervé, J. Janin, M. Karplus, J-C. Patte, D. Perahia, M. Rossmann, and A. Shechter for reading and discussing separate chapters or for stimulating discussions during preparation of the manuscript. We also thank for very helpful suggestions and discussions Drs. R. L. Baldwin and M. Karplus.

We are indebted to Drs. N. Kellershohn and D. Vergé, and also to M. Desmadril, M. Laurent, and M. Tempête-Gaillourdet for their help in carefully reading particular chapters. We want to acknowledge all the authors who sent us original figures, documents and preprints of their manuscripts.

We are indebted to Mrs. Barthélémy and Mrs. Lavorel for typing and retyping the manuscript, Mrs. Clais for providing documentation and for their help during the material preparation of the manuscript.

Introduction: Cellular Environment and Significance of Folding Processes

Dans le processus de structuration d'une protéine globulaire, on peut voir à la fois l'image microscopique et la source du développement epigénétique autonome de l'organisme lui-méme.

J. Monod, "Le Hasard et La Necessité"

1

Folding and Processing: The Last Events in Protein Biosynthesis

1.1 EVOLUTION OF THE MAIN CONCEPTS OF PROTEIN FOLDING

With reference to the tremendous progress in molecular biology this last decade, it might seem that all mechanisms by which genetic information is transmitted have been elucidated. It might also appear that all events involved in the formation of a biologically active protein are well known. In fact, only duplication, transcription, and translation processes which contribute to the formation of a linear polypeptide chain with a known sequence of amino acids are rather well understood. But posttranslational (or cotranslational) events, i.e., the ones that generate globular structure and biological function, are not yet fully elucidated. It is possible to distinguish two different kinds of events:

(a) Covalent processes including limited proteolysis and chemical modifications such as glycosylation, phosphorylation, hydroxylation, lipidation, methylation, ADP-ribosylation, and so on;
(b) Noncovalent processes such as polypeptide chain folding and subunit assembly.

Figure 1.1 summarizes the different events possibly involved in producing an active protein. They are of importance in generating functional properties of proteins, recognition of specific ligands, catalysis, active transport,

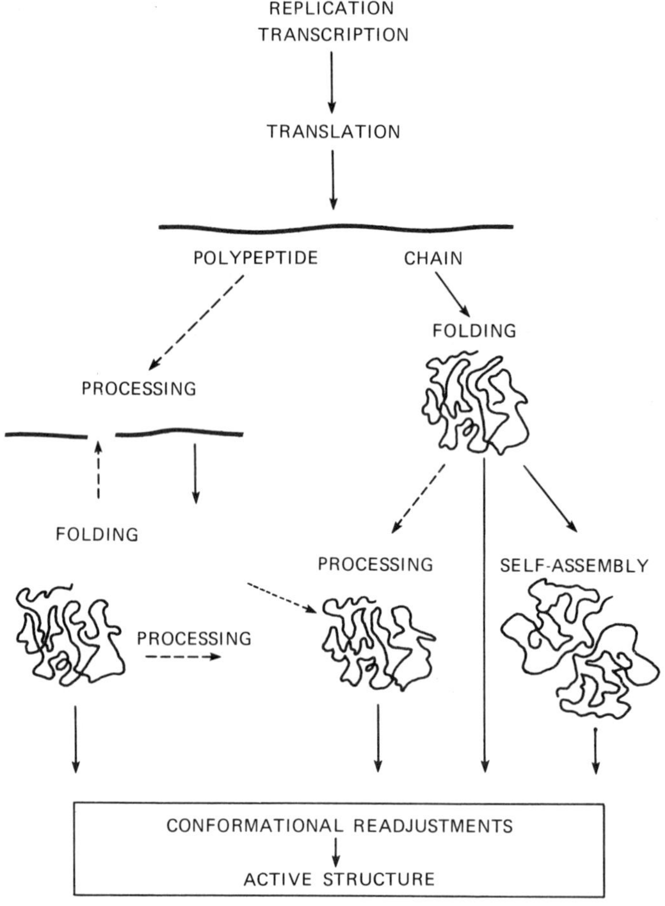

Fig. 1.1. Posttranslational and cotranslational processes.

and hormonal action. Furthermore, the biological function of a protein appears in specific loci in cells and higher organisms, thus allowing very precise and subtle regulations which insure the preservation and function of living organisms.

This chapter is an introduction to the main problems of protein folding. The term protein folding includes all the events occurring from translation to formation of the functional structure of a protein molecule. These problems must be considered at the cellular level taking into account the particular environment, the effect of compartmentalization (Palade, 1975), the sequence of the different processes, with reference to their biological meaning, and the regulatory processes by which functional structure takes place. However, because of the difficulties encountered when attempting to study protein folding accurately under cellular conditions, it is necessary to shift from the cellular to the molecular aspect of these problems. Until now, mechanisms of protein folding were studied on molecules isolated from their biological environment. A rapid historical survey of the evolution of the main concepts is presented.

The processes involved in the formation of the "native structure" of proteins were questioned for a long time before the mechanism of protein biosynthesis was known, and before the three dimensional structure and even the primary structure of proteins were determined. With increasing knowledge in both fields, protein structure and protein biosynthesis, questions concerning protein folding can be asked more precisely. More significant experimental approaches of the problem can now be attempted. Considering the frequency and importance of the papers which appeared in the literature on this topic, one notes three maxima with a periodicity of about 20 years, approximately located around 1935, 1955, and 1975. Indeed, with a certain lag, each of these periods corresponds to a specific discovery. The first significant works appeared around 1935 with the analysis of the denaturation process by Wu (1931), with the experiments of Anson and Mirsky (1934a,b 1935), and those of Northrop (1932). These first works clearly outlined the correlation between the biological activity of a protein and the so called "native structure," defined not only by the activity but also by some physical characteristics (solubility, ability to crystallize, hydrodynamic properties), and later by chemical properties. In fact, very little was known at this time about proteins. This period was principally marked by success in protein crystallization; one has to remember that such an event ended the old arguments of vitalists concerning the mysterious nature of enzymes. The decrease of solubility and the loss of activity were the first parameters used as conformational probes at the beginning of the work on protein denaturation. Thus, the first studies on denaturation remained only qualitative approaches. However, at this period the reversibility of the denaturation–renaturation process was clearly shown but only for few proteins, hemoglobin, chymotrypsinogen, trypsinogen. This represents the first important fact concerning the understanding of protein folding.

In the next 20 years, tremendous progress was made in the study of the physical chemistry of macromolecules and the application of new methodology to the study of protein conformation was extensively developed. A better resolution of optical techniques (UV spectroscopy, spectropolarimetry, etc.) allowed their utilization in following conformational changes of proteins. This period was also marked by important developments in protein chemistry and the first successes in determination of protein sequence with the insulin molecule (Sanger, 1956), followed 4 years later by the sequence of ribonuclease (Hirs *et al.* 1960). At the same time, studies of the noncovalent forces involved in protein structure and the role of water in determining conformation appeared (Kauzmann, 1955; Klotz, 1958). Some attempts to study the thermodynamics of the denaturation–renaturation process were presented (Linderstrøm-Lang, 1950,1952; Pauling *et al.*, 1951; Lumry and Eyring, 1954; Kauzmann, 1959a,b). More quantitative and significant studies were realized by the development of physicochemical approaches.

The third period which spans the last few years is essentially the consequence of precise knowledge in the three-dimensional structure of proteins. The X-ray structure of an increasing number of proteins can now be obtained with high resolution allowing protein knowledge to reach the atomic level. This is probably the most important discovery which allows precision concerning concepts and ideas on protein folding. With the first studies on structural fluctuations of proteins, we have now reached a new and very promising period.

Even if crystallographic studies present only a static approach, nevertheless they provide an important piece of information for understanding how the architecture of a protein molecule could be built from the extended polypeptide chain. Crystallographic studies contribute significantly to improving the hypothesis concerning the ways of folding. Like others, the problem of protein folding has gained significantly by the evolution of the knowledge on protein structure. Quantitative studies at a higher level of precision are now possible. When the identification of each amino acid residue in the three-dimensional space is known by means of X-ray structure, it becomes possible and significant to follow the behavior of some of them located at different positions in the polypeptide chain during the folding process. Of course, kinetic approaches are necessary to elucidate the existence and the number of intermediate states. In this regard, the use of rapid kinetic methods, such as flow and relaxation techniques, to study protein folding has brought and still brings very useful contributions; but they remain only phenomenological and are insufficient to determine the nature of these intermediates.

This very rapid survey of the evolution of the studies of protein folding could be treated in parallel with a survey of the evolution of the studies of protein function, mainly enzymatic activity. The same statistical distribution of significant works will be observed; the same events have marked both kinds of studies. The masterpiece of the evolution of works and ideas in both fields originated from knowledge of protein structure. Because a higher level of knowledge has been reached today, the main questions concerning protein folding can be asked more precisely than before.

1.2. STRUCTURE OF FOLDED PROTEINS IN THE BIOLOGICAL ENVIRONMENT

The success of X-ray crystallography in solving the three-dimensional structure of biological marcromolecules at atomic resolution has triggered the most important progress in molecular biology. But the mechanism by which a polypeptide chain folds to reach its native and functional structure remains an intriguing problem of fundamental interest. The knowledge of a large number of protein structures brings new insight to the study of mechanisms of protein folding. The structural organization of globular proteins displays common patterns arising from similar mechanisms of folding and substructure assembly. However, under the influence of thermal motion, protein molecules fluctuate between different more or less stable conformational states, some of which can be stabilized on binding of specific ligands. Functional and regulatory properties probably arise from this conformational flexibility and fluctuations of protein structure. Stability of the macroscopic state of protein structure is strongly dependent on the environment, pH, ionic strength, and solvent; solvent contributes significantly to the energy of stabilization of protein conformation.

1.2.1. Structural Organization of Proteins

Three-dimensional structures of more than 60 globular proteins, known by X-ray crystallography at atomic resolution, indicate similarities in structural organization. Although a great number of conformations might be expected *a priori* for a polypeptide chain with a given sequence of amino acids, structural patterns which are found in globular proteins are restricted in number. Globular proteins contain segments of ordered structures (α helices and extended β structures) and segments of nonregular structures. These pieces of structure interact to form "building blocks," and the assembly of such elements generates the entire structural folded protein. Larger building blocks are organized into domains of a single polypeptide chain, i.e., covalently linked miniglobular proteins.

Many proteins, mainly intracellular proteins, are oligomeric, consisting of the assembly of identical or different subunits. Thus, the protein architecture displays a kind of hierarchy in which building blocks of different levels of complexity interact to produce the aesthetic design of a native protein and the efficient conformation of a biologically active molecule.

1.2.2. Role of Solvent and Environment in Stabilization of Protein Structure

Conformational stability of a protein molecule in a given environment depends not only on the amino acid sequence of the polypeptide chain, but on the environment. Local conditions existing inside each cell compartment, solvent, pH, ionic strength, viscosity, and concentration can modulate the conformational properties of a protein. The role of solvent is of particular importance for stabilization of protein structure. Therefore, globular proteins soluble in aqueous medium and membrane proteins are expected to have different conformational characteristics and must be considered separately.

1.2.2.1. Soluble Globular Proteins

The structure of a globular protein is determined by the amino acid sequence of the polypeptide chain in water solvent. The sequence represents a well-determined arrangement of polar (hydrophilic) and nonpolar (hydrophobic) amino acids. The proportion of nonpolar to polar amino acid side chains does not vary greatly from one protein to another, at least for small globular proteins. There is a thermodynamic tendency of nonpolar groups to avoid contact with solvent and of polar groups to be in contact with water. Kauzmann (1959) introduced the "oil drop" model to describe this situation and evaluated the thermodynamic properties of hydrophobic interactions of proteins in solution. In protein structure, a substantial number of the nonpolar side chains interact to form hydrophobic cores at the interior of the protein, leaving the polar groups mostly on the surface.

The major contributing factor to the stabilization of hydrophobic interactions is solvent entropy. The tendency of hydrophobic groups to aggregate inside the protein is responsible for the close packing of the atoms characteristic of globular proteins.

However, a significant proportion of hydrophobic groups lie close to the surface of protein molecules while some polar groups are buried in the interior. Indeed, part of polar amide and carbonyl groups of the polypeptide backbone are buried because of the folding of the polypeptide chain; they are buried without loss of energy by forming hydrogen bonds.

The stabilization of protein structure in water represents a delicate balance between conformational entropy, which tends to unfold the protein, and various stabilizing energy contributions such as hydrogen bonds, hydrophobic and van der Waals interactions, and electrostatic forces. The resulting free energy of stabilization is generally small compared with certain stabilizing and destabilizing contributions. Therefore, it clearly appears that fluctuations of the environment (temperature, pH, ionic strength, or ligand binding) might destabilize the protein until partial or complete unfolding, or on the contrary, might provide an extra stabilization of the structure.

1.2.2.2. Membrane Proteins

Membrane proteins are exposed to a multiphasic environment. There are different kind of membrane proteins (Fig. 1.2); because the extrinsic ones, only interacting with the membrane, are not very different from soluble proteins, we only consider intrinsic membrane proteins. Some of these are penetrating proteins, with a great part of the molecule localized outside the membrane and the remainder anchored in the lipid bilayer. A distinction was made between ectoproteins and endoproteins based on whether they protrude at the extracytoplasmic or cytoplasmic side of the membranes, respectively (Rothman and Lennard 1977; Roth and Lodish, 1977). Spanning proteins cross the membrane, with a part outside of the membrane, the other part on the other side and a great portion of the protein molecule interacting with lipids. Included proteins, such as proteolipids are totally inserted in the lipid bilayer and generally interact with the nonpolar part of lipids. Especially important is the heterogeneous environment, with part being aqueous solvent, another part comprising the electric charges of polar heads of phospholipids and the hydrophobic parts of the lipid inside the bilayer. Certainly the environment determines the structural organization of membrane proteins and their function of controlling active transport. The asymmetry of the membrane might play a role by stabilizing the final structure of a membrane protein. The structure of membrane proteins, however, is not very well known. The best documented system is bacteriorhodopsin with its seven helices crossing the membrane (Ovchinnikov *et al.*, 1979; Henderson and Unwin, 1975; Engelman *et al.*, 1980; Blake, 1980), and with the helices being quite parallel each with others. Helices are possibly the preferred structures for spanning proteins having a great part of their residues in interaction with lipids.

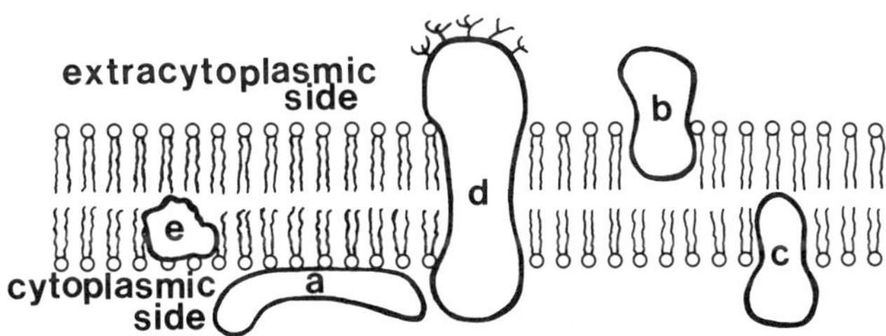

Fig. 1.2. Membrane proteins: extrinsic protein (a); intrinsic proteins: penetrating proteins [ectoprotein (b), endoprotein (c)], spanning protein (d), included protein (e).

Apart from this well-documented system, and perhaps one or two others, there is really insufficient information at this time to discuss folding and even the structure of intrinsic membrane proteins. It remains however an important question for the near future.

1.2.3. Conformational Fluctuations of Proteins

The conformation of a protein in its folded state fluctuates around its equilibrium position in a given environment, since it is subjected to thermal motion and also to fluctuations of the environment. X-Ray diffraction studies, based on a time-averaged structure of a molecule in the crystal provide a rather static picture of protein structure. However, even in their native state, proteins have significant internal motions which cover a large time range varying from picoseconds to seconds, according to the process (Linderstrøm-Lang, 1955; Carreri *et al.*, 1975, 1979; Cooper, 1976; Weber, 1975; McCammon *et al.*, 1976, 1977, 1979, Karplus and McCammon, 1979). Proteins oscillate rapidly between a great number of closely related conformational states. The biological importance of these conformational fluctuations in determining functional properties of proteins was strongly suggested and is emphasized today (Citri, 1973; Viratelle and Yon, 1973; Jencks, 1975; Yon, 1976; Huber, 1979a,b).

Several experimental techniques have have been used to study conformational flexibility in protein molecules. Spectroscopic methods such as NMR, ESR, Mossbauer, and fluorescence techniques provide direct and detailed studies of internal motions in proteins. Hydrogen–deuterium (or tritium) exchange methods are particularly convenient for kinetic studies. Immunochemistry may be also a very helpful method to study conformational fluctuations. These results are discussed in the second part of this volume.

A dynamic model of the protein molecules tends to supersede progressively the rigid view even for crystallographers. In fact, the recent development of refined methods which provide precise atomic coordinates, leads to interpretation of "atomic temperature factors" as structural motion. Frauenfelder and co-workers (1979) have performed X-ray diffraction studies in myoglobin at several temperatures and found mean square displacements larger than expected if they were entirely attributable to vibrations. Artemiuk *et al.* (1979) analyzed atomic displacement in the crystals of hen and human lysozyme and suggested the occurrence of intramolecular motions. Since the active site is located in a region of high displacement, the authors concluded that conformational dynamics play a critical role in enzymatic activity.

Molecular dynamic calculations (Gelin and Karplus, 1975; McCammon *et al.*, 1977; Karplus and McCammon, 1979) promise to provide the theoretical background for understanding motions in folded proteins. Time-dependent structural fluctuations of proteins in the neighborhood of their equilibrium conformation in their native state reveal a time constant of 1 picosecond for the decay of atom fluctuations and the occurrence of concerted motions. Different time fluctuations have to be considered in folded proteins: side-chain rotations, local motion involving the backbone, breathing motions, isomerization transitions corresponding to motions in different time scales. Ligand binding can hinder part of these conformational motions freezing the protein in a more rigid conformation.

The increasing number of papers overlooking the rigid view of protein structure reinforces the importance of conformational dynamics of proteins. After the success in solving the three-dimensional structure of biological molecules, the next step presumably will be the understanding of their structural dynamics and perhaps, by this way, we will understand the genesis of their functional properties.

1.3. POSTTRANSLATIONAL AND COTRANSLATIONAL PROCESSES

1.3.1. Polypeptide Chain Elongation and Protein Folding

The mechanisms by which proteins are synthesized in cells are now quite well explained. Information encoded in DNA is transcribed in messenger RNA, and then mRNA is translated on the ribosome machinery to give a polypeptide chain with a well-defined primary structure. The genetic code (three bases determining one amino acid) is given in Fig. 1.3; the universality of this code is generally accepted. One can remark that when U is the second base of the codon, the amino acid residue is nonpolar and when A is the second base, it is polar (see Dickerson and Geis, 1969).

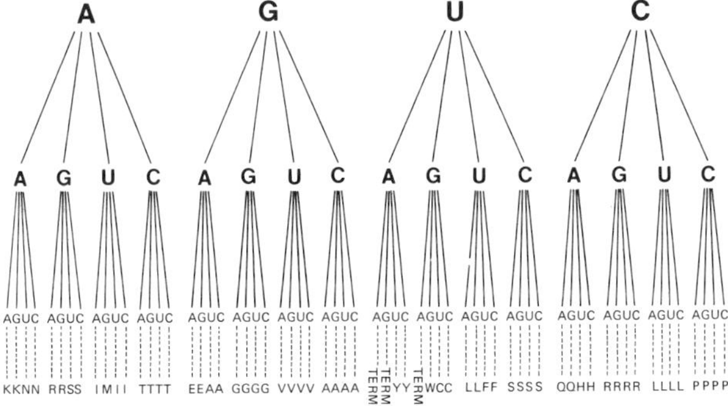

Fig. 1.3. The genetic code.

The translation proceeds by a stepwise sequential addition of amino acids to the growing chain initiated at the amino terminal end and terminated at the carboxyl end. The polarity of translation was clearly demonstrated by the elegant experiments of Dintzis (Dintzis, 1961; Naughton and Dintzis, 1962) on the biosynthesis of hemoglobin by rabbit reticulocytes preceded by the observations of Bishop *et al.* (1960) on the same system. The "polarity" of transcription was then shown for other proteins, bacterial amylase (Yoshida and Tobita, 1961), several proteins from *E. coli* (Goldstein and Brown, 1961), hen egg-white lysozyme (Canfield and Anfinsen, 1963a). Bergmann and Lodish (1979) have presented a quantitative kinetic model showing effect of initiation, elongation, and termination parameters on the rate of formation of a completed polypeptide chain. The model fits quantitatively for the synthesis of α and β globin synthesis in reticulocytes. The total rate of protein synthesis is also affected by a limiting amount of tRNA and by mRNA secondary structure.

Assuming the mechanisms which originate a polypeptide chain are known, nothing is really clear about the events which drive the chain to the three-dimensional structure of a protein on ribosomes in such a very short time. It is generally accepted that this process occurs spontaneously without any additional information. The reversibility of the *in vitro* unfolding–folding process, which can be considered to mimic folding of nascent proteins, although at a slower rate, has reinforced this idea. Crick (1958) proposed that "folding is simply a function of the order of the amino acids." According to this statement, protein folding, which is the formation of the "unique", three-dimensional, "native structure" is entirely governed by the amino acid sequence, in a definite environment.

Many proteins, mainly extracellular proteins, are cross-linked by disulfide bridges which provide the molecule with a greater stability to protect its structure against fluctuations of the environment. These are generally monomeric proteins. Intracellular proteins are devoid of disulfide bridges and most but not all, are oligomeric; staphylococcal nuclease, phage T4 lysozyme, which are monomeric proteins, have no disulfide bridges.

Apart from formation of disulfide bridges, and apart from assembly producing quaternary structure, two questions relevant to all proteins can be raised concerning the folding of nascent protein into ribosome:

(1) Does the folding process occur during the elongation of the polypeptide chain when still attached to the ribosome, or after its termination?

(2) Is the folding process a stepwise sequential process beginning from the amino terminal end and finishing at the carboxylate end, as is the elongation process, or not?

These two questions are not absolutely independent. Indeed, if folding occurs during the elongation process, beginning as soon as enough amino acids are present in the growing chain and continuing until termination of the chain, it is a sequential process. But even if a sequential mechanism was established, it would not lead to the conclusion that the growing chain folds when bound to the ribosome.

These questions must be examined with regard to local circumstances of biosynthesis. In cells, proteins are synthesized either by free polysomes or by rough endoplasmic reticulum (RER) membrane-bound polysomes. For nascent polypeptide chains, these two different situations have to be considered since, in the second case, compartmentalization leads to a vectorial transport and discharge through the membrane. Bound and free polysomes seem to derive from a common precursor pool of ribosomes. Several lines of evidence indicate that membrane-bound ribosomal subunits readily exchange *in vivo* and no major differences in the rRNA sequences and organization of free and bound ribosomes have been found (Fern and Garlik, 1976; see Shore and Tata, 1977).

1.3.2. Proteins Synthesized by Free Ribosomes

In this section, we only want to consider the simplest situation, i.e., protein synthesis by free polysomes. Figure 1.4 refers to this case and illustrates the alternative. The more complex process of protein synthesis by RER membrane-bound polysomes is considered in the next section. The first

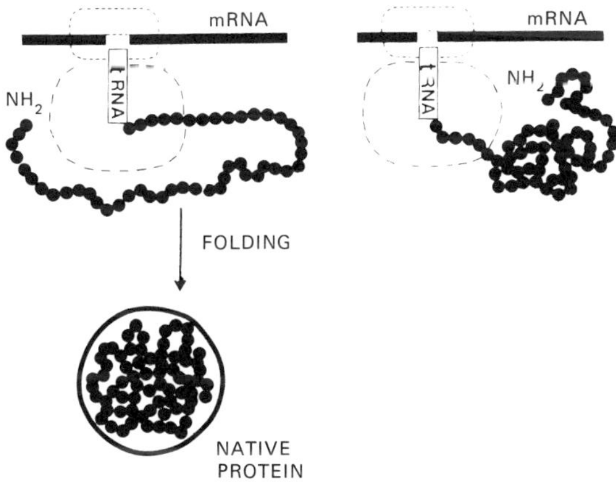

Fig. 1.4. Folding of a nascent polypeptide chain (adapted from Deal, 1969).

question has been discussed by different authors (Chantrenne, 1961; Epstein *et al.*, 1963; Deal, 1969; Wetlaufer and Ristow, 1973). Until now, this problem has received very little experimental support. On the one hand, some papers reported the existence of an enzymatically active form of β-galactosidase attached to the ribosomes (Zipser, 1963; Zipser and Perrin, 1963; Kiho and Rich, 1964). A quite different approach with the same enzyme was reported by Hamlin and Zabin (1972). These authors found immunochemical activity associated with growing chains of β-galactosidase. In these experiments antibodies prepared against the native enzyme were used to measure the binding of anti-β-galactosidase to incomplete chains. Thus, conformational determinants of the native protein were found. The authors concluded that their experiments "suggest that nascent polypeptides begin to assume a conformation that is, by virtue of cross-reaction with anti-β-galactosidase, similar to that attained by the completed, functional protein." By contrast, smaller proteins, such as staphylococcal nuclease and ribonuclease, are unable to fold correctly when a part of the C-terminal extremity is removed by proteolysis (Taniuchi and Anfinsen, 1969). This lead them to conclude that "almost the entire amino acid sequence of nuclease is essential to determine unique folding" (see Part II, this volume). The conclusions of the different approaches are not discussed in this introductory chapter. The intention is to outline what type of information is needed to answer the first question. Experimental studies of the problem require careful determination of conformational properties of the growing polypeptide chain during biosynthesis. Such a study is not easy; it can be performed in proteins whose *in vitro* biosynthesis is well controlled. The immunological activity of nascent fragments can be used to detect the appearance of conformational determinants identical to those of the native protein; however results obtained only by this method might give an ambiguous answer. Binding of antibodies to the nascent chain could induce the right conformation. Therefore, it is necessary to introduce at least a second method, the observation of another conformational parameter to follow the polypeptide folding. The conclusion would be significant only if the results obtained by at least these two different methods agreed. Because of the small quantities of material and the noise provided by the ribosome machinery, the study is certainly a challenge. Significant contributions to determine if protein folding is a stepwise sequential process may be provided by *in vitro* observations. Indeed, the question of a sequential folding beginning from the N-terminal end and finishing at the C-terminal was suggested by different authors (Chantrenne, 1961; Phillips, 1966, 1967; Dunhill, 1967). In the study of isolated molecules, it is possible to follow the behavior of the different parts of the polypeptide chain, and to evaluate their importance in the folding process. Even so, these considerations allow one to question the plausibility of the assumption but will not give a definite proof of the mechanism of protein folding into ribosome.

An additional problem arises for proteins cross-linked by disulfide bridges. How does the formation of S—S bonds proceed *in vivo*? How can one explain their high rate of formation in a cell compared to that observed when isolated in solution? What is responsible for this acceleration?

Another problem is related to subunit assembly in oligomeric proteins. It has been suggested that the formation of quaternary structure would be more efficient if "before release from the ribosome, the individual polypeptide chains were to assume a conformation capable of accepting the other subunits of the protein" (Hamlin and Zabin, 1972). Models were proposed by Deal (1969) to explain folding of another oligomeric enzyme, glyceraldehyde-3-phosphate deshydrogenase, and more precisely the requirement of NAD^+ to achieve the right folding of the molecule (see part II, Chapter 11). However very little experimental evidence exists at this time and the problem of sub-unit assembly in biological environment remains unclear.

Certainly the problem of protein folding *in vivo* remains difficult to evaluate experimentally. It is possible to ask to what extent the *in vitro* studies of totally unfolded proteins is a valuable model leading to an understanding of the mechanism of formation of the tertiary structure of a nascent polypeptide chain. Taking into account the importance of the environmental conditions on the protein conformation, it seems realistic to ask if the ribosome matrix is not able to induce special and well-defined ways of folding. However, on the basis of the ability of a nascent polypeptide chain longer than about 30 amino acids at least to be attacked by proteases, it seems that only the C-terminal part, still attached to the ribosome, is shielded from the aqueous environment. This finding (Malkin and Rich, 1967) was considered as an argument to support the possibility of folding for a nascent incomplete polypeptide chain (Hamlin and Zabin, 1972). It justifies *in vitro* studies of renaturation as a simulation of protein folding in aqueous solution.

1.3.3. Proteins Synthesized by Bound Ribosomes: Preproteins and Signal Peptides

Two distinct classes of proteins are synthesized by membrane-bound polysomes in cells which have a well-defined rough endoplasmic reticulum. They are secreted proteins which are destined for export or for transport to other cellular organelles, and membrane intrinsic proteins which have to be inserted asymetrically toward the noncytoplasmic face of the membrane. The role of the RER for the synthesis of secretory proteins is well established (Palade, 1975). For these proteins, compartmentalization regulates the different processes. They must be transferred across the hydrophobic bilayers of membranes to reach their final localization. Transfer normally occurs during biosynthesis. It seems however that under special physiological conditions, synthesis of proteins which normally occur on bound ribosomes can be made on free ribosomes. An example is provided by albumin from hepatoma (Uenoyama and Ono, 1972). There is no absolute requirement for a specific mRNA to be translated by bound ribosome (for example, see Lowe and Halliman, 1973).

1.3.3.1. Compartmentalization of Protein Synthesis

When synthesis occurs in bound ribosomes, the result is the compartmentalization or segregation of the newly synthesized proteins. It was shown in this case that the polypeptide chain is largely inaccessible to proteases. When proteolysis is performed while ribosomes are attached to membranes (Blobel and Dobberstein, 1975a,b), polypeptide chains remain buried within the membrane. The first direct evidence that secretory proteins are synthesized by membrane-bound ribosomes was obtained from labeling experiments (Siekevitz and Palade, 1960). This aspect was discussed in detail in the reviews published by Palade (1975) and by Shore and Tata (1977). Several secretory proteins have been studied, including immunoglobulins, insulin, α lactalbumin, mellitin from honey bee venom, parathyroid hormone, pancreatic

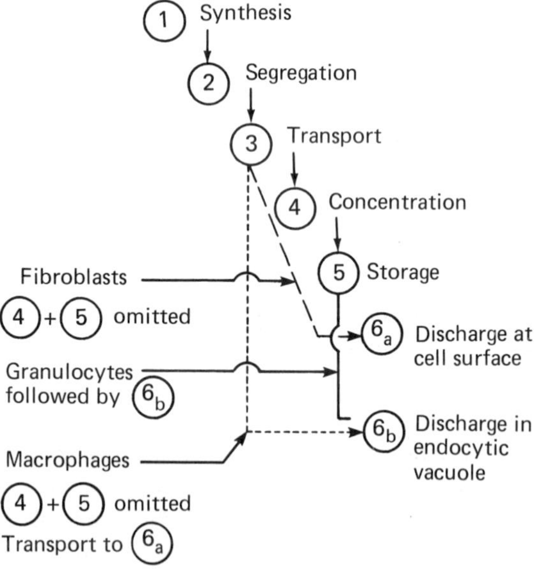

Fig. 1.5. The different steps of secretory process occurring in different compartments within the cell (according to Palade, 1975).

secretory proteins, serum proteins, milk proteins, and proteins from chicken oviduct. For secretory proteins, six successive steps have to occur: synthesis, segregation, intracellular transport, concentration, intracellular storage, and discharge. These steps occur in different compartments within the cell. They are summarized in Fig. 1.5 according to Palade (1975). Even during synthesis, the nascent polypeptide chain has to be discharged either into or across the membrane. The existence of a particular N-terminal sequence referred to as the signal peptide was established by Blobel and Sabatini (1971), Blobel and Dobberstein (1975), Devillers-Thierry *et al.* (1975). The essential characteristic of the signal hypothesis is the presence of a unique sequence of amino acids following the initial methionine (Fig. 1.6), when mRNA translation products have to be transferred across the membrane. It has been suggested that translation of mRNAs containing signal codons begins on free ribosome and that only when the signal peptide of the nascent chain has emerged from the ribosome, does attachement of the ribosome to the membrane occur. Then, on the outer part of the membrane, the first processing occurs upon proteolysis by a membrane-bound enzyme which removes the signal peptide. For this reason, this extrapeptide was ignored until translation products of messengers under acellular conditions revealed larger sequences than *in vivo* secreted proteins. These precursor proteins are proteolytically processed *in vivo* or when ribosomes are bound to the RER membrane.

Other modifications occur to newly synthesized proteins; disulfide bond formation, phosphorylation, glycosylation, hydroxylation, and lipidation. These events often take place within the endocellular membrane network; glycosylation occurs primarily in the endoplasmic reticulum Golgi system; glycosyl transferases are found associated with membranes (see Shore and Tata, 1977).

1.3.3.2. Vectorial Discharge of Nascent Polypeptide Chains Synthesized on Rough Endoplasmic Reticulum-Bound Polysomes

The signal peptide capable of binding on specific sites allows the vectorial transport of the nascent protein across the membrane. Figures 1.6 and 1.7 illustrate the mechanism proposed by Blobel and Dobberstein (1975) to account for vectorial transport. The extra-N-terminal sequences contain up to 30 amino acids (an average of 20) depending upon the protein. They have an unusual amount of hydrophobic residues (see next subsection for the sequences). After initiation of protein synthesis, the growing signal sequence is assumed to be channeled through the 60 S subunit and pass across the membrane; the hydrophobic character of this N-terminal sequence favors the process of driving the protein through the membrane. A model was proposed which assumes the existence of a transient tunnel formed by loose

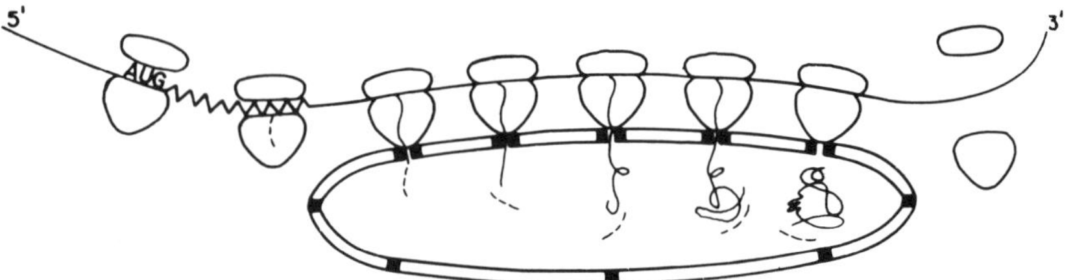

Fig. 1.6. Schematic representation of protein biosynthesis by rough endoplasmic reticulum membrane-bound polysomes (according to Blobel and Dobberstein, 1975a). Codons corresponding to signal peptides after initiation codon AUG are indicated by a zig-zag, the signal sequence is indicated by a dashed line (see text).

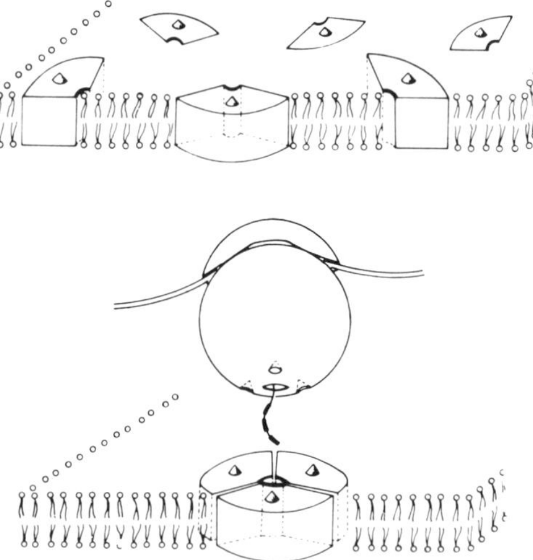

Fig. 1.7. Model for the formation of a transient tunnel for the transfer of the nascent chain (according to Blobel and Dobberstein, 1975b). Specific regions of the signal sequence are recognized by ribophorins allowing their association. Subsequently, ribosomes interact with these proteins at specific sites on both molecules represented by cones on the proteins and notches on the ribosomal subunit.

association of three membrane proteins, each of them possessing a recognition site for a specific region of the signal sequence. These proteins called ribophorins I and II have been identified (Kreibich *et al.*, 1978a,b). Precise synchronization with translation is required for recognition of the signal sequence by membrane proteins. The first step in transfer is the recognition of the signal peptide by the ribophorins I and II which assemble around it by lateral diffusion through the lipid bilayer. Specific binding of the ribophorins to the ribosomal complex then occurs. This probably directs a subsequent assembly of ribophorins to form a more stable and more specific channel through which the polypeptide chain could reach the intracisternal space. Signal peptidase (or clippase) located on the inner face of the RER membrane splits the N-terminal peptide. Proteolytic removal of the signal peptide occurs before completion of the polypeptide chain. After discharge of the completed chain, ribosomes are detached from the membrane. Thus ribophorins are free again to diffuse in the plane of the membrane. The tunnel does not seem to constitute a permanent structure in the RER membrane.

In contrast to secreted proteins, membrane proteins are not discharged into the intracisternal space, but inserted into the membrane. A peptide signal also is required for membrane insertion. Glycosylation can occur during the synthesis, as shown for G protein of vesicular stomatitis virus (VSV) (Rothman and Lodish, 1977). These studies have revealed a precise sequence of events by which the nascent polypeptide chain is glycosylated in two steps. During maturation this protein also binds 1–2 molecules of fatty acid (Schmidt and Schlessinger, 1979).

The existence of an extra peptide is not a characteristic seen only in eukaryotes. Similar mechanisms have been described recently in prokaryotic cells (particularly in *E. coli*), from which a transient amino terminal extension has been identified in several secreted or membrane proteins (Bassford and Beckwith, 1979; Inouye and Halegoua, 1979). The sequence of several peptide signals has been determined: major and minor coat proteins (Sugimoto *et al.*, 1977; Chang *et al.*, 1979), maltose binding protein (Hedgpeth *et al.*, 1980), lipoprotein (Inouye *et al.*, 1977), β-lactamase (Sutcliffe, 1979). The signal sequence of alkaline phosphatase is only partially determined (Sarthy *et al.*, 1979). Identical mechanisms of protein secretion across membranes are described (see review, Davis and Tai, 1980).

Signal hypothesis for the transmembrane transfer of secreted or membrane proteins, appears as a rather general process occurring in prokaryotes as well as in eukaryotes. Furthermore it does not seem specific for proteins synthesized by RER-bound polysomes and has also been described for proteins synthesized by free polysomes especially peroxisomal (Goldman and Blobel, 1978) and mitochondrial proteins (Maccechini *et al.*, 1979, Cote *et al.*, 1979).

Blobel (1980) assumes that parts of a protein sequence bear information for different processes termed as "intracellular protein topogenesis." Topogenetic sequences include signal sequences, transfer sequences which interrupt the translocation process, sorting sequences which determine translocational traffic and insertion sequences which initiate integration of proteins into the lipid bilayer of a membrane (see also Marx, 1980). Thus, beside the genetic code which determines sequence, a so-called zip code determines destination of proteins into the cell. Silhavy *et al.* (1977) succeeded in moving to the outer membrane of *E. coli* a polypeptide coded by a *lamB–lacZ* hybrid gene obtained by gene fusion. Although β-galactosidase is normally cytoplasmic, *lamB* is located in the outer membrane. Its precursor contains a signal peptide of 24 amino acids. However, a protein from a mutant which contains only a part of *lamB* sequence but that includes the signal sequence at the N-terminal and the sequence of β-galactosidase is mainly located in the cytoplasm. This indicates that a signal sequence is not sufficient to lead β-galactosidase out of the cytoplasm (Moreno *et al.*, 1980).

1.3.3.3. Structural Homologies of Signal Peptides from Different Preproteins

A signal sequence has been found recently for several secreted proteins, immunoglobulin light chain (Milstein *et al.*, 1972; Blobel and Dobberstein, 1975a,b Burstein and Schechter 1977, 1978), for insulin (Chan *et al.*, 1976; Duguid *et al.*, 1976 Lomedico and Sauenders, 1976; Lomedico *et al.*, 1977), for pancreatic secretory proteins (Devillers-Thierry *et al.*, 1975; Scheele *et al.*, 1978), for milk proteins (Gaye *et al.*, 1977; Craig *et al.*, 1976; Mercier and Gaye, personal communication), for mellitin from honey bee venom (Suchanek *et al.*, 1975), for parathyroid hormone (Kemper *et al.*, 1974; Habener *et al.*, 1975, 1976). It was found in chicken oviduct proteins, lysozyme, ovomucoid, conalbumin, with an exception for ovalbumin (Palmiter *et al.*, 1977a,b, 1978; Gagnon *et al.*, 1978; Jòlles *et al.*, 1979). Though it is not cleaved during its transfer through the membrane, ovalbumin contains the functional equivalent of a signal sequence which is observed in a tryptic fragment containing residues 229–276 (Lingappa *et al.*, 1978a,b, 1979). The sequences of several signal peptides are given in Table 1.1. It is not an exhaustive presentation of all known signal peptides; several others have been sequenced. This sampling however clearly displays common features in all these sequences. They contain between 15–30 amino acid residues. Most of them have one to three positive charges within the first five residues and also few charged amino acid side-chains at the C-terminal region of the peptide. It has been postulated that these charges interacting with the negative charges of phospholipids favor the anchoring of the signal peptide. Besides these few charged residues, the sequence has a marked hydrophobic character and is composed mostly of consecutive hydrophobic residues. Using the predictive method of Chou and Fasman (1974a,b), Austen (1979) predicted that hydrophobic sequences have helical structure or β-strand structure or both. In most signal sequences, a β turn close to the cleavage site was predicted. Quite similar results were found by Garnier *et al.* (1980).

Austen calculated that about 11 residues in β-strand structure and 23 in α helix would be required to span a membrane of 36 Å dimension, thus possibly explaining the sequence of 11–25 hydrophobic residues found in all signal peptides. Although speculative, these arguments emphasize that common structural features may ensure analogous functions. Table 1.1 does not indicate great sequence homologies among all the signal peptides. Nevertheless, homologies in their conformational preferences, i.e., in their helix or β-strand tendency are shown.

When such a protein is synthesized under acellular conditions in the absence of RER membrane, the signal peptide remains attached to the protein. It is not evident that preproteins fold as proteins without the signal peptide to yield the same native functional protein. This problem has not yet been scrutinized, but contradictory results from one protein to another must be mentioned. Preribonuclease can fold to give active enzyme (Haugen and Heath, 1979). By contrast preamylase does not show any enzyme activity; the extra peptide likely prevents the right folding of the preprotein and therefore the expression of the catalytic activity (Goreki and Zeelon, 1979). Evidence for conformational differences between preprotein and protein was reported for leucine-specific binding protein (Oxender *et al.*, 1980) and for thyroid stimulating hormone (Giudice and Weintraub, 1979).

1.3.4. Covalent Processing

Once the polypeptide chain is folded, further events are often required to permit the protein to reach its functional conformation. Covalent processing, such as further limited proteolysis and chemical modifications is an important event in determining functional properties. These processes induce conformational refinements which can be considered as the last events of protein folding.

1.3.4.1. Role of Proteases in the Formation of the Functional Structure of Proteins

Many physiological reactions are triggered by the proteolytic cleavage of a precursor which regulates formation of active proteins at very specific loci in the organisms (see Neurath and Walsh 1976, 1977; Woodbury and Neurath, 1980). They include activation of pancreatic zymogens, hormone formation, regulatory peptide formation (angiotensins, kinins), blood coagulation, complement, assembly processes (fibrin, collagen, phage heads), fertilization, and development.

To ensure the integrity of cells and organisms, some mechanisms may require that the enzymatic activity or, more generally, the biological activity of a protein be locked and not expressed in determined cellular compartments or biological organs before reaching the final destination. Such is the case of zymogens and prohormones. Pancreatic serine proteases, for example, are

TABLE 1.1

Amino Acid Sequence of Signal Peptides[a]

Precursor	Amino Acid Sequence	↓[b]	Ref.[c]
Eukaryotic proteins			
Pre-ovomucoid	M A M A G V F V L F S F V L C G F L P D A A F G	A E V D	a
Pre-lysozyme	M R S L L I L V L C F L P L A A L G	K V F X	a
Pre-conalbumin	M K L I L C T V L S L G I A A V C F A	A P P K	b
Pre-promelletin	M K F L V X V A L V F M V V Y I X Y I Y A	A P E P	c
Pre-IgG (light chain)			
MOPC41	M D M R A P A Q I F C F L L L L F P G T R C	D I Q M	d
MOPC321	M E T D T L L L W V L L L W V P G S T C	D I V L	d
MOPC104E	M A W I S L I L S L L A L S S G A I S	Q A V V	d
Pre-proparathyrin	M M S A K D M V K V M I V M L A I C F L A R S D G	K S V K	e
Pre-proinsulin 1	M A L W M R F L P L L A L L V L W E P K P A Q A	F V K Q	f
Pre-growth hormone	M A A D S Q T P W L L T F S L L C L L W P Q E A G A	L P A M	g
Pre-proalbumin	M K W V T F L L L L F I S G S A F S	R G V F	h
Pre-trypsinogen 2	M A K L F L F L A L L L A Y V A	F P L D	i
Pre-α_{s1}-casein	M K L L I L T C L V A V A L A	R P K H	j
Pre-k-casein	M R K S I L L V V T I L A L T L P F L I A	Q E Q N	j
Pre-α-lactalbumin	M M S F V S L L L V G I L F X A T Q A	E Q L T	j
Pre-β-lactoglobulin	M K C L L L A L G L A L A C G V Q A	I I V T	j
Pre-opiocortin	M P R L C S S R S G A L L L A L L L Q A S M E V R G	W C L E	k
Pre-lipoprotein	M K A T K L V L G A V I L G S T L L A G	C S S N	l

Protein	Signal sequence	↓	Ref.
Pre-penicillinase	M S I Q H F R V A L I P F F A A F C L P V F A	H P E T	m
Pre-glycoprotein VS Virus	M K C L L Y L A F L F I (H V N) C	K F X I	n
Ovalbumin (not cleaved)	M G S I G A A S M E F C F D V F K E L K V H H A N E N A I M S A L A M	I P Y C P I- V Y L G A K	o
Prokaryotic proteins			
Pre-F_1 major coat protein	M K K S L V L K A S V A V A T L V P M L S F A		p
Pre-F_1 minor coat protein	M K K L L F A I P L V V P F Y S H S A		p
Pre-maltose binding protein	M K K T G A R L L L L A L S A S		q
Pre-lipoprotein	M K A T K L V L G A V I L G T L L A G		r
Pre-β-lactamase	M S I Q H F R V A L I P F F A A F C L P V F A		s
Pre-alkaline phosphatase (partially determined)	M K Q S T . . .		t

[a] Adapted from Austen (1979) and completed.

[b] ↓ Cleavage by the signal protease.

[c] Key to references: (a) Palmiter *et al.* (1977a); (b) Thibodeau *et al.* (1978); (c) Suchanek *et al.* (1978); (d) Burstein and Schechter (1978); (e) Habener *et al.* (1978); (f) Chan *et al.* (1976); (g) Seeburg *et al.* (1977); (h) Strauss *et al.* (1977); (i) Devillers-Thierry *et al.* (1975); (j) Mercier *et al.* (1978); (k) Nakanishi *et al.* (1979); (l) Inouye *et al.* (1977); (m) Ambles and Scott (1978); (n) Lingappa *et al.* (1978a); (o) Blobel and Dobberstein (1975a,b). The authors acknowledge Dr. Gage and Dr. C. Lazdunski for providing information on signal peptides p, q, r, s, t.

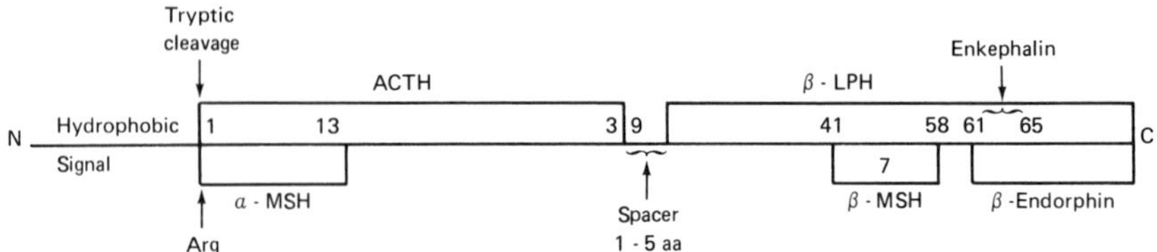

Fig. 1.8. The organization of a multicomponent precursor synthesized in the pituitary and yielding by cleavage different hormones (according to Tata, 1978; Roberts and Herbert, 1977a,b; Roberts *et al.*, 1978). This precursor molecule has a molecular weight of 31,000 and is composed of about 260 amino acids. The hormone ACTH has 39 amino acids and is released by proteolytic cleavage of the N-terminal end; further cleavage of ACTH gives peptide 1–13 which is α-MSH. β-Lipotropin is formed by 91 amino acids in C-terminal position; it contains β-MSH, endorphins and enkephalins which are also obtained by proteolytic cleavage. Peptide 61–76 is the α-endorphin; 61–77 the γ-endorphin; 61–91 which represents the C-terminal fragment is the β-endorphin; and 61–65 is the Met-enkephalin (from Tata, 1978).

secreted as zymogens; therefore, they are inactive and unable to digest pancreatic tissue. Activation occurs in the duodenum where the presence of the enzyme is required (see "The Enzymes" Vol. III, 1971).

The potentially active precursors have conformation either totally or sometimes slightly different from the structure of the functional protein. For these proteins, the expression of biological activity requires the occurrence of structural readjustments produced upon limited proteolysis. First, proteolytic cleavage removes the signal peptide in the preprotein; but for prohormones and zymogens further proteolytic cleavage is required for activation. In several cases, a large fragment of the polypeptide chain is removed in the conversion process. This is exemplified by the transformation of proinsulin to insulin, where possible structural changes are of great amplitude.

Interesting examples of multiprocessing are the multicomponent proteins such as opiocorticotropin and lipotropin precursor. Peptide hormones derived from multicomponent proteins probably have no conformational similarities with their larger precursor. This multicomponent precursor, as illustrated in Fig. 1.8, contains in its sequence a number of polypeptide hormones with different biological activities (Marks, 1979; Krieger and Liotta, 1979; Mains and Eipper, 1979; Liotta *et al.*, 1980), including α and β malanocyte-stimulating hormones (α-and β-MSH), corticotropin hormone (ACTH), α and β lipotropins (α-and β-LPH) with 91 and 58 amino acids, respectively, and smaller fragments with opiate activity (endorphins and enkephalins). Little is known about the way this large precursor is proteolytically processed in different loci of the organism. It has been suggested that such multicomponent proteins represent a selective advantage for all the covalently linked activities and likely have arisen by gene duplication and fusion.

For all proteins which are processed in an ultimate stage, conformational rearrangments occur upon conversion of precursor to active protein. These conformational variations may be of very weak amplitude as in chymotrypsinogen and chymotrypsin whose crystal structures are very similar (Wright, 1973). In serine proteases, limited proteolysis of zymogen leads to the formation of a salt bridge between the negative charge of a carboxylate neighboring the reactive serine and the N-terminal group liberated by processing. It is noteworthy that chymotrypsinogen displays a very weak but significant enzymatic activity, 10^6–10^7 times smaller than the activity of the enzyme (Gertler *et al.*, 1974), indicating that the active site is functional even in the zymogen. Conformational refinements made possible after limited proteolysis, although very small in amplitude, seem decisive for amplification of the enzymatic activity. These kinds of zymogens may be considered as incompletely folded proteins, locked in a conformation very close to, but different enough from that of the functional enzyme. Study of conformational readjustments occurring as a consequence of the conversion of a zymogen into an active enzyme may be of great interest for understanding the mechanisms by which a protein folds to its functional structure. When activation

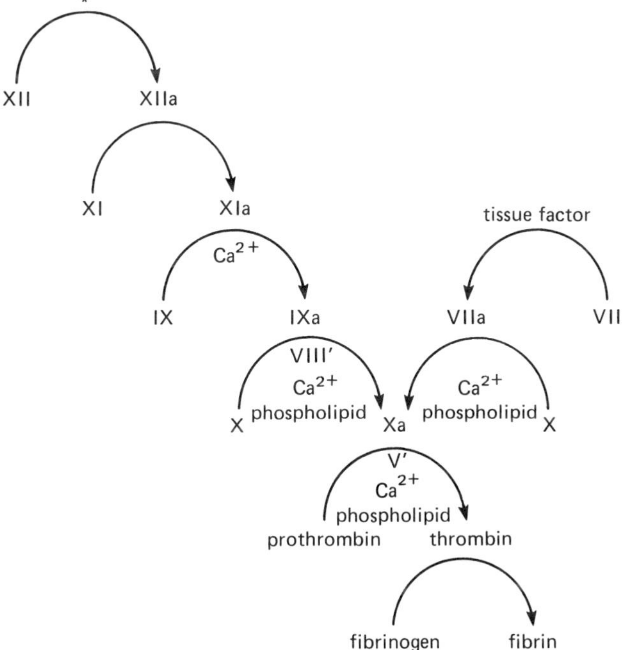

Fig. 1.9. Schematic representation of the cascade system in blood coagulation (from Neurath and Walsh, 1977).

proceeds only through a very limited proteolysis, zymogen structure may be regarded as an intermediate in the folding pathway. Thus, it is easy to study the transformation of this intermediate.

Of particular interest are the cascade systems that are involved in blood coagulation (Fig. 1.9). Such systems are defined by sequential zymogen activation reactions or cascade: the enzyme produced after activation at the first step specifically cleaves the second zymogen and so forth. The intrinsic pathway is controlled by five zymogen activations, the extrinsic one by three. Such processes, which are quite irreversible, represent considerable amplification and permit a very rapid response. At each level, specific noncovalent inhibitors are available to modulate and regulate the system. Generally, covalent regulation by a cascade system allows a more rapid and efficient response than noncovalent regulatory processes such as allostery. It is required for defense against any aggression of the organism, the existence of additional auxilliary noncovalent regulation in the same system, and offers possibility of modulating the process when necessary.

1.3.4.2. Role of Chemical Modifications in the Formation of the Functional Structure of Proteins

Other types of processing such as chemical modifications certainly induce variations in the structural organization of a protein and influence its stabilization in a particular environment. Many proteins undergo further covalent modifications which can determine their stability and functional properties. This aspect will not be dealt with in detail as it is the subject of many recent reviews (Wold, 1981), but the frequently occurring chemical modifications will be discussed briefly.

Hydroxylation of prolyl and lysyl residues occurs mainly in collagen, elastin, and some plasma proteins. Hydroxylysines and hydroxyprolines have been found also in acetylcholinesterase and in the C_1q subcomponent of complement (see review, Kivirikko and Myllylä, 1980). In collagen, the role of hydroxylation seems to consist in stabilization of the triple helix.

Glycosylation is a very common chemical modification leading to covalent linkage of carbohydrate moieties with a protein. Only few amino acid residues are capable of accepting a sugar moiety: serine or threonine, asparagine, and hydroxylysine. They are often located in turns at the surface of the molecule. Beside their biological function specific for glycoproteins (cell recognition, immunity), the presence of sugar certainly modifies protein stability and also may protect against proteolytic attacks (see review, Phelps, 1980).

Carboxylation of glutamyl residues plays an important role in prothrombin activity, and perhaps several other plasma vitamin K-dependent proteins involved in blood coagulation. (see review, Suttie, 1980).

Reversible phosphorylation requiring both a kinase and a phosphatase is implicated in the control of many cellular process. Generally, a phosphate group is linked to serine or threonine and also to tyrosine residues in proteins. Phosphorylation in enzyme molecules generally leads to a modification of the catalytic activity (see review, England, 1980). Other modifications have been also described such as methylations and ADP-ribosylation. Formation of disulfide bonds is also a posttranslational process.

1.3.5. Mechanisms of Protein Folding

In this introductory chapter, the fundamental questions at the molecular level focus on two points. The first point is a discussion of the thermodynamic and kinetic determination of protein folding. In fact, it can be divided in two related questions. Does the native protein correspond to the most stable conformation? Does the native protein corresponds to a unique structure? The second point is related to the pathway of protein folding, the existence, the number, and the nature of the intermediates. In this presentation of the fundamental questions, we will try to delineate what decisive arguments, if any, could be presented, and what significant experiments could be made to prove or disprove the different hypotheses.

1.3.5.1. Thermodynamic and Kinetic Control of Protein Folding

"Does the native protein correspond to the most stable conformation?" This question has been discussed abundantly in this decade and perhaps should be de-emphasized now. But in this introductory chapter, an historical point of view is certainly interesting to consider, even if it only allows us to discover the evolution of the ideas. What does the so-called "thermodynamic hypothesis" (Epstein *et al.*, 1963) mean? It has been clearly defined in a review by Anfinsen and Scheraga (1975):

> "It is currently believed that the three dimensional structure of a native protein in a given environment (solvent, pH, ionic strength, presence of other components, temperature, etc.) is the one in which the Gibbs free energy of the whole system is a minimum with respect to all degrees of freedom, i.e., that the native conformation is determined by the various interatomic interactions and hence by the amino acid sequence, in a given environment."

According to this hypothesis, the native structure represents the most stable structure. On the other hand, the concept of kinetic control of protein folding means that some energetic barrier could hinder the access of the most stable conformation, thus trapping the structure in a metastable state. The idea of kinetic control of protein folding has been defined by Wetlaufer and Ristow (1973):

> "This would mean that in many, perhaps most proteins the native structure is not the structure of the lowest Gibbs free energy (the global minimum). Of course, it would be a structure in a free energy minimum, the lowest free energy minimum of the kinetically accessible structures."

In fact, it was believed for a long time that protein folding is a thermodynamically controlled process. Proposed in the early papers (Mirsky and Pauling, 1936; Eyring and Stearn, 1939), this statement was generally accepted by different authors (Kunitz, 1948; Pauling and Corey, 1951; Linderstrøm-Lang, 1950, 1952; Lumry and Eyring, 1954, Schellman, 1955a; Kauzmann, 1959a,b. Scheraga, 1963; Schellman and Schellman, 1964; Tanford, 1962a,b, 1970; Brandts, 1969). It is at least implicit in all discussions. It seemed justified by the reversibility of the unfolding–folding process observed even in these early studies. However, in the first studies, the doubt often remained as to whether or not the protein was totally unfolded. This ambiguity can be avoided in the most recent studies performed with proteins whose sequence is known. In this case the complete unfolding can be verified by various methods. Thus the ability to refold the denatured and reduced ribonuclease and to restore the native structure with the correct disulfide bridging, even from the scrambled ribonuclease, and to find again fully active enzyme was often considered as a decisive argument in favor of the thermodynamic control of protein folding (Sela *et al.*, 1957; Anfinsen and Haber, 1961; Anfinsen *et al.*, 1961, 1972; Anfinsen, 1962, 1966, 1972, 1973; Haber and Anfinsen, 1962; Epstein *et al.*, 1963). The hypothesis representing the native structure as the one of lowest free energy was assumed by all the authors who attempted prediction of conformation. The rather good agreement between the data provided by X-ray crystallography and predictions of conformation based on calculations that involve energy minimization is often used in favor of thermodynamic control.

Although accepted for a long while, the thermodynamic determination of protein structure was seriously questioned by Levinthal (1968a,b, 1969) who proposed a kinetic control of protein folding. He argued that the time for a random search through all the possible conformations even for a small protein is not compatible with the time biologically needed. In kinetic control, the folding process may occur through a unique, well determined, and specific pathway which drives the protein in a metastable state. The idea of kinetic control was assumed by several researchers (Levinthal, 1966, 1968, 1969; Teipel and Koshland, 1971a,b; Wetlaufer, 1973; Wetlaufer and Ristow, 1973).

The alternative, thermodynamic versus kinetic control, greatly influenced different studies on protein folding several years ago. Several attempts to resolve this problem have been reported in the literature for proteins with and without disulfide bridges. Indeed it is difficult to prove the type of control, thermodynamic or kinetic, involved in protein folding; in most cases, we have no decisive arguments to discard one rather than the other hypothesis. In this respect it is significant to outline the point of view of Anfinsen and Scheraga (1975): "Thus, in our view, it appears likely though by no means proved, that the native conformation of a protein is indeed the one of the lowest free energy."

At this time, thermodynamic and kinetic control of protein folding does not seem mutually exclusive. It might considered a plausible assumption that the native conformation is the one of lowest energy but is reached through kinetically controlled intermediates.

"Does the native protein corresponds to a unique stucture?" The uniqueness of the native conformation from a given sequence in a given environment could be questioned. Does a unique structure for a native protein always really exist? Or is it possible that several exist, although there are a limited number of native conformers? Because of the flexibility of proteins driven by conformational entropy, and the possibility of conformational fluctuations due to rotations of the backbone and side-chains, it does not seem unreasonable to imagine a protein oscillating between at least two states separated only by a small free energy. Some parts of the chain could exist in two conformational states in rapid equilibrium. The existence of allosteric enzymes obeying the model of Monod, Wyman, and Changeux (1965) could be considered as such a possibility. Even in the absence of ligand, two states of the protein R and T do exist. Furthermore in K systems, they have the same catalytic activity, and therefore are native forms. Some ligands are able to stabilize one or the other of these forms. The allosteric enzymes are oligomeric proteins, generally proteins without disulfide bridges. Disulfide bounds considerably reduce the conformational entropy, and therefore the conformational fluctuations, and perhaps restrict the number of possible conformers.

1.3.5.2. Pathway(s) of Protein Folding

The understanding of protein folding requires a precise knowledge of the pathway(s) involved in the process.

Whatever hypothesis may be stated, thermodynamic or kinetic control of the formation of the tertiary structure, it seems realistic to admit a limitation of the number of possible pathways. "Evolution (with thermodynamics dictating the folding), has selected the amino acid sequence to form a biologically active molecule with presumably a limited number of pathways from the unfolded form to a unique native structure of the lowest free energy" (Anfinsen and Scheraga 1975).

A random search for the native conformation among all the possible structures is very improbable since a polypeptide chain can fold in a short time.

Different models have been proposed to describe nucleation and folding of proteins:

(1) The classical nucleation–propagation model involving nucleation followed by a rapid growth of the polypeptide chain to the folded state, nucleation being the limiting step. This model predicts a highly cooperative process;

(2) A dynamic model of diffusion–collision was developed by Karplus and Weaver (1976). In this model nucleation occurs within smaller parts of the molecule forming microstructures. Thus all conformational possibilities may be searched very rapidly but are not stabilized. Several of these micro-domains have to collide and to coalesce to produce a substructure with the native conformation. Folding is achieved by a series of such diffusion–collision steps.

(3) Folding in stages involving several levels of structuration in which several elements are formed and assembled; this model requires a sequence of well defined events with a relative stability of structured elements at each stage of the process.

A possible hierarchy of protein folding corresponding to the hierarchy in protein structure was suggested (Schulz, 1977). It can be summarized by the following scheme for folding in stages with an additional level for oligomeric proteins, the subunit assembly:

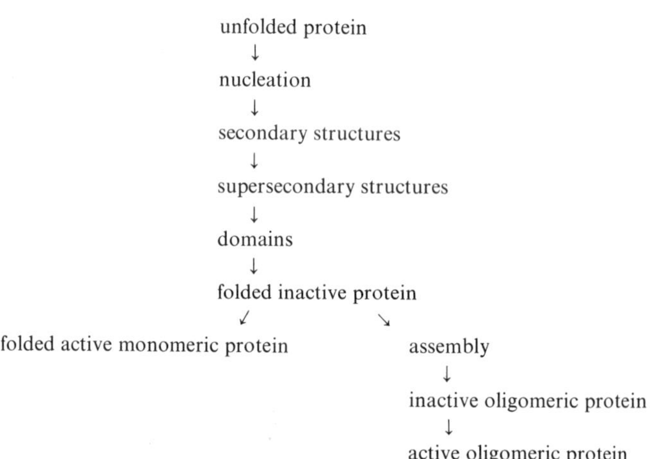

Many experimental studies have been made to determine the pathway of folding of several proteins, and to detect and identify the intermediates.

From earlier studies and over a rather long period of time, native unfolded protein transition was described by the two-state approximation, which is generally applicable for very cooperative transitions. However, as early as 1942 (Neurath *et al.*), the occurrence of intermediates was proposed to explain some discrepancy between kinetics of refolding when different parameters were used to follow the progress of the reaction. The validity of the two-state approximation for protein folding–unfolding will be discussed in great detail in Part II, Chapter 6; but in many cases no evidence for the existence of intermediates was seriously provided. The introduction of fast kinetic methods, stopped-flow and T-jump, to follow protein folding has brought direct evidence for a multi-step process.

Based on the first results of rapid kinetics, it was proposed that protein folding proceeds through a sequential pathway, the first event being presumably a nucleation step. Nucleation, by initiating and directing the folding, would restrict the number of possible structures during the search for the right conformation. The concept of nucleation was developed. Earlier, the nucleation centers were postulated to be mainly helical structures (Lewis *et al.*, 1970, 1971). An α helix tendency or even an ordered structure is no longer considered a necessity for a segment of polypeptide chain to be a nucleation site. Matheson and Scheraga (1978) predicted nucleation sites buried in hydrophobic binding. In their review, Anfinsen and Scheraga (1975 foot note p. 208) used the term of "nucleation site" to designate a portion of the amino acid sequence in which a specific local backbone conformation has a tendency to form because of short and medium range interactions. "Regular structure tendency is not a necessity for a portion of the chain to be a nucleation center. Depending upon the length and the sequence of the polypeptide chain, it is possible that more than one nucleation site occurs". According to these concepts, Anfinsen had proposed a scheme to explain the mechanism of protein folding (Anfinsen 1972, 1973, Anfinsen and Scheraga 1975). The nucleation sites could generate fragments of the polypeptide chain in the so called "native format". In this first step, they are in flickering equilibrium with the random conformation until folding occurs and stabilization by long range interactions drives the equilibrium towards the native form.

Secondary and supersecondary structures could be represented as elementary building blocks folded in the early step of protein folding but only stabilized upon interactions with other structural elements.

The existence of domains in proteins, i.e., miniproteins inside a polypeptide chain, and its implication for protein folding was particularly underlined by Wetlaufer (1973) who distinguished folding in continuous and discontinuous regions. He speculated that for continuous regions, folding could result from a process of nucleation followed by rapid growth, whereas discontinuous regions will require considerably longer folding time. In several cases, structural domains carry on a definite function. As an example, the protomer of several dehydrogenases consists of a coenzyme binding domain and a catalytic domain (Rao and Rossmann, 1973). It was suggested that these proteins arose from gene fusion during evolution. It is proposed by several authors that protein segments coded by exons (expressed sequences of DNA separated by noncoding sequences or introns) correspond to functional domains (Gilbert, 1978; Craik *et al.*, 1980). Domains in multidomain proteins can behave as folding units able to fold and to be stabilized independently.

1.3.6. Assembly Processes

Although the mechanism has not been elucidated, the spontaneous folding of a single polypeptide chain to reach its three-dimensional structure represents the most elementary event in the genesis of living organisms. Self-assembly of subunits forming quaternary structure occurs at a higher level, but proceeds from the same thermodynamic principles as the folding of a polypeptide chain. Oligomeric proteins generally are stabilized through subunit interactions. Quaternary constraints can generate new properties of regulation so that amplification or attenuation processes occur that are directly related to subunit interactions (Citri, 1973). At a higher level, associations of several different molecules lead to complex structural organization and also generate particular properties due to proximity such as channeling effects, vectorial reactions, and so on.

1.3.6.1. Conformational Refinements on Subunit Assembly

Oligomeric proteins result from the assembly of either identical or nonidentical subunits. Since most oligomeric proteins cannot exist in a functional monomeric form, conformational variations, at least localized at the active site, might be driven by subunit interactions. These conformational readjustments occurring through quaternary constraints are not necessarily of great amplitude. Variations in position of the atoms at the active site as small as a fraction of an angström can determine the appearance of activity. Oligomeric proteins, when allosteric, represent a selective advantage in regulating enzymatic activity according to the requirement of cell metabolism by modulating this activity by amplification or attenuation mechanisms in response to the concentration of an effector present in the environment. All oligomeric proteins are not allosteric. Some of them do not display any deviation from Michaelis behavior. In this case, however, oligomeric structure is required for stabilization of functional tertiary structure of each subunit. In all oligomeric proteins, structural variations upon subunit assembly can be expected.

In the last step of protein folding, the one which is decisive for functional properties of oligomeric as well as monomeric proteins, some very small but efficient conformational refinements, perhaps only ensured by side-chain rotations, are certainly required to generate biological activity. One can ask if these mechanisms are very different for oligomeric and monomeric proteins since all proteins are built of smaller substructures.

1.3.6.2. Macromolecular Assembly

Multienzyme complexes, as exemplified by the well-documented system, fatty acid synthetase, result from a higher order of assembly. The fatty acid synthetase system in yeast contains seven different enzymes associated in a complex. It consists of two polypeptide chains, each containing several enzymes. Subunit A contains the acyl carrier protein, the condensing enzyme, and β-ketoacetyl reductase. Subunit B has four enzymatic activities: acetyltransacylase, malonyltransacylase, hydroxyacyldehydratase, and enoylreductase. Each chain is coded by a single gene (Lynen, 1972). This supramolecular organization requires specific interactions at each level to regulate the activity of the whole system. Multienzymatic systems are likely endowed with great efficiency because the components are arranged in the same sequential order as the reactions in the metabolic pathway. Several multienzyme complexes have been described. Presumably such organization represent an economy for the cell, minimizing the entropy effects in metabolic pathways (Gaertner, 1978). In some eukaryotic multienzyme complexes, several enzymes are even linked covalently giving rise to multifunctional proteins. Covalent linkage represents a double advantage in evolution; first, it ensures a coordinated controlled synthesis of each enzyme, and second, it represents a greater stabilization of the complex than a noncovalent association (Stark, 1977).

The self-assembly of membrane structure leads to higher degrees of organization in which lipid and protein components assemble by noncovalent forces. The structures of ribosomes and viruses are also determined by self-assembly. An interesting and now well documented system is the self-assembly of tubulin to form an organization known as microtubules. Assembly processes may be involved in the formation of more complex systems, such as cell organelles, mitochondria, and chloroplasts. Larger supramolecular structures such as organs, arise not only from assembly processes, but also may involve other mechanisms.

Eventually, the elucidation of the mechanisms of protein folding and assembly will lead to an understanding of the orders of organization which are reached in living organisms, since microcosm reflects macrocosm.

I

Considerations of Protein Folding Deduced from Characteristics of Folded Proteins

Since protein research is now at the apex of the third period of protein description marked by a precise knowledge of space parameters, Part I begins with a presentation of the structural characteristics of folded, globular proteins (Chapter 2). The hierarchy of protein organization as revealed by the different levels of structure is detailed. The most basic features, the structural invariants encountered in globular proteins are presented. The occurrence of a small number of topologies is examined from the double point of view of limitations in folding mechanisms and of protein evolution.

Stability of the overall native conformation is ensured by a great number of intramolecular interactions and interactions with solvent. Thermodynamic studies of protein conformation have provided information regarding the importance of different energetic contributions to the folding process and in stabilization of the final native structure (Chapter 3).

These two fundamental pieces of information provided by examination of the final native structures and by determination of all energetic contributions necessary to their stabilization have been used in attempts to deduce rules of protein folding. Predictive rules of secondary and even supersecondary structures, as well as simulations of protein folding are discussed in Chapter 4. However, it should be noted that, at present, native and functional structures of globular proteins of reasonable size have not been unequivocally deduced by computations from the amino acid sequence of the polypeptide chain.

In the last section of Chapter 4, recent information on structural fluctuations of protein structure is presented. It is an important aspect in the understanding of the folding of a protein and of the expression of its biological activity.

2

Structural Characteristics of Folded Proteins

2.1. DEFINITION OF THE LEVELS OF PROTEIN STRUCTURE

The terminology introduced by Linderstrøm-Lang (1952) is generally used to define three levels of protein structure designated as primary, secondary, and tertiary structure. The primary structure is defined by the amino acid sequence of the polypeptide chain which is maintained by covalent links called the peptide bonds and does not describe the spatial arrangement. However, stereoisomerism (L and D forms) is included in this level of structure.

The secondary structure refers to the local spatial arrangement of the backbone without regard to the conformation of the side chains or the overall arrangement of the whole chain. The tertiary structure refers to the spatial arrangement of the entire chain, including the side chain of the amino acids and results from side chain interactions. Thus, the spatial arrangement of the polypeptide chain is frequently described as the secondary and the tertiary structures.

However, the distinction between these two levels of structure is not always evident. Historically, after the discovery of helical structures by Pauling and Corey (1951a,b,c,d, 1952, 1953a,b), Pauling (1940), and Pauling *et al.* (1953, 1962), it was commonly thought that the polypeptide chain first folds into regular structures (mainly α helix) which are stabilized by peptide hydrogen bonds that give the secondary structure, and that these structures then fold and are stabilized by side-chain interactions to form the tertiary structure thus yielding the compact overall shape of the molecule. However, further research has shown that regular structures in globular proteins may be limited locally and that many parts of the chain are irregularly folded, but these are still capable of interacting with the rest of the molecule.

The distinction between these two levels of structure is rather artificial since both structures are stabilized by the same kind of noncovalent interactions. Therefore, it is often not necessary to distinguish these two levels of structure and it is preferable to refer only to the conformation of the protein. This has been already proposed by Wetlaufer (1961), who included in this definition the subunit assembly of oligomeric proteins.

However, a distinction between the two types of structure is useful in some cases, for example the evaluation of certain physical or chemical studies. Some methods used to study the conformation are sensitive mainly to the local arrangements of the backbone, and specifically to the presence of regular structures, but do not reveal the overall folding of the molecule. These methods are considered to be probes of the secondary structure [e.g., optical rotation (ORD) and circular dichroism (CD), infrared (IR) spectrophotometry]. Some techniques are sensitive to the nature of the overall folding of the polypeptide chain and thus to the changes in the conformation, but do not reveal the details of the local conformation of the polypeptide chain. They reflect the changes of the overall conformation and are probes of the tertiary structure (e.g., changes in solubility, or viscosity.) A more precise definition is preferred which terms these two types of structures as backbone and side chain conformations.

Finally, the fourth level of structure is defined by the quaternary structure. As termed by Bernal (1958), this describes the geometry of the subunit assembly. Indeed, many protein molecules are composed of a specific number of subunits which may or may not be identical and which are connected by noncovalent interactions.

An additional definition was proposed by Edelstein who designates as "quinary structure" the interactions between helical fibers that occur in sickle cell hemoglobin or in tubulin units in microtubules (Edelstein, 1980). However, this term is not generally accepted.

2.2 PRIMARY STRUCTURE

Proteins are built of peptide units, the amino acids, whose side chains give the structural feature and therefore the functional properties to the molecule. Among the amino acids naturally occurring in proteins, one can distinguish between polar and nonpolar ones.

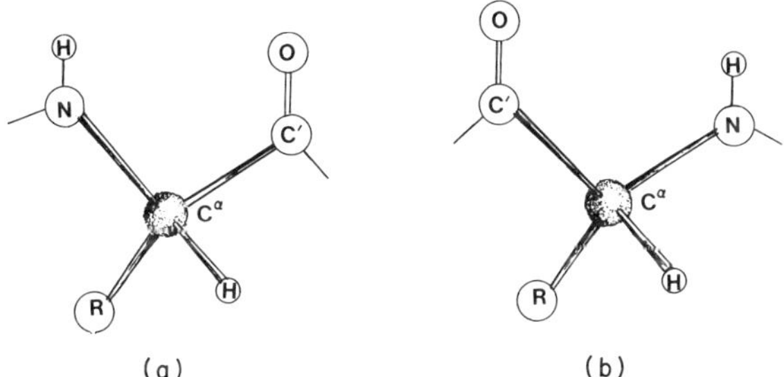

Fig. 2.1. The absolute configuration of amino acids. An easy rule to find the absolute configuration is the following: if one looks from the hydrogen attached to C$^{\alpha}$, to the three other groups in the order R, N, and C′, they appear in clockwise direction for L configuration (a) and in counterclockwise direction for D configuration (b).

With the exception of glycine, all of these amino acids have an asymmetrical carbon. Although there can exist D and L stereoisomers, proteins are composed exclusively of L amino acids. The reasons for this are not evident and are a result of processes associated with the origin of life and earlier stages of the evolution. Some speculative arguments, such as the chirality of the universe, have been proposed. The terms L and D stereoisomers refer to the absolute configuration without regard to rotation of light to the left or to the right respectively. The absolute configurations have been determined by Bijvoet and co-workers (1951) and are shown in Fig. 2.1. Polypeptide chains arise from the formation of covalent peptide linkages by condensation of α-carboxylate and α-amino groups with the removal of water molecules.

The chain length and the amino acid sequence characterize the protein. During the last 20 years, rapid progress has been achieved in sequence analysis. From the earlier successes in protein sequencing with insulin (Sanger, 1956) and ribonuclease by Anfinsen's group *et al.*, (Redfield and Anfinsen, 1956; see also Canfield and Anfinsen, 1963; Hirs *et al.*, 1960), the number of known structures increased rapidly as revealed by the size of the successive issues of the "Atlas of Protein Sequence and Structure" (Eck and Dayhoff, 1966; Dayhoff and Eck, 1967–1968, Dayhoff 1969, 1972, 1973, 1976, 1978). This extensive knowledge of primary structure has encouraged the study of the evolution of proteins by comparison of their primary structure. Most frequently a given protein, which has the same function in different organisms, exhibits small variations among the different species. A remarkable homology is observed: the evolutionary process seems to have been rather conservative. Different categories of amino acid replacements have been observed. From a chemical viewpoint, exchanges are conservative, when one amino acid is replaced by a homologous one (i.e., Glu by Asp, Ser by Thr, Arg by Lys) while the same chemical or ionic characteristic of the side chain is maintained, but not necessarily the same conformational properties (see Section 2.4.1). Radical replacements are those that alter the chemical characteristics of the molecule (i.e., Arg by Glu, Glu by Val, Ser by Gly) (Margoliash and Shejter, 1966). Amino acids are homologous either in their chemical nature or in their conformational preference. For example, Asp or Glu are chemically homologous although they ellicit completely different conformational behavior (Asp is a strong helix breaker and Glu is a strong helix former).

The sequences of some proteins from different species have been determined and this has permitted researchers to construct phylogenetic trees that retrace the history of a protein in the different species from the early stages of evolution to the present day.

2.3. DESCRIPTION OF PROTEIN CONFORMATION

Since confusion in terminology and definitions still persists in the literature in spite of recommendations by the Commission on Biochemical Nomenclature, it is important to start this section by recalling the main definitions used to describe the conformation of the polypeptide chain.

2.3.1. Definitions

The term conformation in general encompasses the spatial arrangement of a molecule as determined by rotations about the single bonds. In the specific case of the polypeptide chain, conformation describes the overall spatial organization. The conformation can be modified without breaking any covalent bond. In the literature on synthetic polymers, this spatial arrangement is often termed the configuration. However, to avoid any confusion at this level, it is preferable to use exclusively the term configuration to designate the spatial arrangement of the polypeptide chain. The term configuration must be restricted to the existence of stereoisomers (D and L amino acids) as is generally accepted in organic chemistry.

2.3.2. Characteristics of the Polypeptide Chain: Rules of Nomenclature

The description of polypeptide conformation involves the specification of bond length, bond angles, and angles of internal rotation about the single bonds. For the sake of convenience a special convention has been adopted to describe polypeptide conformation. The International Union of Pure and Applied Chemistry and International Union of Biochemistry (IUPAC–IUB) Commission on Biochemical Nomenclature has proposed a detailed set of recommendations. The original proposals (Edsall *et al.*, 1966; IUPAC–IUB Commission of Biochemical Nomenclature, 1966a,b) were modified in 1970 (see Edsall *et al.*, 1970, 1971).

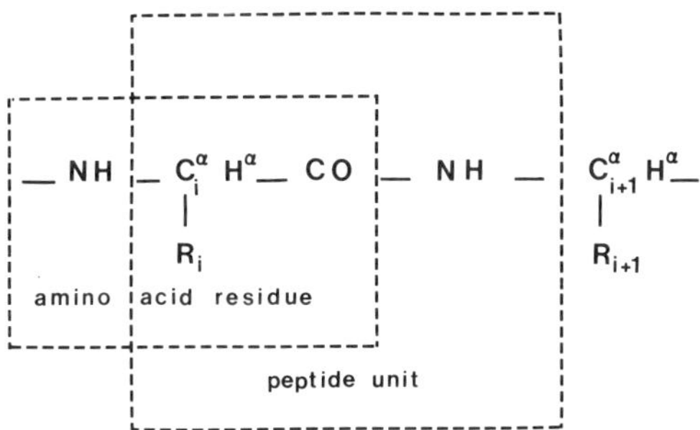

Fig. 2.2. Amino acid residue and peptide unit.

One should note the difference between amino acid residues which refers to —NH—CHR—CO and peptide units which refers to —CHR—CO—NH— (Fig. 2.2). The peptide unit corresponds to the properties of the peptide bond; it has a partial double-bond character because of the existence of two resonance forms (Fig. 2.3a) which strongly restricts the possibility of rotation around the C—N bond. Consequently, this group is generally planar; however, deviations from planarity of the peptide bond have been observed. An evaluation of conformational energy calculations indicated deviation of only 1–3° in low energy minima of Gly–Gly, but as much as 10° in Gly–Pro (Zimmerman and Scheraga, 1976).

The recommended symbols for bond length and bond angles are b and τ, respectively. They are generally found to have fixed values. For the peptide unit these values were originally determined by Corey and Pauling (1953); revised values were published by Momany *et al.* (1975). They are given in Fig. 2.3b. The only degrees of internal freedom are those of rotations around the single bonds. The angles of internal rotation, called dihedral angles or torsion angles, are generally denoted θ. In a system of four atoms A—B—C—D, for a rotation around B—C, the angle between the projection of A—B and the projection of C—D on a plane perpendicular to B—C is referred to as the torsion angle θ. This angle is taken equal to zero in the eclipsed conformation in which the projections of A—B and C—D coincide. It is considered to be positive for a clockwise rotation of the rear group (D),

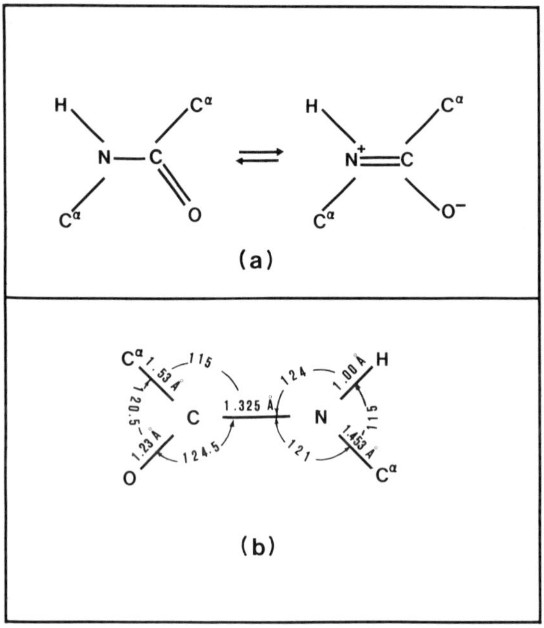

Fig. 2.3. The peptide bond: (a) the two resonance forms of the peptide bond, (b) bond length and bond angles of the peptide bond with values revised by Momany *et al.* (1975).

when the front group (A) is fixed (Fig. 2.4). Angles are measured from $-180°$ to $+180°$ rather than from 0 to 360°. This notation has the advantage that enantiomeric conformations appear with the same absolute value of dihedral

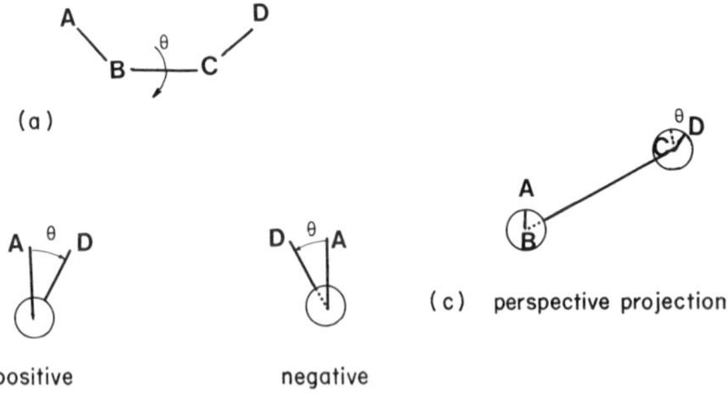

Fig. 2.4. Representations of the dihedral or torsion angle θ. (a) Usual representation; (b) Newman projections; (c) perspective projection.

angle, but with an opposite sign. The values of torsion angles from 0 to 360° were used by several authors before the recommendations of the Commission; this notation is still found in literature. The sequence rule gives formal priority to the peptide chain over the branches; the order of priority is the order of atomic number. If a compound has the geometry

$$\begin{matrix} A & & & & D \\ & \searrow & & \nearrow & \\ P- & B & - & C & -E \\ & \nearrow & & \searrow & \\ Q & & & & F \end{matrix}$$

then the sequence rule gives the priorities A > P > Q and D > E > F, and the principal torsion angle θ refers to the atoms A—B—C—D; the branches beginning in C are numbered as follows:

$$C_1\text{—}D,\ C_2\text{—}E,\ C_3\text{—}F.$$

The important torsion angles in polypeptides are denoted ϕ, ψ, ω, ν and χ (θ is a generic symbol covering all these). The rotation angles around the N—C^α and the C^α—C′ bonds are denoted ϕ and ψ respectively; ϕ_i and ψ_i refer to rotations around bonds which pertain entirely to residue i; ω_i is the rotation around the peptide bond C—N which follows residue i. The dihedral angle ν is a measure of the nonplanarity of the carbonyl oxygen atom; ν^O and ν^H, the dihedral angles of rotation around C'_i—N_{i+1}, measure the nonplanarity of the set of atoms C_{i+1}—N_{i+1}—C'_i—O_i and C_i—C'_i—N_{i+1}—H_{i+1}, respectively; they are defined to be zero for a planar peptide unit (Ramachandran and Sasisekharan, 1968). The atoms of the polypeptide chain are used as references to define the values of the dihedral angles. As indicated in Fig. 2.5, the position of C'_{i-1} and C'_i defines ϕ_i, that of N_i and N_{i+1} defines ψ_i. When the two reference atoms are eclipsed, the angle is taken to be equal to zero. Thus the fully extended chain is described by $\phi_i = \psi_i = \omega_i = 180°$. The planarity of the peptide bond restricts ω to the values of 0° (cis) and 180° (trans) or values very close to them. Except for proline and hydroxyproline, the peptide bond is generally trans. Regular structures correspond to definite values of the ϕ and ψ angles.

Similar notations have been recommended for the description of the side chain. Starting from the C^α of the backbone, the consecutive carbon atoms are denoted C^β, C^γ, etc. In branched side chains, a number is added to the Greek letter to differentiate the branch, as indicated in Fig. 2.6: the general symbol for the dihedral angles is χ with a superscript number to distinguish the various angles, i.e., χ^1 is the torsion angle around C^α—C^β; χ^2 is around C^β—C^γ. In branched chains, a second superscript number is added, for example, the angle around C^β—C^γ_1 is referred to as χ^{21} and around C^β—C^γ_2

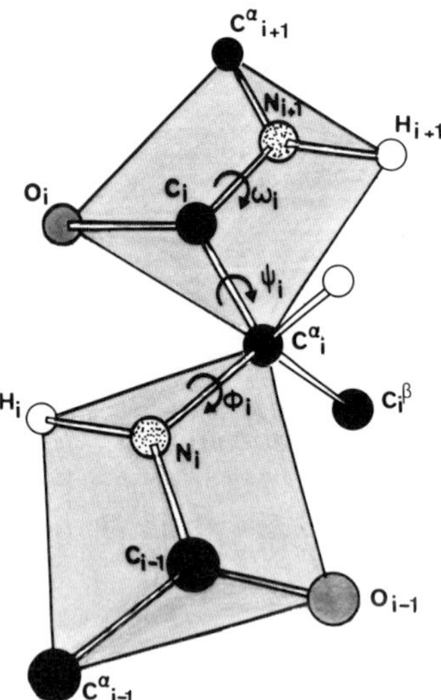

Fig. 2.5. Representation of dihedral angles in protein backbone.

as χ_i^{22}. The subscript i must be added to all symbols for atoms and dihedral angles to designate the residue to which they refer (e.g., χ_i^{22}).

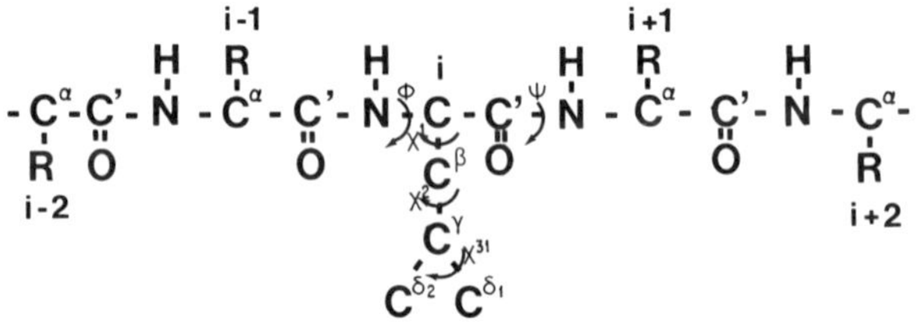

```
            i-1                    i+1
      H     R      H   i      H     R      H
-Cα-C'-N -Cα - C'- N Φ C -C' Ψ N - Cα- C'- N -Cα-
 R   O           O    χ1  Cβ O           O       R
i-2                   χ2                        i+2
                          Cγ
                            χ31
                       Cδ2  Cδ1
```

Fig. 2.6. Dihedral angles in side chain residues of proteins.

Because of the planarity of the peptide unit, the conformation of the polypeptide chain (backbone) is determined by the values of the dihedral angles (ϕ_i, ψ_i). When deviation from planarity occurs, values of ω have to be added $(\phi_i, \psi_i, \omega_i)$. Since there are only two main variables per residue, the local conformation around a residue could be represented by a point (ϕ_i, ψ_i) in a two dimensional diagram. The plot of ψ versus ϕ in rectangular coordinates is called the conformational map, or Ramachandran diagram (Fig. 2.7). This representation is used principally in the studies of conforma-

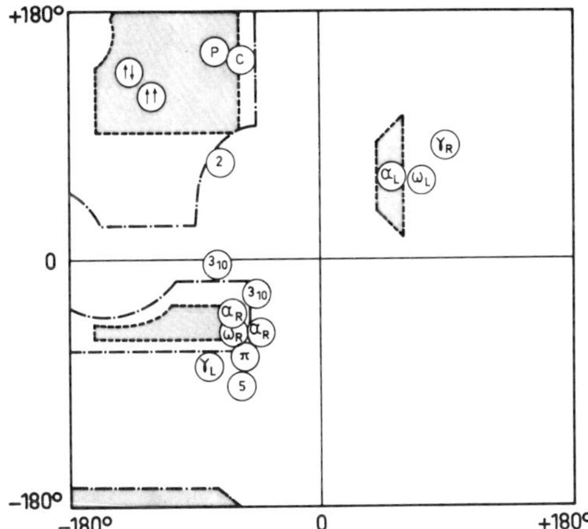

Fig. 2.7. A Ramachandran's diagram indicating the position of the main ordered structures: (α_R), (α_L) are the right and left α helix respectively; (π), (π) helix; (3_{10}), 3_{10} helix; (γ_R), (γ_L), right- and left-handed γ helices respectively; (ω_R), (ω_L), right- and left-handed ω helices respectively; (↑↓), antiparallel β pleated sheet; (↑↑), parallel β pleated sheet; (2), 2_7 ribbon; (P), polyproline helix; (C), collagen coiled coil; and (5), five-membered ring (adapted from Dickerson and Geis, 1969).

tional restrictions and preferences, or to summarize the distribution of local conformations in proteins of known three-dimensional structure. On this diagram the enantiomeric structures are located symmetrically with respect to the origin; for instance, right-handed helix and left-handed helix. Because of serious restriction on folding resulting from unfavorable interactions (see Chapter 3), only a small part of the conformational map corresponds to possible local conformations.

A polypeptide chain organized in an irregular folding is represented by a sequence of points which are more or less scattered on the Ramachandran map. If a polypeptide chain is arranged in regular structure (i.e., the local conformation of each amino acid residue is identical and thus defined by the same ϕ and ψ), then the chain folds into regular arrangement and is specified by a single point on the conformation map.

Proline is more restricted than the other amino acids, because the N—C$^\alpha$ bond pertains to a five membered ring. Consequently ϕ is restricted to a value near $-70°$ leaving only ψ as a variable. Furthermore, the peptide bond preceding the pyrrolidine ring can exist in two conformations, the cis or trans form, and ω_{i-1} can have the value 0° or 180° respectively (Fig. 2.8). For all the other amino acids the trans form is preferred because of more favorable

trans bond with proline cis bond with proline

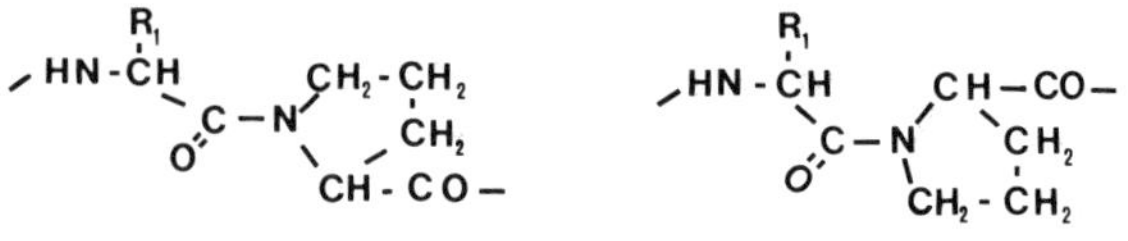

Fig. 2.8. The two conformations of peptide bond preceeding a proline residue.

interaction energies within the peptide chain. The preference for the trans form in peptides having no substituent on the N atom results from the instability of the cis form by ca. 8 kcal/mole over the trans form. This instability is caused by (1) unfavorable interactions of the C_i^{α} and $H_{C_1}^{\alpha}$ with C_2^{α} and $H_{C_2}^{\alpha}$ atoms, (2) favorable interactions present in the trans form which are disallowed in the cis form, and (3) conformational entropy. In N-substituted peptides, as X-Pro peptides, the instability of the cis form is much less than in the unsubstituted peptides, ($\Delta G_{\text{trans}\rightarrow\text{cis}}$ is only 1.87 kcal/mole for Gly–Pro whereas it is 8.33 kcal/mole for Gly–Gly), and a small amount of energy from long range interactions can turn the trans into the cis form (Zimmerman and Scheraga, 1976). In earlier studies, it was proved by infrared spectra that the trans conformation is the one occurring in all open polypeptide chains (Sasisekharan, 1962) except polyproline which can exist in both cis and trans forms. In smaller cyclic peptides, Pro may be in cis form as experimentally observed by ^{13}C NMR., and the cis form content depends on the solvent (Wüthrich *et al.*, 1974). In fact the trans form has been observed in all the naturally occurring proteins or polypeptides, except in small cyclic oligopeptides where the existence of the ring excludes the possibility of the trans form.

2.4 BACKBONE CONFORMATION

As emphasized in Chapter 1, examination of protein three-dimensional structures reveals two main characteristics in polypeptide chain folding: (1) a hierarchy of organization with different levels of structure, and (2) within each level of structure, a small number of structural patterns are encountered in all globular proteins despite the great number of conformations which may be expected (see Schulz and Schirmer, 1979, and the review in Thomas and Schechter, 1980; Richardson, 1981). At the lower level, secondary structures that commonly occur are essentially right-handed α helix, extended structures, and reverse turns. Then, from interactions between adjacent segments of regular structures, arise only few structural motifs which are found in many proteins (see Fig. 2.9).

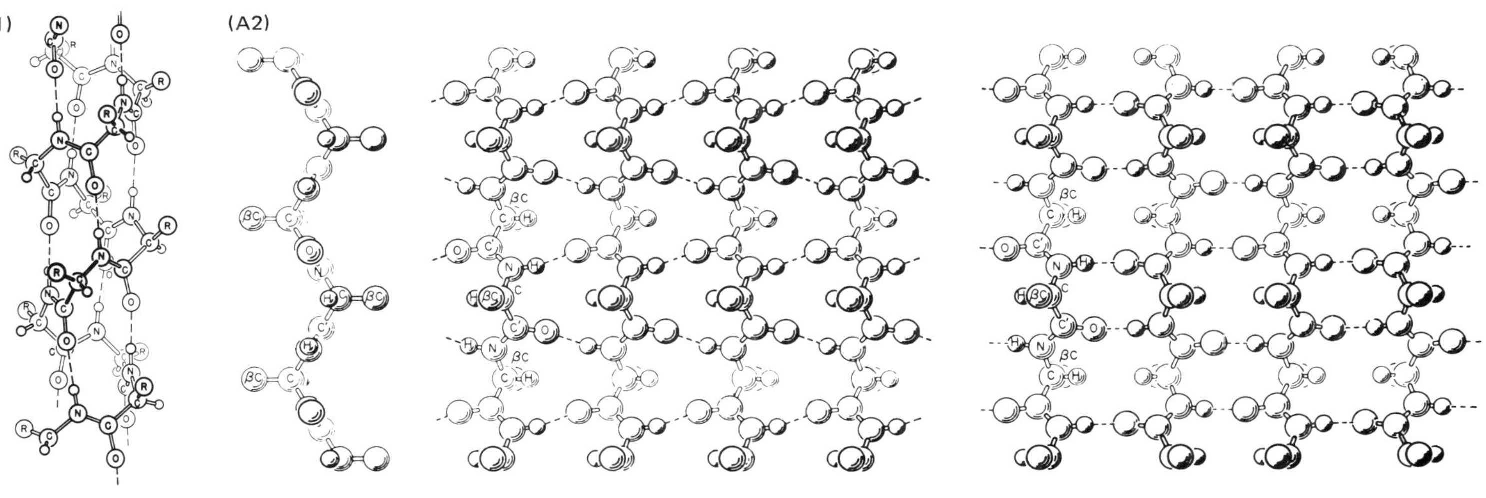

Fig. 2.9. Main structural motifs in globular proteins. (A) ordered structures: (1) α-helix from Cozey and Pauling (1956); (2) segment cf extended structure, parallel and anti-parallel β sheets from Corey and Pauling (1956); (3) β turns type I and II from Lewis *et al.* (1971). (B) Supersecondary structures. (C) Diagrammatic protein structures: left, chymotrypsin with two β barrels; right, triose phosphate isomerase (alternance of helices and β barrels); (from Schulz and Schirmer, 1979) (courtesy of Schulz).

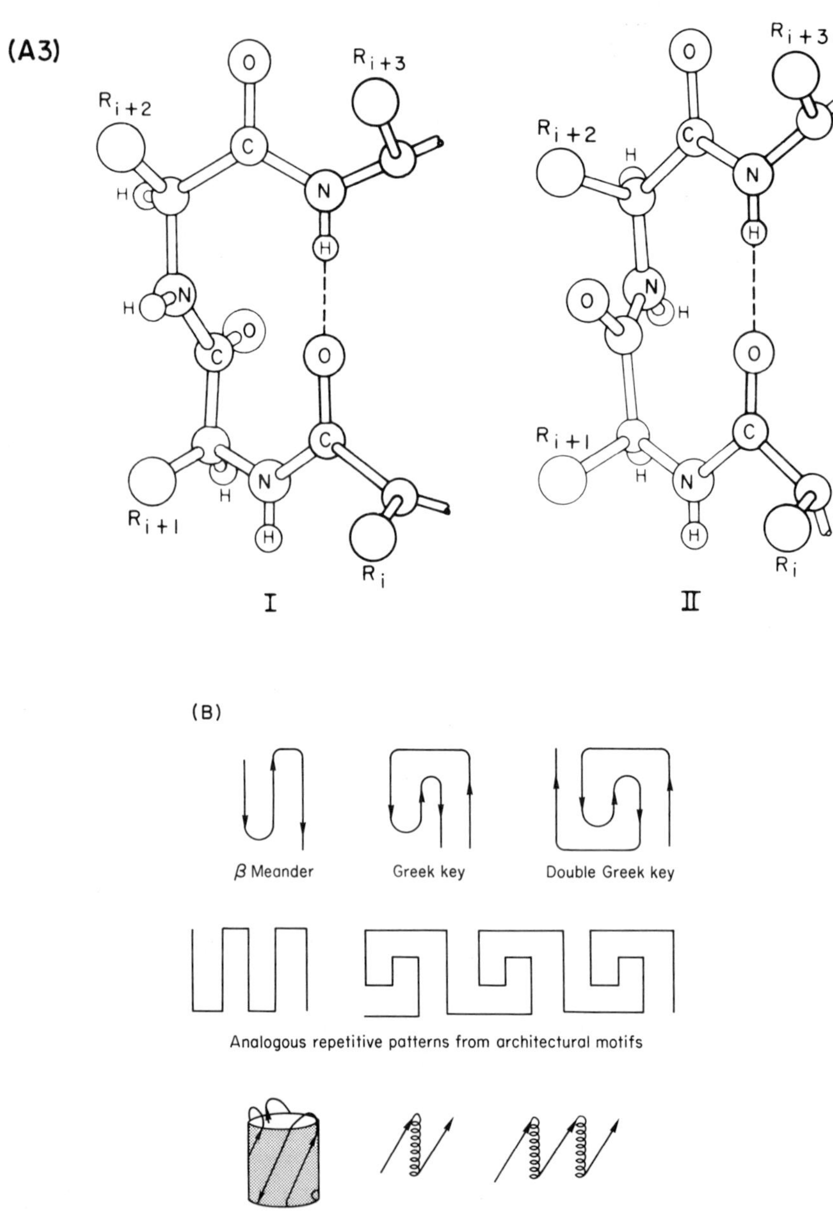

Fig. 2.9 (*continued*)

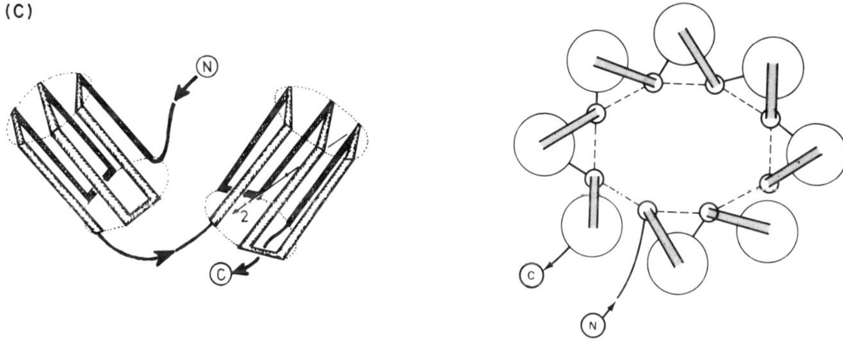

Fig. 2.9 (*continued*)

2.4.1. Characteristics of Ordered Structures in Proteins

The existence of ordered structures for the polypeptide chain was first proposed by Astbury and his group in the 1940s (Astbury, 1940; Astbury and Bell 1940, 1941), by Huggins (1943) and Bragg and co-workers (1950) in connection with the X-ray structures of some fibrous proteins. Expanding these earlier studies, Pauling (1940), Pauling and Corey (1951a,b,c,d,e, 1952, 1953a,b), Pauling and co-workers (1953), and Corey and Pauling (1953, 1956) provided the basis for the current knowledge of protein structural properties. The authors have described a large variety of helical and regular structures. The structural principles which are implied in the formation of regular structures were well-defined and most of them are still considered to be valid. According to Corey and Pauling (1953) five main considerations govern the structure of proteins: (1) bond length and angles are fixed and have standard values (see Fig. 2.3b); (2) the peptide bond is planar, the trans-form is preferred; (3) dihedral angles, ϕ and ψ, were accepted to be restricted to a few favorable orientations; (4) the maximally possible number of NH . . . O=C hydrogen bonds are formed; and (5) the length and angular deformations of the hydrogen bonds vary within narrow limits. Conditions 3 and 4 are no longer considered important for many structures frequently encountered.

Regular structures in proteins (globular and fibrous proteins) are essentially helical structures, extended structures (with the different pleated sheets); reversals or turns of the polypeptide chain are also considered in this section.

2.4.1.1. Helices

When the twists at the α carbon are the same, the polypeptide chain folds into a helix. If the pitch is low enough, hydrogen bonds can form between C=O and N—H groups of adjacent turns of the helix. These hydrogen bonds are roughly parallel to the helical axis. Helices are uniquely defined by the specification of the dihedral angles ϕ and ψ, but, they can be described by parameters characterizing the geometry of the helix. According to the recommended standard notation, these parameters are n, number of residues per turn; h, unit height or translation per residue along the helix axis; $t = 360°/n$, unit twist or angle of rotation per residue along the helix axis. A helix could also be defined in terms of symmetry and hydrogen bond arrangement. In this definition it is referred to as a n_r helix (where n is the number of residues per turn and r is the number of atoms in a ring formed by a hydrogen bond and the segment of main chain connecting its extremities). This notation was introduced by Bragg and co-workers (1950).

Different helices can exist and were described by Pauling and Corey (1952, 1953a,b). However, the most important of the structures proposed by these researchers is certainly the α helix. In this structure each turn of the helix contains 3.6 residues ($n = 3.6$). The pitch (i.e., the distance between adjacent turns) is 5.4 Å, corresponding to a rise of 1.5 Å per residue ($h =$ 1.5 Å). Each N—H group is hydrogen bonded with the C=O group of the fourth preceding amino acid residue, and the number of atoms in the ring formed by each hydrogen bond is 13 (Fig. 2.9). Therefore the helix is also designated as helix 3.6_{13}. The α helix can exist in a right- and a left-handed form. These two forms are mirror images of each other. They are indicated in the Ramachandran diagram (Fig. 2.7). The main parameters for the regular structures are indicated in Table 2.1. The helical parameters for the right-handed α helix are $n = 3.6$, $h = 1.5$ Å, and $t = 100°$.

The specification of dihedral angles (ϕ, ψ) defines uniquely the helix geometry (i.e., it corresponds to unique values of n and h). However, a given value of n and h may not always be sufficient to determine the local conformation (ϕ, ψ); it may correspond to two distinct ϕ, ψ values. The equations for n and h are the following (adapted from Schellman and Schellman, 1964)

$$\cos(180°/n) = +0.817 \sin(\phi + \psi)/2 + 0.045 \sin(\phi - \psi)/2$$
$$h \sin(180°/n) = -2.967 \cos(\phi + \psi)/2 - 0.664 \cos(\phi - \psi)/2$$

The actual numerical values of the constants depend on the values chosen for bond geometry (bond lengths and bonds angles).

Additional information, such as that on hydrogen bonding, is required to distinguish the exact conformation. An alternative conformation to that

TABLE 2.1

Characteristics of Regular Structures in Proteins or Polypeptides[a]

Structure	ϕ	ψ	n	r	h (Å)	t
δ Helix	-98	$-80°$	4.3	14	1.23	85.°4
α Right-handed helix	$-58°$	$-47°$	3.6	13	1.5	100°
α Left-handed helix	$+58°$	$+47°$	3.6	13	1.5	100°
ω Helix	$-64.°06$	$-55.°06$	4	13		90°
π Helix	$-57.°06$	$-69.°6$	4.4	16	1.15	81.°8
3_{10} Helix	$-75.°5$	$-4.°5$	3	10	2.00	120°
α_{II} Helix	$-93°$	$-18°$	3.6	13	1.5	100°
γ Helix	$-84.°64$	$-91.°44$	5.1	17	0.98	70°
2.2_7 Helix	$-78.°1$	59.°2	2.17	7	2.75	
2_7 Helix	$-74.°9$	69.°5	2	7	2.80	
Fully extended chain	180°	180°	2		3.63	180°
β parallel structure	$-199°$	113°	2		3.25	
β antiparallel structure	$-139°$	135°	2		3.5	
Polyproline I ($\omega = 0°$)	$-83°$	158°				
Polyproline II ($\omega = 180°$)	$-78°$	149°				
Polyglycine II	$-80°$	150°				
Collagen	$\begin{cases}-51\\-76\\-45\end{cases}$	$\begin{cases}153\\127\\118\end{cases}$				

[a] ϕ, ψ represent dihedral angles of the backbone; n, number of residue per turn; r, number of atoms in a ring formed by hydrogen bond and the segment of main chain connecting its extremity; h, unit height; $t = 360°/n$, unit twist.

of the α helix is the α_{II} helix ($\phi, \psi = -93°, -18°$) where n and h are identical. The α_{II} helix is organized in such a way that hydrogen bonds cannot be formed, therefore, it is less stable than a regular α helix. It could occur in some proteins at the C-terminus of the α helical segment (Némethy *et al.*, 1967).

Among the possible hydrogen bonded helices, the ω helix, the π helix, and the 3_{10} helix have been reported. The ω helix, which is referred to as a 4.0 helix, has been observed for poly-β-benzyl-L-aspartate by X-ray diffraction and infrared methods (Bradbury *et al.*, 1959). The π helix or 4.4_{16} helix also has a favorable hydrogen bonding between adjacent turns. The π helix and γ helix have never been observed; the ω helix has not been found in globular proteins. Recently another type of helix, the δ helix or left-handed 4.3_{14} helix, was described by Chandrasekaran *et al.* (1979) as only slightly less stable than the right-handed α helix. Short helical segments in several proteins were reported to have a conformation corresponding to that of 3_{10} helix. In some cases, however, when this conformation is reported to occur near the end of an helical region, it is not easy to distinguish it from the α_{II} helical termination, since their (ϕ, ψ) values are very close. Nevertheless, segments of 3_{10} helix have been identified in lysozyme, chymotrypsin, myoglobin, hemoglobin, carbonic anhydrase, cytochrome b_5, and ribonuclease. Segments of α_{II} helix have been found in lysozyme, carbonic anhydrase, and carboxypeptidase (Liljas and Rossmann, 1974). From data obtained with copolypeptides, Takahashi and co-workers (1978) proposed that the C-terminal side of the α helix is more stable than the N-terminal side.

2.4.1.2. Extended Structures

When the polypeptide chain is fully extended, no hydrogen bond can be formed within the chain, because N—H and C=O groups are oriented perpendicularly to the direction of the chain. However, hydrogen bonding with neighboring chains is possible and, by a slight twisting, stable structures can be formed with possible extended structures. Two of them, proposed by Pauling and Corey (1953a,b), are very important and are the β structures or pleated sheet. The parallel and antiparallel β structures correspond to an arrangement in which adjacent peptide chains are directed in the same and in the opposite direction respectively. All N—H and C=O groups are engaged in good hydrogen bonds in both structures, however their orientations are slightly different. Pleated sheets were observed first in fibrous proteins (keratine, silk); it is well known that they also exist in many globular proteins. In the latter, the extention of the sheets is limited often to a couple of short chain segments. For a perfect structure, the characteristics are $n = 2$, $h = 3.25$ Å for the parallel, and $n = 2$, $h = 3.5$ Å for the antiparallel arrangement. An antiparallel β structure formed by a unique polypeptide chain folded back upon itself has been called the cross β structure.

2.4.1.3. Chain Reversals

Backbone connection between β strands can be classified into two main categories. The first category is the hairpin, or plain, or same end connections, while the second is referred as crossover, or cross, or opposite end connections (Chothia, 1973; Rao and Rossmann, 1973; Banner *et al.*, 1975; Richardson *et al.*, 1976; Richardson, (1976, 1977). Figure 2.10 shows the folding of the polypeptide chain according to these two general possibilities.

Reversal of the direction of the chain can be achieved over a distance of two residues. The simplest examples of such reversal of the chain have been termed β fold, or β turn, or β bend. They were analyzed by different researchers (Venkatachalam, 1968; Lewis *et al.*, 1971, 1973a, Kuntz, 1972;

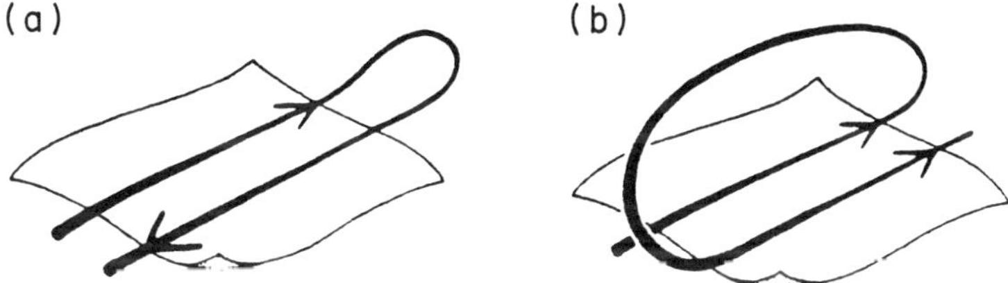

Fig. 2.10. Connectivities of β strands (a) same end or hairpin connection, (b) crossover or opposite end connection (from Richardson, 1976).

Chou and Fasman, 1974a,b, 1977; Crawford *et al.*, 1973; Zimmerman and Scheraga, 1977a,b, 1978a,b,c; Némethy and Scheraga, 1980). A β bend involves a region of four consecutive residues i through $i + 3$ folding back on itself by nearly 180°. A bend can be defined when distance between C_i^α and C_{i+3}^α is less than 7 Å, and when $(i + 1)$ or $(i + 2)$ residue is not in α_R helix. Venkatachalam (1968), was the first to consider different possible types of bends (type I, II, and III). Type III is equivalent to one turn of a 3_{10} helix. As can be seen from Fig. 2.9, type I and type II β turns differ in the orientation of the central peptide unit and in several interactions. Eleven types of bends have been observed in proteins; type N′ corresponds to the mirror image of type N. Each type is defined by the values of $(\phi, \psi)_2$ and $(\phi, \psi)_3$ angles. These values are given in Table 2.2. Several β turns are hydrogen bonded between O_i and N_{i+3}; that requires the distance between those two atoms be less than 3.5 Å. Many distorted bends, similar to β bends, have no hydrogen bond. Of the 135 bends in proteins examined by Lewis and co-workers (1973b), 40% did not possess the hydrogen bond between i and $i + 3$ residues. Némethy and Scheraga (1980), have reconsidered the different bend types and examined the strereochemical requirements for the existence of hydrogen bonds in bends. The possibility of a formation of intramolecular hydrogen bonds depends on the relative orientation of the peptide groups. Hydrogen bonds may be found in bends with dihedral angles different from those of ideal bends. It is especially true in the case of type I β turns.

Most frequent bends have Asn, Cys, Asp in the first position; Asn, Asp, Gly in the second position; Asn, Asp, Gly in the third position; and Trp, Gly, Tyr in the fourth position (Chou and Fasman, 1977). As noted by Finkelstein and Ptitsyn (1976), proline has a great tendency to be in position 2 of β bends type I and III, glycine to be in position 3 of β bends type II, and glycine as well as short hydrophilic side chains also have the tendency of initiating β bends type II and III.

Another possibility of folding for the same end chain reversal is the γ turn, in which the direction of the polypeptide chain reversed over three

TABLE 2.2

Classification of Bend types According to the Values of the Dihedral Angles of Residues $i+1$ (ϕ_2, ψ_2) and $i+2$ (ϕ_3, ψ_3)[a,b]

Bend Type	ϕ_2	ψ_2	ϕ_3	ψ_3	Number of Ideal Bends	Number of Non-Ideal Bends	Total Number of Bends	Number of H-Bended Bends
I	−60	−30	−90	0	130	46	176	99
I′	60	30	90	0	8	5	13	10
II	−60	120	80	0	41	23	64	43
II′	60	−120	−80	0	15	5	20	16
III	−60	−30	−60	−30	66	11	77	45
III′	60	30	60	30	11	2	13	7
IV	A bend with 2 or more angles differing by at least 40° from those given above				0	35	35	5
V	−80	80	80	−80	1	2	3	0
V′	80	−80	−80	80	0	4	4	2
VI	A *cis*-Pro at position 3				8	0	8	6
VII	A kink in the protein chain created by $\psi_2 \simeq 180°$ and $\lvert\phi_3\rvert < 60°$ or $\lvert\psi_2\rvert < 60°$ and $\phi_3 \simeq 180°$				8	0	8	1
Total					283	133	421	234

[a] According to Chou and Fasman (1977).

[b] Frequency of occurrence of ideal and non-ideal bends has been evaluated in 26 proteins.

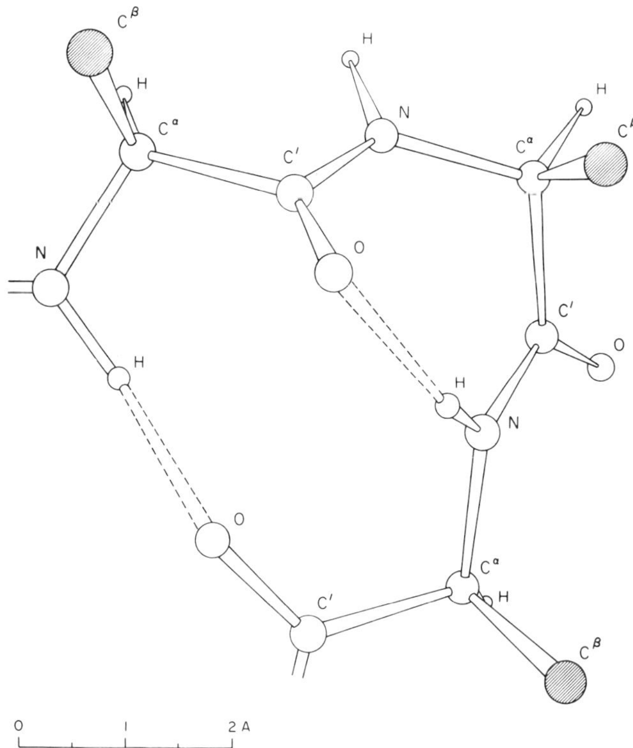

Fig. 2.11. The γ turn according to Némethy and Printz (1972) (Courtesy of G. Némethy). It involves only three residues and is stabilized by two hydrogen bonds. Residue N_{i+1} must be a glycine; for other amino acids, the conformation is of high energy.

residues and not four, as in the β bend. The γ turn was described by Printz *et al.* (1972) and by Némethy and Printz (1972). This conformation, as the β turn, can connect two strands of antiparallel β structure: it requires only three residues. Furthermore, it is stabilized by two hydrogen bonds (Fig. 2.11). The γ turn was reported by Matthews (1972) in the X-ray structure of thermolysin; it involves residues 25–27. Proton and ^{13}C magnetic resonance studies have shown the occurrence of the γ turn in peptides of elastin (Urry *et al.*, 1975). This structure was reported to exist in solution (Khaled *et al.*, 1976).

2.4.1.4. Some Other Regular Structures

Some other intermolecularly, hydrogen-bonded, multiple-chain structures do exist. Polyglycine is known to exist in two forms: (1) polyglycine I which is in an extended structure, and polyglycine II in which the chains are packed hexagonally with hydrogen bonds linking each chain to its neighbors. The parameters are $n = 3$, $h = 3.1$ Å. In collagen, three chains in extended conformation are twisted around each other to form a super coiled structure. The chains are linked by hydrogen bonds. Every third residue is a glycine and the preceding one is generally a proline.

Other types of structures are represented by polyproline I and II. These two structures correspond to the fact that the peptide bond can exist either in the cis ($\omega = 0°$) or in the trans form ($\omega = 180°$). Polyproline I, in which the peptide unit is in the cis form, is twisted in a rigid right-handed three-fold helix ($n = 3.3$, $h = 2.2$ Å). Polyproline II, which corresponds to the trans form of the peptide bond, is a left-handed, three-fold helix ($n = 3$, $h = 3.12$ Å). This conformation is very similar to that of collagen (Dickerson and Geis, 1969; Schellmann and Schellmann, 1964). Of these regular structures, only segments of polyproline conformation can exist in globular proteins.

2.4.2. Characteristics of Nonregular Structures in Proteins

The term nonregular structure in proteins does not mean random conformation of the chain. Among the possible nonregular structures, the random polypeptide chain must be considered to be a description of a denatured protein. However, nonregular but fixed structures do exist in the native conformation of a protein. Even for many globular proteins, they represent the larger part of the molecule, the ordered region being in smaller proportion.

2.4.2.1. Random Polypeptide Chain of Denatured Proteins

The polypeptide chain after denaturation no longer occupies a compact volume; the chain becomes flexible and occupies a large volume. The dihedral angles (ϕ_i, ψ_i) are no longer restricted to certain values, but may vary over all the accessible regions of the Ramachandran map. The conformation of each residue varies in time and can occasionally be in a conformation corresponding to a regular structure. Under conditions where the random structure is stable, the probability for several successive residues to be in the same conformation is very weak. Because of the presence of steric restrictions, the chain is not completely random in the sense of an idealized random coil. However, the rotational freedom around the single bonds is sufficient to permit the loss of correlations along the structure. Random coil is a Boltzmann average over the ϕ, ψ map.

2.4.2.2. *Non-ordered Fixed Structures*

In many native globular proteins large parts of the polypeptide chain are not organized in regular structures as helices or pleated sheets. However, the chain has a fixed and well-defined conformation in these regions of the molecule. The dihedral angles (ϕ_i, ψ_i) have well-defined values. The folding of the chain is as well-determined as that of the ordered regions. Sometimes, these parts of the chain were referred to as the random regions; however, this denomination is rather ambiguous. Indeed, in terms of flexibility and of conformational freedom of the individual residues, they differ very strongly from the random coil. It is recommended that these regions be called non-ordered fixed structures. Some of their physical properties are quite different from those of the random polypeptide chain (e.g., spectroscopic properties of buried chromophores, chemical reactivity of these groups, antigenicity).

2.4.3. Occurrence of Regular Structures in Proteins

The α helices and β sheet strands correspond to repeating structures characterized by the ϕ and ψ values of the torsion angles, and may involve a variable number of residues. Reverse turns do not have this repeating structure and, if they display very precise geometry, they involve a small number of adjacent residues. Nevertheless, they are included in the secondary structure. The problem of characterizing secondary structures in proteins was in fact made difficult by the absence of precise rules. A particular difficulty is locating the end point of helices. Different methods have been used for determining secondary structures in globular proteins, but all of them are based on somewhat arbitrary criteria. The ϕ and ψ torsion angles were used to assign local conformations of each amino acid. More recently used (ϕ and ψ) angles were calculated from atomic coordinates (see, e.g., Burgess *et al.*, 1974; Maxfield and Scheraga, 1976; Tanaka and Scheraga, 1976a,b,c). However, small errors in ϕ and ψ angles can introduce incorrect assignments. Crystallographers prefer to use the hydrogen bonding rather than ϕ and ψ angles for the assignment of regular conformations.

Distances between C^{α}—C^{α} and α torsion angles (i.e., dihedral angles about virtual bonds C^{α}—C^{α}) (see Chapter 4, Section 4.3.1.1) were also used to identify secondary structures (Srinivasan *et al.*, 1975). Levitt and Greer (1977) proposed rules based on combination of three criteria, inter α carbon distances, α torsion angles, and hydrogen bond patterns. The authors have proposed this procedure as the first objective compilation of secondary structures in a large number of globular proteins. The automatic identification procedure is objective, but somewhat arbitrary definitions still remain and their method involves simplifications. It is based on a simplified representation of the polypeptide chain (see Chapter 3). The residues in α helix or in β strand do have different local conformations as measured by the angles; however, the spread in values is rather broad ranging from $-180°$ to $-50°$ for α helical conformation and from $-50°$ to $+150°$ for the β structure. In the overlapping range (i.e., between $-60°$ and $-40°$) the right-handed, α-helical conformation is assigned to residue i provided neighboring residues $i-2$, $i-1$, $i+1$, and $i+2$ have been assigned in the right-handed helix. Residues which have not been assigned to either helix or sheet are tested for reverse turns. These criteria probably lead to an overestimate of secondary structures in proteins. Criterion of reliability of secondary structure assignment is obtained by comparison with the secondary structure derived from the independent set of coordinates available for some proteins. By the combination of methods, the number of extra helical residues is only slightly greater than the number of missed helical residues. However, many more extra β sheets are found than missed.

Levitt and Greer applied their method for analyzing atomic coordinates of 62 proteins. Table 2.3 lists the content of each type of secondary structures that are calculated according to the data of Levitt and Greer for the different categories of proteins.

The authors have considered the four classes of proteins previously defined by Levitt and Chothia (1976): (1) includes all α proteins that have only helices; (2) all β proteins that contain mainly β structure; (3)($\alpha + \beta$) proteins that have both types of secondary structure (these structures do not mix and seem to aggregate separately); (4) α/β proteins that have approximately alternating segments of each structure (see Section 5). In fact, only three of these classes are clearly defined: classes I, II and IV. Proteins which are not included in one of these classes are classified as ($\alpha + \beta$). Those which are mainly α proteins (type I), without any significant amount of β structure, have a high content of helix averaging 80.7% and a relatively low content of reverse turns. The hemoglobins, myoglobins, which are heme proteins, and hemerythin are α proteins. Conversely some proteins such as rubredoxin and superoxide dismutase have very little helix, but have β strands. The average content of β structure in β proteins is 56%. The content of reverse turns is 21.6% which is very close to the mean value obtained for all proteins except the α proteins. Extensive and perfect regions of β strands are found in carboxypeptidase, carbonic anhydrase, and concanavalin A.

Some proteins contain both types of structures, they are either ($\alpha + \beta$) proteins or α/β proteins. The proportion of α helices and β strands is more balanced in this last category than in ($\alpha + \beta$) proteins. Besides these regular structures, the content of nonregular structure in proteins is rather large.

TABLE 2.3

Evaluation of the Different Secondary Structures in the Four Classes of Proteins[a]

(A) All α Proteins

Protein	Number of Amino Acids				Percentages		
	Total	α Helix	β Sheet	Turns	% α Helix	% β Sheet	% Turns
Calcium-binding protein B (carp)	108	63	6	18	58.3	5.6	16
Azomyohemerythrin (*Thermiste pyroides*)	113	90	0	12	79.6	0	10.6
Hemerythrin (*Phascolopsus gouldii*)	118	93	0	14	78.8	0	11.9
Carboxyhemoglobin (*Glycera dibranchiata*)	147	116	0	23	78.9	0	15.6
Cyanmethemoglobin (sea lamprey)	148	116	0	23	78.4	0	15.5
Metmyoglobin (sperm whale)	153	134	0	10	87.6	0	6.5
Aquomethemoglobin (horse)	287	226	0	22	78.7	0	7.6
Deoxyhemoglobin (horse)	287	246	0	22	85.7	0	7.6
Deoxyhemoglobin (human)	287	246	0	24	85.7	0	8.3
Total	1,648	1,330	6	168	80.7	0	10.2

[a] The secondary structures are those estimated by Levitt and Greer (1977) in the classes of proteins defined by Levitt and Chothia (1976).

TABLE 2.3 (*continued*)

(B) All β Proteins

Protein	Total	α Helix	β Sheet	Turns	% α Helix	% β Sheet	% Turns
Rubredoxin at 1,5 A (*Clostridium pasteurianum*)	54	0	22	17	0	40.7	31.5
Variable part of Bence–Jones dimer (human myeloma REI)	214	0	153	42	0	71.5	19.6
Immunoglobulin G F'ab (human myeloma NEW)	426	11	286	76	2.5	67.2	17.8
Bence–Jones dimer Meg (human)	430	29	296	74	6.7	69.4	17.2
Prealbumin dimer (human)	246	14	148	46	5.6	60.2	18.7
Copper–Zinc superoxide dismutase (bovine)	151	0	78	53	0	51.7	35.1
Concanavalin A (Argonne) (jack bean)	237	6	150	52	2.5	63.3	22
Concanavalin A (Rockfeller) (jack bean)	237	6	153	52	2.5	64.6	22
Alkaline protease B (*Streptomyces griseus*)	185	21	96	29	11.35	52	15.7
Trypsin-DIP (bovine)	224	19	125	54	8.4	55.8	24.1
α-Chymotrypsin A (MRC) (bovine)	236	17	129	54	7.2	54.7	22.9
α-Chymotrypsin A (Michigan) (bovine)	236	23	118	58	9.7	50	24.5
Chymotrypsinogen A (bovine)	245	27	121	52	11.02	49.4	21.2
Elastase (porcine)	240	19	109	68	7.9	45.4	28.3
Total	3,361	192	1,884	727	5.7	56	21.6

TABLE 2.3 (*continued*)

(C) ($\alpha + \beta$) Proteins

Protein	% α Helix	% β Sheet	% Turns	Total	α Helix	β Sheet	Turns
Insulin dimer (porcine)	60.8	14.7	10.8	102	62	15	11
Trypsin inhibitor (bovine)	25.9	44.8	8.8	58	15	26	9
Ferredoxin (*Peptococcus aerogenes*)	24	29.6	22.2	54	13	16	12
Ferricytochrome *b* 5 (bovine)	55.2	25.3	10.3	87	48	22	9
Oxidized high-potential iron protein (*Chromatium vinarum* D)	8.2	10.5	31.8	85	7	1	27
Ferrocytochrome *c* (tuna)	47.6	10.7	22.3	103	49	11	23
Ferricytochrome *c* "outer" (tuna)	48.5	9.7	17.5	103	50	10	18
Ferricytochrome *c* "inner" (tuna)	48.5	12.6	18.4	103	50	13	19
Ferrocytochrome *c* (bonito)	47.6	5.8	31	103	49	6	32
Ferricytochrome *c* 2 (*Rhodospirillum rubrum*)	46.4	14.3	26.8	112	52	16	30
Cytochrome *c* 550 (*Paracoccus denitrificans*)	46.3	14.9	20.1	134	62	20	27
Ribonuclease A (bovine)	22.6	46	18.5	124	28	57	23
Ribonuclease S (bovine)	25	53.2	15.3	124	31	66	19
Lysozyme (chicken)	45.7	19.4	22.5	129	59	25	29
Nuclease (*Staphylococcus aureus*)	26.8	38.8	24	14,2	38	55	34
Papain (papaya)	27.8	29.2	24.5	212	59	62	52
Thermolysin (*Bacillus thermoproteolyticus*)	39.5	31.3	18.7	316	125	99	59
Total	38.11	25.2	20.7	2,091	797	528	433

TABLE 2.3 *(continued)*

(D) α/β Proteins

Protein	% α Helix	% β Sheet	% Turns	Total	α Helix	β Sheet	Turns
Oxidized thioredoxin (*Escherichia coli*)	40.7	30.5	17.6	108	44	33	19
Oxidized flavodoxin (*Clostridium MP*)	45	34	13	138	62	47	18
Semiquinone flavodoxin (*Clostridium MP*)	45.6	34	13	138	63	47	18
Adenylate kinase (porcine)	61.9	18.9	14.4	194	120	36	28
Phosphoglycerate mutase (yeast)	44	23	21.6	218	96	50	47
Triose phosphate isomerase dimer (chicken)	49.4	24.3	12	494	244	120	59
Carbonic anhydrase B (human)	16	45.3	25.4	256	41	116	65
Carbonic anhydrase C (human)	15.6	44.1	27.3	256	40	113	70
Subtilisin BPN′ (*Bacillus amyloliquifaciens*)	32.4	37.1	21.1	275	89	102	58
Subtilisin novo (*Bacillus amyloliquifaciens*)	30.2	37.1	20.4	275	83	102	56
Carboxpeptidase A (bovine)	39.7	30.3	20.5	307	122	93	63
Carboxpeptidase B (bovine)	37.4	36.8	15.4	299	112	110	46
Malate dehydrogenase (porcine)	43	25.9	24.3	325	140	84	79
Lactate dehydrogenase (dogfish)	41.6	26.1	19.1	329	137	86	63
Lactate dehydrogenase–NAD-complex (dogfish)	42.9	25	21	329	141	82	69
D-glyceraldehyde-3-phosphate dehydrogenase (lobster) (green)	31.5	38.4	22	333	105	128	73
D-glyceraldehyde-3-phosphate dehydrogenase (lobster) (red)	30	36.4	21.6	333	100	121	72
Alcohol dehydrogenase (horse)	29.1	45	18.7	374	109	150	70
Phosphoglycerate kinase (yeast)	29	21.7	20.9	373	108	81	78
Phosphoglycerate kinase (horse)	42	33.4	13.8	407	171	136	56
Hexokinase B III (yeast)	40.4	16.2	24	450	182	73	108
Total	37.2	30.8	19.6	6,211	2,309	1,910	1,215

TABLE 2.3 (*continued*)

(E) All Proteins

All Proteins	Total Amino acids	α Helix	β Sheet	Turns	% α Helix	% β Sheet	% Turns
Total	13,311	4,628	4,328	2,543	34.8%	32.5%	19.1%

Some deviations from ideality in the regular structures have been observed. Although great regularity is observed in helices which deviate very little from ideality (−57°, −48°), distorsion may be observed near the C-terminus of an α helix. There is a tendency for the carbonyl oxygen to tilt outward from the helix. These distorsions cause helices near the C-terminus to appear to be 3_{10} or α_{II} helices. The left-handed helix is sterically impossible.

The cis-conformation of the peptide bond has been observed in globular proteins for a small number of proline residues. It was reported to occur in ribonuclease (Wyckoff *et al.*, 1970), erythrocruorin (Huber *et al.*, 1971), subtilisin (Alden *et al.*, 1973), thermolysin (Matthews *et al.*, 1974), the Bence–Jones dimer REI (Huber and Steigemann, 1974), and carbonic anhydrase (Kannan *et al.*, 1975). One nonproline cis conformation was reported for carboxypeptidase involving the bond between Ser 197 and Tyr 198 (Hartsuck and Lipscomb, 1971). The rare occurrence of cis peptide units is only partially explained by the intrinsic energy of the cis over the trans conformation which is small in proteins (Zimmerman and Scheraga, 1976; see also Section 2.3.2). As reported by Ramanchandran and Mitra (1976), an additional factor is the conformational restriction imposed by the occurrence of a cis peptide unit in a chain of trans units.

In addition to α helices and β strands, reverse turns occur quite often in proteins with an average frequency of about 20%. The 29 proteins analyzed by Chou and Fasman (1977) revealed 408 β turns. They are now considered to be another type of secondary structure since they are common structural components of all globular proteins (Crawford *et al.*, 1973). Their directing influence in folding of the polypeptide chain have been emphasized by several authors (Lewis *et al.*, 1971, 1973a; Kuntz, 1972; Crawford *et al.*, 1973; Chou and Fasman, 1974a,b, 1977). As previously mentioned, 11 bend types are found to occur in proteins (see Fig. 2.12). Some of them are hydrogen bonded, others are not. The frequency of occurrence of these different kinds of bends is summarized in Table 2.3; type I is the most frequent. Some deviations from ideality are observed for the different types of turns. Indeed the distribution of dihedral angles is spread over a rather wide range as shown

Type I. Subtilisin BPN′;
23–26, Gly-Ser-Asn-Val

Type I′. Staphylococcal Nuclease;
94–97, Ala-Asp-Gly-Lys

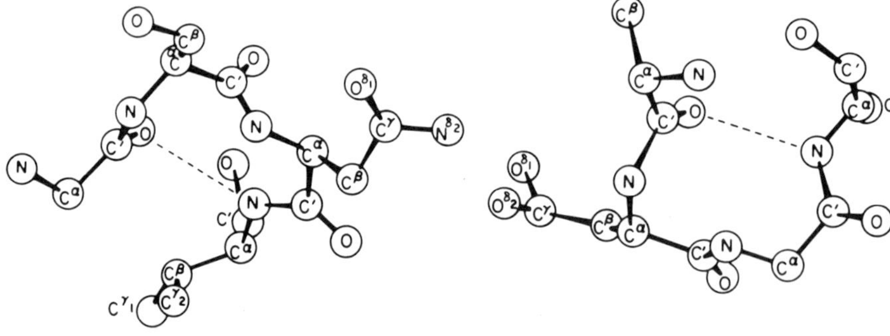

Type II. Hen Egg White Lysozyme;
115–118, Cys-Lys-Gly-Thr

Type II′. Carboxypeptidase A;
277–280, Tyr-Gly-Phe-Leu

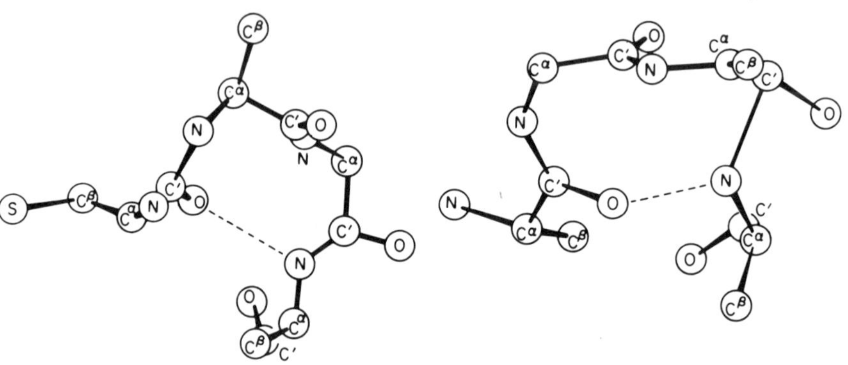

Type III. Sperm Whale Myoglobin;
46–49, Phe-Lys-His-Leu

Type III′. Subtilisin BPN′;
210–213, Pro-Gly-Asn-Lys

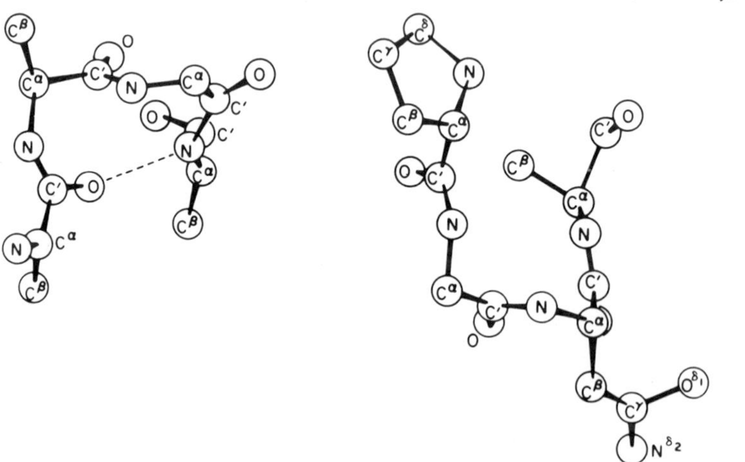

Fig. 2.12. Examples of the different types of β turns found in proteins (From Anfinsen and Scheraga, 1975).

Type IV. α - chymotrypsin ;
99 - 102 , Ile - Asn - Asn - Asp

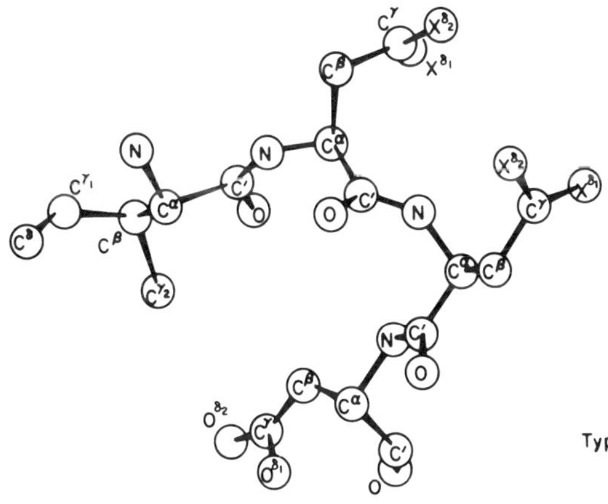

Type V. Horse Ferricytochrome c ;
43 - 46 , Ala - Pro - Gly - Phe

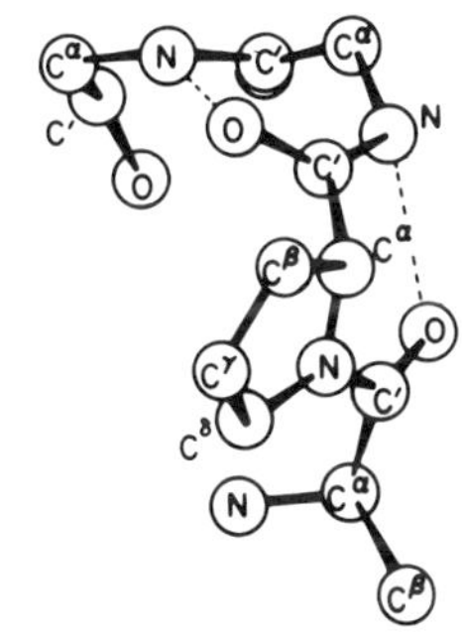

Type VI. Bovine Ribonuclease S ;
91 - 94 , Lys - Tyr - Pro - Asn

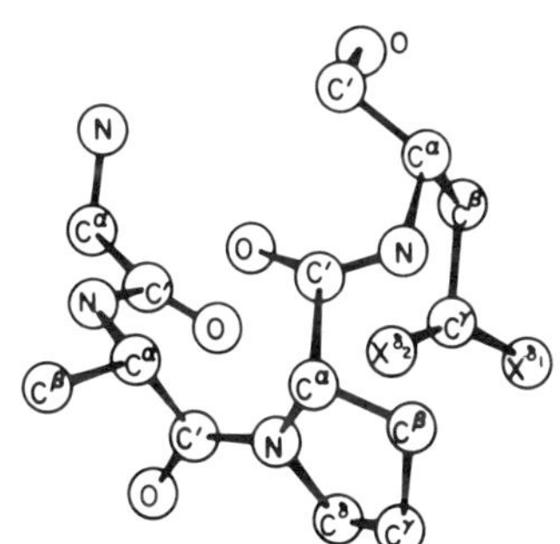

Type VII. α - chymotrypsin ;
67 - 70 , Val - Ala - Gly - Glu

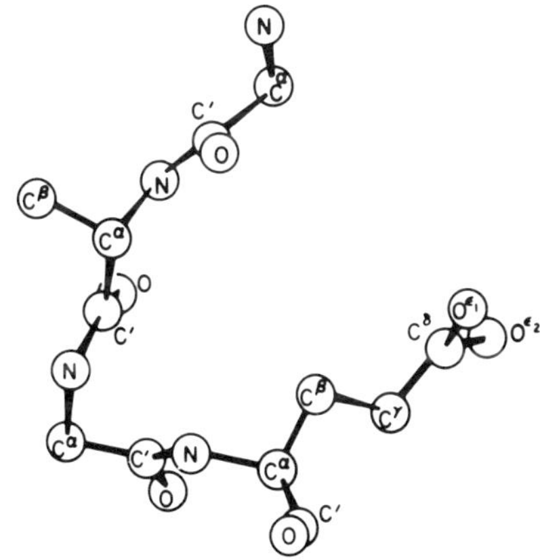

Fig. 2.12 (*continued*)

by Némethy and Scheraga (1980) for residues in bends of type I, II, and III of 23 proteins. As remarked by Kuntz (1972), bends occur on the surface of the protein accessible to solvent. Since most of the bends are associated with the presence of polar residues such as Asp, Asn, Ser, Thr, Tyr, it is not surprising to find bends at the surface of proteins and thus accessible to the solvent. Figure 2.13, according to an adaptation of Chou and Fasman from Matthews *et al.* (1972), shows this fact for cytochrome b_5.

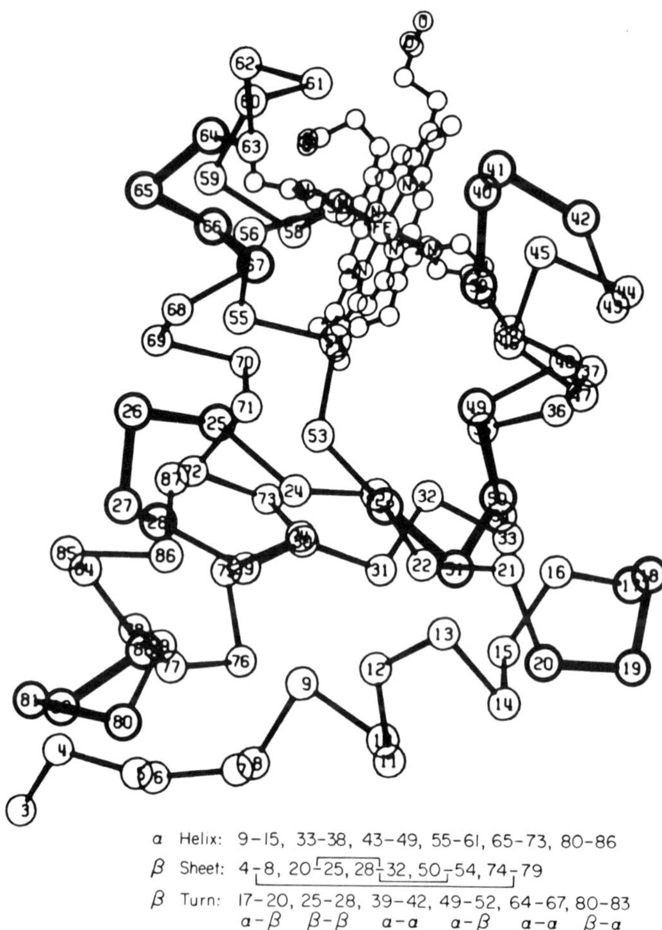

Fig. 2.13. Example of β turns in cytochrome b_5 (an adaption of Chou and Fasman, 1977; from Matthews *et al.*, 1972; with courtesy of G. Fasman). The amino acids forming β turns are indicated by thick circles. They are 17–20. Asn–Ser–Lys–Ser, type III; 25–28, Leu–His–Tyr–Lys, type III; 39–42, His–Pro–Gly–Gly, type I; 49–52, Gln–Ala–Gly–Gly, type II; 64–67, Ser–Thr–Asp–Ala, type III; and 80–83, His–Pro–Asp–Asp, type I.

Rose (1978) proposed some rules to predict the occurrence of chain turns in globular proteins on a hydrophobic basis. Profiles of hydrophobicity were drawn for lysozyme, ribonuclease, and pancreatic trypsin inhibitor in which free energy of transfer is plotted against the residue number. Local minima correspond to the chain turns. A linear relationship between the number of turns in a globular protein and the molecular weight has been reported (Rose and Wetlaufer, 1977). A rather good agreement is found between predicted and experimentally observed secondary structures. About 70% of secondary structures are correctly assigned in several predictive methods (see Chapter 4) (Burgess *et al.*, 1974).

The β bends appear mainly to depend on short range interactions, and this structure has been detected in short oligopeptides. For example, Deslauriers *et al.* (1979) have synthesized a cyclic compound containing dipeptide L-Ala–Gly cyclisized with ε-aminocaproic acid and also the open chain analog. A strong preference for the type II β-bend structure was suggested by both calculation and experimental data in the cyclicized dipeptide. Even the open-chain analog in solution possessed a significant amount of type II β-bend structure among an ensemble of conformations. In methionine enkephalin, (3 Gly–4 Phe) residues were described to form a type II β bend, on the basis of conformational analysis using empirical energy calculation in agreement with available NMR data (Isogai *et al.*, 1977).

Reversal of the chain direction corresponding to β turns are common features of globular proteins. They connect different types of structured or nonstructured segments of the polypeptide chain. Reverse chain folding is stabilized by antiparallel β sheets, helix–helix, and helix–β-strand interactions. Isogai and co-workers (1980) described a new type of structure, referred to as the multiple bend.

In addition to these structures, short pieces of polyproline II structure have been identified in pancreatic trypsin inhibitor (Huber *et al.*, 1971) and in cytochrome c_{551} (Almassy and Dickerson, 1978). Helical segments quite similar to polyproline have been identified in chymotrypsin (Srinivasan *et al.*, 1975).

2.5. SUPERSECONDARY STRUCTURES

Segments of secondary structure occurring in a continuous fold may interact to form some regular aggregates termed for the first time by Rao and Rossmann (1973) as supersecondary structures, and later designated by Levitt and Chothia (1976) as folding units. They represent folding units at a higher level than secondary structure in the hierarchy of protein organization. Packing of the atoms in globular proteins results from interactions between segments of regular or nonregular structures within the polypeptide chain. Several combinations arising from the packing of α helices and β structures, or from α helix–β structure interactions, or from α helix–coil interactions and β structure–coil interactions, have been observed and described. However, a restricted number of such arrangements is found in all known proteins, forming more or less regular topologies. The principles which determine these arrangements have been tentatively rationalized by different researchers (Crick, 1953; Chothia, 1973, 1978; Chothia *et al.*, 1977; Chothia and Janin, 1978; Ptitsyn and Rashin, 1973; Ptitsyn, 1973, 1978; Ptitsyn and Finkelstein, 1978; Ptitsyn *et al.*, 1979; Richardson, 1976, 1977, 1980, 1981; Richardson *et al.*, 1978; Efimov, 1979; Lim and Efimov, 1977; Richmond and Richards, 1978; Cohen *et al.*, 1979; Sternberg and Thornton, 1976, 1977a,b; Nagano, 1973, 1974, 1977a,b; Rackovsky and Scheraga, 1978; Levitt and Chothia, 1976). It is just the beginning of the study of supersecondary structures and most attempts to solve the problem remain crude approaches, since they do not take into account specific interactions, and the number of samples is still insufficient for statistical analysis.

Classifications of proteins have been proposed according to the structural patterns encountered in the molecules.

2.5.1. Characteristics of Supersecondary Structures in Proteins

Some rules have been suggested to determine the packing of secondary structures. Most approaches assume two main principles: (1) residues that become buried in the interior of a protein close pack and they occupy a volume similar to that which they occupy in crystals of their amino acids; (2) associated secondary structures retain a conformation close to the minimum free energy conformation of the isolated secondary structures.

The first principle, which governs the amino acids packing in the whole protein, is presented in more detail in Section 2.7. The second principle relies on the observation that quite all the torsion angles ϕ and ψ in proteins are located in the allowed region of the Ramachandran diagram. Based on these principles, models for describing the packing of α helices, pleated sheets, and the interactions between helices and pleated sheets have been presented.

2.5.1.1. Packing of Helices

Crick (1953) introduced the first packing scheme for α helices using a graphical construction referred to as the knobs into holes model. He concluded that helices of the same sense can pack together about 20° away from parallel, or about 70° away from parallel in another type of arrangement. In their analysis of helix–helix interactions, Chothia *et al.* (1977) started from this simplified model which is represented in Fig. 2.14. All residues of an α helix ($r = 3.6$, radius = 5 Å) are assumed to have the same size and shape. Two helices are slit down one side in a direction parallel to their axis. Each is opened up to obtain a flattened projection. The surface of a helix is represented by rows of adjacent amino acid side chains. One row is formed by the residues i, $i \pm 3$, $i \pm 6, \ldots i \pm 3n$, another by i, $i \pm 4$, $i \pm 8, \ldots i \pm 4n$. These alignments mean that the $i \pm 3n$ or $i \pm 4n$ residue rows form a ridge and that grooves lie between parallel rows. The model of helix–helix packing consists of the arrangement of the surface ridges of the first helix into the grooves of the second one. The arrangement differs slightly from that proposed by Crick (1953). According to this model, called the ridges into grooves model, three classes of interactions are described as summarized in Fig. 2.14.

Chothia and co-workers (1977), analyzed the interactions between helices in different proteins (hemoglobin, thermolysin, lysozyme, calcium binding protein, subtilisin, and staphylococcal nuclease) and found 26 cases of helix–helix packing where three or more residues from one helix are in contact with three or more residues from the other one. Among these pairs of interacting helices, 3 were identified as class I with observed values of Ω* equal to $-80°$, $-85°$ and $-95°$; 16 were identified as class II with angle values ranging from $-20°$ to $-70°$ averaging near $-50°$; and 6 were class III some of which had Ω very close to $0°$ ($+20°$), i.e., parallel. Class I, the less populated one, has been recently reinterpreted (Chothia *et al.*, 1980; Janin and Chothia, 1980). Strongly interacting helices are not always parallel, especially when they are short, but, in some cases, a parallel arrangement of helices is observed. This occurs in myohemerythrin, purple membrane protein, and coat protein of Tobacco Mosaic Virus (TMV). However, this graphic model is oversimplified and does not take into account the effect of individual side chains, their size, their polar or nonpolar character, which is of importance for all intramolecular interactions in proteins.

Different authors have examined the packing of α helices. Ptitsyn and Rashin (1973, 1975) used the string and sausage model in which helices are represented by cylinders and applied their model to myoglobin. Starting from the same model, Richmond and Richards (1978) examined associations of helix pairs in sperm-whale hemoglobin; they considered the approach of two helices along the perpendicular line connecting their axes. Such an approach produces a solvent exclusion effect equivalent to a large hydrophobic contribution to the free energy of interaction. For each helix in

* The angle between helix axes is defined by Ω. This angle is negative if the near helix is rotated in a clockwise direction relative to the far helix. It is positive if it is rotated in the counterclockwise direction (Chothia *et al.*, 1977). Richmond and Richards (1978) do not use sign convention.

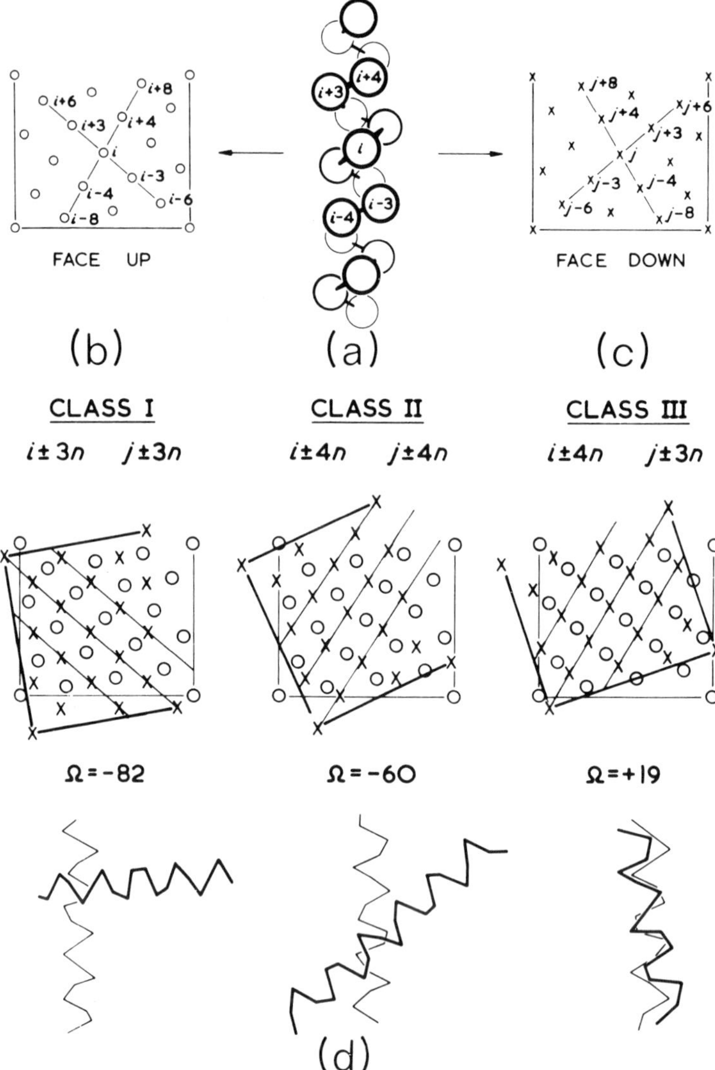

Fig. 2.14. The projections b and c of α helix (a). The residues of an α helix are represented by white circles, i, $i + 3$, $i + 4$, etc. The helix is opened and flattened projections are represented the one face up (b), the other face down (c). These projections will allow the packing of two helices. The ridges into grooves model for helix–helix packing generates three models of interactions (d). In class I, residues forming rows are $i \pm 3n$, $j \pm 3n$ for the first and the second helix respectively. The value for ideal helices is $-82°$. In class II, residues forming rows are $i \pm 4n$ and $j \pm 4n$ for each helix respectively: the resulting angle is $-60°$. In class III, the residues in rows are $i \pm 4n$, $j \pm 3n$ for the first and second helix respectively: the angle Ω is $+19°$. The helix–helix interactions are schematically represented for the three classes (from Chothia *et al.*, 1977).

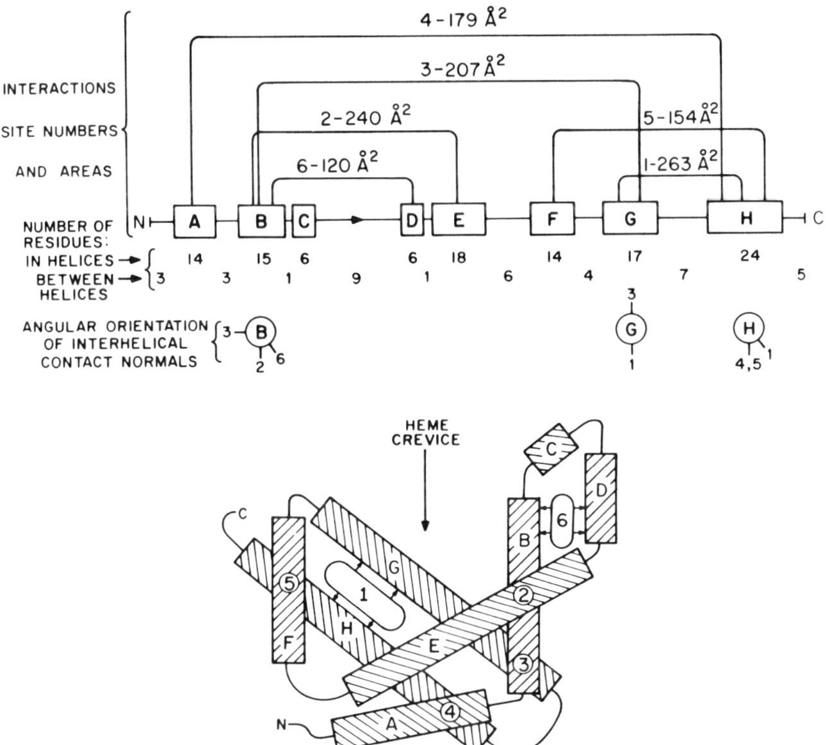

Fig. 2.15. The helix–helix contacts in myoglobin (from Richmond and Richards, (1978). The number of interaction sites and variations of accessible surface area upon interactions are indicated in the upper diagram for each pair of interacting helices (i.e., G–H, B–E, B–G, A–H, F–H, and B–D). The lower drawing is a schematic representation of the three-dimensional structure of the molecule.

myoglobin, the solvent accessibility of all the atoms was evaluated. For the six pairs of helices, it is clear that hydrophobic interactions are involved in the helix–helix packing. Among the six pairs, four have interhelical angles between 65° and 85° (referred to as perpendicular contacts) and two have angles smaller than 30° (referred to as parallel contacts). In perpendicular contacts, distances of separation tend to be about 2 Å smaller than in the parallel contacts. Figure 2.15 shows the contact between helices in myoglobin. The six main interaction sites correspond to 40% of the total area change from extended sausage model to folded, native sperm-whale myoglobin.

In this approach, specifications for central residues i are based on size and polarity. Three classes of α–α unit were also defined according to the

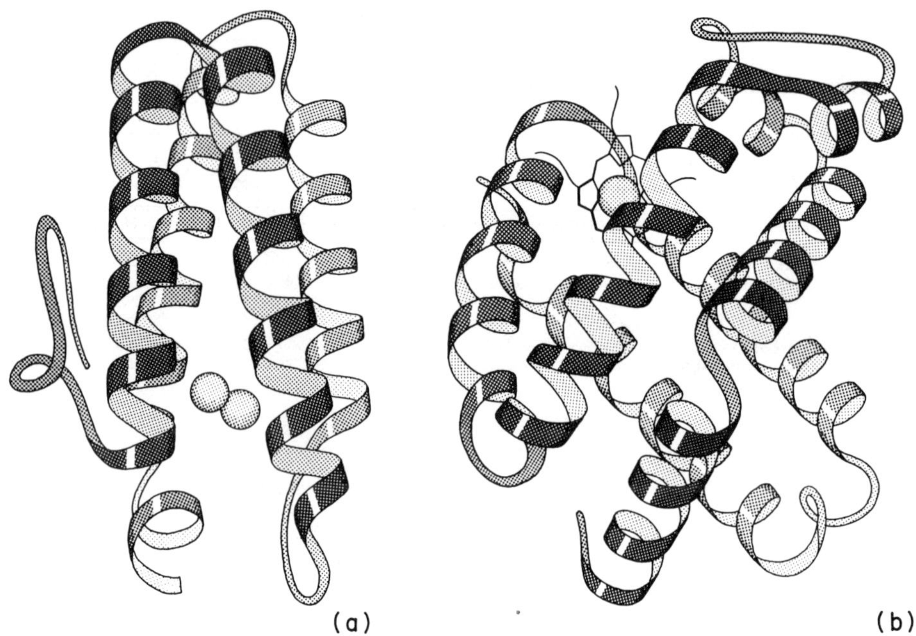

Fig. 2.16. Examples of helix–helix interactions (a) in myohemerythrin and (b) in β subunit of hemoglobin (courtesy of Richardson, 1980).

type of interactions. In class I, the interhelical angle is $80^\circ \pm 10^\circ$ and the central residue can only be glycine. In class II, the central residue may be Ala, Val, Ile, Ser, Thr, or Cys and the interhelical angle is smaller but variable. In class III, the interhelical angle is small, $20^\circ \pm 10^\circ$, and the central residue may be Leu or Met as well as one of those of class II. Different possibilities of helix–helix interactions are shown for myohemerythrin and for the β subunit of hemoglobin in Fig. 2.16 (a and b respectively) according to the clear and aesthetic representation of Richardson (1980, 1981). Using the geometry involved in the close packing of helices according to all these specifications, Richmond and Richards (1978) and Cohen and co-workers (1979) proposed an algorithm for predicting strong helix–helix interaction sites in proteins of known sequence and a computer program to fold a protein consisting only of helical segments and connecting parts of polypeptide chains of known length. They have applied their procedure to myoglobin (see Chapter 3). This approach represents an attempt to include more specific interactions in the formation of supersecondary structures.

Efimov (1979) from Poustchino's group also emphasized the polarity or hydrophobicity of side chains in the packing of α helices. He developed stereochemical rules to predict rotational isomers of hydrophobic side chains of α helices. The analysis of side-chain dihedral angles in 2536 residues from 19 protein structures indicates a small number of conformations observed for all types of side chains. Observed dihedral angles χ^1 and χ^2 remain within 15° to 18° from the values predicted on the basis of energy calculations. They do not depend on the position of the residue on the surface or inside the protein (Janin *et al.*, 1978). It was shown that close packing of hydrophobic side chains on the surface of individual helices or helix–helix structures is obtained for definite combinations of rotational isomers. The author's have introduced apolar and polar packing of helices in the search for different combinations. Apolar packing of helices approximately coincides with the packing described earlier by other researchers (Crick, 1953; Chothia *et al.*, 1977; Richmond and Richards, 1978), with certain differences in the values of dihedral angles.

Argos and co-workers (1977) have proposed a stable four-helical super-secondary structure composed of four roughly parallel α helices. This folding unit occurs in several proteins: hemerythrin, TMV disk protein, and tyrosyl tRNA synthetase. On the basis of an X-ray diffraction study however, Blow and co-workers (1977) did not accept the existence of such a structure in tyrosyl tRNA synthetase.

In cytochrome *c′*, α helices pack with an overall left-handed twist along the direction of the helix axes. Each protomer is thus organized as a left-twisted, 4-α-helical bundle (Weber *et al.*, 1980).

Considering the connectivity of helical segments, Richardson (1980, 1981) distinguished in antiparallel α-helical structures those which are formed of up-to-down helix bundles (each helix connecting to its nearest neighbor) and those which are formed of Greek key helix bundles. The first category which includes myohemerythrin (see Fig. 2.16a), TMV protein, and cytochrome b_{562} contains four helices in the bundle. The second category, with thermolysin domain 2, T_4 lysozyme domain 2, and hemoglobin, consists of 5–7 helices in the bundle (Fig. 2.16b).

2.5.1.2. Packing of β Structures

β Pleated sheets are major features in a large number of globular proteins of known three-dimensional structure. Similar topologies with determined regularities have been noted.

The β strand arrangements depend on the type of connection. They are generated either by hairpin or by cross over connections. Two connected β strands may be nearest neighbors, or one or more other strands may lie between them. They were classified by Richardson (1976, 1977) in type $\pm n$ according to the separation of the connected strands (i.e., the number of intervening strands) for the same hand connections. When the connection

TABLE 2.4

Frequency of Occurrence of Different Connection Types in β Sheets[ab]

	⇅	Mixed	↑↑	Total
± 1	57	48		105
± 1x		12	32	44
± 2		9		9
± 2x	7	8	7	22
 ± 3	8	4		12
 ± 3x		7	5	12
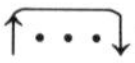 ± 4		2		2
 ± 4x	4			4
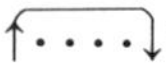 ± 5		3		3
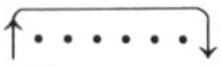 ± 7		1		1
 ± 7x		1		1
 ± 8x		1		1

[a] From Richardson (1977).

[b] The values for antiparallel, parallel, and mixed β sheets are tabulated separately.

crosses to the opposite end of the strand, the connected strands are designated as type $\pm nX$ (summarized in Table 2.4). The occurrence frequencies of each connection type in known proteins is also presented in Table 2.4. Among the different connections analyzed by Richardson (1977), 70% are either ± 1 or $\pm 1X$.

As emphasized by Richardson (1977), the commonly encountered topological patterns of sheets resemble some geometric motifs found in Greek or Indian pottery and weaving. The simplest pattern is the up-and-down β-meander (+1, +1, +1) (Fig. 2.9). Frequent motifs are the greek key (+3, −1, −1) and the double greek key (+3, −1, −1; +3) which are found in staphylococcal nuclease, chymotrypsin, immunoglobulins, superoxide dismutase, and pre-albumin. These elementary motifs often arrange to form barrel structures, as for example chymotrypsin which is roughly composed of two β barrels (Fig. 2.9). Figure 2.17 shows the topological connections of the β sheet in different proteins of known three-dimensional structure. Close similarities appear in β sheet topologies in globular proteins. Richardson (1977) suggested that these similarities had happened by chance rather than by evolutionary relationship since these β sheet topologies are among the most probable and that the number of highly favorable folding patterns is

Fig. 2.17. Schematic representation of topological connectivities of β pleated sheets found in proteins of known three-dimensional structures (courtesy of Richardson, 1977). From top to bottom, the examples are presented according to the increasing number of β strands forming the sheet (from left to right according to the percentage of antiparallel connections). Arrows indicate the strands of β sheet in the plane of the paper. Cross-over connections which are

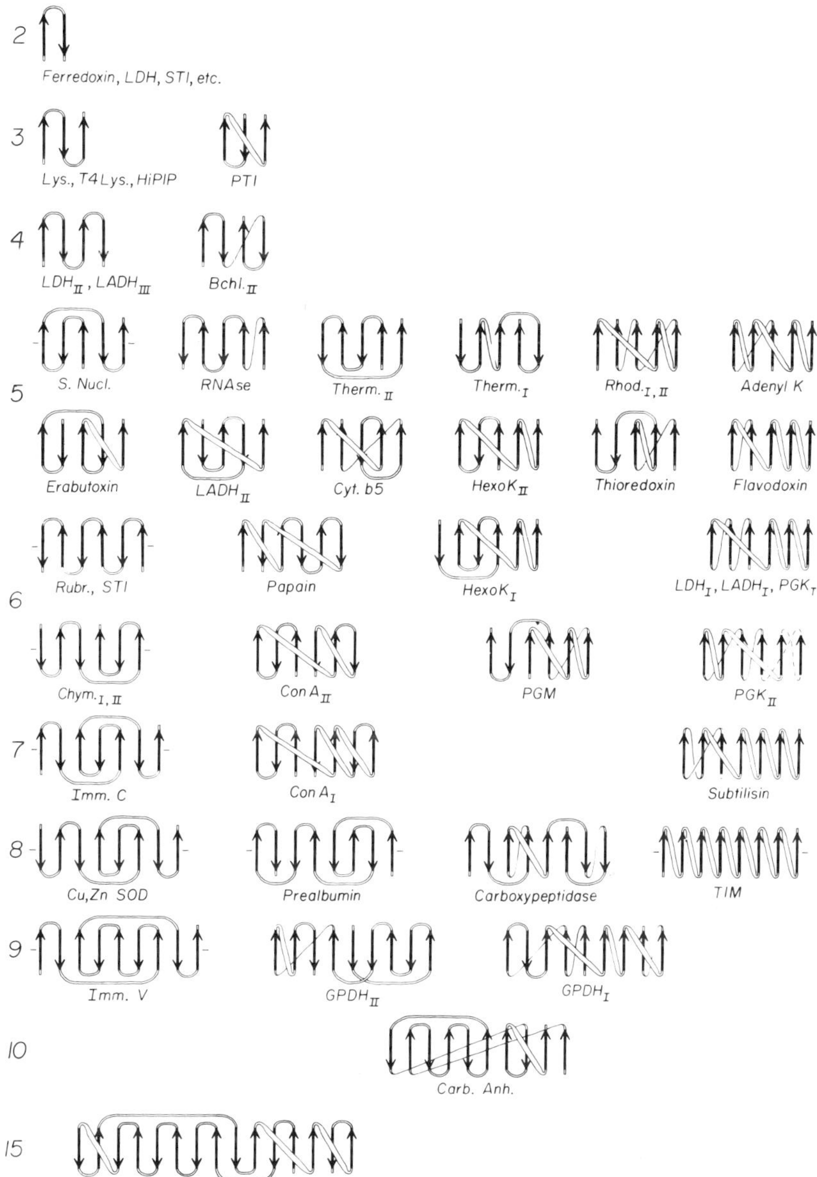

above the sheet are indicated by wide lines, those which are below are indicated by thin lines. The abbreviations for the different proteins are the following: LDH, lactate dehydrogenase; STI, soybean trypsin inhibitor; Lys, lysozyme; PTI, pancreatic trypsin inhibitor; LADH, liver alcohol dehydrogenase; Bchl, Bacteriochlorophyll protein; S. nucl., Staphylococcal nuclease; Rnase, ribonuclease; Therm., thermolysin; Rhod., rhodanese; Adenyl. K., Adenylate kinase; Cyt. b_5, cytochrome b_5; Hexo. K., hexokinase; PGK, phosphoglycerate kinase; Chym., chymotrypsin; Con. A, concanavalin A; PGM, phosphoglycerate mutase; Imm. C., immunoglobulin constant domain; SOD, superoxide dismutase; TIM, triose phosphate isomerase; GPDH, glyceraldehyde-3-phosphate dehydrogenase; Carb. Anh., carbonic anhydrase. For proteins having more than one β sheet, roman numeral subscript are added.

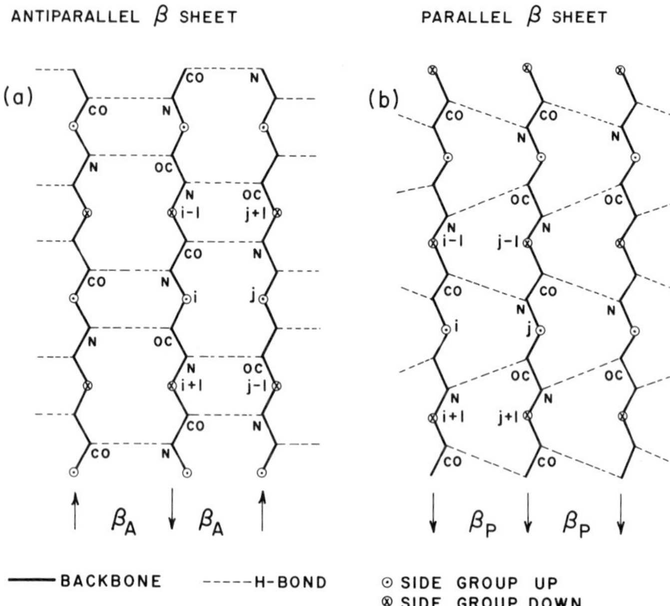

Fig. 2.18. Antiparallel and parallel β sheets (courtesy of Lifson and Sander, 1979).

quite limited. The predominance of right-handed connections was reported (Richardson *et al.*, 1976; Sternberg and Thornton, 1977a,b). As an example, of 66 cross-over connections 64 were right-handed and only two were left-handed. One of the left-handed connections was in subtilisin and the other in hexokinase, which has since been revised (Anderson *et al.*, 1978), therefore there is actually only one example.

The β strands may be parallel or antiparallel according to the connections, forming parallel or antiparallel β sheets. They differ in the arrangement of the strands and also in the hydrogen-bonding pattern (Fig. 2.18). Parallel β strands are not adjacent along the sequence. When they are not formed by an α helix, they are not stable and associate to form sheets which consist of at least five β strands. Therefore, parallel β sheets occur only in relatively large protein domains. Stabilization of parallel β structure probably requires a large cooperativity, whereas antiparallel β structure often occurs with only two β strands. In every protein analyzed, Richardson (1977) noted a marked preference for pure parallel or pure antiparallel structure over mixed β sheets. The reason seems easy to understand; a strand with antiparallel hydrogen bonding on one side is set up for antiparallel hydrogen bonding on the other side. 138 Strands were found with the same type of hydrogen bonding on both sides over only 33 with opposite bonding type on each side.

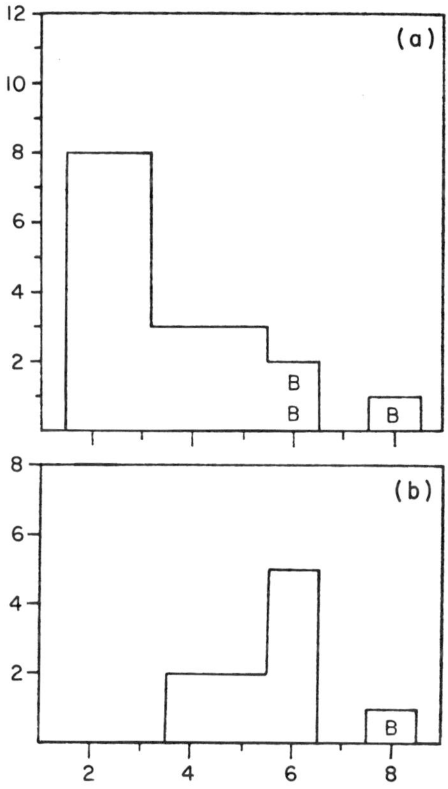

Fig. 2.19. Distribution of the number of strands per sheet (according to Sternberg and Thornton, 1977b). The observed distributions are different, when the connections are antiparallel (a) or when they are parallel (b). The letter B indicates β barrel structures (courtesy of Sternberg).

Furthermore, parallel β sheets and also mixed sheets are buried and are therefore inaccessible to the solvent suggesting that this structure has too low a stability to tolerate solvent access to the hydrogen bonds. Antiparallel β sheets have one side accessible to the solvent (Richardson, 1977). Of the 57 β-sheet structures examined by Sternberg and Thornton (1977a,b), 86% contained two to six strands with an average number of 4.7 strands per sheet. The distribution of the number of strands per sheet in parallel connections and antiparallel connections is given in Fig. 2.19. The observed distributions are different and illustrate the tendency for parallel β structures to be formed by a greater number of strands than antiparallel β structures. They have at least four strands, but generally not more than six. The larger sheets are most often mixed as in glyceraldehyde-3-phosphate dehydrogenase or in carboxypeptidase (Quiocho and Lipscomb, 1971), which contains a central β sheet composed of eight strands, the ones with antiparallel connections, the others with parallel connections connected with α helices. If nonmixed, they are involved in barrel structures.

As reported by Lifson and Sander (1979, 1980a,b), residue contact in antiparallel and in parallel β strands differs in amino acid preference. The correlations were investigated between nearest neighbors (i,j) in adjacent strands (see Fig. 2.18). Pair correlation among groups of similar polarity has been reported by von Heijne and Blomberg (1977, 1978). Parallel β strands impose more severe constraints on amino acid content than antiparallel β strands.

In globular proteins, β sheets are twisted rather than straight as in fibrous proteins, they have a right-handed twist when viewed along the strand. Right-handed strands viewed along the direction of the polypeptide chain give rise to a sheet which has left-handed twist viewed perpendicular to the peptide chain.* Regular and twisted β structures are characterized by the helical parameters $n = 2$ and $n \neq 2$ respectively. The twist of sheets has been studied by several researchers (Chothia 1973; Richardson, 1977; Richardson *et al.*, 1978; Weatherford and Salemme, 1979; Raghavendra and Sasisekharan, 1979; Salemme and Weatherford, 1981a,b; Salemme, 1981).

Chothia (1973) emphasizing for the first time this structural feature suggested that polypeptide chains with a right-handed twist are stabilized by a greater entropy than those with a left-handed twist, rather than by a difference in interchain energy. However, this was deduced from considerations of allowed conformations in the hard-sphere Ramachandran (ϕ, ψ) map. Chothia did not present any energy calculations and did not clearly mention whether he used the coiled coil or the crossing model of β structures. Nishikawa and Scheraga (1976) used the model of two stranded antiparallel β structure in α-coiled coil form for their determination of energy and geometry parameters. They found neither geometrical nor energetic preference for either a right- or a left-handed twist. However Raghavendra and Sasisekharan (1979) found that regular as well as left-handed or right-handed twisted antiparallel β sheets are stereochemically possible but a preference for the right-handed twisted β structure is indicated by their energy calculations. Weatherford and Salemme (1979), considering only parallel β sheets but analyzing both coiled coil and crossing models, attributed right-handed twist to distorsions in the peptide bonds that confer local left-handed helical character to each β strand. The nonplanarity of peptide bonds results from the tetrahedral deformation of the peptide nitrogen atom. These nonplanar distorsions may be attained with low energy variation, and they are frequently observed in peptide crystals (Ramachandran, 1974). Positive values of peptide bond dihedral angle deviation ($\Delta\omega = \omega - 180^\circ$), and negative values of θ_N (θ_N being the angle between the C_i, C_i, N_{i+1} plane and the C_i, N_{i+1}, C_{i+1} plane) impart left-handed local character to the polypeptide chain. Distribution of $\Delta\omega$ values for peptide bonds in 24 crystallographically

* Since the handedness of the twist has been described by either one or the other definition in literature, we use the first definition.

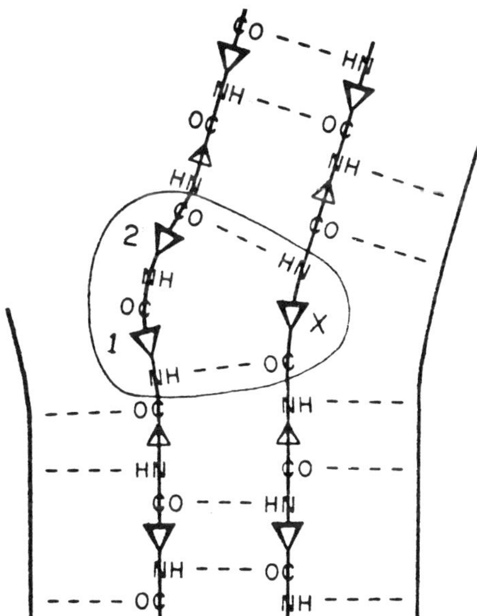

Fig. 2.20. The β bulge (according to Richardson *et al.*, 1978). A β bulge indicated by the outlined region is at the edge of an antiparallel β sheet. Smaller triangles represent side chains that are below the sheet, larger triangles those that are above it.

determined structures indicate a 70% preference for $+\Delta\omega$, $-\theta_N$ distorsions. Thus, according to these authors, nonplanarity of peptide bond, by conferring local left-handed character to each β strand, gives rise to right-handed twisted β structures which favor optimal formation of hydrogen bonds.

However, Sielecki and co-workers (1979) have not found deviations from planarity, or any preference for the sign of $\Delta\omega$, in the 179 trans peptide bonds considered in the refined three-dimensional structure of *Streptomyces griseus* serine protease A at 1.8 Å resolution. They concluded that deformation of the peptide bond does not significantly determine chirality of β sheets.

A small unit of nonrepetitive protein structure involving two strands of sheet, the β bulge, was described by Richardson and co-workers (1978, 1981), (Fig. 2.20); 91 examples were listed. The β bulge is a region between two consecutive hydrogen bonds including two residues on one strand opposite to a single residue on the other strand. This structure, by introducing a slight bend in the sheet, locally accentuates the right-handedness of the twist. It does not exist in β sheets interacting with α helices in α/β proteins (Janin and Chothia, 1980).

Therefore, in all proteins, sheets have a tendency to form a right-handed twist, but the extent of the twist varies for each sheet. A schematic representation of carboxypeptidase, according to Richardson, shows the twist of a

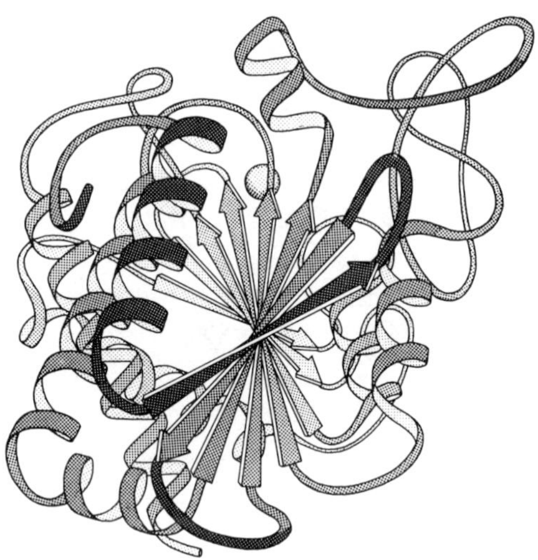

Fig. 2.21. Schematic drawing of carboxypeptidase displaying the twist of β sheet (with courtesy of Richardson, 1980).

β sheet (Fig. 2.21). However the origin of the right-handed twist is not yet well understood. Energetic preference for such β structure is not generally accepted and distorsion of the peptide bond, if plausible, does not appear to be a general feature in proteins.

Once the sheets are formed, they are also able to associate. Chothia and co-workers (1977) have emphasized some principles for sheet–sheet packing. They distinguished two classes of packed pleated sheets: the ones which are formed by two large and independent pleated sheets and the others which consist of the folding over of single sheets. The association of two independent pleated sheets is dependent on the twist of each of them and also on the amino acid residue composition. Two sheets with the same degree of twist are closely packed when their strands are parallel (Fig. 2.22). Two sheets with different twists (T_1 and T_2) pack more closely when the rotation angle between the sheets is given by the expression:

$$\Omega_c = -2(|T_1 - T_2|)$$

The packing of the β sheet was investigated by Chothia and co-workers (1977) in four proteins: immunoglobulin domains, concanavalin A, superoxide dismutase, and prealbumin. All these proteins contain two large antiparallel pleated sheets of different twist which are packed together (Fig. 2.22). The overall twist of each of the sheets was measured. The relative orientation was calculated using the foregoing expression for Ω_c obtained by computer model building. The values of Ω_c were compared to the observed

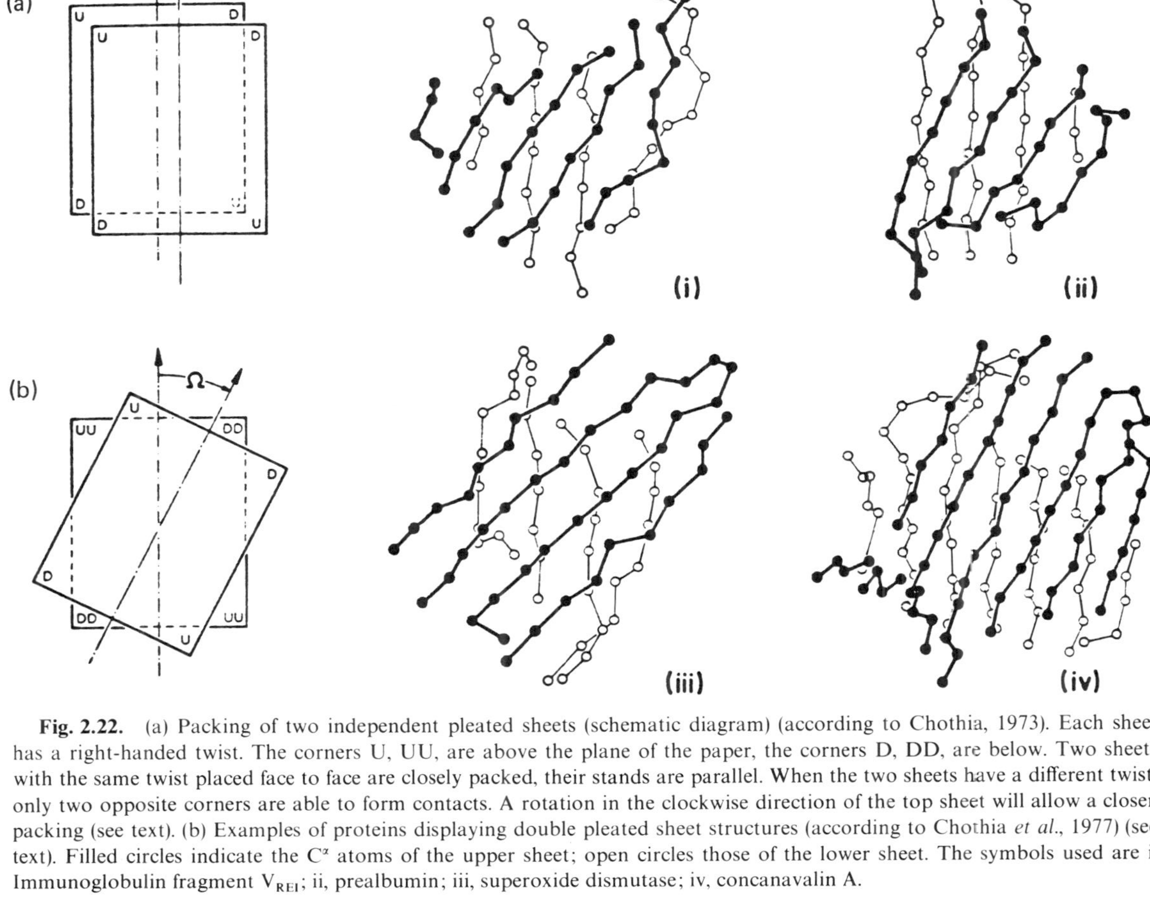

Fig. 2.22. (a) Packing of two independent pleated sheets (schematic diagram) (according to Chothia, 1973). Each sheet has a right-handed twist. The corners U, UU, are above the plane of the paper, the corners D, DD, are below. Two sheets with the same twist placed face to face are closely packed, their stands are parallel. When the two sheets have a different twist, only two opposite corners are able to form contacts. A rotation in the clockwise direction of the top sheet will allow a closer packing (see text). (b) Examples of proteins displaying double pleated sheet structures (according to Chothia *et al.*, 1977) (see text). Filled circles indicate the C^{α} atoms of the upper sheet; open circles those of the lower sheet. The symbols used are i, Immunoglobulin fragment V_{REI}; ii, prealbumin; iii, superoxide dismutase; iv, concanavalin A.

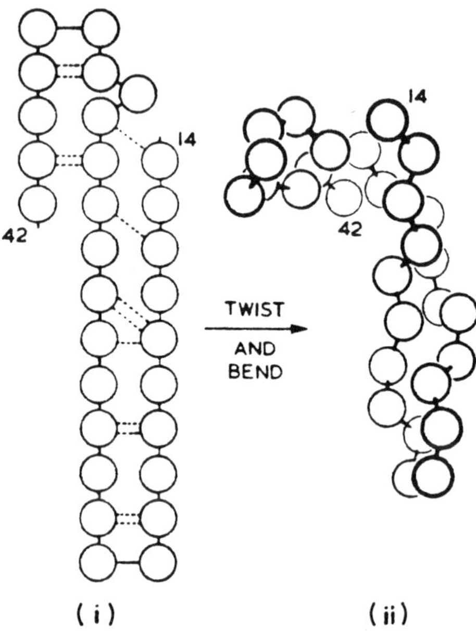

Fig. 2.23. The packing of pleated sheets with a right-handed supertwist resulting from interactions over a single sheet; such a topology is actually occurring in elastase where residues 14–42 form a triple stranded β sheet (Chothia *et al.*, 1977).

ones. Except for the case of concanavalin A, the expected values were in good agreement with the observed ones. For concanavalin A, the discrepancy was attributed to the pleated sheets being bent as well as twisted.

The interactions over a single sheet involve a folding through a right-handed supertwist which can retain interchain hydrogen bonds and the right-handed twist of each part of the pleated sheet. Such an arrangement is shown in Fig. 2.23 for the first domain of elastase.

2.5.1.3. Interactions between α Helices and β Structures

The recognition, in certain globular proteins, of a ($\beta\alpha\beta$) folding unit, (see Fig. 2.9), in which two parallel β strands are connected by an α helix, is the result of the observation of Rao and Rossmann (1973). The ($\beta\alpha\beta$) folding unit may be regarded as a substructure of the mononucleotide binding fold. It has a remarkable preference for the right-handedness; 57 of the 58 $\beta\alpha\beta$ units analyzed by Sternberg and Thornton (1976, 1977a,b) were found to be right-handed; only one left-handed unit was encountered in subtilisin. The right-handed sense which is certainly the most striking characteristic of these structures was explained by Chothia (1973) as being a consequence of the twist of the β sheet which creates steric constraints.

Sternberg and Thornton (1977a,b) have shown that it can be understood by the interactions within the individual strands of the sheet.

The importance of the interactions between α helices and β strands has been particularly emphasized in recent years. These interactions are expected to occur from the hydrophobic side chains. Nagano (1974, 1977a,b) suggested that nucleation occurs from medium range interactions between regions having a high helical potential (α candidate) and a high β structural potential (β candidate). Moreover the determination of the polypeptide conformation is presented as resulting from interactions between nuclei. In insulin B for example, the hydrophobic side chains of the helix (10–18) are fixed on one side whereas the β candidates are flexible enough to interact with those of the helix leading to the formation of a flickering nucleus. Such interactions may be stabilized by the existence of disulfide bridges. Thus in pancreatic trypsin inhibitor the interaction between an α candidate (47–54) and a β candidate is stabilized by the Cys 30–Cys 51 bond.

The ($\beta\alpha\beta$) folding unit occurs in many proteins. It is the essential structure of α/β proteins, a common structure of dehydrogenases (alcohol dehydrogenase, lactate dehydrogenase), some kinases (hexokinase, adenylate kinase, phosphoglycerate kinase), and phosphoglycerate mutase. It is also found in carbonic anhydrase, carboxypeptidase A, cytochrome b_5, flavodoxin, papain, rhodanese, thioredoxin, and subtilisin.

The occurrence of the β strand–α helix–β strand–α helix–β strand structure, ($\beta\alpha\beta\alpha\beta$), was recognized for the first time by Rao and Rossmann (1973) in lactate dehydrogenase, malate dehydrogenase, and flavodoxin. That is the reason why it is commonly referred to as the "Rossmann fold." More generally, this structure has been found in nucleotide binding proteins such as dehydrogenases and kinases. It is found also in several other proteins, subtilisin and flavodoxin (as previously mentioned) which contain only one ($\beta\alpha\beta\alpha\beta$) structure. A part of glycogen phosphorylase also contains this structure (Fletterick *et al.*, 1976). In dehydrogenases (lactate dehydrogenase, alcohol dehydrogenase, glyceraldehyde-3-phosphate dehydrogenase, malate dehydrogenase), two Rossmann folds are involved in forming the binding site of NAD^+. Figure 2.24 is a diagrammatic representation of the two Rossmann folds involved in the NAD^+ binding site of dehydrogenases (Rossmann *et al.*, 1974). It shows the connections of the strands. In Fig. 2.25 a schematic representation of the NAD^+ binding domain of horse liver alcohol dehydrogenase with the two Rossmann folds is presented (Eklund *et al.*, 1976).

A recent analysis of the structure of triose phosphate isomerase (TIMase) compared to that of lactate dehydrogenase (LDH) was reported by Phillips and co-workers (1978). Triose phosphate isomerase has a striking regularity with the eightfold repeat of β–α unit $[(\beta\alpha)_2\beta]$. The eight parallel β strands

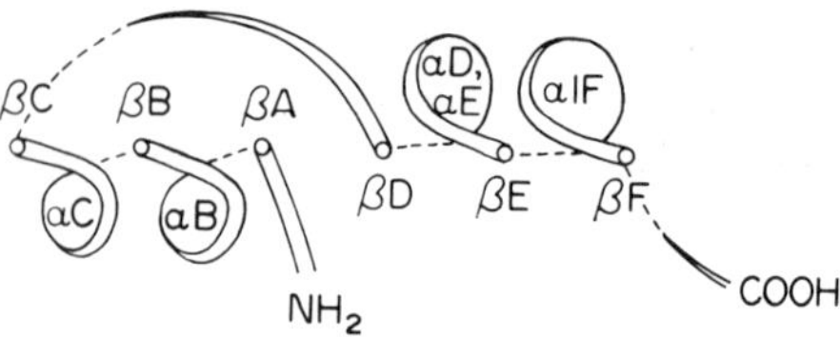

Fig. 2.24. Diagrammatic representation of the NAD$^+$ binding fold in dehydrogenase indicating the connectivities of the strands from the N- to the C-termini (according to Rossmann *et al.*, 1974; courtesy of M. G. Rossmann).

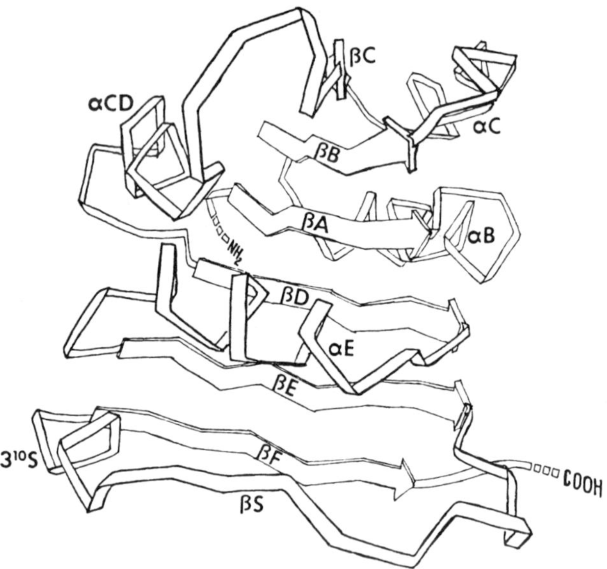

Fig. 2.25. Diagrammatic representation of the polypeptide chain fold of the NAD$^+$ binding domain indicating the arrangement of the segment S of secondary structures in alcohol dehydrogenase (according to Eklund *et al.*, 1976).

form an internal barrel, each pair of strands being connected by one or two of the 12 helices (11 α helices and one 3_{10} helix), approximately antiparallel to the corresponding strand. This protein contains 51% of α helix, 23% of β sheet, and 3% of β bends. Gene duplication in TIMase seems rather unlikely. Although there is an eightfold repeat of $\beta\alpha$ unit, there is no evidence of repeating sequence within the polypeptide chain. However, several three-dimensional and sequence similarities in TIMase and in LDH have been found which suggests that some sections of the molecule are related (see Section 2.11).

In all the proteins considered, α helices may pack onto either parallel or antiparallel pleated sheets. Three different possibilities have been described by Chothia and co-workers (1977), using their simplified model for helices and for sheets. In the simplest case, any α helix packs onto a pleated sheet with its axis parallel to the strands of the sheet. In such a model the helix residues in contact with the sheet are i, $i+1$, $i+4$, $i+5$, $i+8$, $i+9$ and the angle between the helix axis and the sheet is null. However, because of the twist of the sheet, the orientation of the helix away from the parallel position ($\Omega = 0°$) is more likely. If it occurs in the negative direction ($\Omega < 0°$), then only the center of the helix is capable of close packing (the ends move away from the sheet). Conversely, when the helix is oriented in the positive direction ($\Omega > 0°$), only the ends pack. The center of the helix is above the sheet creating an internal cavity in the protein. Helices packed on sheets with a large twist have a negative Ω. As for helix–helix packing, Ω defined the relative orientations of the two pieces of secondary structures.

Chothia and co-workers (1977) reported that in some proteins, such as flavodoxin, carboxypeptidase, subtilisin, and triose phosphate isomerase, the predominant structure is a central pleated sheet flanked by α helices. These researchers noted that in these proteins, 19 helices each have at least four residues (i.e., in total 129 residues) in contact with the pleated sheet. Among the observed values for the helix–sheet angle Ω, 13 of the 19 range from $-10°$ to $+10°$.

A single model for the packing of helices into pleated sheets in α/β proteins was proposed by Janin and Chothia (1980). Their model assumes that (1) the helix axis is parallel to the β strands, (2) the β sheets have a regular geometry, and (3) side chains in the contact region are selected to produce a smooth surface. When helices pack against each other on the face of a β sheet, helix–helix arrangement corresponds to class II (see Fig. 2.14). Eight proteins have been considered: flavodoxin, adenylate kinase, triose phosphate isomerase, subtilisin, rhodanese, carboxypeptidase, the NAD^+ binding domain of lactate dehydrogenase, and glyceraldehyde-3-phosphate dehydrogenase.

2.5.1.4. Interactions between α-Helices or β-Structures and Coil

The α helices or β structures may interact with or may be connected by coil (nonregularly structured) segments (c) of the polypeptide chain. The $\beta c \beta$ units have been observed in proteins. The 13 $\beta c \beta$ units examined by Sternberg and Thornton (1977a) were right handed. The loop regions of these units had length varying from 5 to 24 residues. Four series of folding units, helix–coil (αc), helix–coil–helix ($\alpha c \alpha$), extended-coil (βc), and extended-coil-extended ($\beta c \beta$) have been considered by Rackovsky and Scheraga (1978) in their computational procedure to determine the role of ordered structures and nonregular parts of the chain in protein folding. Their approach which is presented in detail in Chapter 4 emphasizes the importance of the interactions between structured strands and adjacent nonregular segments in the folding process.

2.5.2. Structural Patterns and Protein Folding

To compare protein structures, several diagrammatic representation or topology packing diagrams have been used by the different researchers (Schulz and Schirmer, 1974; Rossmann *et al.* 1974; Levitt and Chothia, 1976; Sternberg and Thornton, 1977a,b; Nagano, 1977a,b). For the sake of simplicity, only the one proposed by Levitt and Chothia (1976) is used here. The symbolism of the main folding units including direction of the polypeptide chain is explained in Fig. 2.26.

As already mentioned, classification of globular proteins according to the structural patterns, has been introduced by Levitt and Chothia (1976). The proteins which contain $\beta\alpha\beta$ units pertain to class IV; the proteins which pertain to one of the three first classes do not have any $\beta\alpha\beta$ folds. Figure 2.27 displays several examples of proteins which pertain to the different categories according to the simplified representation. (Examples of all $-\alpha$ proteins in

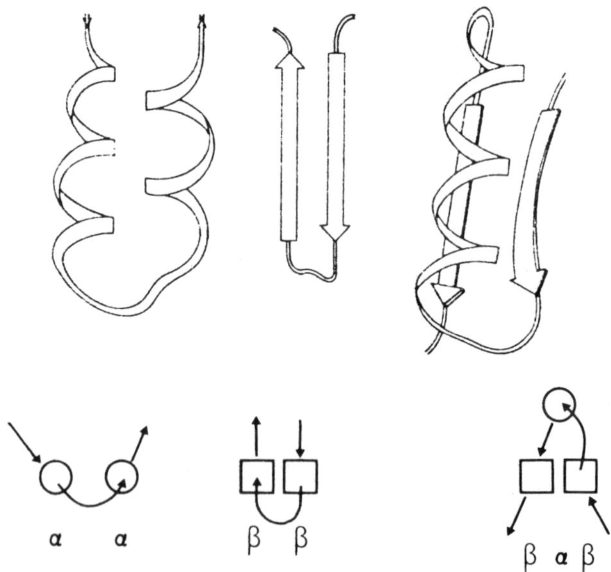

Fig. 2.26. Schematic representations of different folding units (from Levitt and Chothia, 1976).

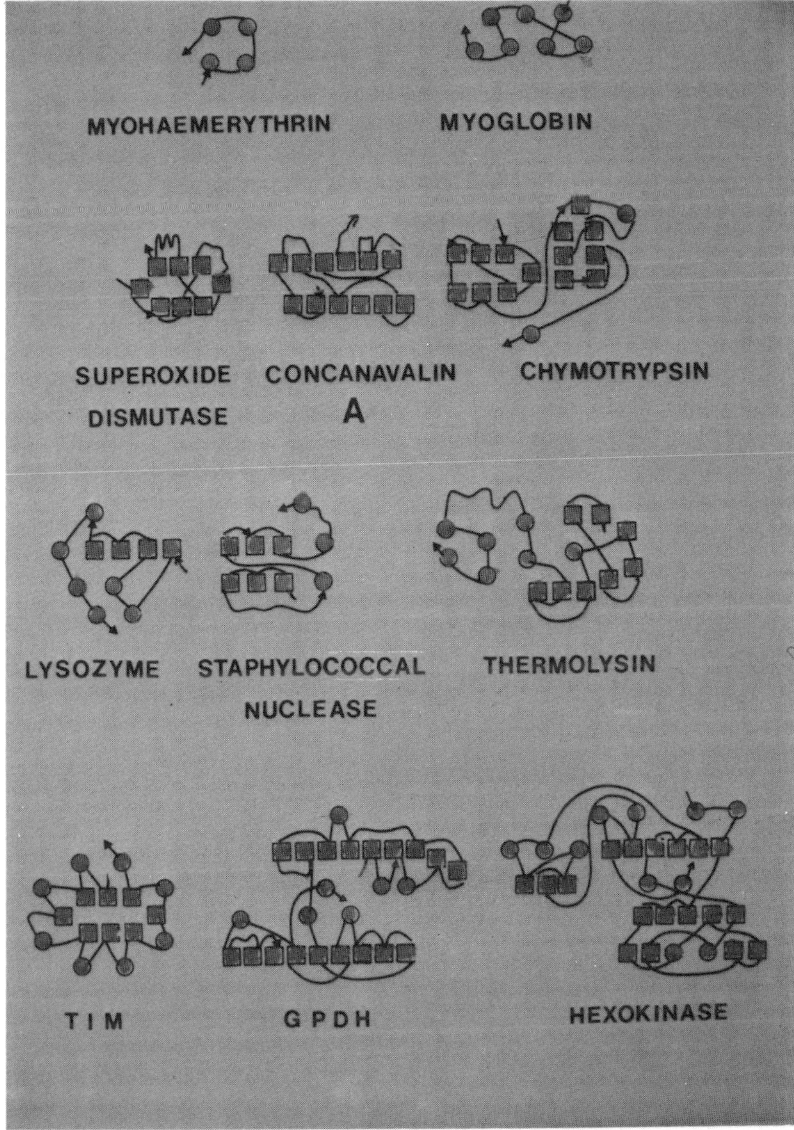

Fig. 2.27. Examples of proteins from the different classes (adapted by from Levitt and Chothia, 1976).

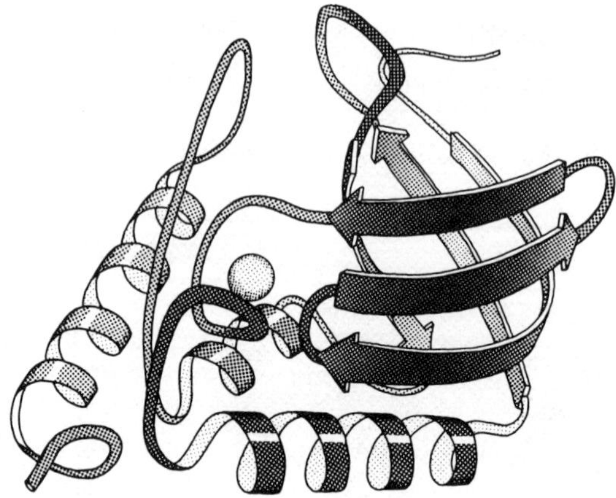

Fig. 2.28. Schematic drawing of Staphylococcal nuclease (courtesy of Richardson, 1980).

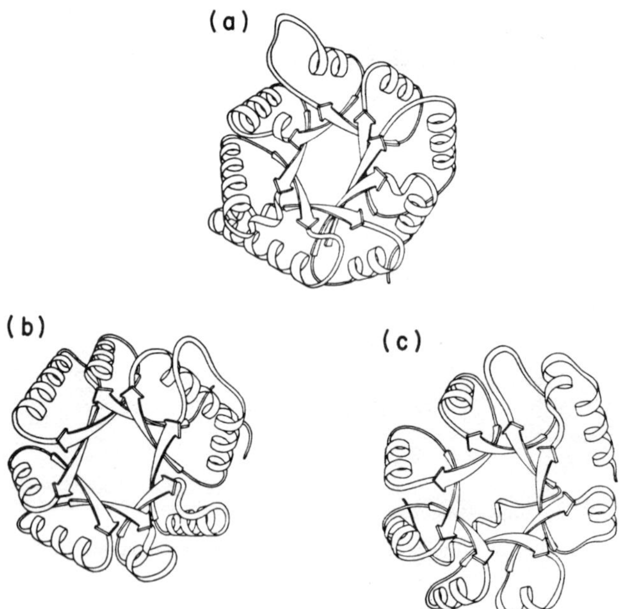

Fig. 2.29. Schematic drawings for (a) a triose phosphate isomerase subunit; (b) the first domain of pyruvate kinase; (c) a reconnected 2-keto-deoxy-6-phosphogluconate aldolase subunit (from Richardson, 1979).

the schematic three-dimensional representation used by Richardson are presented in Figure 2.16.) In the same representation Fig. 2.28 shows staphylococcal nuclease which is a typical example of ($\alpha + \beta$) proteins with a region of interacting β strands distinct from a region of α helices. Figure 2.29 shows typical examples of α/β proteins.

Although the number of samples is insufficient to allow significant statistical analysis, a crude correlation indicates that most of all α and α/β proteins have no disulfide bonds. Disulfide-linked proteins are generally found in all β and in ($\alpha + \beta$) proteins. In this last class disulfide bonds do exist between any kind of structured or nonstructured segments.

Richardson (1980, 1981) proposed a slightly different classification for protein structures. The author also distinguished four major classes. The first one is the antiparallel α structure with two subgroups, the up-and-down helix bundles, and the Greek key helix bundles (these kinds of structures are shown in Fig. 2.16). The second category is the parallel α/β proteins, the first subgroup being the singly wound parallel β barrels, the second one the doubly wound β sheets (as in lactate dehydrogenase, gluthathione reductase), the third subgroup has a central mixed β sheet flanked by helical layer as in carboxypeptidase (see Fig. 2.21). The third category is the antiparallel β structures and has several subgroups according to the topology (up-and-down β barrels, greek key β barrels). While the two first categories of structures are quite identical to all β and α/β proteins defined by Levitt

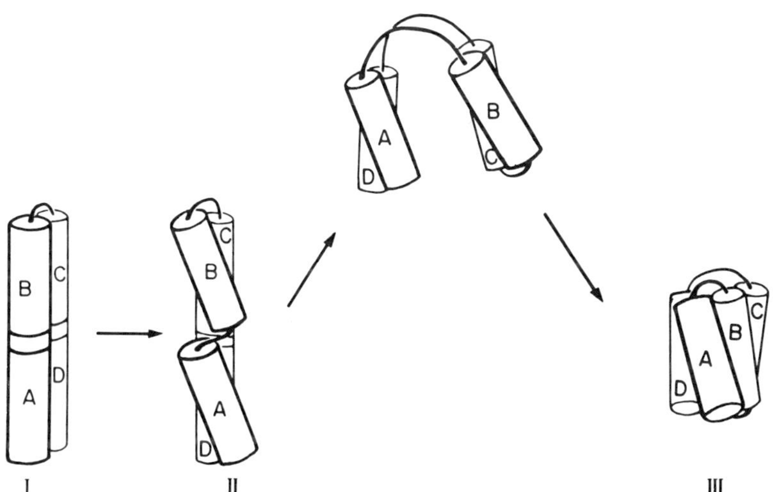

Fig. 2.30. Mechanisms of protein folding for all-α proteins (according to Ptitsyn and Finkelstein, 1978; courtesy of Ptitsyn).

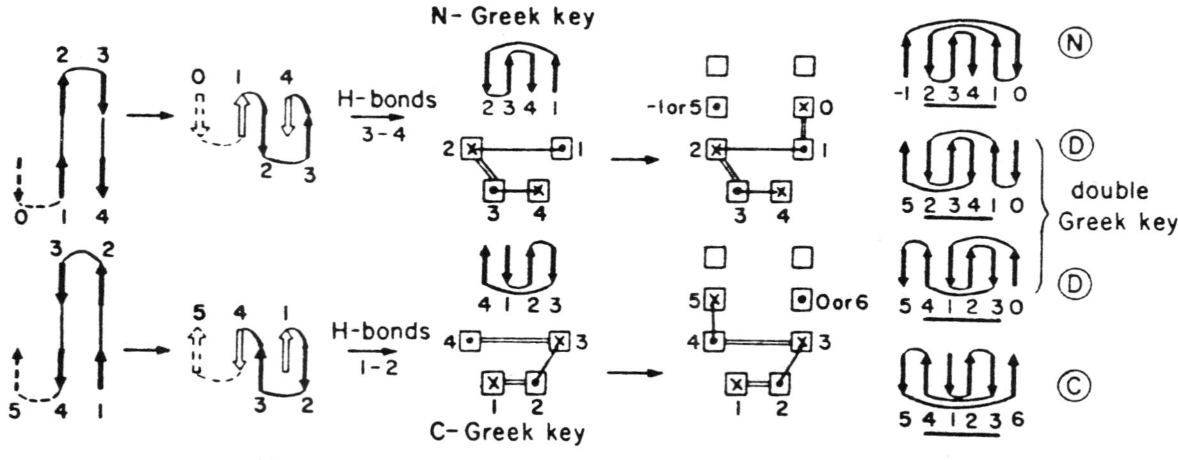

Fig. 2.31. Mechanisms of protein folding for all-β proteins (according to Ptitsyn and Finkelstein, 1978; courtesy of Ptitsyn).

and Chothia (1976), the third class not only includes all β proteins, but also proteins which contains α helix segments (for example, staphylococcal nuclease). The last category defined by Richardson (1980, 1981) includes the small disulfide-rich or ligand-rich structures such as pancreatic trypsin inhibitor, cytochrome *c*, rubredoxin. A group of small proteins structured around a four disulfide-bonded core including several neurotoxins and domains of wheat germ agglutinin have been described by Drenth and co-workers (1980).

Examination of these different patterns of protein structure had suggested possible mechanisms for protein folding. It was proposed that folding occurred by stages, in which pieces of secondary structures are first folded, then diffuse and interact to form folding units. These small building blocks also associate to form larger pieces of structure and so forth until the native conformation is formed (Ptitsyn and Rashin, 1975; Levitt and Chothia, 1976; Lim and Efimov, 1977; Ptitsyn and Finkelstein, 1978; Ptitsyn *et al.*, 1979). Ptitsyn and Finkelstein (1978) distinguished pathways for the formation of the different classes of proteins. For all α proteins, they suggest formation of one long initiating hairpin formed from long adjacent helices which then break off one of the helices. Similar breaking of the other helix gives a tetrahedral complex with an optimal packing (Fig. 2.30). On the same principle, starting from a long hairpin β structure and then subsequent breaking, Ptitsyn and Finkelstein explained formation of all β proteins (Fig. 2.31). Such a representation remains very speculative; it is supported neither by experimental nor by sufficient theoretical arguments. Richardson (1980, 1981) suggested a concerted folding of rather long segments in β structure.

2.6. STRUCTURAL DOMAINS

There are different levels of substructure in proteins. The folding units or supersecondary structures which are an assembly of secondary structures have been considered. The term domain defines larger assemblies having the characteristics of complete globular protein, i.e., compactness and stability (see Rossmann and Argos, 1980).

2.6.1. Structural Characteristics of Domains

Domains may be considered to be independent globular regions covalently linked into a protein with the side chains forming an hydrophobic core. The concept of domain was first introduced to describe the structure of immunoglobulins (Edelman, 1970). Structural domains were even separated by limited proteolysis in the case of the IgG molecule (Porter, 1959). The existence of domains in the structure of immunoglobulins was later confirmed when the three-dimensional structure was solved (Schiffer *et al.*, 1973; Amzel *et al.*, 1974; Poljak *et al.*, 1973; Davies *et al.*, 1975; Amzel and Poljak, 1979). The IgG immoglobulin molecule is made of 12 domains (4 in the two light chains and 8 in the two heavy chains) (Fig. 2.32). In each light chain, one domain consists of the variable region and the other of the constant region. In the heavy chain, one domain is also built from the variable region and the three others from the constant region (Edelman and Gally, 1968; Edelman, 1970). Detailed three-dimensional structure of the

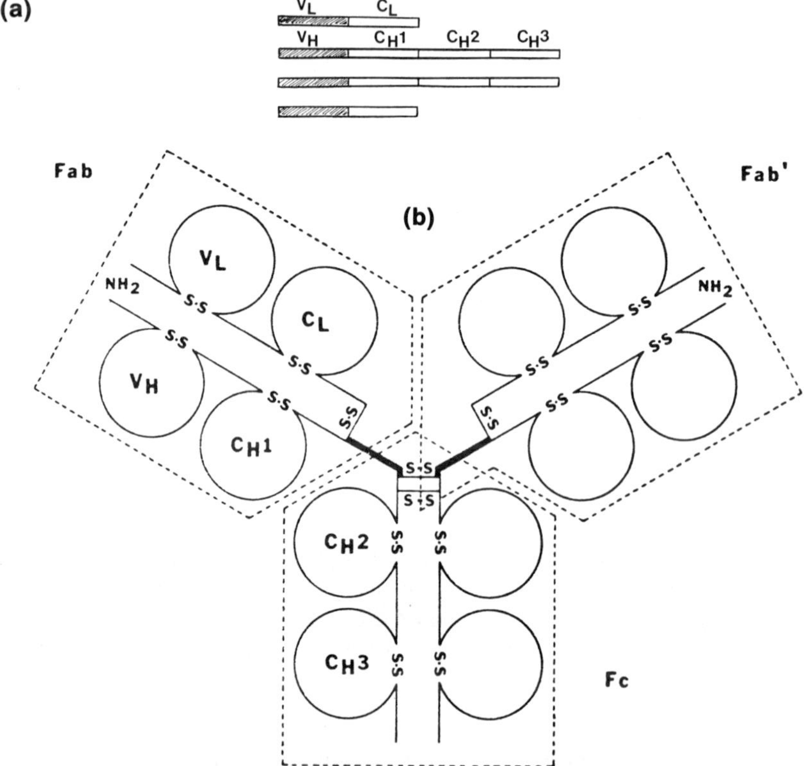

Fig. 2.32. Structure of immunoglobulins (a) schematic representation of the two light chains and the two heavy chains with the variable and constant regions of the sequence (Edelman, 1970); (b) schematic three-dimensional representations (according to Amzel and Poljak, 1979); (c) three-dimensional structure of immunoglobulins (from computer drawings, with courtesy of Feldmann).

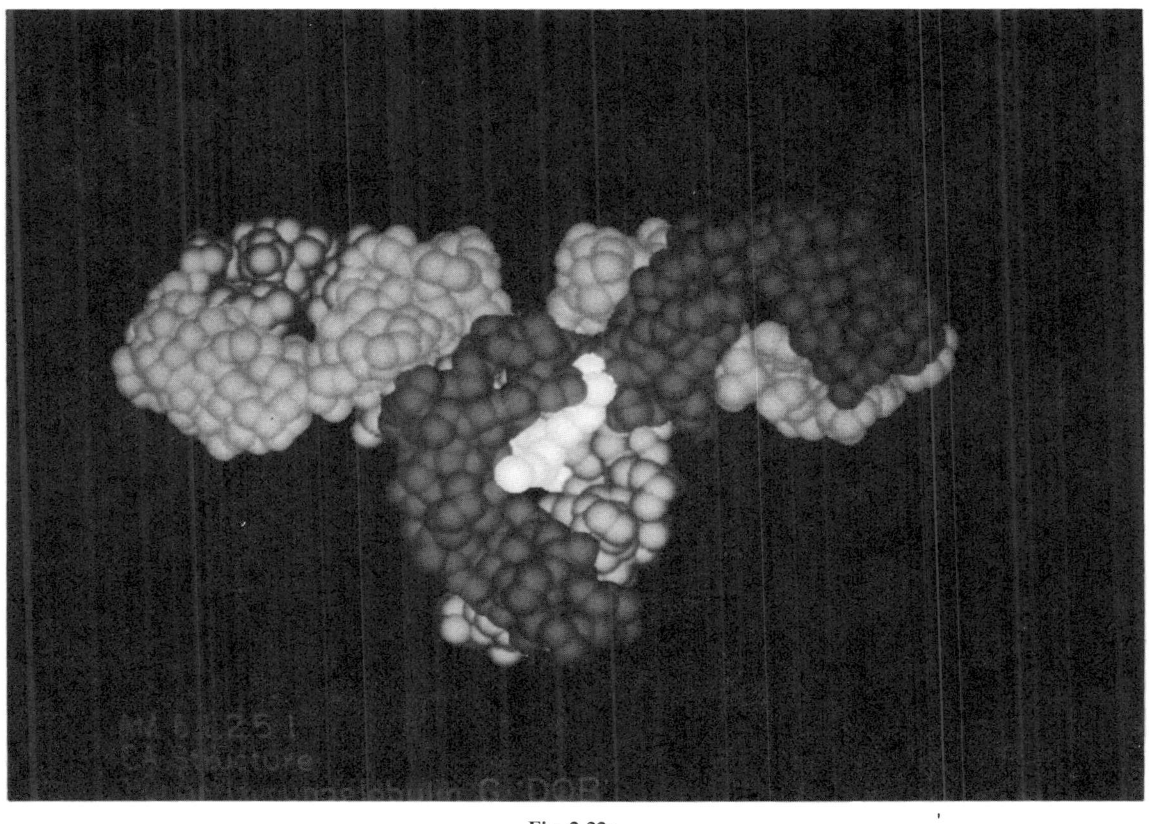

Fig. 2.32c.

F_{ab} fragment (two variable domains V_L and V_H and two constant domains C_L and C_H1) is known (Poljak *et al.*, 1973). Some other details of the three-dimensional structure are available in the literature (e.g., C^α coordinates, domain interactions, hydrogen bonding) (Schiffer *et al.*, 1973; Davies *et al.*, 1975; Amzel *et al.*, 1974; Richardson *et al.*, 1976).

The existence of domains as structural patterns commonly encountered in proteins of molecular weight larger than 16,000–20,000 was emphasized by Wetlaufer (1973). Wetlaufer distinguished between continuous domains formed by adjacent segments of the polypeptide chain and discontinuous domains which are formed from nonadjacent parts of the polypeptide chain. He proposed some rules for the folding process that is expected from such a structural organization. For proteins folded into several continuous domains, folding is expected to occur independently in each domain by formation of one nucleation center followed by the rapid growth of the structure. For proteins built up of discontinuous domains, a longer time of folding is predicted.

With several exceptions, most proteins of large molecular weight are generally organized into domains. The size of a domain ranges from 40 to 150 amino acids and the size depends on the type of structure. Richardson (1977) indicates that for domains built around a parallel β sheet, the average size is 179 residues with a minimum of 135, whereas domains built around an antiparallel β sheet have an average size of 128 residues with the minimum being 54. Carboxypeptidase A and thermolysin have the same molecular weight of about 34,600. Both are zinc proteins with a requirement of Zn^{2+} for the catalytic activity. Thermolysin is organized into two domains. Unexpectedly, despite its size, carboxypeptidase A does not fold into domains. The same situation is observed for concanavalin A (27,000 for the protomer) and for carbonic anhydrase (30,000). All these proteins are characterized by a high content of β structure.

The recognition of structural domains in proteins may be of importance for understanding the assembly of the structural elements of a molecule. A method based on the distance map, as proposed by Nishikawa *et al.* (1972), Ooi and Nishikawa (1973), and Nishikawa and Ooi (1974), was used by Rossmann and Liljas (1974) and Liljas and Rossmann (1974) for comparison and detections of structural similarities of domains. For these maps, the distances between C^α atoms were plotted as a square matrix with the sequence numbers arranged horizontally and vertically. Contours were drawn for various C^α—C^α distances. Moreover, in such a plot the different ordered structures display typical representations. Helices are represented by short interactions along the diagonal. Parallel β pleated sheets are represented by a line parallel to, but distant from the diagonal. Antiparallel β strands are shown as interactions in a line perpendicular to the diagonal.

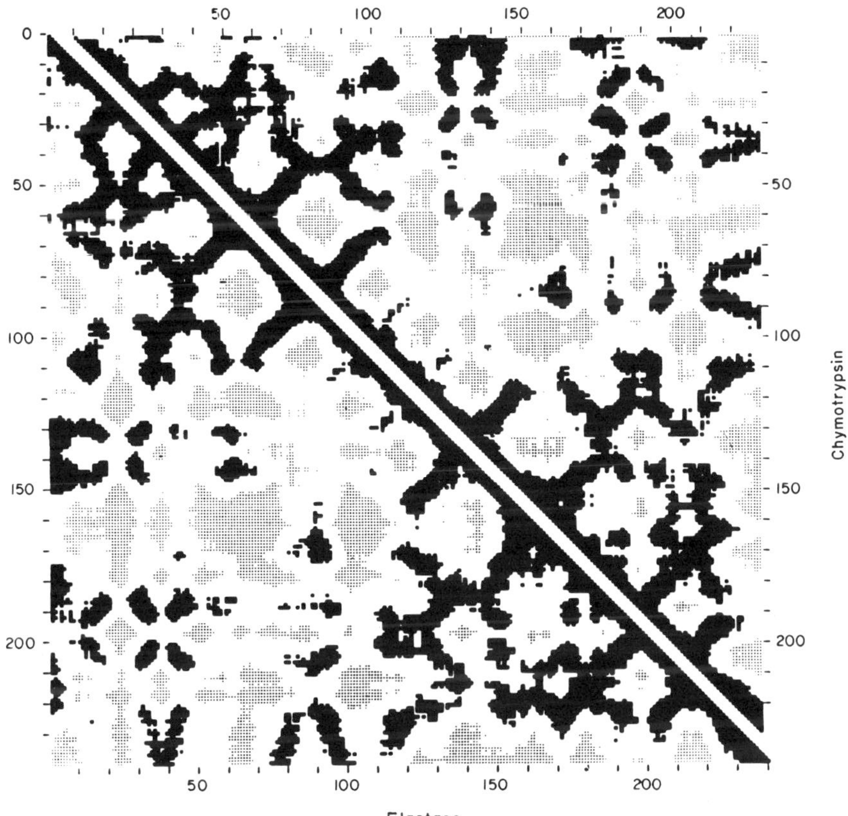

Fig. 2.33. Interatomic distance matrix (according to Ooi and Nishikawa, 1973) for elastase (lower left) and chymotrypsin (upper right). Black areas correspond to separation less than 15 Å; dotted areas to separation greater than 30 Å. The plot clearly indicates: (1) similarities in both structures, (2) the organization of the molecules into two distinct domains (which correspond to the two regions of high density of black areas) (from Sawyer *et al.*, 1978). Black areas correspond to separation less than 15 Å; dotted areas to separation greater than 30 Å.

Figure 2.33 shows the map for elastase and chymotrypsin. Rose (1979) developed an algorithm approach that requires only X-ray atomic coordinates. In his approach, the three-dimensional problem is reduced to an analysis in a plane by projection of the C^{α} position onto a disclosing plane. According to this procedure, domains are iteratively decomposed into subdomains. A method for recognition of domains in proteins was also developed by Wodak and Janin (1980). Two continuous domains were observed in the structure of serine proteases, trypsin, chymotrypsin, and elastase (Hartley and Shotton, 1971; Blow and Steitz, 1970; Birktoft and Blow, 1972;

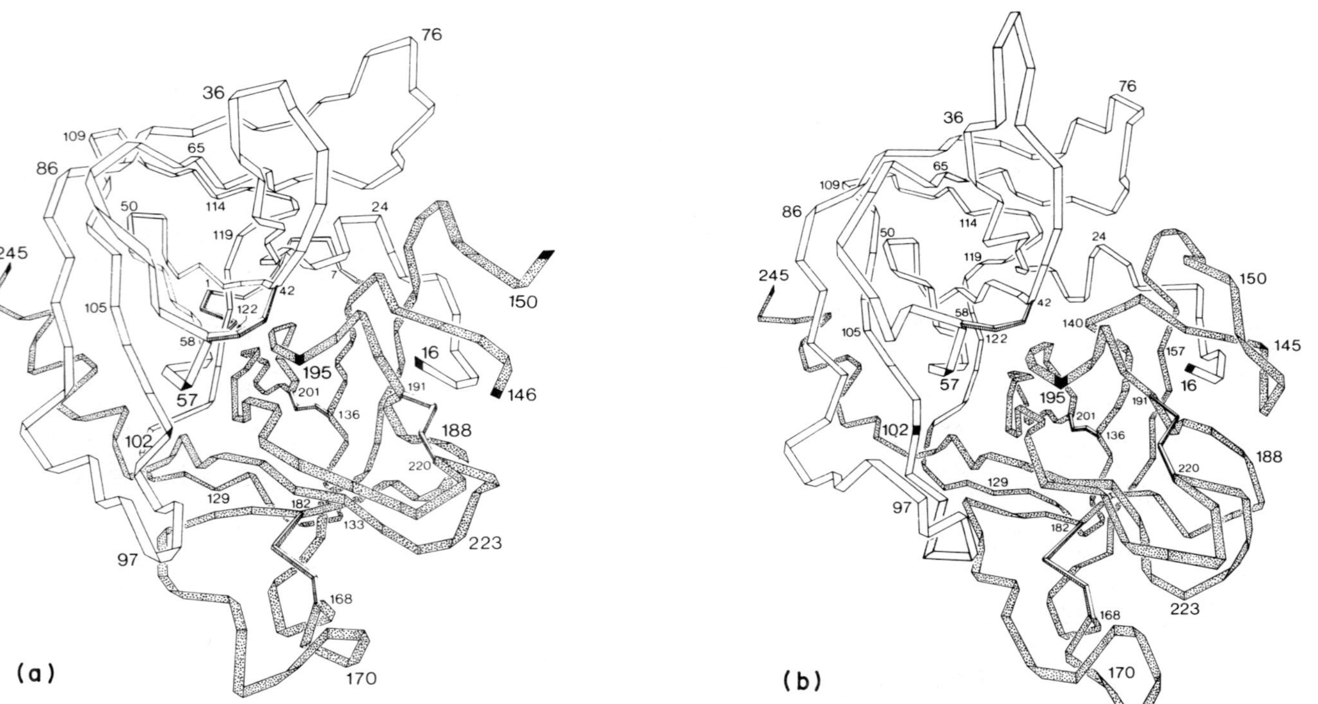

Fig. 2.34. Ribbon diagrams showing the three-dimensional structure of (a) chymotrypsin, (b) elastase (from Hartley and Shotton, 1971). The organization of these molecules into two domains is indicated in the drawings which also display the structural similarities between both enzymes.

Bode and Schwager, 1975; Sawyer *et al.*, 1978). Amino acids 16–130 form the first domain, residues 131–245 the second one (amino acid numbers are those of chymotrypsinogen). Figure 2.34 shows by ribbon diagrams the folding into domains of chymotrypsin and elastase. The structural organization of these two proteins in domains clearly appears in the interatomic distance matrix shown in Fig. 2.33. The catalytic site of these enzymes is composed of residues located in both domains. His 57 and Asp 102 are in the first domain and Ser 195 is in the second domain. The binding site of substrates is situated in the second domain. The cleft is located between the two lobes. Similar organization into two domains with an active site cleft between them has been found in acid proteases (Subramanian *et al.*, 1977; Tang *et al.*, 1978). Papain is structured as two well-separated domains: segments 1–11 and 112–207 (Drenth *et al.*, 1971). This is also the case for phosphoglycerate kinase (Blake *et al.*, 1972; Banks *et al.*, 1979) and for hexokinase (Steitz *et al.*, 1973; Anderson *et al.*, 1978, 1979). Thioredoxin has two continuous domains. Subtilisin is folded into three continuous regions: segments 1–100, 100–176, and 177–275. Thermolysin offers a remarkable example of a two domain protein: consisting of segments 1–157 and 158–316. Lysozyme has one continuous and one discontinuous region (Wetlaufer, 1973). Each protomer of glutathione reductase can be divided into four domains. The two nucleotide-binding domains have similar chain folds (Schulz, 1980).

The three-dimensional structure is now well established for several dehydrogenases. These enzymes are oligomeric proteins; each subunit is folded into two distinct globular lobes: the NAD^+ binding domain (two Rossmann folds) and the catalytic domain. Figure 2.35 shows these structures for *Bacillus stearothermophilus* glyceraldehyde-3-phosphate dehydrogenase (GPDH) (Biesecker *et al.*, 1977). In dehydrogenase the average size of the NAD^+ binding domain is about 140 amino acids. It consists of two units: the main structural elements are six strands of parallel sheet (βA, βB, βC, βD, βE, and βF) and four helices (αB, αC, αE, and α1F). Two helices are located on each side of the sheet. The comparison of the NAD^+ binding structures in the different dehydrogenases (i.e., LDH, GDPH, and ADH) indicates a correct superposition of the individual residues and a quite constant angle for the twist between the strands (Rossmann *et al.*, 1975).

2.6.2. Interactions among Domains in Proteins

The globular character of domains in proteins is marked by the close packing of the atoms (see Section 2.7). These closely packed domains interact with each other in a protein burying a part of the accessible surface area

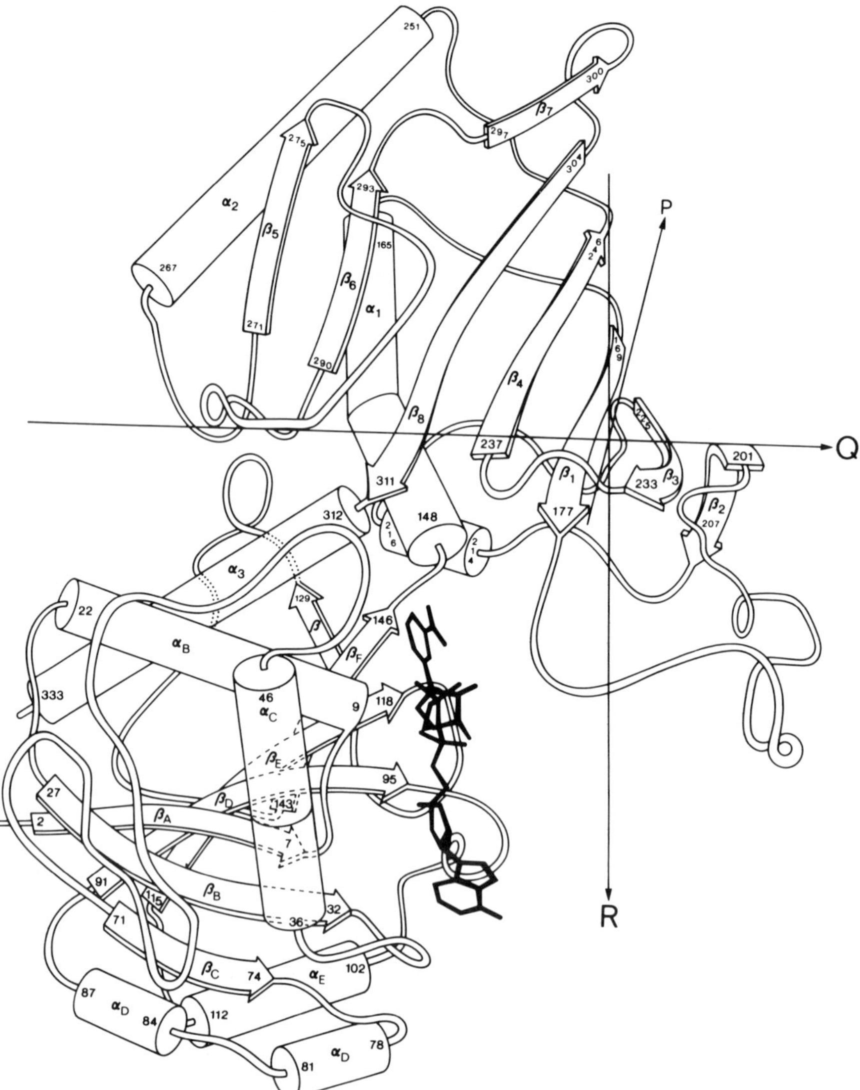

Fig. 2.35. Schematic diagram of the structure of *Bacillus stearothermophilus* glyceraldehyde-3-phosphate dehydrogenase (according to Biesecker *et al.*, 1977; courtesy of J. Walker) The first domain, NAD^+ binding domain, consists of residues 1–149 organized in two Rossmann folds; the position of NAD^+ is indicated. In the second domain (residues 150–311), an extensive region of antiparallel β sheet form a subunit interface; the S loop is formed by residues 178–201; the C-terminal helix (residues 312–333) fits into a groove in the first domain.

(see Section 2.7 and Chapter 3). The area buried between the two domains in thermolysin was calculated by Chothia (1976). It is rather small since it represents only 7% of the total buried surface. It is somewhat larger in elastase where 11% of the total buried surface is involved in the interdomain contacts. The surface between domains is largely nonpolar. This is also true for the contact between protomers in oligomeric proteins. For thermolysin, 73% of the interdomain surface is nonpolar and 66% of the interdomain surface of elastase is nonpolar.

As an exception, a small number of isolated water molecules can be trapped inside a protein. They are often located on the surface of the domains (see Section 2.9).

Symmetry may be apparent in the arrangement of domains. As emphasized by Rossmann and Argos (1976) similar, but not identical, domains existing within a polypeptide chain may associate together with a pseudosymmetry. It has been noted for the association of domains in dehydrogenase and also for the association of the two cylinders in chymotrypsin. In chymotrypsin, both domains have a symmetrical structure consisting of β barrels formed of six antiparallel β strands paired around a dyad axis. A twofold symmetry was also described for copper–zinc superoxide dismutase and for protease B from *Streptomyces griseus* (Mc Lachlan, 1979a,b, 1980a,b), and TMV coat protein (Mc Lachlan *et al.*, 1980). Pseudosymmetry with an approximate dyad axis between domains within a single chain has been observed in other proteins such as carp parvalbumin (Kretsinger and Nickolls, 1973), rhodanese (Bergsma *et al.*, 1975), bacterial ferredoxin (Adman *et al.*, 1973), and acid proteases (Subramanian *et al.*, 1977; Tang *et al.*, 1978). Soybean trypsin inhibitor has a threefold axis (Mc Lachlan, 1979b).

2.7. OVERALL CONFORMATION OF PROTEINS

The conformational description of proteins and peptides has been reviewed extensively by various researchers (Schellman and Schellman, 1964; Dickerson, 1964; Davies, 1965, 1967; Harrington *et al.*, 1966; Ramachandran and Sasisekharan, 1968; Scheraga, 1971; Lee and Richards, 1971; Richards, 1974; Chothia, 1974, 1975; Anfinsen and Scheraga, 1975; Chothia, 1975; Chothia and Janin, 1978; Némethy and Scheraga, 1977). The overall conformation of the polypeptide chain is entirely described by the values of dihedral angles (ϕ_i, ψ_i). For a complete description of the protein conformation, the dihedral angles χ_i^{jk}, corresponding to the side chains must also be known. Crystallographers generally describe protein conformation by the atomic coordinates in an adequate system. They have been collected in the Protein Data Bank (Brookhaven, 1977). Theoretical methods to express the effect of ϕ_i and ψ_i rotations in the standard coordinate system have been developed (Némethy and Scheraga, 1965; Ramachandran *et al.*, 1966; Ramakrishman, 1964).

The main rules which govern the overall conformation of globular proteins were mainly deduced from the knowledge of three-dimensional structure at atomic resolution, which displays structural invariants. These main rules may be expressed as follows: (1) the absence of knots in its topology; (2) the close packing of the interior of the molecule (or of the domain in the case of a multidomain protein), which prevents the presence of water inside the molecule; (3) the tendency of hydrophobic groups to be located at the interior of the molecule and of the polar residues to be at the surface; and (4) the presence of bound water molecules at the surface in interactions with the polar groups including CO and NH of the backbone.

These two last characteristics of protein structure are discussed in Chapter 3 rather than here, because they are directly related to the thermodynamic properties of protein molecules in water solvent.

2.7.1. Absence of Knotted Topology in Proteins

It is a striking fact that proteins do not fold into a knotted conformation. The absence of knots (which occur only very rarely at the extremity of a polypeptide chain) indicates some restrictions which must prevent or hinder a polypeptide chain from achieving this kind of topology. It was particularly emphasized by Richardson (1977) for the connectivities of sheet. However, some parts of a polypeptide chain located either at the N-terminal or at the C-terminal ends can be tucked in a loop. Such a situation has been described for carboxypeptidase where the polypeptide chain forms a knot containing the first 30 residues and the last 20 residues (Quiocho and Lipscomb, 1971). In pancreatic trypsin inhibitor, the N-terminal part of the polypeptide chain passes through the covalent loop formed by the 30–51 disulfide bond forming a threaded topology (Deisenhofer and Steigemann, 1975). Penetrated loops (which are one segment of a protein encircled by another, but not necessarily in a knot) were described for colipase and *Androctonus* neurotoxin (Klapper and Klapper, 1980).

2.7.2. Compactness of Globular Proteins

The examination of the three-dimensional structure of proteins interpreted at atomic resolution indicates that the various atoms and functional groups pack together particularly well. This aspect of protein structure was studied by Lee and Richards (1971), Richards (1974, 1977, 1979), and Chothia (1975). The close packing of the atoms in proteins was quantitatively approached

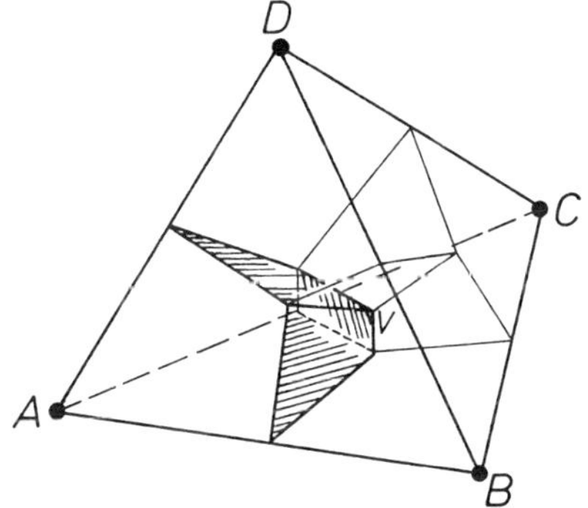

Fig. 2.36. Illustration of Voronoi polyhedra procedure (see text) (from Schulz and Schirmer, 1979).

by Richards, who evaluated packing density which represents the standard van der Waals volume of an atom or a group of atoms divided by the volume actually occupied in space. This last volume was determined by an adaptation of the Voronoi polyhedra procedure with the atomic group centers being taken from atomic coordinates. Figure 2.36 shows a normal Voronoi polyhedra construction, which is built from the center of four atoms A, B, C, and D.

Space is divided into polyhedra around atoms in a way adapted from the Voronoi procedure. For a set of atoms, space is divided into polyhedra with one atom at each corner. Bissecting planes are drawn on the plane vector at a position which divides the distance between atoms A and B according to the ratio of the van der Waals radii if they are noncovalently bound, and according to the ratio of covalent radii if they are covalently bound. The four bissecting planes subdivide the tetrahedron uniquely into four parts. The part which belongs to atom A is represented by a hatched surface in the figure. The atom A may belong to different tetrahedra. To determine the total polyhedra around A, the same construction has to be performed for all tetrahedra, and all parts belonging to A have to be added. This procedure has to be repeated for each atom. When it is, space is totally divided.

For closely packed spheres, the packing density is 0.74 while it is 0.91 for closely packed cylinders of infinite length. For crystals of many organic molecules this dimensionless number ranges from 0.70 to 0.78. The average packing density for lysozyme and ribonuclease S is close to 0.751 (Richards, 1974). Thus, the volume occupied by a particular amino acid in the interior of a protein is constant and the same as the volume occupied by this residue in a crystal form. Thus the description of molecular crystals is accurate (Liquori, 1966, 1969; Chothia, 1975). The total volume of these two proteins derived by this procedure is larger by 7–10% than that estimated from measured partial specific volume. The average density in the protein interior varies in different parts of the same protein. In lysozyme and ribonuclease, packing densities range from 0.5 to 1.0. Indeed, some parts have low densities, with packing defects which allow motion, as for example, at the active site of an enzyme. High-density regions, which are not compressible in globular proteins, may transmit or correlate motion over long distances (Lee and Richards, 1971).

Occasionally, the presence of holes inside the molecule was described. One hole in myoglobin can accomodate a xenon atom (Schoenborn *et al.*, 1965). Lee and Richards (1971) have determined and listed cavities in three proteins, myoglobin, lysozyme, and ribonuclease S. In myoglobin, the larger cavities have respective volumes of 2.85 $Å^3$ and 2.29 $Å^3$. The heme bound water molecule forms parts of the boundary of a cavity of 1.781 $Å^3$. Another cavity of 0.853 $Å^3$ can accomodate a xenon atom. One or two cavities in ribonuclease S appear to be unoccupied. In many cases holes inside proteins are filled with solvent. In lysozyme one large and two smaller cavities were reported from the electron density map, and three buried water molecules were found (Lee and Richards, 1971; Richards, 1974). In chymotrypsin, water molecules are inserted into cavities.

2.8. QUATERNARY STRUCTURE OF PROTEINS

While globular proteins are commonly built by the assembly of domains, an ultimate stage of folding is often required to generate functional properties of a molecule. This is the selfassembly of subunits in oligomeric proteins.

2.8.1. Definition

Many proteins (mainly endocellular ones) are composed of several subunits. This subunit assembly is termed the quaternary structure, according to Bernal (1958). A subunit does not necessarily coincide with a single chain. For instance, α chymotrypsin is built of three chains, A, B, and C, but does not contain three subunits. Similarly, the A and B chains of insulin correspond to a unique subunit. Mostly, the number of subunits in proteins is rather small and corresponds to a well-defined arrangement. For this reason Monod *et al.* (1965) proposed the term oligomer meaning a protein composed of a small number of identical subunits. The term polymer refers only to the indefinite association of a large number of subunits. The subunits in an oligomer are termed protomers. The term subunit may be used to refer to any identifiable submolecular entity within a protein, whether identical to or different from the other components of the molecule. For example, in hemoglobin, α and β and even $\alpha\beta$ are subunits, while $(\alpha\beta)$ is a protomer. Monomers designate proteins with a unique structured polypeptide chain.

Most oligomeric proteins are made of identical protomers. The association of protomers to give an oligomeric protein can occur according to different geometries, each of them involving well-defined contacts between protomers. The driving force resulting from interprotomer interactions tends to stabilize the structural organization of the overall conformation. Some restrictions might be imposed which limit the number of possible arrangements. In this section, the possible modes of association of protomers are considered first, then the possible geometries of oligomeric proteins are discussed with references to the architectural patterns deduced from X-ray diffraction studies. Finally, some examples of protomer interfaces are examined, (i.e., the contact regions between protomers in some oligomeric proteins as determined by high-resolution, crystallographic studies).

In addition to the earlier paper by Monod *et al.* (1965), some reviews of the quaternary structure of proteins have pointed out the importance of the oligomeric structure for the stability and the functional properties of the molecule as well as for its advantage along molecular evolution (see Sund and Weber, 1966; Hanson, 1966, 1968; Klotz *et al.*, 1970; Matthews and Bernhard, 1973; Friedman and Beychok, 1979). Klug (1967) has discussed the possible geometry for protein oligomers that pertain specifically to virus structure.

2.8.2. Mode of Associations of Protomers

The geometry of oligomers is determined by the subunits contacts. The stability of the overall conformation is plausibly the main reason why nature has selected oligomeric proteins. In such proteins the protomer itself is less stable than the whole protein. The driving force for the association of isolated subunits certainly originates from the stabilization resulting from the interface bonding between protomers. The larger should be the number of contacts between protomers, the more stable should be the oligomer. Most oligomeric proteins are not linked by disulfide bridges which provide an extrastabilization to the molecule. One can imagine that subunits without disulfide bridges need to be associated to gain this increment of stability.

The nature of interprotomer contacts and the strength of the binding may vary according to the geometry of the oligomer. The region of a protomer which is involved in the binding with another protomer is termed by Monod *et al.* (1965) as binding set. The domain of bonding defines the region that encloses the complementary binding sets of two protomers. The domain of bonding between protomers could be made of identical binding sets. In such a situation the association is said to be isologous. If the domain of bonding involves two different binding sets, the association is heterologuous. Figure 2.37 shows these two modes of association. In their review, Matthews and

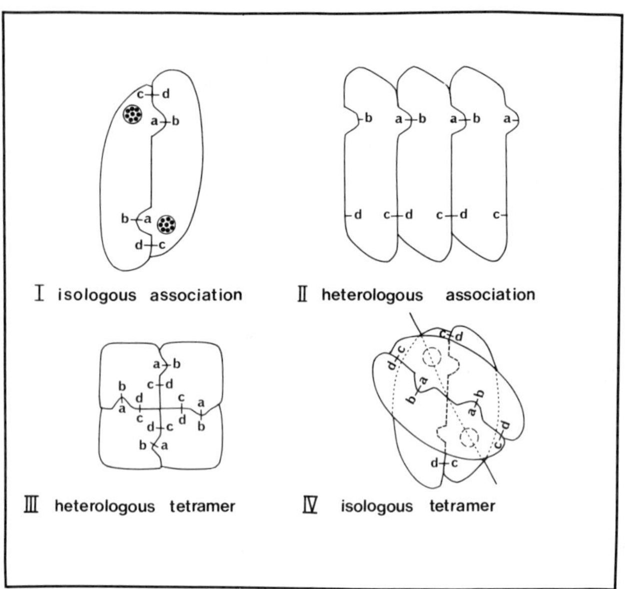

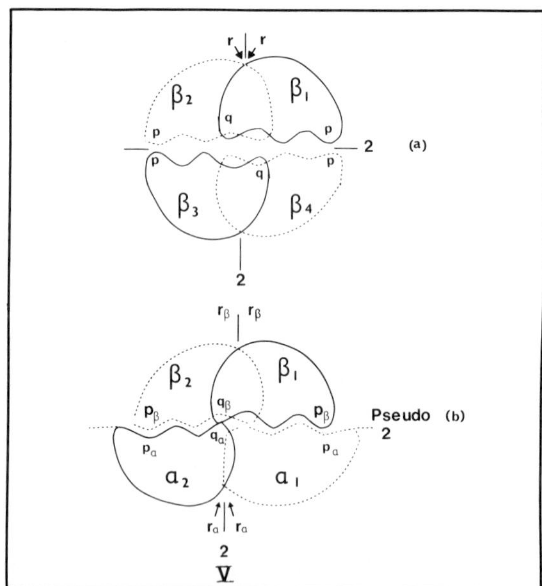

Fig. 2.37. Modes of association of protomers in oligomeric proteins (I) isologous associations; (II) heterologous associations: (III) heterologous tetramer; (IV) isologous tetramer (pseudotetrahedral); and (V) exact and pseudoisologous associations in a tetramer I, II, III from Monod *et al.*, 1965) (IV), V from Matthews and Bernhard, 1973).

Bernhard (1973), taking into account the detailed structure of oligomeric proteins then available, added two additional classes of association. The one is referred to as pseudoisologous association and appears when the domain of bonding is composed of two almost identical binding sets. The other denominated axial association corresponds to a situation where the domain of bonding includes interactions across an axis of symmetry.

In an isologous association the domain of bonding has a twofold axis of rotational symmetry. Furthermore, this kind of association leads to the formation of closed structures. Heterologous associations can generate helical structures; if the pitch of the helix is zero, a closed cyclic structure will be formed. By this kind of cyclic arrangement, closed oligomers can be built, containing any desired number of subunits larger than two. Isologous bonding is the only possible type of association for a dimer. Isologous interactions can generate oligomers containing only an even number of protomers. Heterologous bonding represents the only possibility of subunit associations for an oligomer with an odd number of protomers, while in an oligomer with an even number of protomers, the association may be either isologous or heterologous. Pseudoisologous associations correspond to a geometry in which the interacting protomers are to a high degree of approximation twofold related, but there are some small shifts from exact dyad symmetry. Examples of axial interactions are provided by insulin and by hemoglobin. In the insulin hexamer, each of three subunits has a histidine group which interacts with the same Zn^{2+} located on a threefold symmetry axis. In hemoglobin a molecule of diphosphoglycerate situated on the symmetry axis mediates the interaction between the two dyad related β chains. Since these interactions are not meaningfully termed isologous or heterologous, they were referred to by Matthews and Bernhard (1973) as axial.

2.8.3. Geometry of the Oligomer

For determination of the geometry of oligomeric proteins, Klotz and coworkers (1975) assumed rules which impose certain restrictions to the arrangements. The first assumption admits that all protomers in an oligomeric protein are in an identical (or pseudoidentical) environment. Second, taking into account that oligomeric proteins are built up of a small number of protomers, they assumed that binding regions must be saturated. Therefore, arrangements of protomers in the whole protein are closely restricted. If a protein is a tetramer, the formation of a hexamer or an octamer would be unlikely. As a consequence of these restricting conditions, the protomers could be regularly arranged around a central point of the oligomer. A line symmetry is ruled out by the first condition since in helical arrangement the subunits at the extremities are not in the same environment

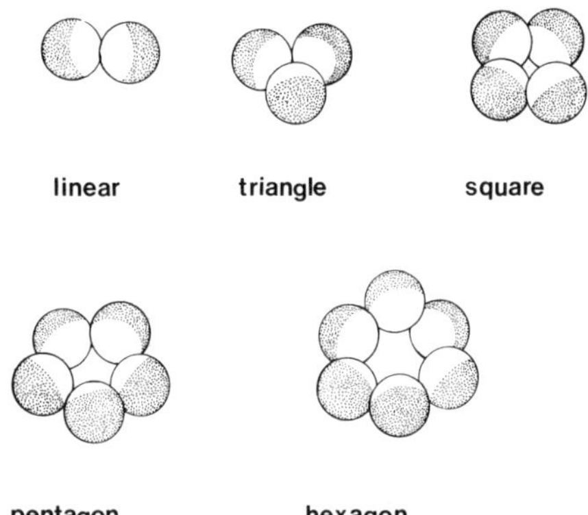

Fig. 2.38. The arrangements of protomers in cyclic symmetry (adapted from Klotz *et al.*, 1975).

as the others. All linear arrangements of protomers are disregarded. Only point group symmetries are considered as possible. The point group symmetry of an object may be defined as an operator or set of operators which, when applied about some point, transpose the object into itself. For oligomeric proteins, only point group symmetries characterized by rotation axes are possible. The possible geometries are cyclic, dihedral, or cubic symmetries.

In cyclic symmetry, there is a *n*-fold rotational axis, *n* being the number of subunits. Figure 2.38 shows the arrangements of protomers for dimer, trimer, tetramer, pentamer, and hexamer. The dimer has the C_2 symmetry (each protomer coincides to the other by a rotation of 360°/2), the trimer has the C_3 symmetry (rotation of 360°/3) and is required for the protomer transposition. The cyclic symmetry is referred to as C_n in the Shoenflies notation, *n* being the number of protomers. In the "International Hermann–Mauguin notation" it is referred to as *n*.

Dihedral symmetry refers to oligomers which have *n* twofold axis at right angles to any single *n*-fold axis. The number of subunits must be 2*n*. Dihedral symmetry is termed D_n in the Schoenflies notation. For example, a tetramer arranged in dihedral geometry is designated to have a D_2 symmetry (222 in International notation). Figure 2.39 indicates the arrangement in dihedral symmetry for several oligomers. For each oligomer, two possible geometries can be considered. For example, a tetramer could be formed of a square arrangement of the protomers. This would require a threefold axis of symmetry which is unlikely, since protomers are asymmetric. Structures with D_3, D_4, D_5, D_6, . . . D_n may exist for proteins with 6, 8, 10, 12, . . . 2*n* protomers respectively.

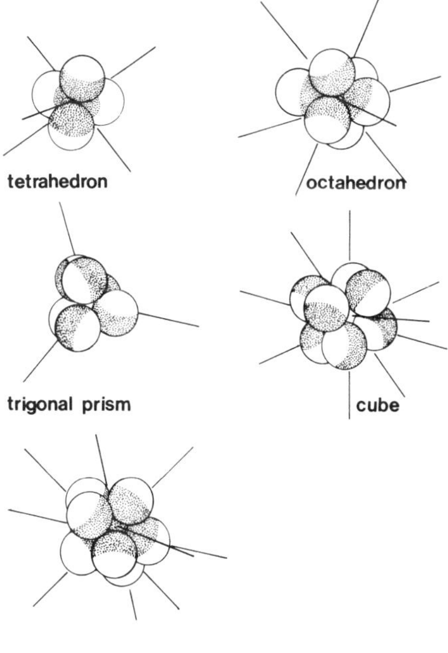

Fig. 2.39. The arrangements of protomers in dihedral symmetry (adapted from Klotz *et al.*, 1975).

Another arrangement is provided by the cubic symmetry: the cubic point groups. This kind of arrangement does not exist for smaller proteins. Only oligomeric proteins with $12n$ protomers may possess cubic symmetry. This symmetry designated as T (or 23) is characterized by more than one rotational axis greater than twofold. For example, the tetrahedron possesses four threefold axes. The threefold symmetry as previously mentioned is not possible when asymmetric protomers of a protein are at the vertices of a tetrahedron. However, it can be obtained if a group of three subunits are located at the four vertices of a tetrahedron. This involves a number of 12 asymmetric protomers.

Another arrangement involving a point symmetry of three subunits placed at each of the vertices of a cube is designated as O symmetry. For this arrangement the protein must contain at least 24 protomers.

Table 2.5 summarizes the different types of symmetry possible for proteins and the types of associations which are involved in the geometry of the molecule, according to the number of protomers.

TABLE 2.5

Different Types of Symmetry in Protein Oligomers[a]

Number of Protomers	Schoenflies Notation	International Notation	Type of Interactions	Geometry
2	C_2	2	Isologous	Linear
3	C_3	3	Identical heterologous	Triangle
4	C_4	4	Identical heterologous	Square
4	D_2	222	Two different isologous	Square
4	D_2	222	Three different isologous	Tetrahedron
5	C_5	5	Identical heterologous	Pentagon
6	C_6	6	Identical heterologous	Hexagon
6	D_3	32	Two different isologous	Hexagon
6	D_3	32	Isologous and heterologous	Trigonal prism
6	D_3	32	Isologous and heterologous	Octahedron
7	C_7	7	Identical heterologous	Heptagon
8	C_8	8	Identical heterologous	Octagon
8	D_4	422	Two different isologous	Octagon
8	D_4	422	Isologous and heterologous	Cube
8	D_4	422	Isologous and heterologous	Square antiprism
n	C_n	n	Heterologous	
$2n$	D_n	n 22 (n even) n 2 (n odd)	Isologous, or Isologous and heterologous	
12	T	23	Isologous and heterologous	
24	0	432	Isologous and heterologous, or Heterologous	
60		532		Icosahedron

[a] Adapted from Klotz *et al.* (1975).

2.8.4. Occurrence of Oligomeric Proteins and Their Structural Characteristics

In their listing of subunit constitution of proteins, Darnall and Klotz (1972) considered more than 500 proteins. From this table, the frequency of occurrence of the different oligomeric proteins according to the number of subunits was plotted by Raibaud (1977) (Fig. 2.40). Clearly the dimcric proteins are the most frequent, and to a lower extent the tetramers. The frequency of occurrence rapidly decreases with the increasing number of subunits.

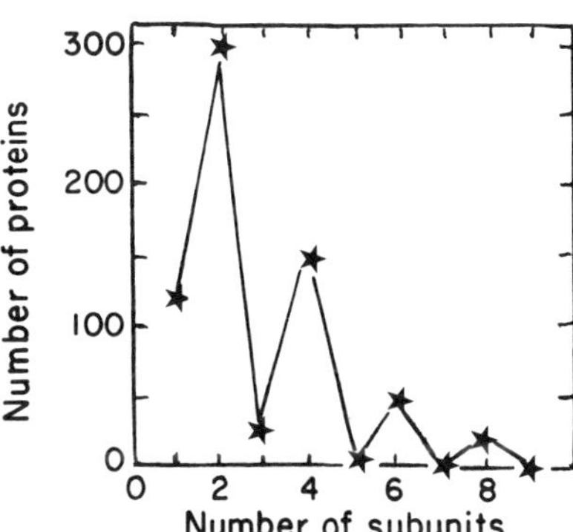

Fig. 2.40 Frequency of occurrence of the different oligomeric proteins (according to Raibaud, 1977; from the listing of Darnall and Klotz, 1972) (see text).

When $n = 2$, 3 or any odd number, only one type of geometry is possible involving a cyclic symmetry. For an oligomer containing $2n$ protomers, cyclic and dihedral geometries are possible. Dimeric proteins have essentially a C_2 symmetry and tetrameric proteins are mostly with a D_2 symmetry. The dihedral symmetry is found rather frequently in oligomeric proteins. No crystallographic data exist at the present time to support the existence of tetramer with cyclic symmetry and there is no evidence for the existence of tetramers with a C_4 symmetry or pseudo-symmetry. Table 2.6 gives some examples of symmetries found in proteins. The T symmetry appears to be involved in the geometry of L- aspartate-β-decarboxylase, an enzyme composed of 12 subunits according to the data of electron microscopy (Bowers *et al.*, 1970). For dihydrolipoyl transuccinylase, an enzyme composed of 24 subunits, both electron microscopy and X ray diffraction indicate an O symmetry (de Rosier *et al.*, 1971).

Oligomeric proteins are composed of either identical or nonidentical subunits. The structures of the type $\alpha_2\beta_2$, $\alpha_4\beta_4$, are not exceptions. These proteins may possess axes of pseudosymmetry as does hemoglobin ($\alpha_2\beta_2$).

TABLE 2.6

Examples of Symmetries Found in Proteins[a]

Dimer	C_2	Hexokinase B (Steitz, 1971)[b] Insulin (Adams *et al.*, 1969)[b] Rhodanese (Drenth and Smith, 1971) β-Lactoglobulin (Aschaffenburg, *et al.*, 1965) Triose phosphate isomerase (Banner *et al.*, 1972) Alkaline phosphatase (Hanson *et al.*, 1970) Alcohol dehydrogenase (Zeppezauer *et al.*, 1967) Leucine tRNA synthetase (Chirckjian *et al.*, 1972) Lysine tRNA synthetase (Lagerkvist *et al.*, 1972) Phosphoglucose isomerase (Campbell *et al.*, 1971) Phosphorylase b (Eagles *et al.*, 1972)
Trimer	C_3	2-Ketodeoxy-6-phosphogluconate aldolase (Hammerstedt *et al.*, 1971)
Tetramer	D_2	Aldolase (Penhoet *et al.*, 1967) Lactate dehydrogenase (Rossmann *et al.*, 1967) Pyruvate kinase (Campbell *et al.*, 1971) Prealbumin (Blake *et al.*, 1971) Glyceraldehyde-3-phosphate dehydrogenase (Watson and Banaszak, 1964) Phosphoglycerate mutase (Campbell *et al.*, 1972) Concanavalin A (Hardman *et al.*, 1971) Phosphorylase (Kiselev and Lerner, 1971) L-Asparaginase (Epp *et al.*, 1971) Tryptophanase (Morino and Snell, 1967)
Hexamer	D_3	Aspartate transcarbamylase (Wiley and Lipscomb, 1968)
Dodecamer	T	L-Aspartate β-decarboxylase (Bowers *et al.*, 1970)
24 Subunits	O	Dihydrolipoyl transacetylase (de Rosier *et al.*, 1971)

[a] Adapted from Klotz *et al.* (1975).
[b] With a distorted axis of symmetry.

Determination of symmetry is in general obtainable by electron microscopy and by X-ray crystallography. Some other elegant but indirect methods are also available. Studies of dissociation, when possible, yield valuable information on the structure of oligomeric proteins (Guidotti *et al.* 1963). In appropriate solvents, oligomers are often dissociable into subunits without denaturation. The size and the composition of these subunits reflect the symmetry of the oligomer. The fact of dissociation into dimers is a diagnostic for dihedral symmetry. For instance, a tetramer having a D_2 symmetry would be expected to form stable dimers under favorable conditions: indeed, it contains one type of dimer bonds and two types of tetramer bonds. If the tetramer has a C_4 symmetry, since all the contacts are identical, it would be expected to produce only monomers dissociation. The intramolecular structural symmetry of oligomeric proteins can be approached by chemical investigations. Chemical labeling studies by substrate analogue reagents which form covalent intermediates on the one hand, binding studies on the other hand, can give indications about the geometry of oligomeric proteins. The existence of half-site reactivity and the determination of two classes of binding sites in a molecule was interpreted by Matthews and Bernhard (1973) in tetrameric enzymes as resulting from a $D_2 \rightleftarrows C_2$ transition.

2.8.5. Interfaces between Subunits in Oligomeric Proteins

The stability of a geometric arrangement depends on the energy of binding (i.e., the number of contacts and the strength of the bonds). The bond strength is not well established. The interprotomer contacts are all identical in cyclic symmetry, while they are different in dihedral geometry. A part of the surface of protomers is buried as a result of their association. Generally the area buried in interface upon formation of quaternary structure consists of a larger proportion of nonpolar contacts. Accessible surface area (see Chapter 3) buried after formation of quaternary structure was calculated for methemoglobin and deoxyhemoglobin by Chothia and co-workers (1976). They found a decrease in the accessible surface area by about 6,000 Å^2 upon the association of the four subunits. Consequently one-fifth of the subunit surface area is buried in subunit interfaces since the accessible surface area of each subunit is about 7,000 Å^2. In horse methemoglobin 60% of the area buried in interfaces are found in $\alpha_1\beta_1$ and $\alpha_2\beta_2$ contacts. The surfaces involved in contacts are mostly hydrophobic. Only 32–41% of the contacts result from the polar residues except for the contacts between subunits which are more polar due to the salt bridges between the two protomers of the same type. The interfaces between dimers are closely packed; the area buried in the interface upon formation of quaternary structure consists of a larger proportion of nonpolar contacts.

Considering that rough recognition of interacting sites is provided by distribution of hydrophobic groups on the surface, a tentative approach for predicting quaternary structure of proteins was proposed by Wodak and Janin (1978) and Rashin and Yudman (1979).

The contacts between subunits have been detailed for several oligomeric proteins, for example in several dehydrogenases. A schematic representation was proposed first for describing subunits contacts in lactate dehydrogenase (Liljas and Rossmann, 1974). Each subunit is represented on a two-dimensional diagram by a surface which has the same number of straight edges as it has intersubunit contacts. In LDH, each subunit is a triangle schematically designated by a color code, red, green, yellow, or blue as for

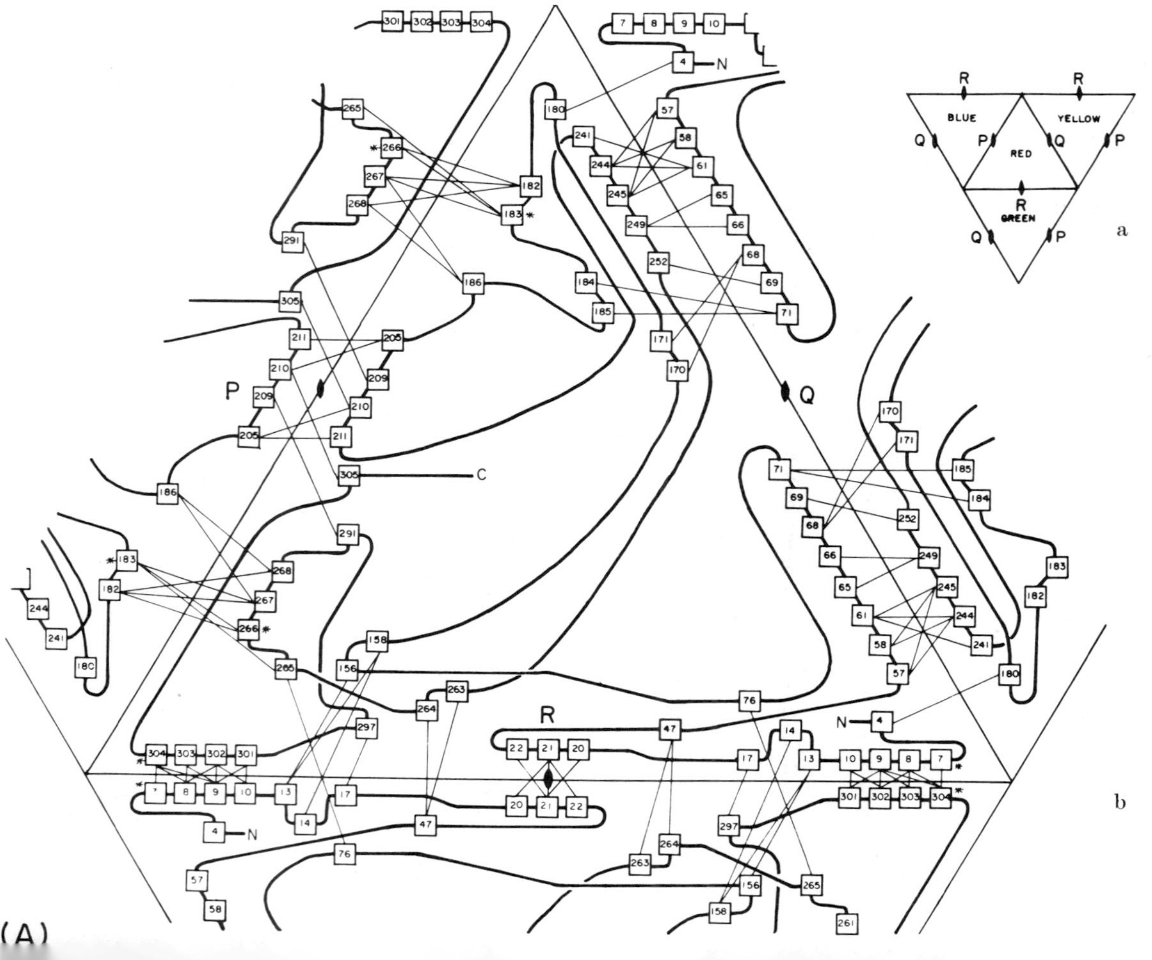

BLUE
RED
YELLOW
GREEN
P
Q
R
N
C
a
b
(A)

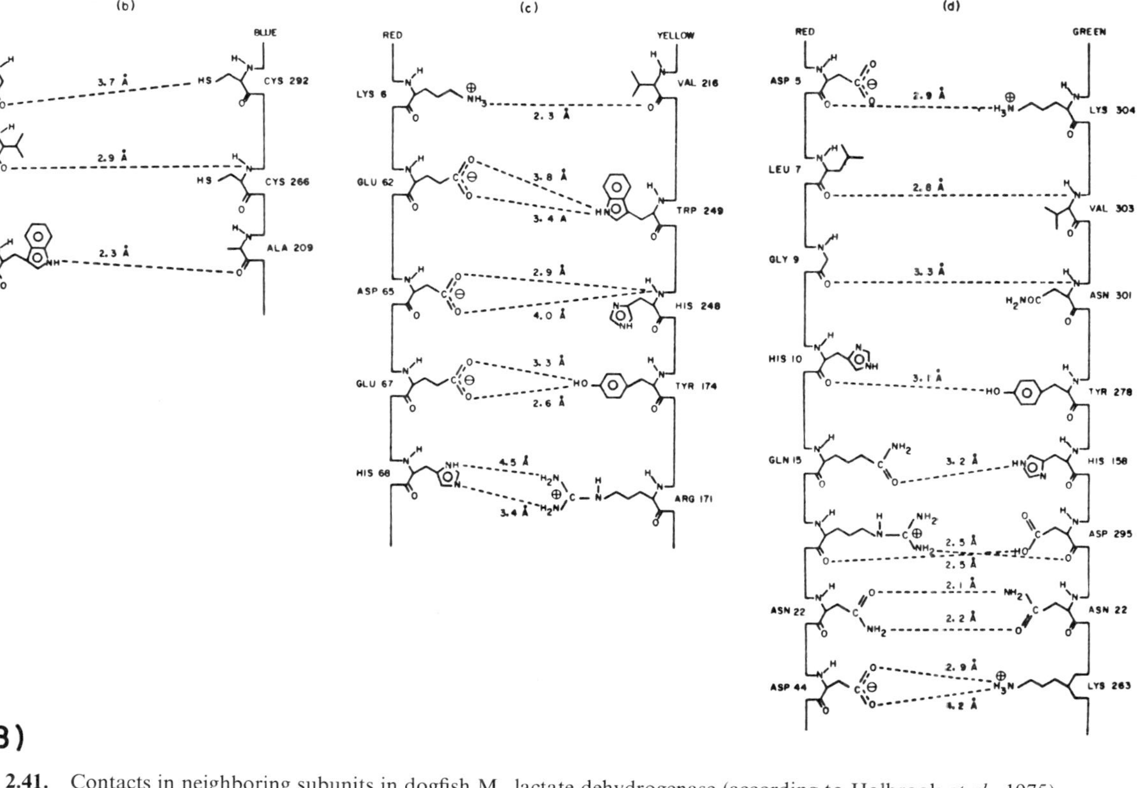

Fig. 2.41. Contacts in neighboring subunits in dogfish M_4 lactate dehydrogenase (according to Holbrook *et al.*, 1975). (A) Schematic representation of the different subunit contacts (see text), (B) detailed examples of some hydrogen bonds involved in the subunit interactions.

other dehydrogenases. The red subunit interacts with the blue, yellow, and green ones by the P, Q, R, axes respectively. Some of these contacts are nonpolar, some others involve hydrogen bonds or salt bridges. Figure 2.41 shows such a representation of the contacts in neighboring subunits in dogfish M_4 lactate dehydrogenase.

Subunit boundary is constitued by the pleated sheet structure in prealbumin, concanavalin A, and glyceraldehyde-3-phosphate dehydrogenase.

For isologous dimers, Morgan *et al.* (1979) defined the concept of self-complementary surfaces and the symmetry they require. He illustrated this concept by the Chinese symbol of T'ai ki (Fig. 2.42). A detailed description of these surfaces indicates that concanavalin A and the chymotrypsin dimer have this symmetry, whereas α chains of deoxyhemoglobin have not and hence do not dimerize.

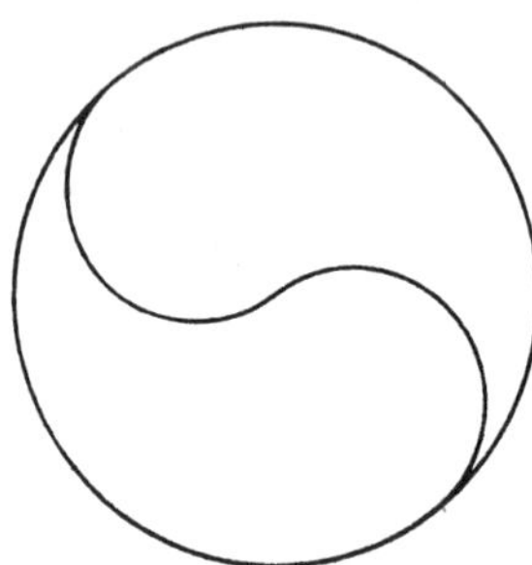

Fig. 2.42. The Chinese symbol T'ai ki illustrates the symmetry of self-complementary surfaces (according to Morgan *et al.*, 1979).

Molecular structure of glyceraldehyde-3-phosphate dehydrogenase has been determined for the enzyme from different species, particularly for the lobster enzyme and for the enzyme from *B. stearothermophilus.* Comparison of the interactions in thermophilic versus mesophilic protein is important for understanding thermostability. In fact, *B. stearothermophilus* GPDH retains structural integrity and full activity at temperatures up to 60°–70°C and the *thermus aquaticus* enzyme is still stable at 95°C (Biesecker *et al.*, 1977). A great conservatism is observed in the residues located at the inter-protomer interfaces. However, small differences have been observed. Contrary to prediction, protomer interfaces are no more hydrophobic in the *B. stearothermophilus* GDPH than in the lobster enzyme. Furthermore the subunit interfaces contain three extra buried salt bridges in the thermophilic enzyme. One is formed between Arg 194 and Asp 293 from the *P* axis related subunits. The others involve Arg 281 of one subunit forming a double salt bridge with Glu 201 from both the *R* and *P* axis related subunits. These additional salt bridges which occur between protomers stabilize the structure of the thermophilic oligomeric enzyme and contribute, at least in part, to the stabilization energy.

Hemoglobin A_2 is more resistant to heat denaturation than hemoglobin A. Analysis of the amino acid variations indicates that a replacement in the $\alpha_1\beta_1$ contact introduces an extra hydrogen bond between the guanidium group of Arg 116 β_1 and the carbonyl oxygen of Pro (114) α_1 in hemoglobin A (Perutz and Raidt, 1975). Perutz and Raidt have compared the amino acid substitution in these hemoglobins and in bacterial ferredoxins from mesophilic and thermophilic species and observed very few substitutions. They concluded that these changes can be accomodated without significant modification in the backbone conformation. Similar conclusions were reached by Argos and co-workers (1979) who compared sequences of thermophilic and mesophilic molecules of glyceraldehyde-3-phosphate dehydrogenase, ferredoxin, and lactate dehydrogenase. They have shown that Gly, Ser, Lys, and Asp in mesophiles are generally substituted by Ala, Thr, Arg, and Glu in thermophiles. Perutz (1978) emphasized that a few salt bridges would be sufficient to account for the stability of thermophilic molecules.

2.8.6. Structural Variations Related to Functional Characteristics

Most of the oligomeric proteins, but not all, deviate from a Michaelis behavior with regard to the effectors or substrates. Quaternary structure for oligomeric proteins which obey the Michaelis law, may be only a conformational requirement with extrastabilization occurring through the subunit interactions. For allosteric proteins, deviations from hyperbolic law may be either a cooperativity which is an amplification, or, on the contrary, an anticooperativity which is an attenuation of the functional properties (binding or catalytic activity). This allows a delicate regulation of the functional properties of the protein (Monod *et al.*, 1965). These properties occur through interactions between binding sites mediated by the protein which insures the transmission of the signals from one subunit to the others. Such a transmission involves a right coupling of subunits in the oligomer. The failure of this coupling upon desensitization drives the protein toward Michaelis behavior.

The role of subunit interfaces in the allosteric mechanism was analyzed by Chothia and co-workers (1976) for hemoglobin by comparison of the subunit interfaces of the deoxyhemoglobin and of methemoglobin. Indeed, binding of oxygen on hemoglobin is a cooperative process. Deoxyhemoglobin and methemoglobin represent respectively the T and R states of the protein in the allosteric model proposed by Monod and co-workers (1965). The surface area buried in the subunit interface is larger in the deoxy- than in the methemoglobin structure by about 800 Å^2, as a consequence of the structural change. This corresponds to an energy of 20 kcal/mole mainly because of hydrophobic interactions. Therefore 5 kcal per protomer is used to

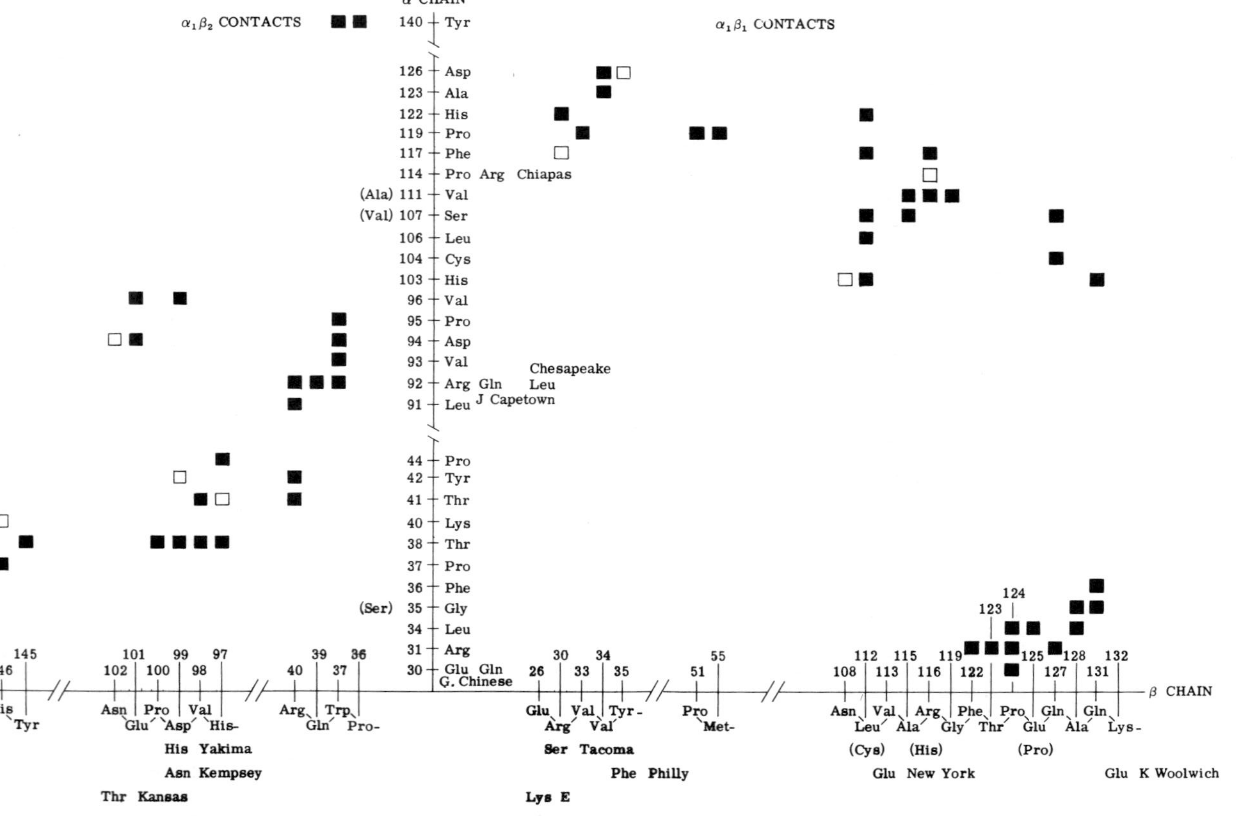

Fig. 2.43. Variations of $\alpha_1\beta_2$ subnit contacts in horse deoxyhemoglobin and oxyhemoglobin (according to Perutz *et al.*, 1968; Klotz *et al.*, 1975) Filled squares indicate van der Waals contacts: open squares indicate the contacts involving a hydrogen bond (courtesy of Klotz).

stabilize deoxyhemoglobin compared to methemoglobin. The authors conclude that an increase of hydrophobic contacts stabilized the *T* form in which the protomer conformation is in an unfavorable energetic conformation. The variations of the contacts as determined from X ray crystallography data by Perutz and co-workers (1968a,b) are shown in Fig. 2.43.

Detailed investigation of the structural changes in hemoglobin induced by ligand binding using computer graphics was reported by Baldwin and Chothia (1979). Tertiary and quaternary changes are coupled.

Variations of quaternary structure resulting from ligand binding were studied for various proteins such as for example hexokinase (Bennett and Steitz, 1978, 1980a,b; Shoham and Steitz, 1980).

2.9. LOCALIZATION OF WATER MOLECULES INTERACTING WITH PROTEINS

The close packing of proteins does not favor the presence of water molecules at the interior. However, water molecules can be trapped inside the protein but only in small quantity. Internal cavities inside a protein molecule were identified by the procedure developed by Lee and Richards (1971) for the estimation of accessibility to solvent molecules when they are large enough to contain at least one solvent molecule.

The bound water molecules are all attached to polar groups located at the surface of the molecule (CO and NH groups of the backbone included). Icelike structure has never been reported to occur around the hydrophobic side chains exposed to the solvent.

The possibility of hydrogen bond formation by separate water molecules with polar side chains must be considered, as principally emphasized by Birktoft and Blow (1972) and Lim (1974a,b). Three situations are of interest for the energetics of protein conformation: (1) side chain-H_2O-side chain, (2) side chain-H_2O-backbone, and (3) backbone-H_2O-backbone. They may contribute appreciably to the stabilization of protein conformation. The importance of water molecules in stabilization of irregular structures has been emphasized by Kuntz (1972), Berendsen (1972), and Lim (1974a,b). These researchers assumed an increase of the number of water molecules forming hydrogen bonds from type 1, 2, and 3 in irregular regions of proteins.

Solvent molecules directly interacting with the proteins have been localized from electron density maps for several proteins. Two kinds of solvent molecules, the external and the internal water molecules, should be considered. The external water molecules generally form hydrogen bonds with the surface polar groups. Some of them form single hydrogen bonds. Others bind to at least two groups of the enzyme. In elastase, 90 external water molecules are

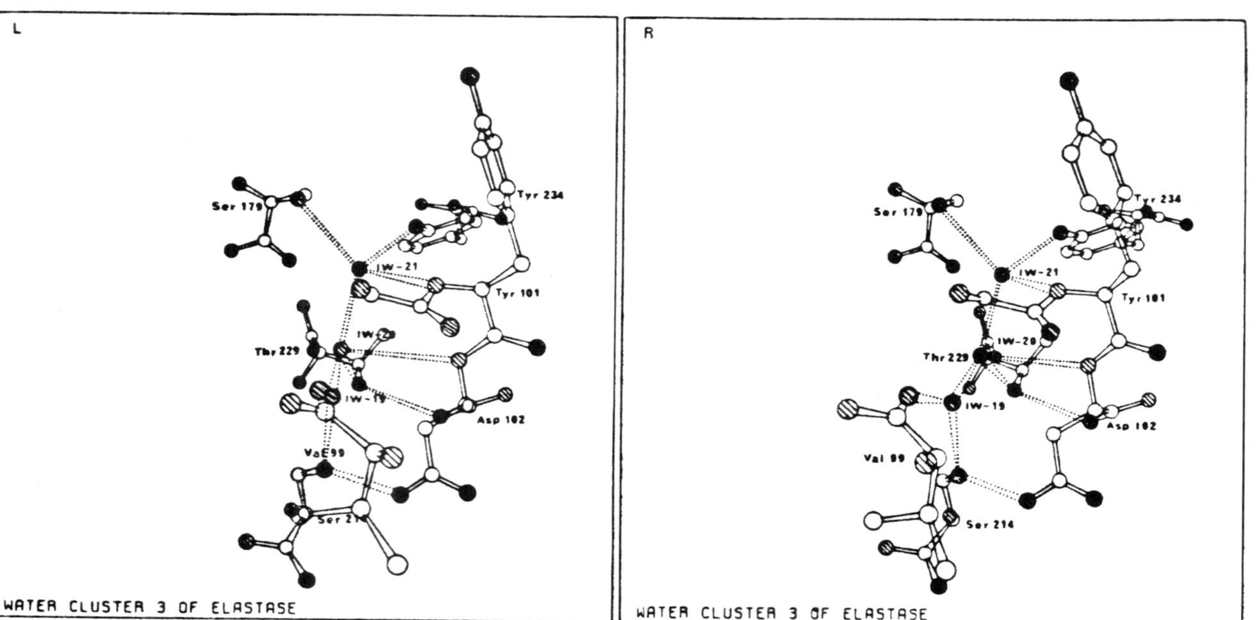

Fig. 2.44. Localization of the internal water molecules forming cluster 3 in elastase. These solvent molecules stabilize the interactions between the aspartate loop (in the first domain) and the C-terminal domain (see text) (from Sawyer *et al.*, 1978).

bound to the protein by a single hydrogen bond whereas 21 are bound by two or more groups of the protein.

The internal water molecules are localized into cavities or around the domains, some being sandwiched in the interfaces between domains, 3 were identified in lysozyme, 10 in carboxypeptidase (Matthews, 1977). In chymotrypsin, of the 13 water molecules trapped into the protein, 6 are located at the contact region between the two domains and form hydrogen bonds between both lobes. In fact the 13 water molecules were located at the surface of the two domains (Birktoft and Blow, 1972). In this protein, water molecules were supposed to stabilize the charged N- and C-terminal residues by charge delocalization. Generally, internal water molecules are isolated ones. In chymotrypsin and still more in elastase, they have a noticeable tendency to occur in clusters. These clusters certainly help to stabilize some parts of the tertiary structure of the protein by locally increasing the dielectric constant. There are 25 water molecules trapped inside the molecule that are inaccessible to the solvent. Three of the four clusters are very similar to those formed in chymotrypsin. Two of them, cluster 2 with three water molecules, and cluster 3 with three water molecules IW 19, IW 20, and IW 21 (as shown in Fig. 2.44), are located between the two domains. This last cluster which is homologous in chymotrypsin, probably stabilizes the interaction between a part of the aspartate loop and the C-terminal domain of the molecule by delocalizing the negative charge of the carboxylate of Asp 102. Cluster 1 (7 water molecules) is close to the end of the first cylinder which forms the first structural domain. Cluster 4 lies near the end of the second domain (Sawyer *et al.*, 1978).

2.10. STRUCTURAL SIMILARITIES BETWEEN MONOMERIC AND OLIGOMERIC PROTEINS

It is worthwhile to emphasize the structural similarities between the architectural arrangements of domains in multidomain proteins and of subunits in oligomeric proteins. The only difference is that domains are covalently linked whereas subunits are associated by exclusively noncovalent interactions. Domains are characterized by a close packing of the atoms forming independent globular building blocks inside a protein, such as subunits do in oligomers. Their assembly within a monomeric unit may be compared to the assembly of protomers in oligomeric proteins. Symmetry exists in the arrangement of subunits in oligomeric proteins. Symmetry or at least pseudosymmetry is often observed in the arrangement of domains. The interdomain contacts as well as the intersubunit contacts are mainly hydrophobic. Water molecules are located at the surface of the domains as at the surface of protomers. Some of them are trapped at the interface between domains or subunits.

There are striking similarities between monomeric and oligomeric proteins which both result from assembly of smaller subunits. The interaction of these units probably induce structural readjustments which may be of small amplitude but which are decisive for the full expression of biological activity. Therefore, by reference to oligomeric proteins, one can assume that the right conformational coupling of domains in multidomain proteins is an important and last event in protein folding, the one which induces conformational refinements allowing transmission through all the molecule, thus amplifying or generating biological activity.

2.11. EVOLUTIONARY ASPECT OF PROTEIN STRUCTURES

Comparison of protein structures from different living organisms has allowed construction of phylogenetic trees for several families of proteins. Until recently, only few proteins have been subjected to this analysis. Evolutionary trees derived either from protein sequences or from morphological considerations are qualitatively similar. As Margoliash (1972) stated: "We are still largely at the threshold of the new biological era of protein taxonomy." Different theories have been proposed to account for molecular evolution. The Darwinian theory is based on natural selection whereas neutral theory of molecular evolution accepts that variability is simply the result of chance and that replacement amino acids are selectively equivalent (Kimura, 1979). Another question in comparison of proteins exhibiting similarities is whether these similarities result from divergent or convergent evolution. In the present section, evolution of proteins derived from sequence alignments is considered first, and then the evolution of three dimensional structures is discussed. In a special subsection the existence of domains and its evolutionary significance is discussed with regard to structural and functional aspects.

2.11.1. Evolution of Primary Structures

There are a great number of possibilities for a change of amino acid sequence. From consideration of the genetic code, only by single base substitution, there are nine possibilities of changing a triplet codon. Therefore, there are $9 \times 61 = 549$ possibilities for change in amino acid sequence since 61 of 64 codons specify amino acids. There are 134 silent mutations because of code degeneracy, and 23 others because of a change of termination codon. Of the remaining 392, 132 lead to conservative changes, 160 to radical changes (see Section 2.2), and 100 represent intermediate changes (for example Val/Gly, Leu/Pro.). Considering the amino acids, there are 190 possible interchanges for the 20 amino acids of which 75 are attainable only from single base substitution, 101 are attainable from substitution of two bases in appropriate codon position, and 14 can only occur when all three bases of the codon are changed (Doolittle, 1979).

There are several ways to compare two homologous amino acid sequences. One can determine the percent identity after optimization of the alignment. Another way consists of counting the differences to express them in terms of necessary base substitution. Two categories of methods permit the construction of phylogenetic trees from the information regarding amino acid sequences of proteins. One consists in deriving the tree from a reconstructed ancestral sequence; the other is the matrix method. According to the first method used by Dayhoff (1972), the ancestral sequence is derived from alignments of amino acid sequences, at each branch point (or node), and thus the tree is shown in the direction of the most similar sequence. The best topology in terms of branch points and length is the one which involves the minimum number of mutations. The matrix method (Fitch and Margoliash, 1968) consists of building the matrix of percent of differences between sequences. Such a matrix is given for cytochrome *C* in Fig. 2.45. In the topology of the tree, the connections represent the order of divergence of the proteins and therefore of the different species. The length of the branches are related to the number of changes between divergence points. All these methods require computer processing of the information. Their limitation comes from the fact that it is not possible to know the exact number of changes which have occurred at one site, especially if several mutations have restored the same amino acid.

The matrix method was used by Fitch and Margoliash (1968) to construct the phylogenetic tree of the cytochrome *C* family (Fig. 2.46). Cytochrome *C* is a small protein with a molecular weight of 12,400. It is a polypeptide chain which contains 103–113 amino acids, depending on the origin. This respiratory protein is found in the mitochondria of all eukaryotes and contains one heme per molecule. The important work of different groups (Margoliash and Shejter, 1966; Margoliash, 1971, 1972; Margoliash and Smith, 1965; Boulter *et al.*, 1972) makes available the primary structure of 67 different cytochromes *C* (see review in Dickerson and Timkovitch, 1975). A great conservatism is observed. Fully 30% of the residues are identical in all species including vertebrates, insects, fungi, yeast, and wheat. Most replacements are conservative.

Relationship between geological time and the events of phylogenetic trees constructed from protein sequences have been evaluated, assuming that divergence between fish and man represents about 400 million years. Each

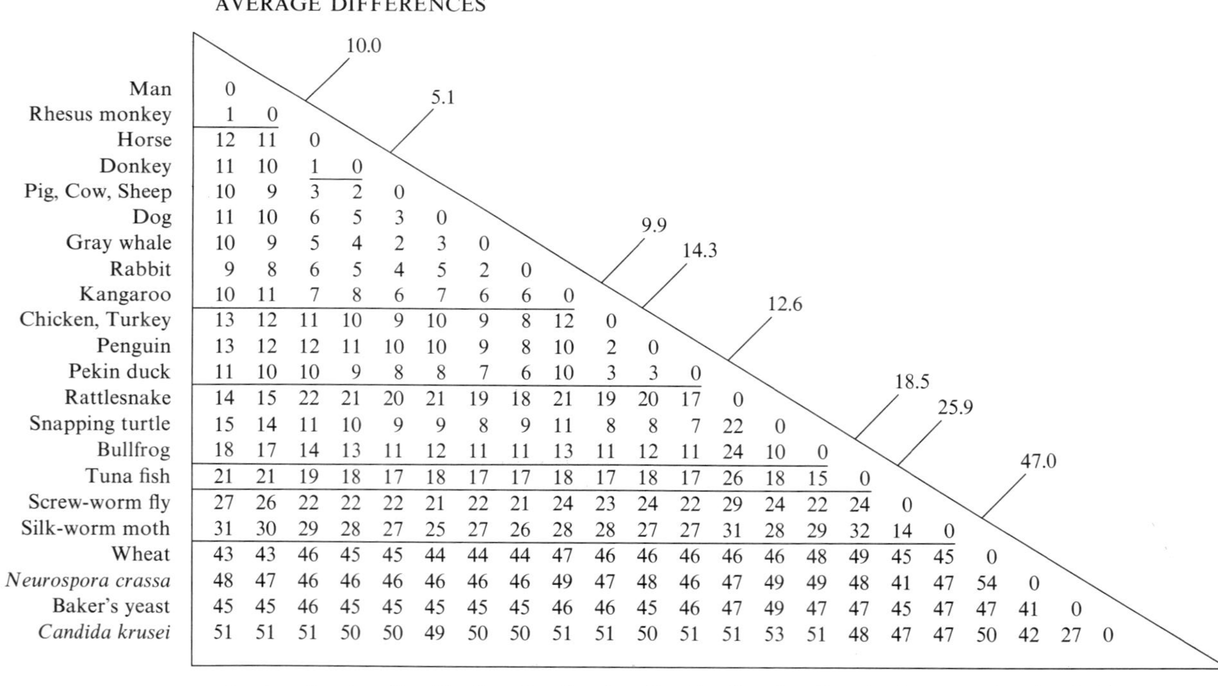

	Man	Rhesus monkey	Horse	Donkey	Pig, Cow, Sheep	Dog	Gray whale	Rabbit	Kangaroo	Chicken, Turkey	Penguin	Pekin duck	Rattlesnake	Snapping turtle	Bullfrog	Tuna fish	Screw-worm fly	Silk-worm moth	Wheat	*Neurospora crassa*	Baker's yeast	*Candida krusei*
Man	0																					
Rhesus monkey	1	0																				
Horse	12	11	0																			
Donkey	11	10	1	0																		
Pig, Cow, Sheep	10	9	3	2	0																	
Dog	11	10	6	5	3	0																
Gray whale	10	9	5	4	2	3	0															
Rabbit	9	8	6	5	4	5	2	0														
Kangaroo	10	11	7	8	6	7	6	6	0													
Chicken, Turkey	13	12	11	10	9	10	9	8	12	0												
Penguin	13	12	12	11	10	10	9	8	10	2	0											
Pekin duck	11	10	10	9	8	8	7	6	10	3	3	0										
Rattlesnake	14	15	22	21	20	21	19	18	21	19	20	17	0									
Snapping turtle	15	14	11	10	9	9	8	9	11	8	8	7	22	0								
Bullfrog	18	17	14	13	11	12	11	11	13	11	12	11	24	10	0							
Tuna fish	21	21	19	18	17	18	17	17	18	17	18	17	26	18	15	0						
Screw-worm fly	27	26	22	22	22	21	22	21	24	23	24	22	29	24	22	24	0					
Silk-worm moth	31	30	29	28	27	25	27	26	28	28	27	27	31	28	29	32	14	0				
Wheat	43	43	46	45	45	44	44	44	47	46	46	46	46	46	48	49	45	45	0			
Neurospora crassa	48	47	46	46	46	46	46	46	49	47	48	46	47	49	49	48	41	47	54	0		
Baker's yeast	45	45	46	45	45	45	45	45	46	46	45	46	47	49	47	47	45	47	47	41	0	
Candida krusei	51	51	51	50	50	49	50	50	51	51	50	51	51	53	51	48	47	47	50	42	27	0

Fig. 2.45. The matrix of the differences between sequences of cytochrome *c* in the various species. For example ten mutations have occurred between penguin and dog.

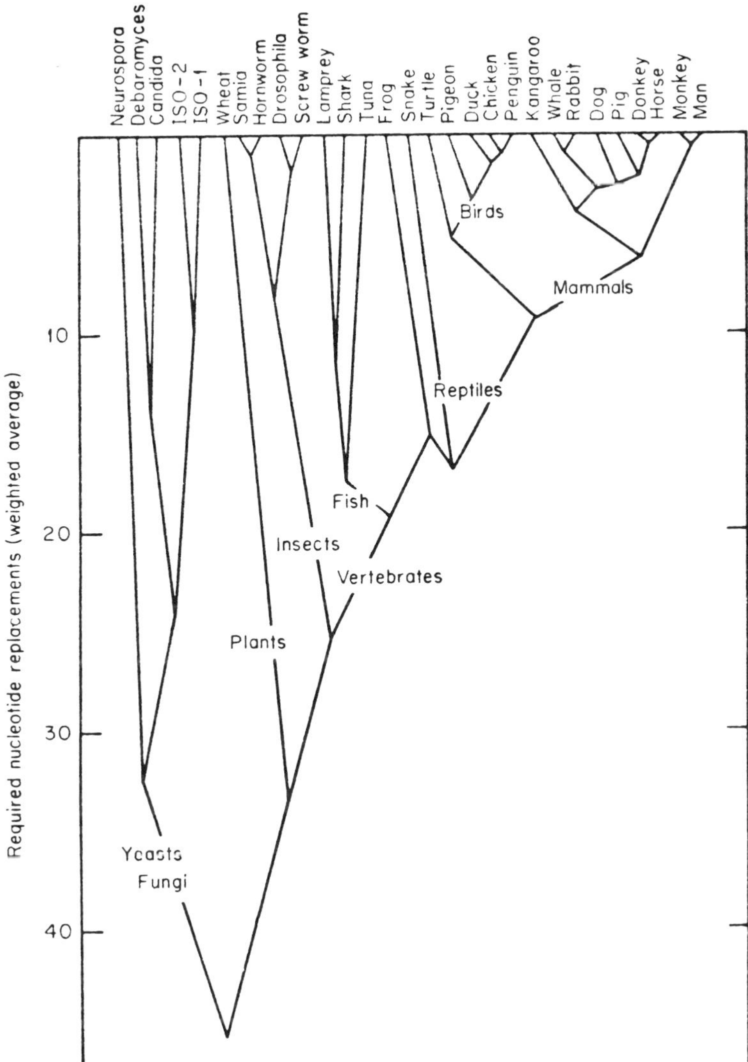

Fig. 2.46. Phylogenetic tree of cytochrome *c* families (according to Dickerson and Timkovitch, 1975).

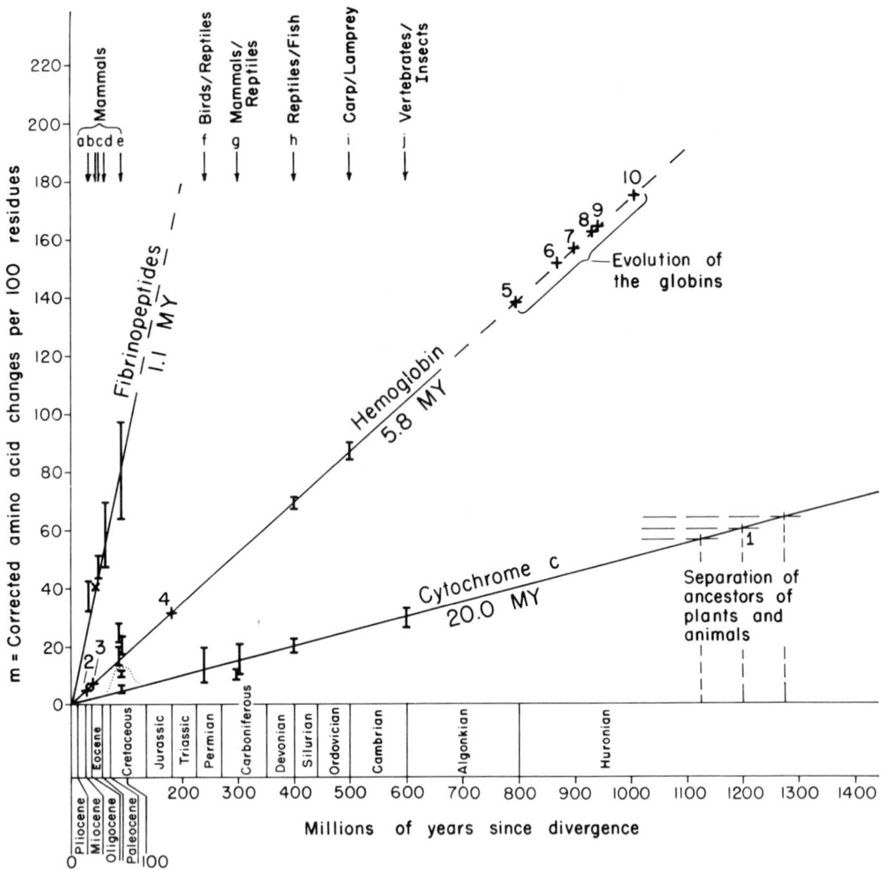

Fig. 2.47. The rate of evolutionary changes in amino acid sequence of three different proteins, cytochrome *c*, hemoglobin, and fibrinopeptides. In abscissea, the time needed for a 1% change in sequence between two diverging branches of the phylogenic tree (in million years), in ordinates the number of corrected amino acid changes for 100 residues. The slope allows one to obtain the rate of mutation (in million of years) (according to Dickerson, 1971).

family of protein is accepting mutation at different rates. The rate of evolutionary change in protein sequences is illustrated in Fig. 2.47 for three families of proteins. It indicates that each protein has a characteristic unit of evolutionary period. This rate of mutation is generally expressed in PAMs units (i.e., accepted point mutation per 100 links, per 100 million years). It clearly appears that the cytochrome *c* family represents one of the most conservative proteins.

2.11.2. Three-Dimensional Structure and Evolution

The analysis of structural and functional similarities of proteins (Rossmann and Argos, 1976, 1977, 1980; Ohlsson *et al.*, 1974; Haber and Koshland, 1970) indicates a great conservation of three-dimensional structure which is not necessarily supported by sequence homologies. Comparison of sequences and three-dimensional structures in different families of proteins, might explain (1) why certain amino acids of the sequence are constant during evolution, (2) what the necessary requirements are to insure similarities in three-dimensional structure, and (3) what the minimum requirements are to determine and to conserve a definite function.

Methods have been developed to compare the folding of three-dimensional structure of proteins (Rossmann and Argos, 1975, 1976, 1977; Mc Lachlan, 1972a,b, 1979a,b; Remington and Matthews, 1978, 1980). Superposition of two chains and minimization of distances between corresponding C^{α} atoms are used to detect similarities between C^{α} backbones of proteins.

Rossmann and Argos (1977) defined structural equivalence indicating that any residue or structural elements (α helix, β pleated sheet) is coincident with that of another molecule by superposition. It is called positive equivalence when similarly directed and negative equivalence when oppositely directed. Topological equivalence refers to a run of positive equivalence. According to these methods, a taxonomy of protein three-dimensional structure was developed.

Among the known proteins, several families are characterized by similarities in function, amino acid sequence, and three-dimensional structure. However, certain families of proteins display similarities of tertiary structure and function without any significant memory of ancestral primary structure. Examples are provided by the heme binding fold in cytochrome b_5 and in the globins (Rossmann and Argos, 1975), in polysaccharide binding fold in hen egg white and phage lyzosyme (Rossmann and Argos, 1976), and in nucleotide binding fold in NAD^+ dehydrogenases and some kinases (Ohlsson *et al.*, 1974; Rossmann *et al.*, 1974). For other proteins a similarity in fold is observed, but neither sequence homology nor common function are obvious. Superoxide dismutase and immunoglobulin domain are typical illustrations of such a situation. During evolution the three-dimensional structure of a protein seems to be much better conserved than the amino acid sequence. The most conservative parts of protein molecules are generally in structural folds essential for function. Modification of only one amino acid in a polypeptide chain may modify the function of a protein, as illustrated by hemoglobins.

Homology in amino acid sequence leads to analogy in three-dimensional structure. The conservatism of the evolution imposed by the function is reflected by the maintenance of persisting sequences in the polypeptide chain that determine the folding of the active protein. Different proteins which exhibit a high degree of similarity in their primary sequence have similar three-dimensional structures. Thus in the serine proteases family, chymotrypsin, trypsin, and elastase have great similarities in both sequence and structure. In hemoglobin and cytochrome *c* families, homologies and similarities in three-dimensional structure have also been observed. In cytochrome *c* there is a conservatism on the part of the structure which is in close contact with the heme group, especially the groups which attach the heme (Fig. 2.48). Probably some other conservative regions could be related to the area in interaction with cytochrome oxydase and reductase. There is also an invariant segment at the left side of the protein designated by Dickerson as the weak side of the protein, because it undergoes conformational changes during the function of the protein. The location of the invariant amino acids in functional regions of proteins emphasizes the fact that the function imposes some serious constraints to the evolution.

The heme-binding domain of several proteins [cytochrome b_5, flavocytochrome b_2, and sulfite oxidase designated as the cytochrome b_5 fold (Guiard and Lederer, 1979)] have also great sequence homologies. These three proteins are probably the products of divergent evolution from a common ancestor.

A conservatism is also observed in the amino acids located at the interior of proteins forming the hydrophobic core. Most replacements occur at the surface where greater permissivity seems possible. In 24 homologous ribonucleases tested, which differ in 34% of their amino acid residues, Lenstra and co-workers (1977) noted the invariability of the hydrophobic residues.

Differences in amino acid sequence do not necessarily lead to variations in three-dimensional structure. There are some structural similarities which do not necessarily result from identical or analogous amino acid sequences.

Different polypeptide chain folds are remarkably conserved in various proteins and families of proteins, leading to these similarities of structural patterns as previously described.

Lesk and Chothia (1980) and Chothia and Lesk (1980) have analyzed and compared the structure of nine different globins. They have remarkably similar secondary and tertiary structure despite homologies as low as 16% for the most distantly related molecules. The helices are particularly well conserved. These researchers discuss the method by which different amino acid sequences can generate very similar three-dimensional structures and the mechanism by which proteins adapt to evolution. Lesk and Chothia examined the packing of helices in all the globins. Variations in sequence lead to change in the volumes of buried residues and therefore in the geometry of helix packing. The relative positions and orientations of homologous pairs

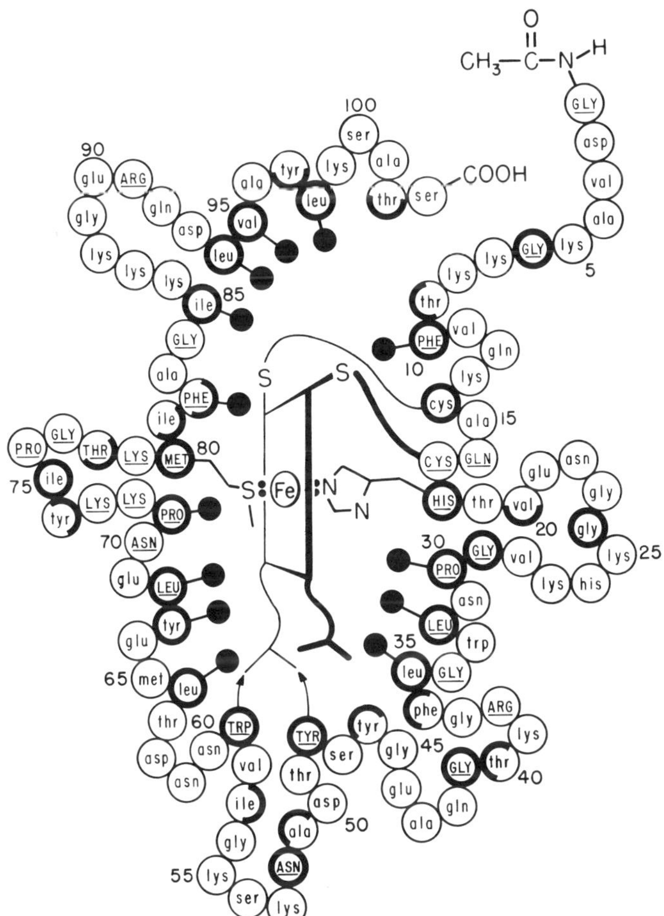

Fig. 2.48. Structure of cytochrome *c* displaying the binding of the heme to the apoprotein (adapted from Dickerson and Timkovitch, 1975). Many of the groups in close contact with the heme group are invariants (shadowed circles).

of helices are shifted, but these shifts are coupled so that the geometry of the heme pocket is similar in all globins. Thus the active site structure is retained by coupling the shifts. Coupling the local structural variations represents a mechanism by which biological function is preserved during the evolution.

The structure of β-turns is also conservative. Many of the chain reversal regions in proteins are conserved independently of the conservations of the residues. For example, 21 of the 27 β turns of elastase were found in chymotrypsin at the same position and with the same bend type. However, only 37 of 84 β turn residues are conserved indicating that sequence homology does not play the main role in determining conformational similarities in globular proteins (Chou and Fasman, 1977).

Topologies of β sheets, particularly six-stranded parallel β sheets are identical in different proteins.

Most of the secondary and supersecondary structures are conservative independently of the variability in sequence. A comparison of the sequence in α/β proteins indicates that residues involved in helix–sheet contacts have the same variability as the rest of the molecule (Janin and Chothia, 1980).

The Rossmann fold is also a conservative structure. Such a structure, which consists of parallel strands of β sheets flanked by two helices on each side which are antiparallel with regard to the sheets, is found in NAD^+ dehydrogenases, in several kinases (e.g., adenylate kinase, phosphoglycerate kinase), and in all enzymes which have a coenzyme with a nucleotide part, NAD^+ or ATP.

A well documented example of similarity in three-dimensional structure with a lack of sequence homology is displayed by the coenzyme binding domain in dehydrogenases. A careful analysis by Ohlson and co-workers (1974) of liver alcohol dehydrogenase (LADH), lactic dehydrogenase (LDH), and glyceraldehyde-3-phosphate dehydrogenase (GDPH) revealed that there are only four invariant residues among the 94 positions compared: These are Gly 199, Gly 204, Asp 223, and Gly 270 (corresponding to liver alcohol dehydrogenase).

Such a conservatism is not so evident for kinases. The Rossmann fold has been described in adenylate kinase and phosphoglycerate kinase. However in hexokinase, the region which binds ATP has a folding pattern different from that which binds NAD^+ in lactate dehydrogenase (Fletterick *et al.*, 1975; Steitz *et al.*, 1976; Anderson *et al.*, 1978, 1979). Adenosine binds at the carboxyl end of a β pleated sheet that is composed of three parallel and two antiparallel strands.

A great similarity of part of structure of subtilisin and lactate dehydrogenase has been noted (Rao and Rossmann, 1973). In subtilisin and in carboxypeptidase four parallel central strands of β pleated sheets are superposable (Rossmann and Argos, 1977). Since these similar structural patterns form structural domains in the corresponding proteins, their evolutionary relationship is discussed in the following paragraphs.

It has been remarked that proteins containing the nucleotide binding pattern, have the property of binding Cibacron blue-3GA the dye moiety of blue dextran sepharose (Rossmann *et al.*, 1975; Thompson *et al.*, 1975; Thompson and Stellwagen, 1976; Stellwagen, 1977; Stellwagen *et al.*, 1975; Witt and Roscovsky, 1980). The binding of Cibacron blue F3GA to liver alcohol dehydrogenase was studied by X ray crystallography and compared to the binding of NAD$^+$. When similarities in the binding of the dye with the binding of coenzyme NAD$^+$ were observed for rings B, C, D, then differences were observed for ring A (Biellmann *et al.*, 1979). However, despite the similarity of Cibacron blue to coenzyme NAD$^+$, the binding of the dye to a protein does not appear to be a specific probe of the existence of the NAD$^+$ binding fold. Beissner and co-workers (1979) reported several examples that contradict such a specific binding and have shown that *E. coli* arabinose binding protein which has a dinucleotide fold does not bind Cibacron blue. Phospholipase A_2, which is not able to bind nucleotides has a good affinity for Cibacron blue (Barden *et al.*, 1980). Barden and co-workers (1980) suggested that binding, when it occurs, is due to the hydrophobic character of a β sheet flanked by hydrophilic helices. In cytochrome b_5 reductase, cibacron blue binds not only to the NADH site but also to the flavin site (Pompon *et al.*, 1980).

Similar conformations have also been noted in (1) short polypeptide hormones such as insulin, insulinlike growth factor, and perhaps relaxin (Blundell, 1979), and (2) neutrotoxins which are composed of a central core of four disulfide bridges and which are characterized by extensive antiparallel pleated sheet structure (Drenth *et al.*, 1980; Drufton and Hides, 1980). Similar function may be produced by completely unrelated proteins. For instance, there is a remarkable degree of homology between the active site of α chymotrypsin and subtilisin BPN′ as revealed by X ray crystallography (Kraut *et al.*, 1972; Robertus *et al.*, 1972; Alden *et al.*, 1971. However, the respective sequence and tertiary structure of these two enzymes are completely different. Moreover, there is no resemblance in the amino acid composition of their respective sites which do, have a similar charge relay system at the catalytic site and homologous binding site. It seems very unlikely that these two enzymes have descended from a common ancestor; both proteins during the evolution have been selected for an analogous function.

2.11.3. Evolutionary Significance of Structural Domains

The domain hypothesis (Edelman, 1970) states that the single polypeptide chain in a multidomain protein was formed by gene fusion or gene duplication. As a consequence, one should expect different functional properties to be associated to the different domains. Continuous domains result from gene fusion and discontinuous domains from gene insertion (Janin, 1979) (Fig. 2.49).

The earliest and the best illustration of domain proteins is provided by immunoglobulins. The 12 domains of the IgG molecule have been already described and are shown in Fig. 2.32. There exist internal homologies in the

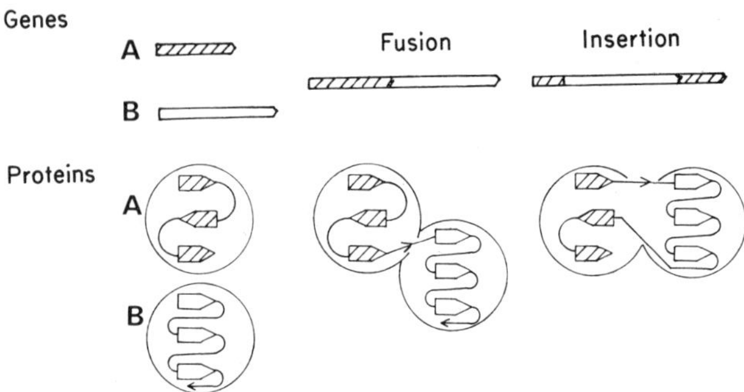

Fig. 2.49. Gene fusion and gene insertion (from Janin, 1979). The gene coding for protein A is fused head-to-tail with the gene coding for protein B; a fused protein made of two continuous domains is produced. The gene coding for protein B is inserted in the A gene; in this case a protein made of two discontinuous domains is produced (courtesy of Janin).

immunoglobulin molecule; the variable regions of the light chain and the heavy chain are homologous. In addition, in the constant region of the heavy chain, three domains which have similar sequence are homologous to the constant region of the light chain. These homologies have suggested that antibodies have evolved from gene duplication and diversification.

The binding site for antigen is formed by the association of two domains V_L and V_H, F_C fragment has the function to bind the complement.

Domains have similar three-dimensional structure. They are composed of antiparallel β strands forming a kind of bilayer structure, the so-called basic immunoglobulin fold (Poljak *et al.*, 1974). A striking resemblance of three structures has been observed between the immunoglobulin domains and a functionally unrelated protein, the copper–zinc superoxide dismutase subunit (Richardson *et al.*, 1976; Rossmann and Argos, 1976) (Fig. 2.50). The same topological pattern of antiparallel β sheets is observed in both proteins while there is no sequence homology between them. Moreover, the external loops of this protein are located in a position equivalent to the hypervariable loops of IgG. The close resemblance in the folding pattern of these two proteins is neither related to the amino acid sequences which differ greatly, nor to functional properties which are completely different. Consequently no evidence supports the hypothesis of their evolutionnary relationship.

In the dehydrogenase subunit, the two domains support different steps of the same function. Since one is the coenzyme binding domain, then the other is the catalytic one which also binds the specific substrate. In fact, it is artificial to assign two different functions in this case since the binding of NAD^+ and the catalysis are sequential and necessary steps of the same enzymatic function. Nevertheless a partition of the work between two different folded regions of the molecule may represent an economy during the evolution. The

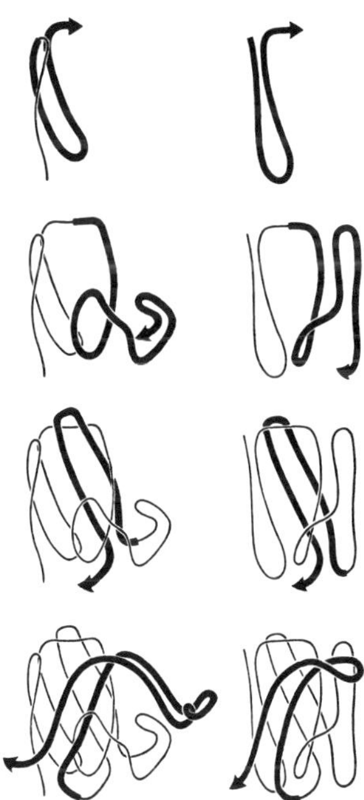

Fig. 2.50. Comparison of step-by-step build-up of backbone configurations for superoxide dismutase (down the left side) and an immunoglobulin variable domain (down the right side). New backbone added at each step is shown by heavy arrows (from Richardson *et al.*, 1976).

coenzyme binding domain does not have the specificity requirements of the other domain whose function is to recognize and to transform the specific substrate. The NAD^+ binding domain may be a building unit common to many dehydrogenases.

The occurrence of the Rossmann fold in many nucleotide binding proteins raises the question of its significance with regard to the evolution. This problem was the subject of debate as to whether it represents either convergent or divergent evolution. Buehner and co-workers (1973) and Rossmann and co-workers (1974) suggested that all the proteins which possess a Rossmann fold have diverged from a common ancestor the function of which was probably to bind a nucleotide (Fig. 2.51) and which was determined by one gene. A gene fusion had occurred with another polypeptide chain which had provided the specific substrate binding ability and the catalytic function. The structure of the different dehydrogenases can be explained by this single scheme (Fig. 2.52). The nucleotide binding domains which consist of the two Rossmann folds A_1 and A_2, occupy the N-terminal part of the molecule of LDH, MDH, and GPDH and the catalytic domain C, a position close to the C-terminal. The catalytic domains in LDH and MDH have great similarties. However, the significance of the nucleotide binding fold has been discussed by Blake (1977). For example, a part of phosphorylase *b* is folded into a structure which resembles the NAD^+ binding pattern of lactate dehydrogenase. The binding site of AMP, the allosteric effector of the enzyme

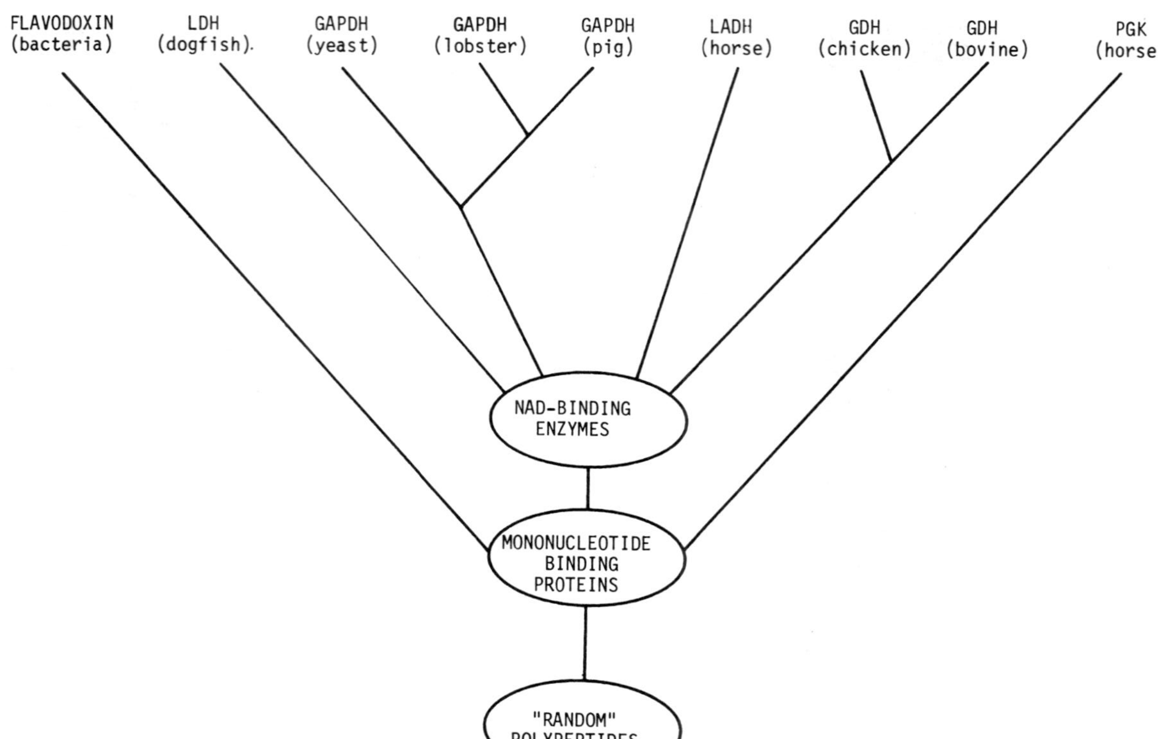

Fig. 2.51. The evolutionary model for dehydrogenases and other nucleotide binding proteins with the hypothesis of divergence from a common ancestor, (adapted from Rossmann *et al.*, 1975; by Doolittle, 1979; courtesy of Doolittle).

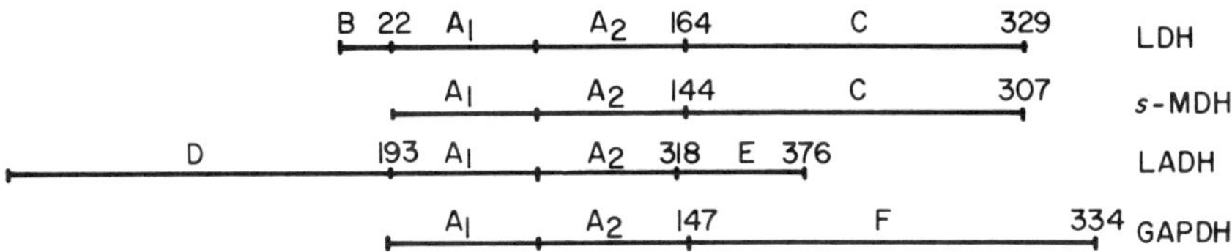

Fig. 2.52. The organization of the polypeptide chain in dehydrogenases in two distincts domains (courtesy of Rossmann *et al.*, 1975). The position of domains with respect to the sequence is indicated: For three dehydrogenases (LDH, MDH, and GPDH) the binding domain is located at the N-terminal end and the catalytic one at the C-terminal. In LADH the catalytic domain is formed by both N-terminal segment *D* and segment *E*.

is not located in this region of the molecule and the function of the NAD^+ binding pattern in phosphorylase is unknown. In the case of alcohol dehydrogenase from horse liver, after binding of NADH, catalytic domains rotate and narrow the clefts between the domains shielding the active site from the solution (Eklund and Branden, 1979).

While there are great similarities among the dehydrogenases, the situation seems quite different for kinases which also bind specific nucleotides. Adenylate kinase has a part similar to a part of phosphoglycerate kinase (PGK). Phosphoglycerate kinase contains one or two copies of the nucleotide binding site and also binds similarly the nucleotide. Hexokinase seems to have a unique structure (Blake, 1977). An interesting situation is found in pyruvate kinase (PK). Each subunit is composed of three domains. The largest domain, *A*, has a similarity with triose phosphate isomerase (TIM) suggesting that domain *A* in PK and TIM have evolved from a common ancestor. A part of domain *C* is similar to other nucleotide binding proteins, but the mechanism of binding is quite different from that found in dehydrogenases. The three domains appear to have different evolutionnary histories. Whatever it was, kinases do not form as closely a related group of structures as do dehydrogenases (Levine *et al.*, 1978). However, certain characteristics are found in many of them. Several kinases consist of two lobes separated by a deep cleft which binds substrates. Upon substrate binding, these enzymes undergo a conformational change which is a hinge motion of the two domains closing the cleft between them. It has been shown for hexokinase (Bennett and Steitz, 1978), phosphopyruvate kinase (Muller *et al.*, 1972), phosphoglycerate kinase (Banks *et al.*, 1979; Pickover *et al.*, 1979), and appears to be a rather general situation for kinases.

The proposed alternative was an evolution from different sources which, through energetic or functional requirements or both, has converged to a similar folding unit. Convergent evolution would be expected if the resulting topology is energetically favored. Divergent evolution could signify that nature has conserved a stable structure. Both, convergent and divergent theories involve a particular stability of the ($\beta\alpha\beta$) folding unit. More plausible than the convergence theory seems to be the one assuming that building blocks (domains) through gene fusion, during evolutionary time, have built proteins: these proteins have diverged and this mechanism has led to differences of function. This is more likely to account for the existence of domains in proteins.

Some speculative, but necessarily incomplete comparison may be made for serine proteases. It is possible to distinguish a binding domain in proteases, the one located toward the C-terminus (*C* domain), but it is not possible to refer to the other domain as the catalytic one since it contains only two of the three catalytic residues of the charge relay system. Ser 195, which keeps a part of the substrate covalently bound in the acyl–enzyme intermediate pertains to the binding domain.

In acid proteases, which are also organized in two topologically similar domains, there is a remarkable internal homology which indicates that these enzymes have evolved by gene duplication of an ancestral protein of about 150 residues having a fold similar to one domain of pepsin (Tang *et al.*, 1978; Subramanian *et al.*, 1977).

In some proteins different parts of the polypeptide chain have clearly different functions. Multifunctional proteins represent interesting examples. In some cases limited proteolysis allows two separate, different fragments of an enzymatic molecule still endowed with some of the functional properties. Aspartokinase–homoserine dehydrogenases (either AK I–HDH I or AK II–HDH II) from *E. coli* K 12 have two enzymatic activities. In AK II–HDH II, the larger fragment (2 × 37,000) obtained from limited proteolysis retains full homoserine dehydrogenase activity (Dautry-Varsat and Cohen, 1977). The situation is analogous to that found in AK I–HDH I (Veron *et al.*, 1972). Veron and co-workers concluded that the polypeptide chain of aspartokinase–homoserine dehydrogenases is organized in two different domains. In this case three-dimensional structure is not known. Although it supports a unique enzymatic function (homoserine dehydrogenase) the molecular weight of the nicked subunit (37,000) allows one to suppose that the functional fragment is itself organized in more than one structural domain of smaller size. In other words, structural domain and functional unit inside a polypeptide chain are not necessarily identical.

Anyhow, globular proteins are commonly built by the assembly of globular domains. It is a plausible assumption, although very difficult to prove, that at the early stage of protein evolution, domains were independent molecules endowed with specific functional properties. Later, after the different events of gene duplication or fusion, they appeared as compact building blocks of protein structure whose assembly gave rise to the function with various arrangements of these domains allowing a greater differentiation of function and specificity.

In conclusion, one can emphasize that the important set of data today available on protein structure allows some generalization as to topological patterns of globular proteins. It appears that these are some invariants in protein folding. The same rules apply to all globular proteins. A hierarchy in the organization of protein molecules may be observed with (1) the same structural elements of secondary structures, (2) the occurrence of a limited number of possible arrangements of these pieces of secondary structures giving supersecondary structures, (3) the existence of domains, (4) the frequent association of either quite similar or even nonidentical domains in protein with a pseudosymmetry, and (4) a higher level of structuration in oligomeric proteins, the assembly of subunits with a definitive symmetry. Thus protein organization results from microstructure interactions at different levels of complexity and size.

At each level, association of the corresponding building blocks stabilizes the structure. At each stage, or perhaps only at the last one, conformational readjustment might occur which generate biological function. Domains plausibly have evolved independently. Later, following gene fusion, different proteins with various specificities appeared.

3

Energetics of Protein Conformation Conditions Restricting the Allowed Conformations

3.1. INTRODUCTION: THE RANDOM SEARCH FOR THE NATIVE STRUCTURE

It is commonly accepted that the primary structure, i.e., the amino acid sequence, determines the unique conformation of a native protein (in a given environment) and that this unique conformation is the most stable one. However, the theoretically possible number of conformations for a polypeptide chain with a defined amino acid sequence is extremely large. This is because of the possibility of rotations around the single bonds of the backbone (N—C^{α}, C^{α}—C′). On the basis of only two conformations per peptide unit, that is of course an underestimation, a crude calculation indicates that a polypeptide chain containing 150 residues can have 10^{45} possible conformations (Anfinsen and Scheraga, 1975; Wetlaufer, 1973) (see Chapter 1). Severe restrictions must occur to reduce this large number of allowed conformations to uniqueness. The laws governing these restrictions are presented here in terms of energy of interactions. The driving force which directs the folding process could represent very precise and rather specific interactions, since the native biologically active three-dimensional structure is either unique, or at most represents a very limited number of conformations (i.e., a narrow distribution of conformations around an equilibrium). Even if a biologically active protein can exist in several distinct conformations, the number of these conformations must be small and each of them is stable under given conditions. This point is discussed later.

Except for proteins with disulfide bridges, the interactions which stabilize the conformation of globular proteins are noncovalent. Némethy and Scheraga (1965) stated that "The question may be asked, how many specific noncovalent interactions are required in a given polypeptide chain in addition to steric restrictions, in order to determine a unique or nearly unique conformation of the chain." The conformation of a native protein is determined by a large number and a large variety of noncovalent interactions. All the noncovalent interactions which are involved in smaller molecules do exist in proteins. Various functional groups contribute to one or more different kinds of interactions. These groups can belong to the side chains or to the backbone. Noncovalent interactions occur not only between various groups of the protein molecule, but also between various groups of the protein and the solvent molecules.

A thermodynamically stable conformation arises from the minimization of the overall free energy of interaction resulting from all intramolecular and intermolecular contributions. Whatever the control of the folding process may be, thermodynamic or kinetic, the conformation must be thermodynamically stable with respect to small perturbations of the molecular geometry. Of course, it is also stable relatively to the unfolded state. In spite of controversies, the thermodynamic hypothesis represents a necessary working hypothesis for all current attempts of a theoretical analysis of protein stability and for determination of the allowed conformations of a polypeptide chain.

The potential energy representing the overall conformation energy of a protein is composed of several contributions. Some of these contributions are intrinsic factors depending only on the nature of the macromolecule; others arise from interactions between some parts of the molecule and the solvent.

In this section each of these contributions is described in more detail and they are presented in the following order: (1) intramolecular interactions arising from factors intrinsic to the protein; (2) intramolecular interactions influenced by the solvent; (3) intramolecular interactions mainly determined by the solvent (represented by hydrophobic interactions in aqueous solution); and (4) direct interactions between protein molecule and solvent. This problem has been analyzed by different researchers (Schellman and Schellman, 1964; Némethy and Scheraga, 1965, 1977; Ramachandran and Sasisekharan, 1968; Némethy, 1969, 1972; Yon, 1969; Scheraga, 1971, 1977; Lee and Richards, 1971; Hagler *et al.*, 1973; Hopfinger, 1973; Richards, 1974, 1977, 1979).

3.2. INTRAMOLECULAR INTERACTIONS ARISING FROM FACTORS INTRINSIC TO THE PROTEIN MOLECULE

In this section only the noncovalent and covalent (S–S bridges) interactions which are not influenced by the presence of the solvent are discussed. The conformation of a polypeptide chain is determined by the values of the angles of rotation (ϕ_i, ψ_i) around the single bonds, if one assumes first that planar peptide units are a good enough approximation to eliminate the angle ω and the dihedral angle of rotation around $C_i—N_{i+1}$ (which are measures of nonplanarity of peptide bond) and second that bond length and bond angles are fixed. The range of dihedral angles (ϕ_i, ψ_i) is restricted by several factors intrinsic to the protein molecule which include (1) steric effects arising from van der Waals contacts between atoms in adjacent peptide residues, (2) van der Waals contacts between atoms in side chain (attractive nonbonded interactions), (3) barriers of internal rotation around the single bonds, (4) formation of specific noncovalent bonds between residues of side chains (or of backbone segments), and (5) formation of loops closed by covalent bonds such as disulfide bridges.

If the geometry is slightly altered, some strain can occur in bond length and bond angle and these constraints change the potential energy. The contribution associated both with bending and stretching of bonds can be evaluated.

The same types of interactions restrict the possible conformations of the side chains. Nevertheless, rotation barriers restrict the dihedral angles of side chains to a limited range of variation whereas ϕ_i and ψ_i have much larger ranges of variation.

3.2.1. Noncovalent Interactions

3.2.1.1. Nonbonded Interactions

Attractive forces between neutral molecules may include three contributions according to the nature of the molecule. These three interaction energies are dipole–dipole or Keesom interactions, dipole–induced dipole or Debye interactions, and induced dipole–induced dipole interactions or London dispersion forces (Table 3.1). Since all of them depend on the inverse of the sixth power of the intramolecular distance, they are generally combined in only one term, representing the total van der Waals attraction and this term is the sum of the three energies:

$$E_a = -A/r^6,$$

TABLE 3.1

The Three Contributions to the Total van der Waals Attractive Interaction Energy[a]

Type of Interaction	Equation
Dipole–Dipole or Keesom interactions	$E - -2/3(\mu_1\mu_2/\varepsilon kTr^6)$
Dipole–Induced dipole or Debye interactions	$E = -(-\alpha_2\mu_1^2 + \alpha_1\mu_2^2)/r^6$
Induced dipole–induced dipole or London interactions	$E = 3/2[\alpha_1\alpha_2 I_1 I_2/r^6(I_1 + I_2)]$

[a] Legend: μ is the dipole moment, α the polarizability, I the ionization energy of the molecule, ε the dielectric constant, k the Boltzmann constant, T the absolute temperature, and r the distance between the centers of the dipoles.

r is the distance and A depends on the polarizability of the interacting pair of the atoms. The energy decreases very rapidly with the distance and becomes negligible beyond 6–8 Å.

In most cases the contribution of the London dispersion interactions is the largest. These attractive forces arise between neutral atoms at a distance where the overlap between the electronic wave function of the atoms is not appreciable. When two atoms approach each other at a shorter distance, a rapid increasing repulsion due to the overlap tends to separate them. While it is known that this energy of repulsion increases very rapidly with the distance, the exact relationship of this energy as a function of the distance is not known:

$$E_r = B/r^n$$

various values were suggested for n ranging from 9 to 15, or even higher. The different approximations leading to different formulations of the potential function can now be discussed. The total nonbonded interaction may be expressed as

$$E = E_a + E_r = -(A/r^6) + (B/r^n)$$

A represents either the London attraction only, or the sum of the three energies mentioned previously, and can be evaluated.

Different forms of potential functions have been suggested with different approximations of the repulsion term: the Lennard–Jones 6–12 potential, the Buckingham or 6-exp potential, and the universal function of Kitaigorodsky (1961, 1965). Only the two more usual are presented. The last one has not been used since 1967 (Venkatachalam and Ramachandran, 1967).

The potential function referred to as the Lennard–Jones 6–12 potential is given by the expression

$$E = E_a + E_r = -(A/r^6) + (B/r^{12})$$

The constant B can be evaluated by considering the potential to be minimum when the distance r is equal to R the sum of van der Waals radii of the two atoms ($B = 1/2AR^6$). This function was used in many studies (see e.g., Scott and Scheraga, 1966; Brant *et al.*, 1967). The total nonbonded interinteraction is the sum of the energy of each interacting pair of atoms:

$$E_{\text{nonbonded}} = \textstyle\sum_{i,j}[-(A_{i,j}/r_{i,j}^6) + (B_{i,j}/r_{i,j}^{12})]$$

The summation is taking over all pair of atoms i and j in the molecule ($i \neq j$). In order to reduce the terms of the summation, CH_3 and CH_2 groups often are considered as a single unit (united atom) in the calculations with the constants adjusted accordingly (Dunfield *et al.*, 1978). Table 3.2 give values of parameters for nonbonded Lennard–Jones 6–12 potential corresponding to types of interactions most frequently occurring in proteins.

The Buckingham 6-exp potential gives the following expression for the total nonbonded interaction:

$$E = E_a + E_r = -(A/r^6) + B\exp^{-\mu r}$$

the constant μ can be evaluated from collisions between rare gases and similar atoms. Best estimates vary from 4.35 to 4.70. The value of B is adjusted to give a potential minimum for $r = R$, R being the sum of the van der Waals radii of the interacting atoms. Brant and Flory (1965a,b) used a constant value of μ equal to 4.6 for all types of contacts.

The use of the different nonbonded potential functions for the prediction of polypeptide conformation has been critically reviewed and compared by Venkatachalam and Ramachandran (1967) and also by Ramachandran and Sasisekharan (1968). Not only do the potential functions differ by the repulsive term, but also by the different values of parameters.

3.2.1.2. "Hard Sphere" Approximation

An approximation in the evaluation of the repulsive term is given by the hard sphere approximation. Indeed, because of the rapid rise of the repulsive branch of the nonbonded interaction energy curve, it was assumed in some calculations that two interacting atoms can approach to a fixed distance R_0; that corresponds to an infinite repulsion when $r < R_0$. The atoms are considered to be hard spheres. The choice of the values of R_0,

TABLE 3.2

Parameters of the Lennard–Jones 6–12 Nonbonded Potential

Atom Pair[a]	C^{kk} (kcal Å^6/mole)	$A^{kk} \times 10^{-4}$ (kcal Å^{12}/mole)	$-\varepsilon^{kk}$ (kcal/mole)	(r_0^{kk}) (Å)
$H_1 \cdots H_1$	45.5	1.410	0.037	2.92
$H_2 \cdots H_2$	45.5	0.843	0.062	2.68
$H_3 \cdots H_3$	45.5	1.439	0.036	2.93
$H_4 \cdots H_4$	45.5	1.169	0.044	2.83
$C_6 \cdots C_6$	370.5	90.603	0.038	4.12
$C_7 \cdots C_7$	766.6	104.898	0.141	3.74
$C_8 \cdots C_8$	370.5	47.530	0.073	3.70
$N_{13} \cdots N_{13}$	363.1	73.254	0.045	3.99
$N_{14} \cdots N_{14}$	401.0	37.494	0.107	3.51
$O_{17} \cdots O_{17}$	369.0	17.019	0.200	3.12
$O_{18} \cdots O_{18}$	217.2	12.563	0.094	3.24
$S_{20} \cdots S_{20}$	249.0	36.318	0.043	3.78

[a] From Momany *et al.* (1974a). The values are parameters of the equation:

$$E(r_{ij}) = \sum_{i=1}^{n} \sum_{j=1}^{m} [(A^{k1}/r_{ij}^{12}) - (C^{k1}/r_{ij}^{6})]$$

with $A^{k1} = -\varepsilon^{k1}/r_g^{k1}$, ε^{k1} being the energy at the minimum of potential and r_o^{k1} the minimum position; superscript kk indicates that both atoms are of the same type. Subscripts are defined as follows: H_1 (aliphatic hydrogen), H_2 (primary and secondary amine or amide hydrogen), H_3 (aromatic hydrogen), H_4 (hydroxy or carboxylic hydrogen), C_6 (aliphatic carbon), C_7 (carbonyl, carboxylic acid or peptide bond carbon), C_8 (aromatic carbon), N_{13} (primary or secondary amide nitrogen), N_{14} (uncharged primary or secondary amide nitrogen), O_{17} carbonyl or carboxylic acid (C=O) oxygen, O_{18} (hydroxyl or carboxylic acid C—O—H oxygen), S_{20} (sulfur).

represents the major problem in this approximation. It would correspond to the sum of the van der Waals radii of the two interacting atoms. However, the concept of this radius itself is not exact and it corresponds to contact distances found in crystals. Occasionally contacts with lower values are found which may arise from steric constraints caused by favorable interactions elsewhere in the molecule. This situation indicates that a choice of the normal van der Waals radii would be too restrictive, since it would exclude many atomic contacts which actually may occur. The sum of the van der Waals radii of a pair of atoms would represent only the equilibrium average distance between these atoms. For these reasons, some researchers and particularly Ramachandran and co-workers (1963a,b) proposed two sets of distances: a normal limit or normally allowed limit which corresponds to the values found in crystals, and an extreme limit which corresponds to

TABLE 3.3

Limiting Distances (in Ångstroms) for Various Interatomic Distances[a]

Type of Contact	Normal Limit	Extreme Limit
H · · · H	2.0	1.9
H · · · O	2.4	2.2
H · · · N	2.4	2.2
H · · · C	2.4	2.2
O · · · O	2.7	2.6
O · · · N	2.7	2.6
O · · · C	2.8	2.7
N · · · N	2.7	2.6
N · · · C	2.9	2.8
C · · · C	3.0	2.9
C · · · (CH)[b]	3.2	3.0
(CH) · · · (CH)[b]	3.2	3.0

[a] According to Ramachandran and Sasisekharan (1968).
[b] (CH) stands for a CH_2 or CH_2 group in which the hydrogens have not been definitely located.

smaller values and allows all observed contacts (Table 3.3). If, in a conformation, some contact distances are between the normal and the extreme limit, the conformation is called partially allowed: these conformations are less stable than the fully allowed ones. The conformations involving contacts within a distance smaller than the extreme limit are unlikely and are called disallowed conformations.

The hard sphere approximation is certainly less precise than the nonbonded potential functions; however, it may serve as a first approximation allowing refinement of the analysis, by the use of energy expressions. As an example, Fig. 3.1 shows an energy curve drawn for a C · · · C interaction according to the parameters given by Scott and Scheraga (1966), and according to the hard sphere approximation.

The hard sphere approximation does not allow one to evaluate the relative stability of the various possible conformations. For such an evaluation, nonbonded potential functions including both repulsive and attractive interactions, must be used. Use of the hard sphere approximation is superseded by more complete potential functions. Quantitative evaluation of the variations of conformational energies are given by lines of equal potential energy on the Ramachandran diagram as shown in Fig. 3.2. In the same conformational map, the allowed regions as determined by the hard sphere approximation are indicated by sharp boundaries. The allowed regions of

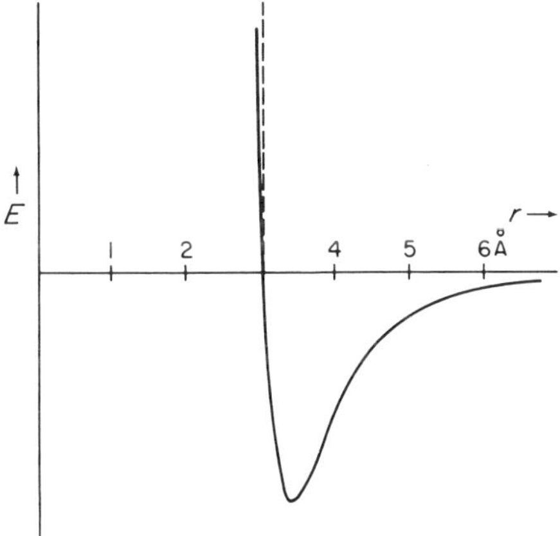

Fig. 3.1. Example of a van der Waals potential energy curve for a C · · · C interaction according to the parameters given by Scott and Scheraga (1966; from Némethy 1969) compared to the hard sphere approximation curve (dashed line).

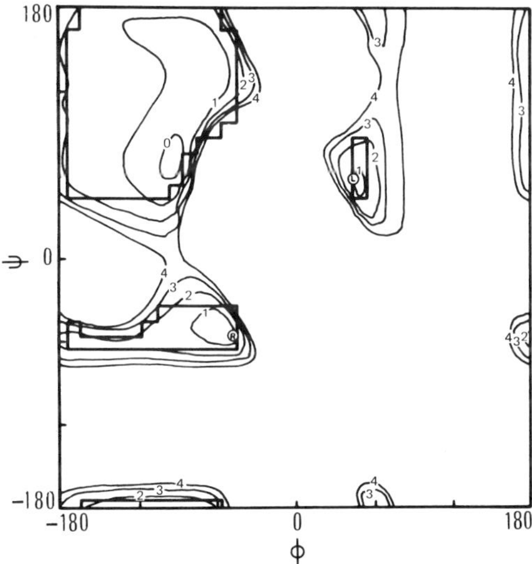

Fig. 3.2. Ramachandran map for a dipeptide with an alanyl side chain (according to Scheraga *et al.*, 1967). The contour lines of equal potential energy are drawn at 1 kcal/mole intervals (from 0 to 4 kcal/mole). The regions of allowed area for hard sphere approximation are indicated.

this steric map coincides reasonably with the low energy regions of the diagram indicating that the hard sphere is a good first approximation for the evaluation of allowed conformations.

3.2.1.3. Potential Barriers to Internal Rotations: Torsional Potential

Internal rotation is not completely free and is limited by some barriers. The allowed angles of internal rotation around the single bonds are determined by periodic potential functions which result in part from the interactions between electron orbitals belonging to the atoms linked by the single bond, and in part from the van der Waals repulsions between substituents on these atoms. The general form of the torsional energy for a single bond of a tetrahedrally bonded atom is the following:

$$E = 1/2E_1(1 - \cos 3\theta) + 1/2\mathrm{E}_2(1 - \cos 6\theta) + \ldots$$

where E_1, E_2 are the height of the torsional barriers which have a threefold or sixfold symmetry respectively. For the rotations ϕ and ψ in a polypeptide chain, Mizushima and Shimanouchi (1961) have taken into account only the barriers of a threefold symmetry. Indeed, the barriers of the sixfold symmetry are generally very small (0.006 kcal/mole in nitromethane) whereas the threefold symmetry barriers have a torsional potential value of about 2.8 kcal/mole in ethane. In hydrocarbons such as ethane, the potential function has three equivalent minima corresponding to staggered positions of the atoms ($\pm 60°, 180°$): The maxima correspond to eclipsed positions of the substituents and have equivalent height of the torsional barrier.

In proteins, the barriers about N—C$^{\alpha}$ and C$^{\alpha}$—C′ have been estimated by comparison with the values obtained in analogous model compounds. A value near 0.6 kcal/mole seems to be a good estimation for E_{ϕ_1}, whereas the intrinsic barrier for ψ can be estimated to be 0.2 kcal/mole (Scott and Scheraga 1966). Brant and Flory (1965a,b) used values of 1.5 and 1.0 kcal/mole for E_{ϕ_1} and E_{ψ_1} respectively.

The minima occur when $\psi = \pm 60°$ and 180° and $\phi = 0$ and $\pm 120°$. While the data regarding these barriers are uncertain, it is evident that they are much lower than those corresponding to rotations around C—C bonds in hydrocarbons. They play a minor role in the determination of the conformation of a polypeptide chain. Momany and co-workers (1975) have shown that these backbone intrinsic torsional terms are not needed.

In side chains, the rotational barriers for bonds between tetrahedral carbon atoms are nearly the same as those in hydrocarbons. For bonds next to an aromatic ring or a charge carboxylate group, they are nearly zero.

The torsional potential around the S—S bridge has two minima ($\psi \pm 90°$) and a barrier of 12 kcal/mole.

For double bonds such as C=C, the barrier is very high and rotation around such bonds is quite impossible. However, for the C ··· N of the peptide bond, which has a character of partial double bond, the torsional potential for the ω rotation must be considered. Its variation is given by the expression:

$$E_{\text{tors}} = 1/2E_{\omega}(1 - \cos 2\omega)$$

and for very small variations around the minimum, the variation of energy is

$$\Delta E_{\omega} = \Delta E_{\omega 1}(\Delta\omega)^2$$

The torsional barrier can be evaluated by different methods, mainly infrared spectroscopy (IR), nuclear magnetic resonance (NMR), and optical rotation studies. In spite of uncertainties in the determinations, the value of E_{ω} is near 20 kcal/mole. This would lead to an increase of only 0.65 kcal/mole for a nonplanar distorsion of 10°.

The out-of-plane distorsion of the O of the carbonyl group and of the H of the N—H group, are more difficult to evaluate.

3.2.1.4. Strain Energy of Bond Length and Bond Angle

Distorsions of structural geometry leading to stretching of bond length and bending of bond angles may occur in strained structures, particularly in cyclic peptides and in closely packed regions of a protein molecule. The increase in energy associated with stretching of a bond is given by

$$E_l = 1/2K_l(\Delta l)^2$$

where Δl is the change in length (measured in angströms), K_l is the force constant (in kcal/mole/Å) associated with the deformation.

A similar expression accounts for the energy of bond bending:

$$E_{\tau} = 1/2K_{\tau}(\Delta\tau)^2$$

$\Delta\tau$ being the change in angle (measured in radians), K_{τ} the bending force constant (in kcal/mole). The K_l and K_{τ} values are determined by IR spectroscopy.

As an example, a change of 0.1 Å in a bond length leads to an increase in energy of 5 kcal/mole. Deformations greater than 0.05 Å are not expected to occur and generally bond stretching is negligible in the conformational energy. Similarly an increase of 0.3 kcal/mole corresponds to a change of 5° in bond angle. Such variations can occur in the different angles in polypeptides and proteins, since possible variations of about 30% around the mean value are evaluated for the angles in the backbone.

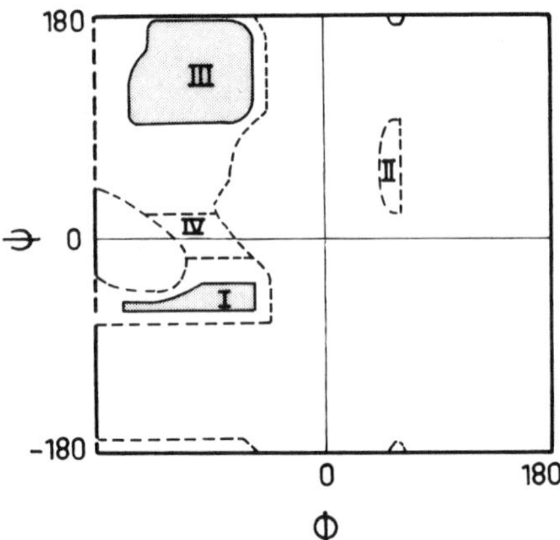

Fig. 3.3. Ramachandran diagram of a system of two peptide units linked at an alanyl α carbon (see text). Hard sphere approximation was used: (—) boundaries of fully allowed regions; (---) boundaries of partially allowed regions (from Ramachandran *et al.*, 1963).

3.2.1.5. Restrictions Imposed by Intrinsic Noncovalent Interactions

The different contributions to conformational energy by noncovalent interactions mentioned previously impose severe restrictions for the folding of a polypeptide chain. One must consider restrictions in backbone folding (ϕ_i, ψ_i) and restrictions in side chains geometry (χ_i^{jk}).

Only a small part of the Ramachandran diagram corresponds to the allowed conformations of the backbone. The nature of the amino acid residues attached to the C^α greatly influences restrictions imposed to the backbone conformation. In Fig. 3.3 a typical Ramachandran diagram shows the allowed conformation of a system of two linked-peptide units when a C^β is present on the C^α. This steric map was determined with the hard sphere approximation and delineates four regions of the diagram. Regions I and III correspond to fully allowed conformations, regions II to partially allowed. It is calculated with the extreme limit values of the interatomic contacts. Region IV is, in fact, a disallowed region; however, it accounts for the fact that few contacts are slightly smaller than the extreme limits, for example $N_1 \cdots N_2 = 2.68$ Å (extreme limit value 2.60 Å), therefore conformation will be expected to occur in this region also. Fully allowed conformations represent only 7.5% of the total area, whereas partially allowed regions extend to 22.5% (Ramachandran *et al.*, 1963a). Regular structures are included in the allowed regions: region I contains the right-handed helices, α, 3_{10} and Π; region II, which is only partially allowed, contains left-handed helices; region III contains β structures. The good accuracy

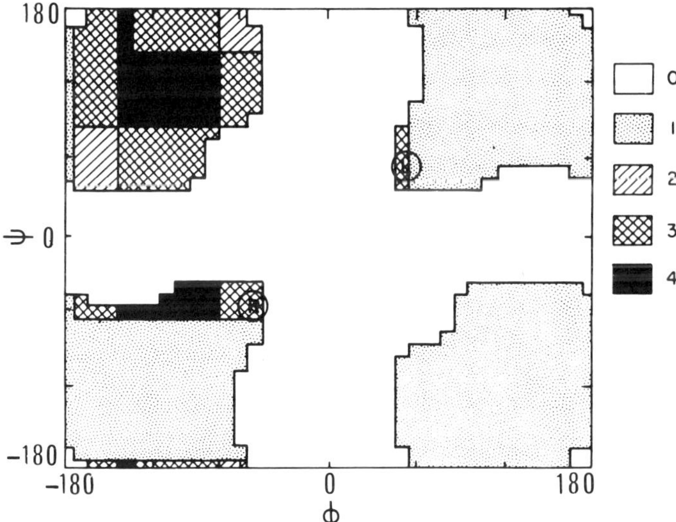

Fig. 3.4. Ramachandran diagram showing the allowed conformations for different side chains. The increase of the length of the side chain and the branching on the C^{β} restricted the allowed regions of the diagram. All conformations in area 0 are excluded; conformations in areas 1–4 are allowed for glycine only; areas 2–4 are allowed for alanine; areas 3–4 are allowed for higher homologs with no branching on the C^{β}; area 4 is allowed for valine and isoleucine. [Reprinted with permission from Némethy *et al.*, 1966, *J. Phys. Chem.* **70**, 999. Copyright © (1966) American Chemical Society.]

between such a diagram, the steric map, and the conformational map using the nonbonded energy potentials, has already been mentioned (Fig. 3.2). It is interesting to point out that the α_R helix corresponds to a region of low potential energy. This leads to the conclusion that the α_R helix is the most stable helical structure, even when only the nonbonded interactions are considered, and the stabilization by hydrogen bonding is not taken into account, although it was originally proposed that this structure is mainly stabilized by hydrogen bonds.

The restrictions are less severe for a glycyl C^{α} atom. The allowed conformations are less restricted when there is no C^{β} attached to the C^{α}. In the glycyl case, as shown in Fig. 3.4, the fully allowed conformation represents 45% of the total area and the partially allowed conformation 61% (Némethy *et al.*, 1966; Scheraga *et al.*, 1967). It is interesting to note the total symmetry of the allowed regions for a glycyl C^{α} which reflects the absence of chirality of the molecule. When longer side chains are present, more severe restrictions are imposed to the geometry (Fig. 3.4) of the backbone. Although this

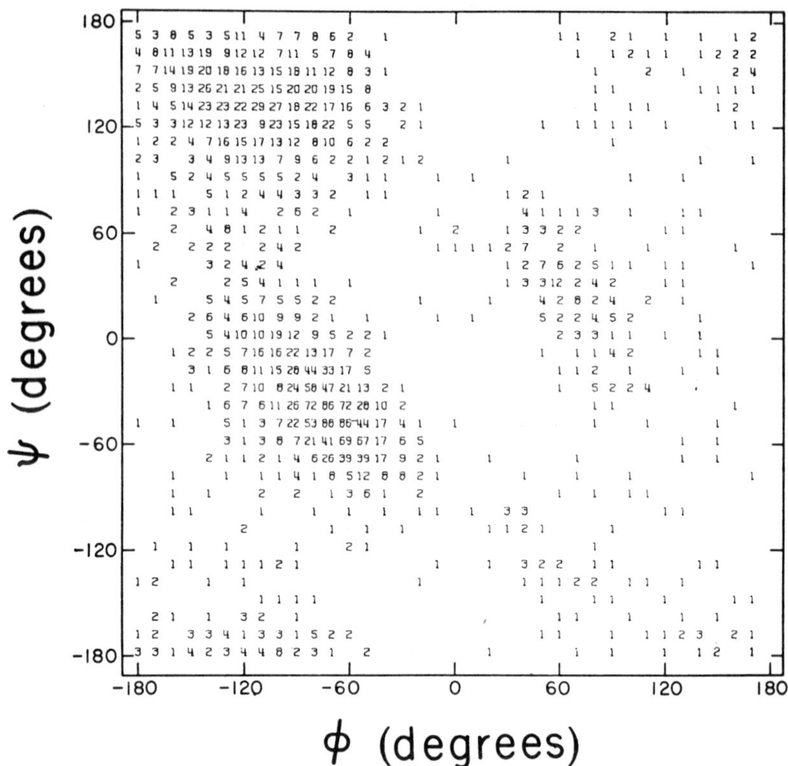

Fig. 3.5. Distribution of ϕ and ψ dihedral angles for all residues in 20 proteins with known three-dimensional structure. The diagram was constructed by Némethy and Scheraga (1977) from listings of coordinates obtained from the Protein Data Bank (Brookhaven National Laboratories). It includes the following proteins: myoglobin, thermolysin, cytochrome b_5, tosyl elastase, tosyl α-chymotrypsin, carboxypeptidase A, papain, bovine pancreatic trypsin inhibitor, concanavalin A, high potential iron protein, hen egg white lysozyme, carp myogen, subtilisin BPN′, ribonuclease S, staphylococcal nuclease, D-glyceraldehyde-3-phosphate dehydrogenase, clostridial flavodoxin, α and β chains of deoxyhemoglobin, sea lamprey hemoglobin, and rubredoxin. Numbers indicate the number of occurrences of conformations within 10° intervals of dihedral angles.

evaluation is not recent and was obtained from hard sphere approximation, Fig. 3.4 is the simplest and clearest illustration of conformational restriction imposed by different side chains.

The location on the Ramachandran diagram of all observed conformations in proteins whose structure is known from high-resolution X-ray crystallography, clearly indicates the preference for the allowed regions of the map. Figure 3.5 reproduces the distribution of ϕ and ψ dihedral angles corresponding to backbone conformations for all residues in 20 proteins with known X ray structures, according to Némethy and Scheraga (1977). Several conformations are present in region IV. Few points fall within the excluded regions.

The conformational features of the side chains are also restricted. The rotations about the single bonds corresponding to dihedral angles χ^{jk} are more restricted than for the backbone rotations (ϕ, ψ). This problem has been largely discussed by Lakshminarayan and co-workers (1967) and the allowed values of dihedral angles χ^{jk}, have been determined for all amino acid residues. As an example χ^1 for a C^α—C^β is sharply distributed around the values $-120°$ and $+120°$.

3.2.2. Covalent Interactions: Disulfide Bridge

Most of the covalent side-chain interactions in naturally occurring proteins are disulfide bonds. These cross-links mostly occur in extracellular proteins providing an extra stabilization to the molecules, which allows the maintenance of the native conformation in spite of fluctations in the environment. According to Kauzmann (1959b), the protein stability increases as the number of cross-links increases. Disulfide bridges producing loops closed in specific sites in proteins impose restrictions on the folding of the polypeptide chain. In fact, they serve mainly to reduce the conformational fluctuations in the denatured form, thus stabilizing the native form relative to denatured one. Thus, Scheraga (1963) and Némethy and Scheraga (1965) using Flory's (1953) theory calculated a decrease of 8.5 eu for the denatured form if one disulfide bridge occurs at every 28 residues. Disulfide bridges also reduce the conformational fluctuations of the native state; however, the largest effect was reported to be on the entropy of the denatured state. The number of amino acids between two cysteine residues has little influence on their ability to form disulfide bonds above the lower limit which may be evaluated to four residues, allowing closure of a loop without steric hindrance (Hardy *et al.*, 1971; Creighton, 1974c).

Restrictions imposed on protein conformation are due to the properties of disulfide bonds. The potential for rotations around the S—S bond, which correspond to the dihedral angle χ^3 or χ^s, has a minimum value of about $\pm 90°$; $\chi^s = +90°$ corresponds to a right-handed twist, $\chi^s = -90°$ corresponds to a left-handed twist. Both are observed for the L-cystine and are found in proteins. In lysozyme, for example, of the four disulfide bridges, two are right handed, two are left handed. The stereochemistry of the disulfide bond is shown in Fig. 3.6. The rotational barrier is 12 kcal/mole.

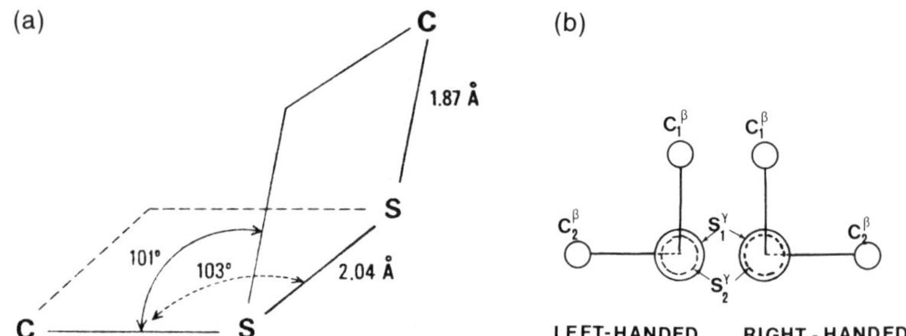

Fig. 3.6. Stereochemistry of disulfide bond. (a) Bond length and angles [reproduced from Yon (1969), "Structure et Dynamique Conformationelle des Proteines." Hermann, Paris.]; (b) left- and right-handed disulfide conformation.

The rather high rotational barrier explains the constraint imposed on the polypeptide chain by the closure of a loop with a disulfide bridge. However, the restrictive conditions arising from the presence of side chains contribute to a greater extent than restrictions imposed by disulfide bridges. These relative effects have been estimated by Némethy and Scheraga (1965), who selected an octapeptide sequence (from residue 65 to residue 72) in ribonuclease. This sequence represents the smallest loop closed by a disulfide bridge in the protein:

65 72
Cys–Lys–Asn–Gly–Thr–Asn–Cys

The allowed conformations were computed taking into account the relative effectiveness of various restrictive conditions. In order to compare the effect of closing the loop by a disulfide bridge to the restrictions imposed by the presence of side chains, two different kinds of computations were performed. On the one hand, when computation did not take into account the presence of side chains, the allowed conformations of the backbone were reduced to 20% by only closing the loop with a disulfide bridge. On the other hand, the effect of steric restrictions arising from the presence of side chains compared with those arising from the backbone only (including the C^β atoms) was computed for the cyclic octapeptide. The estimations indicate that the number of allowed conformations is reduced to about 2%, emphasizing the importance of side chains in restricting the possible conformations of the polypeptide chain. The fact that folding is mainly determined by all the intramolecular noncovalent interactions has also received some experimental support (see Part II).

3.3. INTRAMOLECULAR INTERACTIONS INFLUENCED BY THE SOLVENT

3.3.1. The Hydrogen Bond

3.3.1.1. General Properties

The hydrogen bond arises from electronegative atoms linked by a hydrogen atom, the hydrogen atom being covalently bound to one of the electronegative atoms. It can be schematically represented as D–H $\cdots$ A, D and A denote the donor and the acceptor respectively. In most cases, D and A are oxygen, nitrogen or one of the halogens. The contact involved in hydrogen bonds is much closer than the usual van der Waals contact distances for the corresponding atoms. The violation of the outer contact distance criteria is caused by the strong attractive electrostatic interactions between donor and acceptor. Hydrogen bonds are strongest when the three atoms, D, H, and A are colinear. Furthermore, optimally one of the lone electron pairs of A must be directed toward the hydrogen. However, the angular preferences are not very strong; for example, the angle H—O—C in peptide hydrogen bonds N—H $\cdots$ O=C varies from about 90° to 180° in the plane of the lone electron pair of the oxygen and the hydrogen may be located by as much as 40° away from this plane.

The strength of the hydrogen bond ranges between the nonbonded interactions and covalent bonds. The intrinsic energy of hydrogen bonds is usually between -2 and -8 kcal/mole. The hydrogen bond might be considered in first approximation as a usually strong dipole–dipole interaction between the strong dipole D—H and the dipole formed by the center of the lone electron orbital and the nucleus of the acceptor A. However, more detailed studies indicate that there is a covalent contribution to the strength of the hydrogen bond attributable to electron delocalization. This contribution enhances the directional character of hydrogen bonds. The detailed studies concerning the hydrogen bond have been reviewed and discussed by Pimentel and McClellan (1960) and by Hamilton and Ibers (1968).

The strength of the hydrogen bond decreases as the distance H $\cdots$ A increases; the shape of the energy curve as a function of this distance is similar to that of the energy curve for nonbonded interactions (Fig. 3.1). Various approximations have been proposed to evaluate the hydrogen bond energy as a function of distance, but none of them has been universally accepted. The first attempts by de Santis and co-workers (1965) to include hydrogen bond energies in the potential function for the calculation of polypeptide conformations used only the dipole–dipole approximation. A more accurate treatment of hydrogen bond energies was presented by Lippincott and Schroeder (1955) and by Schroeder and Lippincott (1957) and used by Scott and Scheraga (1966) in their conformational calculations. Some quantum mechanic approaches have been reported for hydrogen bonds (e.g., Dreyfus *et al.*, 1970). Energy parameters of hydrogen bonds between different coordinations of hydrogen binding atoms were recalculated by Momany and co-workers (1975) using the second version of Complete Neglect of Differential Overlap (CNDO/2) method and crystal data of model compounds.

The net energy of formation of intramolecular hydrogen bond can be considerably weakened in water or in another polar solvent, because the DH and A groups can form hydrogen bonds with the solvent. The latter may compensate for the energy required to break the $D—H \cdots A$ hydrogen bond. This effect, referred to as specific hydration by Némethy and co-workers (1978) and Hodes and co-workers (1979a,b), can be important in proteins and is discussed later.

In proteins, hydrogen bonds such as N—H ··· O and O—H ··· O are seen most frequently. The main hydrogen bonds in proteins are shown in Fig. 3.7 (according to Scheraga, 1963). The case of the $—NH_3^+ \cdots {}^-OOC—$ interactions is included by Scheraga in hydrogen bonds, although it can be considered as an ion pair. Very weak hydrogen bonds can be also formed with an SH group, acting as a donor. The electron system of an aromatic ring may behave as a hydrogen bond acceptor.

3.3.1.2. The Peptide Hydrogen Bond

The N—H ··· O=C hydrogen bond between peptide groups of the backbone is the most frequently observed in proteins. It stabilizes regular structures such as α helices, β structures, and turns. Many such bonds also occur between irregularly folded portions of the polypeptide chain. The energy of the peptide hydrogen bond varies as a function of the N ··· O distance (R) and the hydrogen bond angle θ (Fig. 3.8), with a minimum of -4.5 kcal/mole (for $\theta = 0°$) (Ramachandran, 1973). The net energy of the hydrogen bond can vary according to its location and especially water access, because of the possible competition with water hydrogen bonds (Schellman, 1955a):

$$\rangle N—H \cdots O=C\langle + H_2O \cdots H_2O \rightleftharpoons \rangle N—H \cdots OH_2 + HOH \cdots O=C\langle$$

The thermodynamic parameters corresponding to the global reaction represent a balance for the rupture and formation of the different hydrogen bonds. These parameters have been determined by Schellman (1955a) on the basis of a helix random-coil transition. The treatment neglected the influence of the side chains, but accounted for the end effect. Therefore the energy for a helical polypeptide chain of n residues was given by the equation

$$\Delta G = n\,\Delta G_{res} + C$$

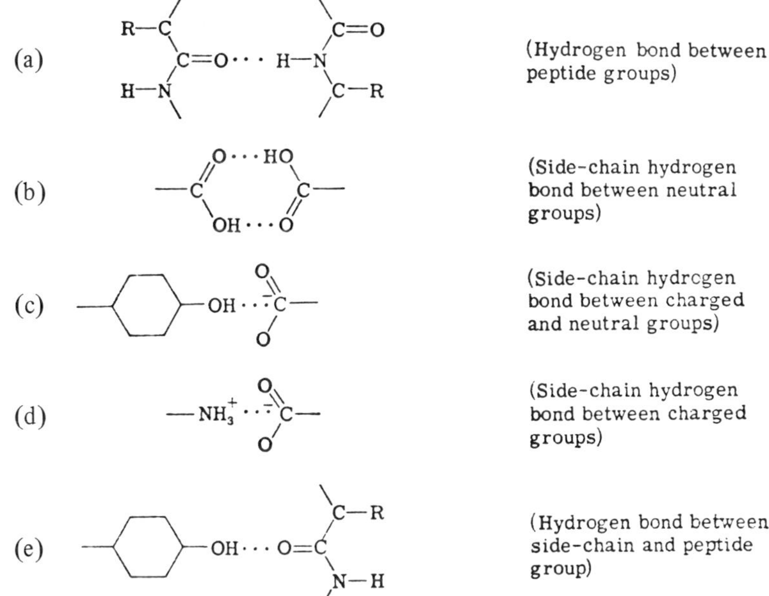

Fig. 3.7. Hydrogen bonds in proteins (from Scheraga, 1963).

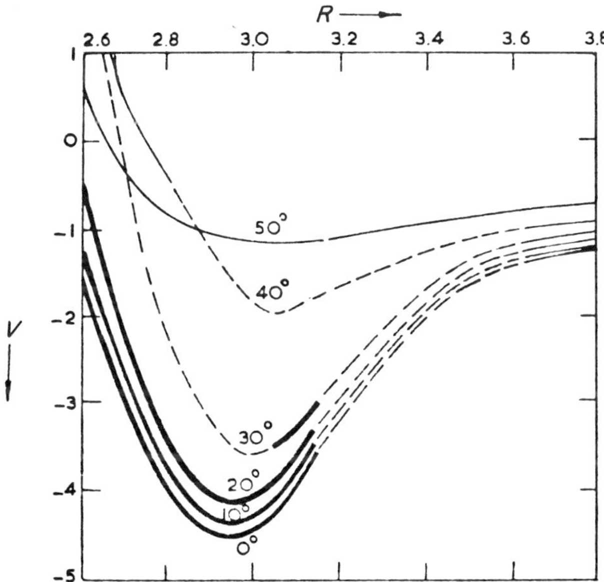

Fig. 3.8. Variation of the energy of the peptide hydrogen bond with N ⋯ O distance for different values of θ, the hydrogen bond angle (from Ramachandran, 1973).

with C corresponding to the end effect which becomes negligible when n is large enough, and ΔG_{res} representing the energy of a hydrogen bond breaking since there is one hydrogen bond per internal residue. The values of thermodynamical parameters have been evaluated with different model compounds. Accounting for different investigations and theoretical estimations, values of $\Delta H_{res} = 1.5$ kcal/mole and $\Delta S_{res} = 4.2$ eu were used by Scheraga (1963). However, in water various degrees of hydrogen bonding can occur and especially multiple bonding as follows:

$$\overset{\cdot H_2O}{\overset{\vdots}{>N}}\!-\!H\cdots \overset{\cdot H_2O}{\overset{\vdots}{O}}\!=\!C\!< + H_2O\cdots H_2O \rightleftharpoons \overset{\cdot H_2O}{\overset{\vdots}{>N}}\!-\!H\cdots OH_2 + HOH\cdots \overset{\cdot H_2O}{\overset{\vdots}{O}}\!=\!C\!<$$

Such a situation described by Némethy and co-workers (1963) has been found already in myoglobin (Kendrew, 1962). As previously mentioned energy parameters of hydrogen bonds have been reevaluated (Momany *et al.*, 1975).

3.3.1.3. Side Chain Hydrogen Bonds

The various categories of hydrogen bonds are presented in Fig. 3.7. The side chain polar groups at the interior of the proteins without access to the solvent are energetically unfavorable. They tend to form hydrogen bonds with nearby polar groups. Many polar groups are located at the outside of the molecule and therefore, they are able to form hydrogen bonds with neighboring water molecules. In general, except in some particular cases, the position of outside amino acid side chains are not well defined in X ray structures because of their flexibility and mobility and the same is true for the water molecules.

Hydrogen bonding may modify the physical properties of groups involved in the bond, specially their ionization behavior. This was already discussed by Laskowski and Scheraga (1954) and by Scheraga (1963). The following example was particularly discussed: hydrogen bond between a tyrosyl and a carboxylate group can exist as long as the tyrosyl group is not ionized and the carboxylate is not protonated. However, the dissociation of the proton from the tyrosyl group is more difficult when it is engaged in a hydrogen bond. This effect leads to a higher value of the observed pK of the tyrosyl group, that is, it is higher than expected for free tyrosyl groups. At the opposite, the dissociation of the proton from the carboxylate group is facilitated when this group acts as a hydrogen bond acceptor. The occurrence of such types of hydrogen bonds in proteins was revealed by X-ray crystallographic data. In deoxyhemoglobin, for example, Tyr 42 of a α subunit forms a hydrogen bond with Asp 99 of a β subunit. This hydrogen bond is shifted in oxyhemoglobin when Asp 42 contracts a hydrogen bond with Asp 102 of the β chain. In hemoglobin, as in myoglobin, hydrogen bonds exist between tyrosine groups and the oxygen of the carbonyl of the backbone, specially in *F* and *H* helices (Perutz, 1962; Kendrew, 1962; Watson, 1968; Perutz and Ten Eyck, 1972). Such hydrogen bonding between tyrosine groups and the oxygen of the peptide bond is frequently encountered in proteins. It has also been reported in pancreatic trypsin inhibitor among other proteins (Huber *et al.*, 1972). Salt bridges, which are included by Scheraga in the type of side chain hydrogen bonds, often occur in globular proteins. The typical example of the salt bridge in serine proteases occurs between the α amino group of the N-terminal, which becomes free after activation of the corresponding zymogen (Ile 16 in trypsin and chymotrypsin, Val 16 in elastase), and the carboxylate of Asp 194 (i.e., the amino acid adjacent to the active serine). Such a salt bridge provides the active conformation of these enzymes, which is lost by disruption of the bridge either upon protonation of the carboxylate or upon deprotonation of the amino group.

These effects were rationalized by Laskowski and Scheraga (1954) and Scheraga (1963). The formation of a hydrogen bond between two side chain amino acids corresponds to the equilibrium

$$P_{(DH,A)} \rightleftharpoons P_{(DH \cdots A)}$$

defined by the equilibrium constant K_h:

$$K_h = \frac{P_{(DH \cdots A)}}{P_{(DH,A)}}$$

which represents the strength of the hydrogen bond; $P_{(DH \cdots A)}$ and $P_{(DH,A)}$ represent species in which the hydrogen bond does exist and does not exist, respectively. The fraction of molecules with the hydrogen bond is given by the equation

$$x_h = P_{(DH \cdots A)}/[P_{(DH \cdots A)} + P_{(DH,A)}] = K_h/(1 + K_h)$$

If K_h is a very large number, then x_h approaches unity and the hydrogen bond is very strong. Conversely, if K_h is a small number, the hydrogen bond is weak. The foregoing equation applies under conditions of pH where the ionization of the acceptor and of the donor are most favorable. Besides these optimal conditions, the variations of x_h can be represented by the equation

$$x_h = K_h/[1 + K_h + (K_1/\mathrm{H}^+) + (\mathrm{H}^+/K_2)]$$

which accounts for the ionization of the donor group (ionization constant K_1) and the acceptor (ionization constant K_2). Therefore, outside of the optimal conditions, pH variations weaken the strength of the hydrogen bond. However, as previously mentioned, the observed pK of both donor and acceptor groups engaged in a hydrogen bond is modified. For the donor group, the observed pK is given by the following expression

$$K_{obs} = (D, A)(H^+)/[(DH, A) + (DH \cdots A)]$$

and upon substitution, this gives

$$K_{obs} = K_1/(1 + K_h)$$

showing quantitatively that $K_{obs} < K_1$, meaning that the dissociation of the proton from the donor group is more difficult when engaged in a hydrogen bond as predicted qualitatively. By a similar argument, it was shown that the increase of the observed dissociation constant of the acceptor (i.e., the lowering of the pK_{obs}) is given by

$$K_{obs} = K_2(1 + K_h)$$

However, the hydrogen bonding is not the only explanation of an abnormal pK which is always observed when a polar group is buried inside the protein and is thus not freely accessible to the solvent. In this later case, the ionization requires a conformational change, thereby an additional energy. The salt bridge of serine proteases provides a typical example. In this case, the additional energy is reflected by a significant shift of the pK of the implicated groups. The pK of the amino group is shifted to 10.1 in trypsin (d'Albis and Béchet, 1967; Béchet and d'Albis, 1969; Chevallier *et al.*, 1969), 8.5 in chymotrypsin (Oppenheimer *et al.*, 1966; Ghélis *et al.*, 1970; Garel and Labouesse, 1970), and 10 in elastase (Kaplan *et al.*, 1971; Ghélis, unpublished data); the apparent pK of the carboxylate is 3.5 in trypsin (Béchet and d'Albis, 1969) and 3 in chymotrypsin (Garel *et al.*, 1974). The whole energy of the salt bridge in chymotrypsin was evaluated to be about 10 kcal/mole (Garel and Labouesse, 1970).

When the donor and the acceptor groups are accessible to water, they can participate in more than one hydrogen bond with water as follows:

$$\begin{matrix} H_2O & H_2O \\ \vdots & \vdots \\ DH & +A \\ \vdots & \vdots \\ H_2O & H_2O \end{matrix} \rightleftharpoons \begin{matrix} H_2O & H_2O \\ \vdots & \vdots \\ DH \cdots & A \end{matrix} + H_2O \cdots H_2O$$

and the degree of hydrogen bonding may vary according to the existence of steric effects. Therefore the energy of side chain hydrogen bonds may be quite different according to the environment and conditions. Nevertheless, Scheraga (1963) suggested average values for the thermodynamic parameters of side chain hydrogen bonds accessible to water ($\Delta G = 0 \simeq 0.6$ kcal/mole; $\Delta S = -5$ eu).

Today in empirical energy computations on conformation of oligopeptides, free energy terms introduced to account for hydration are included by use of a hydration shell model. The effect of water on hydrogen bond is represented by an additional energy due to the formation of hydrogen bonds between water molecules of the hydration shell and acceptor or donor atoms of the solute. This effect called specific hydration is evaluated by assuming that only one water molecule can bind specifically to each donor or acceptor atom (Hodes *et al.*, 1979a,b; Némethy *et al.*, 1978).

Several physical properties are influenced by the formation of hydrogen bonds; UV and IR spectra, fluorescence behavior, and NMR spectra can be used to detect the presence of hydrogen bonds. Hydrogen bonds are identified in proteins whose three-dimensional structure is known at atomic resolution. As an example Fig. 3.9 shows hydrogen bonding in a protomer of lactate dehydrogenase (Rossmann *et al.*, 1972; Holbrook *et al.*, 1975), in both holoenzyme and apoenzyme.

3.3.2. Electrostatic Interactions

Electrostatic interactions may play an important role in protein conformation. Many amino acid residues are charged at a given pH. Even at the isoelectric point, where the net charge of the molecule is zero, there is a great number of positively and negatively charged side chain groups. These charges interact with each other and are also able to interact with free ions in the solvent. These electrostatic interactions depend on the number and on the distribution of the charges, and therefore differ in folded and unfolded states of a protein. The energy of such interactions is described by Coulomb's law.

Besides the entire charge of ionized residues, there are partial charges on atoms, even in their covalently bonded state, and these charges contribute to the electrostatic energy of protein conformation. This contribution has to be included in the total potential energy of protein conformation. To account for these partial charges, the calculation may be made by two possible ways, either by the evaluation of dipole–dipole interactions, or by the calculation of monopole interaction energy, if the values of partial charges are evaluated.

3.3.2.1. Charge–Charge Interactions

The interactions between entire charges are potentially stronger than the nonbonded van der Waals interactions, and decrease less rapidly with distance. The energy between two charges q_1 and q_2 located at a distance r is given by the following expression

$$E = -q_1 q_2 / \varepsilon r$$

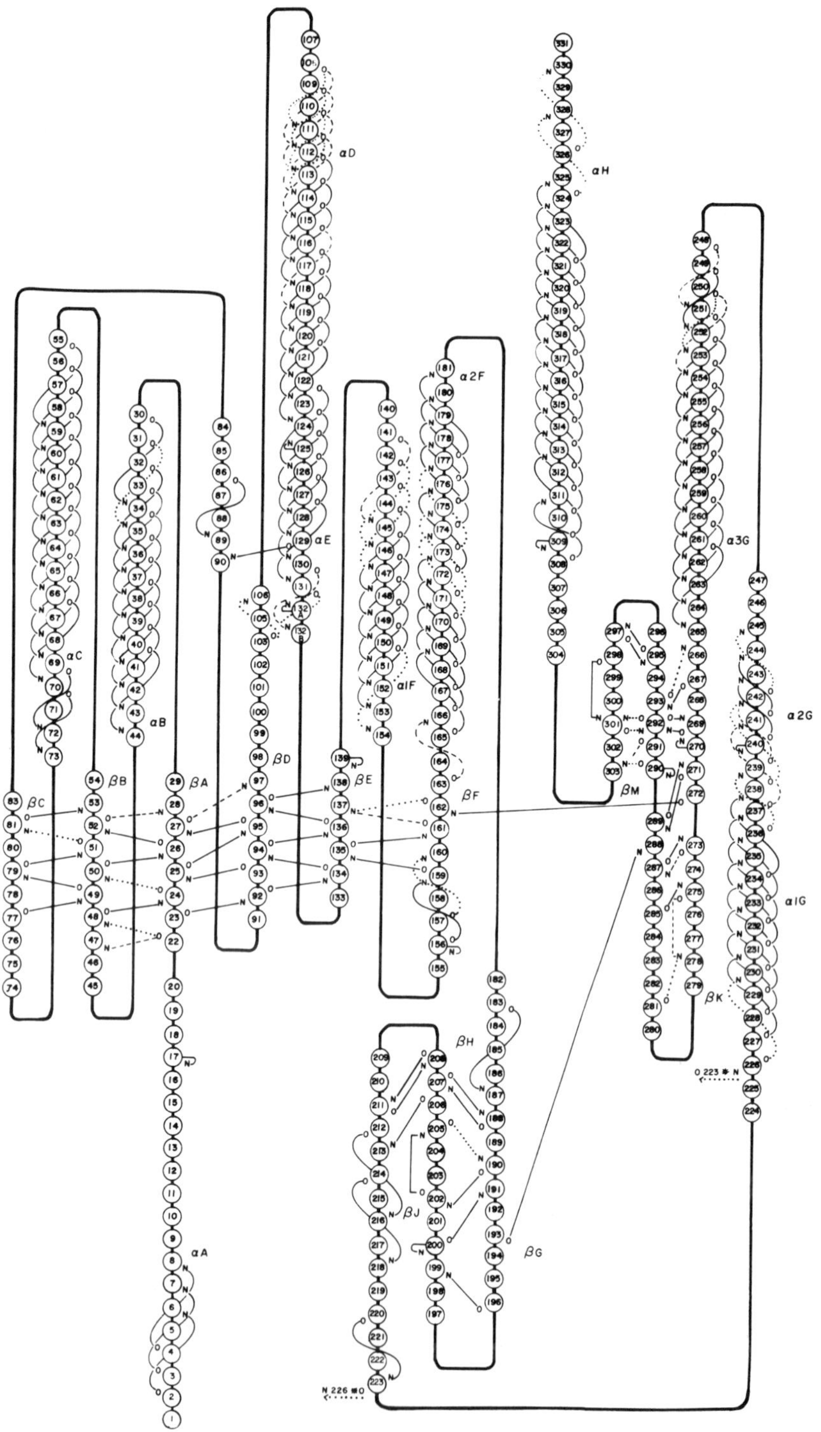

Fig. 3.9. Hydrogen bonding of lactate dehydrogenase (according to Holbrook *et al.*, 1975). Hydrogen bonds are indicated for apoenzyme and for the ternary complex: LDH–NAD$^+$–pyruvate. Dotted bonds occur only in the ternary complex. (Courtesy of Rossmann.)

where ε is the dielectric constant of the medium. In aqueous solution, because of the high dielectric constant of water, the attractive energy between two charges is weakened. For example, the interaction energy between two monovalent ions at a distance of 5 Å in vacuum is -66.3 kcal/mole, whereas this value is reduced to -0.85 kcal/mole in water. This last value is comparable to the thermal kinetic energy.

Almost all charged groups in proteins are located on the outside completely exposed to water. They may be strongly hydrated and their contribution to the potential energy of stabilization is very small. Salt bridges, i.e., attractive interactions between oppositely charged side chains, are not stable in aqueous environment because of the interactions of charged groups with water. Conversely, charged groups buried inside the proteins in an environment with a low dielectric constant form strong salt bridges. The importance of such salt bridges between a carboxylate and an α-amino end group in proteolytic enzymes such as chymotrypsin, trypsin, and elastase have been discussed previously. In these enzymes the salt bridge is able to lock the right conformation of the active center. The distinction between this kind of interactions and an amino–carboxyl hydrogen bond is rather artificial and Scheraga includes them in hydrogen bonds (Fig. 2.8).

At extreme pH values, there is a large number of charges of the same sign; therefore, the repulsions may become strong enough to provoke unfolding of the protein. Acid and alkaline denaturation of proteins are examples of these effects (Kauzmann, 1955; Tanford, 1958, 1961; Scheraga, 1961, 1963).

3.3.2.2. Electrostatic Interactions Caused by Partial Charges: Monopole and Dipole Approximations

As already mentioned, to account for the electrostatic energy arising from partial charges of the atoms, two ways of calculation that are approximations can be used, either the monopole or the dipole approximations. The first one describes the electrostatic interaction by Coulomb's law similar to that given for interactions between two charges in which the residual charges of the atoms are considered. These charges are regarded as point charges located at the geometric center of the atoms of the molecule. They must reproduce the known bond moments of the overall dipole moment of the entire group. There are different methods to calculate the residual charges.

The other way of calculation uses the dipole model and considers the dipole moment μ of bonds and groups of atoms. The dipole–dipole interaction energy is given by the following equation:

$$E = -(\mu_1\mu_2/\varepsilon r^3)\cos\theta$$

The peptide bond has a dipole moment of 3.7 (Brant and Flory, 1965a,b) located at the center of the peptide group. The energy between dipoles of adjacent peptide units can give a large contribution to the overall interaction energy.

For the evaluation of interactions between atoms at distances of a few Angströms, the monopole approximation gives, however, more accurate values of the potential than the dipole moment method.

The main difficulty is the assignment of a value for the dielectric constant. Brant and Flory (1965a, b) used 3.5 and Scott and Scheraga (1966) employed a value of 4. The evaluation of the dielectric constant near the surface of a macromolecule is complicated by the heterogeneity of the environment of the charges (or dipoles). However, the solvent can no longer be considered as a structureless continuous fluid. These effects are not very well understood and are discussed in Sections 3.4 and 3.5. Most studies of electrostatic interactions in polypeptides and proteins use values of dielectric constant ranging from 3 to 4.

Dipole interactions due to the peptide bonds make a significant contribution to the cooperative stabilization of α helices. Furthermore, dipole interactions between the backbone peptide groups and side chains can be decisive in determining the sense of the α helix.

3.4. INTRAMOLECULAR INTERACTIONS DETERMINED MAINLY BY THE SOLVENT

Water is the natural environment of globular proteins; it is their solvent par excellence, except for membrane proteins (the problem of membrane proteins is discussed later). The special properties of liquid water greatly influence structure and reactivity of proteins; particularly they determine hydrophobic interactions which arise from nonpolar groups interacting in water. The free energy of such interactions is essentially due to the favorable entropy changes of water structure. The overall native conformation of globular proteins is strongly dependent on the solvent. Various methods have been proposed to account for the role of solvent in determining intramolecular interactions for proteins in aqueous solutions. Some use models and their treatment is by statistical mechanical methods. Others derive properties of aqueous solutions from empirical potential energy function. The theory and thermodynamics of hydrophobic interactions have been extensively worked out by various groups. From the earlier studies by Franck and Evans (1945), Franck and Wen (1957), Klotz (1958, 1960), and Kauzmann (1955, 1959a) detailed treatments have been proposed (Némethy and Scheraga,

1962a,b,c; Scheraga *et al.*, 1962). The problem has been extensively reviewed (Scheraga, 1963, 1979; Némethy, 1967, 1968, 1969, 1972; Yon, 1969). More refinements of the theory have been published (Hagler *et al.*, 1972, 1973; Magat and Reinish, 1973; Lentz *et al.*, 1974; Owicki *et al.*, 1975). More recently, other approaches have been used to investigate interactions in aqueous solutions. They are derived from empirical potential function, either using models based on molecular dynamics (Stillinger and Rahman, 1972, 1974a,b; Rahman and Stillinger, 1973) or Monte Carlo simulations (Owicki and Scheraga, 1977a,b,c, 1978; Barker and Watts, 1969; Lie and Clementi, 1975; Lie *et al.*, 1976).

The study of hydrophobic bonding cannot be separated from the examination of the structure of liquid water. Thus the theories on the structure of liquid water and the thermodynamics of hydrophobic interactions are summarized briefly here.

3.4.1. Structure of Liquid Water

Many of the physical properties of liquid water, especially those related to its behavior as a solvent, differ from the properties of most other simple liquids. The unusual properties of water are due to the ability of a water molecule to form relatively strong hydrogen bonds with its neighboring molecules. Each water molecule can form four hydrogen bonds, two as a proton donor, two as proton acceptor. Thus, a regular hydrogen bonded crystalline structure forming an unlimited three-dimensional network is realized in ice. It is generally accepted that liquid water also is structured. Hydrogen bonding exists to a considerable extent in the liquid, but this structure is not as regular as in ice. The structure of liquid water is not really known and a large number of theoretical models has been proposed to account for the properties of liquid water. In first approximation, the theoretical interpretations can be classified into two groups: the continuum theories and the mixture theories. The continuum theories assume that each water molecule in the liquid is always in the same state as the others. This involves some distorsions in the hydrogen bonds, as well as a uniform arrangement of hydrogen bonding which persists throughout the liquid. Such a model was proposed by Pople (1951).

By contrast, in the mixture models, emphasis is placed on differences between molecules with hydrogen bonds and molecules with those bonds broken. Because of the cooperativity of hydrogen bonding, water molecules tend to aggregate. Thus, clusters of water molecules connected by hydrogen bonds are rapidly formed and destroyed due to the thermal motion and local energy fluctuations in the liquid. Such a model of flickering clusters was first proposed by Franck and Wen (1957). Figure 3.10 shows these models.

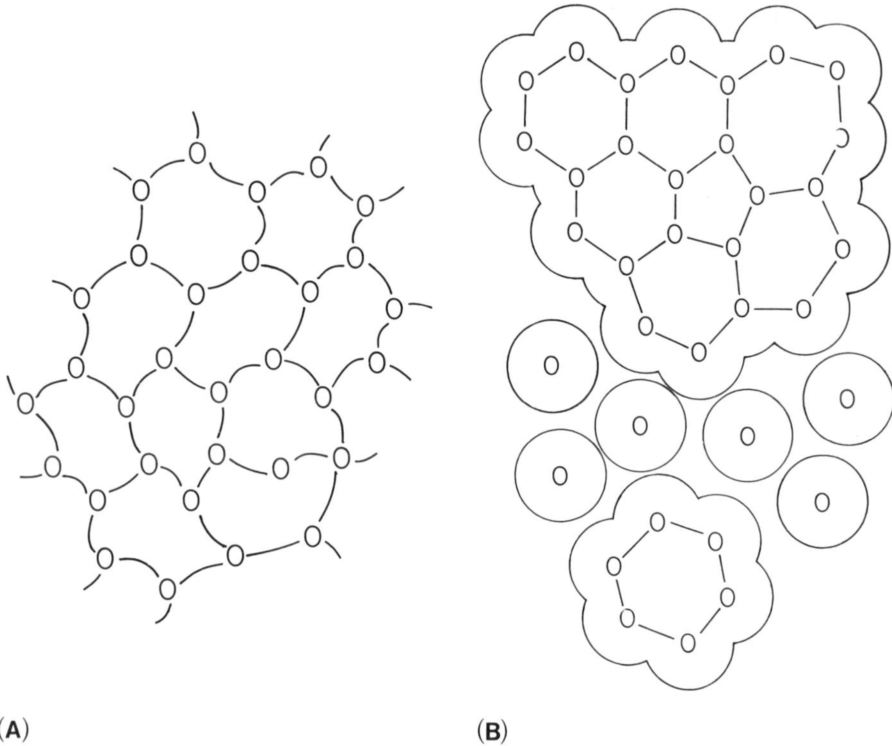

Fig. 3.10. Two different models of liquid water. (A) Continuum model; (B) mixture model (courtesy of Némethy, 1972).

The theory of Némethy and Scheraga (1962) was based on the model of Franck and Wen (1957); furthermore, these researchers assumed that liquid water is composed of a mixture of monomers and clusters of uniform size. The size of the clusters, their proportion in the liquid, and the thermodynamic properties of liquid water were computed as a function of temperature by a statistical thermodynamic treatment. In a more recent analysis, Hagler and co-workers (1972, 1973), Lentz and co-workers (1974), Owicki and co-workers (1975) used a similar model of liquid water, but they assumed that water consists of a continuous distribution of all possible cluster sizes in equilibrium. They calculated a median cluster size of 11.2 at 0°C and they did not find significant numbers of large clusters.

Calculations for liquid water and also for aqueous solutions have been performed by other methods. Rahman and Stillinger (1971, 1973), Rahman *et al.* (1975), and Stillinger and Rahman (1972, 1974a,b) used molecular dynamics simulation techniques. Owicki and Scheraga (1977a,b,c) have proposed a Monte Carlo simulation of liquid water in the isothermal–isobaric ensemble at 298°K and atmospheric pressure. They have calculated different thermodynamic properties. Difficulties in computing free energies by the Monte Carlo method were discussed by these researchers. The intent here is not to detail this aspect, but rather only to mention that similar conclusions were reached by molecular dynamics and by Monte Carlo methods. A smooth distribution of hydrogen bond energy was found. These results are better described by a continuum of bent and stretched hydrogen bonds rather than by populations of made and broken hydrogen bonds. Using molecular dynamic simulation, Stillinger (1980) described liquid water as "a macroscopically connected random network of hydrogen bonds with frequent strained and broken bonds continually undergoing topological reformation." Hydrogen bonds tend to be organized in polygones. This model accepts in fact a continuum of possibilities between the two models described in Fig. 3.10.

3.4.2. Hydrophobic Interactions

Nonpolar solutes have very slow solubility in water. In the neighborhood of nonpolar groups the number of hydrogen bonds per mole of water increases, thereby increasing the structural order of the liquid and thus greatly decreasing the entropy.

The thermodynamics of solution of nonpolar solutes in water was investigated by Némethy and Scheraga (1962a,b,c). It is characterized by (1) a high positive free energy of solution (i.e., a low solubility) resulting from a negative enthalpy of solution counterbalanced by a large excess of negative entropy of solution; (2) a very high partial molal heat capacity; and (3) a decrease of partial molal volume.

Hydrophobic interaction can be considered to be essentially the reversal of the solution process for a nonpolar solute in water. This process is energetically favorable since it is accompanied by a large negative free energy change ($\Delta G_{\mathrm{H}\emptyset}$) due to the large positive entropy contribution ($\Delta S_{\mathrm{H}\emptyset}$); the enthalpy contribution ($\Delta H_{\mathrm{H}\emptyset}$) although positive, is very small for amino acid side chains pairing as calculated by Némethy and Scheraga (1962a,b,c).

The free energy of hydrophobic interaction, $\Delta G_{\mathrm{H}\phi}$, is the sum of several contributions, the free energy corresponding to the structural change of water ($\Delta G_{\mathrm{W}}^{0}$) and the free energy corresponding to the change of states of the side chains ($\Delta G_{\mathrm{S}}^{0}$):

$$\Delta G_{\mathrm{H}\phi} = \Delta G_{\mathrm{W}}^{0} + \Delta G_{\mathrm{S}}^{0}$$

The first contribution is from the structural change of water on forming a hydrophobic bond between two side chains. Molecules of water, ΔY_{S}, (with free energy $G_{\mathrm{W}}^{\mathrm{c}}$) in the first shells around the two side chains are removed.

The resulting energy variation is

$$\Delta G_{\mathrm{W}}^{0} = \Delta Y^{\mathrm{S}}(G_{\mathrm{W}}^{0} - G_{\mathrm{W}}^{\mathrm{c}}),$$

where G_{W}^{0} is the free energy of a molecule of liquid water.

When two nonpolar side chains interact, ΔY^{s} water–nonpolar side chain interactions are broken corresponding to an energy loss of ΔY^{S}. The energy of water solute interaction is E_{RW}. The van der Waals interactions, Z_{R}, result from the pair contact, with a total energy gain of $Z_{\mathrm{R}}E_{\mathrm{R}}$ (E_{R} being the attractive energy for a pair contact). Thus, Z_{R} depends on the extent of contact between the nonpolar side chains. The formation of such an interaction introduces some restrictions of internal bond rotations and limits the degree of freedom. The corresponding increase of free energy (ΔG_{rot}) which arises from the decrease in entropy due to the restrictions of rotations, is generally small (Scheraga, 1963). The total contribution to the free energy is

$$\Delta G_{\mathrm{S}}^{0} = -1/2 E_{\mathrm{RW}} \Delta Y^{\mathrm{S}} + Z_{\mathrm{R}}E_{\mathrm{R}} + \sum \Delta G_{\mathrm{rot}}$$

Therefore, the free energy of formation of hydrophobic interactions is the sum of these different contributions:

$$\Delta G_{\mathrm{H}\phi} = \Delta Y^{\mathrm{S}}(G_{\mathrm{W}}^{0} - G_{\mathrm{W}}^{\mathrm{c}}) - 1/2 \Delta Y^{\mathrm{S}} E_{\mathrm{RW}} + Z_{\mathrm{R}}E_{\mathrm{R}} + \sum \Delta G_{\mathrm{rot}}$$

Thermodynamic parameters for the formation of hydrophobic interactions between all possible pairs of nonpolar side chains in proteins have been calculated at various temperatures by Némethy and Scheraga (1962c). They depend on the nature of side chains, their size, and the extent of the contact. At 25°C the parameters have the ranges: $\Delta G_{\mathrm{H}\phi}$ from -0.2 to -1.5 kcal/mole; $\Delta H_{\mathrm{H}\phi}$, from $+0.3$ to $+1.8$ kcal/mole; and $\Delta S_{\mathrm{H}\phi}$, from $+1.7$ to 11 eu. Some values for specific paired hydrophobic bonds are given in Table 3.4

TABLE 3.4

Thermodynamic Parameters[a] for the Formation of Pairwise Hydrophobic Bonds of Maximum Strength between Different Amino Acid Side Chains at 25°

Side chains	Thermodynamic parameters			Structural parameters	
	$\Delta G_{\mathrm{H}\phi}^{0}$ (kcal/mole)	$\Delta H_{\mathrm{H}\phi}^{0}$ (kcal/mole)	$\Delta S_{\mathrm{H}\phi}^{0}$ (eu)	ΔY^{s}	Z_{R}
Ala–Ala	−0.3	0.4	2.1	4	2
Ile–Ile	−1.5	1.8	11.1	10	5
Phe–Leu	−0.5	0.9	4.7	6	3
Phe–Phe	−1.4	0.8	7.5	12	6

[a] According to Scheraga (1963). Structural parameters ΔY^{S} and Z_{R} (see text) are also given.

TABLE 3.5

Thermodynamic Parameters for the Transfer of Nonpolar Amino Acid Side Chains from Water to the Inside of a Protein[a]

	Parameters			
Amino acid side chains	ΔG^0_{tr} (kcal/mole)	ΔH^0_{tr} (kcal/mole)	ΔS^0_{tr} (eu)	ΔG (kcal/mole)
Ala	−1.3	+1.5	+9.4	−0.73
Val	−1.9	+2.2	+13.7	−1.69
Leu	−1.9	+2.4	+14.3	−2.42
Ile	−1.9	+2.4	+14.5	+2.97
Met	−2.0	+2.7	+16.0	−1.30
Pro	−2.0	+2.2	+14.0	—
Phe	−0.3	+2.7	+10.1	−2.65
	−1.8	+1.0	+9.5	—

[a] According to Scheraga (1973).

The character of hydrophobic interactions is revealed by their temperature dependence. At low temperature, their stability increases in temperature, contrary to what is observed for hydrogen bonds (Némethy and Scheraga, 1962a,b,c, Némethy *et al.*, 1963; Scheraga *et al.*, 1962). The variation of free energy of hydrophobic interactions as a function of temperature is described by the following equation:

$$\Delta G_{H\phi} = a + bT + cT^2,$$

a, b, and c vary for each pair of side chains.

The possibility of interactions with the peptide backbone has also been considered. Interactions involving more than two side-chain amino acids can occur, the limiting case being the formation of an entire hydrophobic region. The thermodynamic parameters for the transfer of nonpolar groups from water to the inside of the protein considered to be a nonpolar region were calculated and are listed in Table 3.5.

A somewhat different approach was used by Tanford (1962a,b, 1964) and by Nozaki and Tanford (1963, 1965, 1970, 1971). These researchers determined experimentally the energy of transfer of a hydrophobic side chain from water into ethanol, with the assumption that this solvent can reflect the inside of the protein. The use of ethanol with the presence of a OH polar group may explain some discrepancies between the calculated values and the experimental determinations (Table 3.5).

Hildebrandt (1979a,b) objects to hypothesis of a hydrophobic effect on the basis of free energy of interfaces between water and n-hexane and water and perfluorotributylamine indicating no antipathy between hydrocarbon and water. Tanford (1979), from interfacial free energies, shows that the antipathy between hydrocarbon and water rests on the strong attraction of water for itself.

Approaches using empirical potential functions can account for interactions of both nonpolar and polar groups with water. Therefore, they are considered in Section 3.5.

3.4.3. Importance of Hydrophobic Interactions in Globular Proteins

3.4.3.1. Role of Hydrophobic Interactions in Stabilization of Protein Structure

Approximately 40–50% of the amino acids in most globular proteins have a totally or partially nonpolar side chain and therefore can participate in hydrophobic interactions. Thus, they may be expected to make a large contribution to the stability of protein structure. Contrary to hydrogen bonds, hydrophobic interactions are more stable when temperature increases up to a certain value; above that, their stability decreases. Since thermal unfolding is a process generally occurring for most globular proteins, the contribution of hydrogen bonding is predominant. The temperature range in which a folded protein is stable depends on the balance of these two contributions.

Hydrophobic interactions are energetically determined by the entropy change that arises from the change of the water structure. Consequently, the nature of the solvent may be expected to have an important role in protein folding. The effects of the solvent can be correlated to its ability to form hydrogen bonds (Singer, 1962; Némethy, 1972). It has been established that hydrogen bonding plays an important role in determining the properties of a solvent and the solvent–solute interactions through the formation of hydrogen bonds. It is also the case for solvents such as ethylene glycol. The action of different solvents on protein conformation was investigated in the 1960s. It was shown that some solvents, such as 2-chloroethanol, favored the formation of helical structures (Doty, 1959), whereas other solvents induced the formation of β structures. Hydrophobic interactions may also influence the stability of hydrogen bonds. In particular, the hydrophobic bond can contribute to the formation of a hydrogen bond and the hydrogen bond is stronger in the presence of nonpolar groups participating in the formation of hydrophobic bonds, if the two processes are stimultaneous. There is some mutual influence between polar and nonpolar side-chain interactions in proteins. A quantitative evaluation was presented by Némethy and co-workers (1963). The equilibrium constant

TABLE 3.6

Thermodynamic Parameters for the Formation of Tyrosyl–Glutamyl Hydrogen Bond Stabilized by the Simultaneous Formation of a Glutamyl–Leucyl Hydrophobic Interaction[a]

	Parameters		
Situation	ΔH^0 (kcal/mole)	ΔS^0 (e.u.)	ΔG^0 (kcal/mole)
Hydrogen bond in absence of hydrophobic bond	−1.5	−5	0
Hydrogen bond in presence of hydrophobic bond	−0.95	−0.03	−0.94

[a] According to Némethy *et al.* (1963).

for the formation of a hydrogen bond with the simultaneous formation of hydrophobic bond is

$$K = K_h(1 + K_{H\phi})$$

K_h is the equilibrium constant for the formation of a hydrogen bond in the absence of hydrophobic bond; $K_{H\phi}$ the equilibrium constant corresponding to the formation of the hydrophobic bond.

The thermodynamic parameters corresponding to the formation of a tyrosyl–glutamyl hydrogen bond in the presence of a leucyl–glutamyl hydrophobic bond are given as examples in Table 3.6.

3.4.3.2. Accessible Surface Area

Taking into account the data derived from X-ray atomic coordinates, different investigators have attempted to rationalize the problem of accessibility to solvent of residues in globular proteins (Lee and Richards, 1971; Richards, 1974, 1977, 1979; Shrake and Rupley, 1973; Finney, 1975, 1978; Chothia 1974, 1975, 1976; Janin, 1976). The concept of accessibility to solvent by residues in proteins was first introduced by Lee and Richards (1971). The accessible surface area, as defined by these researchers describes the extent to which atoms on the surface of a protein can form contacts with water. It represents the locus of the center of the solvent molecule as it rolls along the protein, making the maximum allowed contacts (van der Waals contacts) without penetrating any other atom. More recently Richards (1977) introduced additional concepts, those of contact area and of re-entrant area, their sum giving the molecular area. Figure 3.11 shows the

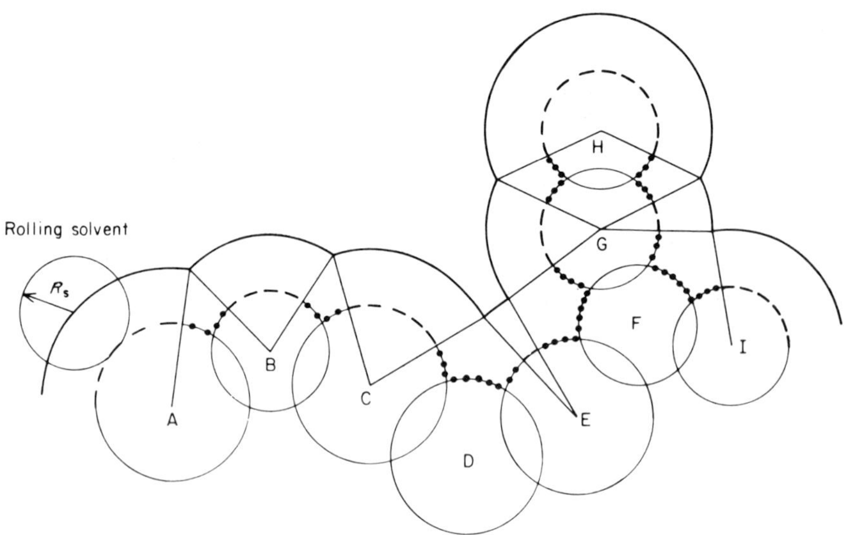

Fig. 3.11. Schematic representation of the accessible surface area in a section of the protein. The symbols A, B, C, . . . I are the atoms of the protein, S the solvent molecule; accessible surface area is defined by the locus of the center of the solvent molecule rolling on the surface of the protein (—); contact area (– – –) and re-entrant area (· · ·) are also indicated in the figure; atoms D and F are inaccessible to solvent (they have no accessible surface) (Finney, 1978) according to definitions given by Richards (1977).

situation for a section of the protein. Finney (1978) estimated the surface roughness by the ratio of molecular area (A_M) to accessible area (A_A).

For the nonpolar side chains of amino acids, Ala Val, Leu, and Phe, there is a linear relationship between hydrophobicity and accessible surface area. The slope is equivalent to 22 cal $Å^{-2}$. For the side chain residues with hydroxylic groups, the slope is 26 cal $Å^{-2}$ (Chothia, 1974). This linear dependency is shown in Fig. 3.12. Chothia (1974, 1975, 1976) and Janin (1976) determined that the loss of accessible surface area in monomeric proteins after folding is proportional to the hydrophobic energy and is a simple function of their molecular weight; it is proportional to the two-thirds power of their molecular weight (Fig. 3.13) (Janin, 1976).

The surface area buried as the result of folding and the hydrophobic contribution to the stability of the folded protein may be easily deduced from the simple rules derived by Chothia and Janin from the X-ray crystallographic data of several proteins. Since on the one hand, the accessible surface area of the unfolded chain is proportional to its molecular weight (Chothia, 1975)

$$A_u \sim 1.44\, M$$

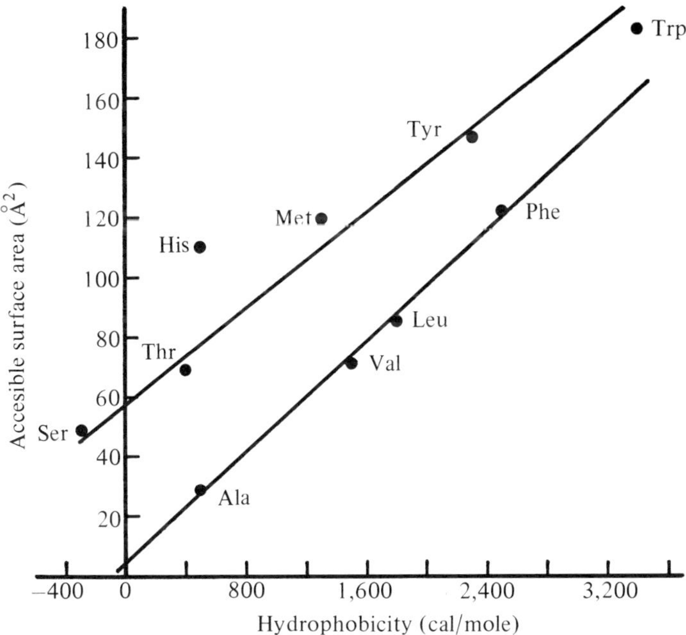

Fig. 3.12. The plot of the accessible surface area versus hydrophobicity for different residue side chains. It indicates the two linear dependencies (see text). [From Chothia (1974) reprinted by permission from *Nature* (*London*) **248**, 338. Copyright © 1974 Macmillan Journals Limited.]

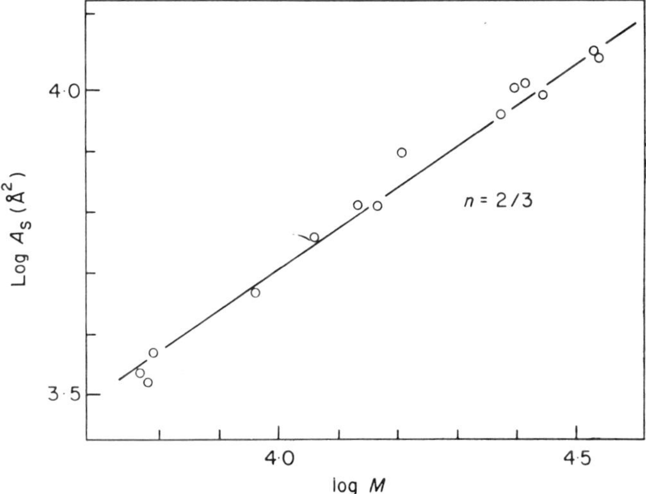

Fig. 3.13. Proportionality of the loss of accessible surface area to the molecular weight of proteins (from Janin, 1976). Different proteins: insulin, rubredoxin, pancreatic trypsin inhibitor, HIPIP, calcium binding protein, ribonuclease S, lysozyme, staphylococcal nuclease, papain, chymotrypsin, concanavalin A, subtilisin, thermolysin, carboxypeptidase A.

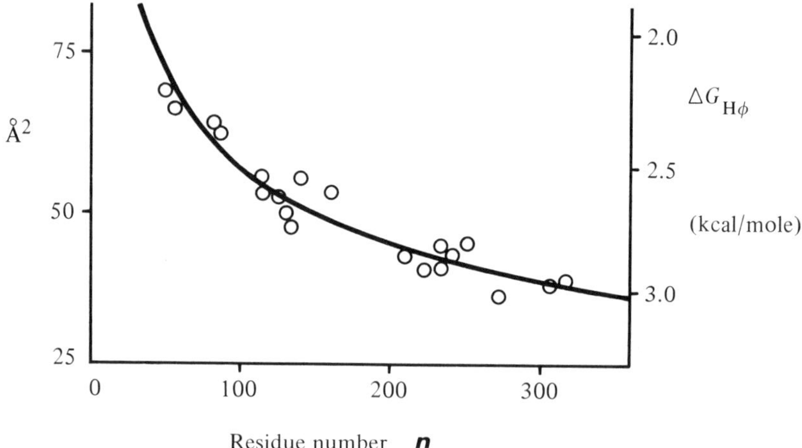

Fig. 3.14. Average residue accessibility (A_n/n) as a function of the number of residues in the polypeptide chain measured on 20 protein structures. On the right ordinate, the hydrophobic contribution to the free energy of the folded protein. [Reprinted with permission from Janin and Chothia, *12th FEBS Meeting Dresden* **52**, 232. Copyright (1978), Pergamon Press, Ltd.]

and, on the other hand, the accessible surface area of the native proteins is correlated to its molecular weight by the relation (Janin, 1976),

$$A_n = 11.1\ M^{2/3}$$

then the surface area buried upon folding (Janin and Chothia, 1978) is:

$$A = A_u - A_n = M(1.44 - 11.1\ M^{1/3})(\text{Å}^2)$$

The contribution of hydrophobicity may also be derived from these simple rules; the contribution per residue is given by

$$\Delta G_{H\phi} = \gamma(A_u - A_n)/n.$$

Figure 3.14 summarizes these rules and indicates that the hydrophobic contribution per residue varies from 2 kcal/mole for chains of about 40 residues to 3 kcal/mole for chains of 340 residues. The values were measured for 20 protein structures.

From these relationships, Janin (1976) concluded that the actual shape of the protein has only minor effect on the accessible surface area. Gates (1979), reconsidering this problem, has shown that the shape of a protein molecule can play a significant role in determining its accessible surface area. A sphere model is not valid for all protein molecules; larger proteins being more aspherical or more textured than smaller proteins.

3.4.3.3. Location of Hydrophobic and Polar Residues

The idea, based on thermodynamic considerations, that globular proteins fold with all nonpolar residues inside and polar residues at the surface accessible to the solvent, was revealed to be oversimplified when enough crystallographic data were available to localize the different side chain residues.

There is a tendency for hydrophobic side chains to be buried at the interior of the proteins. However, if the interior of the protein is composed of nonpolar residues, all the hydrophobic residues are not inside. Whereas hydrophobic groups, such as Val, Leu, Ile, and Phe have a tendency to be buried, some are often fully exposed to the solvent. Shorter side chains, such as Gly or Ala, are especially found at the surface of proteins. Almost all the polar side chains are located on the surface of the proteins. The tendency for polar residues to be outside seems to be more pronounced than for hydrophobic residues to be inside.

From the atomic coordinate of three proteins, lysozyme, myoglobin, and ribonuclease S, Lee and Richards (1971) found that ca. 40–50% of the surface area in each protein is occupied by nonpolar groups. Furthermore, the reduction in the polar surface area following protein folding is about the same as the reduction in the nonpolar surface (Lee and Richards, 1971; Shrake and Rupley, 1973; Chothia, 1976). Chothia (1976), after determining in 12 proteins the occurrence of a residue to be buried, found, however, a strong negative correlation with the polarity of the side chain: 44% of the nonpolar residues bury 95% or more of their surface, whereas only 22% of residues with one polar side chain atom and 11% with two polar side chain atoms do so. The buried surface is more nonpolar (61–74%) than the accessible surface (50–63%).

During the folding of a protein, the effect of the formation of secondary structures on the reduction of the accessible surface area was determined by Chothia (1976). He found a reduction in the polar accessible surface (47–63%) in all cases greater than the reduction in the polar surface (26–51%). This is because of the large contribution of hydrogen bonded main chain atoms to the buried surface. The surface buried within the secondary structures is more polar (50%) than that which remains accessible (38%). Figure 3.15 indicates the character, polar or nonpolar, of the surfaces which remain accessible or which become buried by formation of secondary structure. The proportion of polar groups forming intramolecular hydrogen bonds is constant and close to 50% (Chothia, 1975). The polar surface buried between secondary structures is constituted by atoms that form intramolecular hydrogen bonds (80%), which occur within the same piece of secondary structure; 20% of the intramolecular hydrogen bonds are

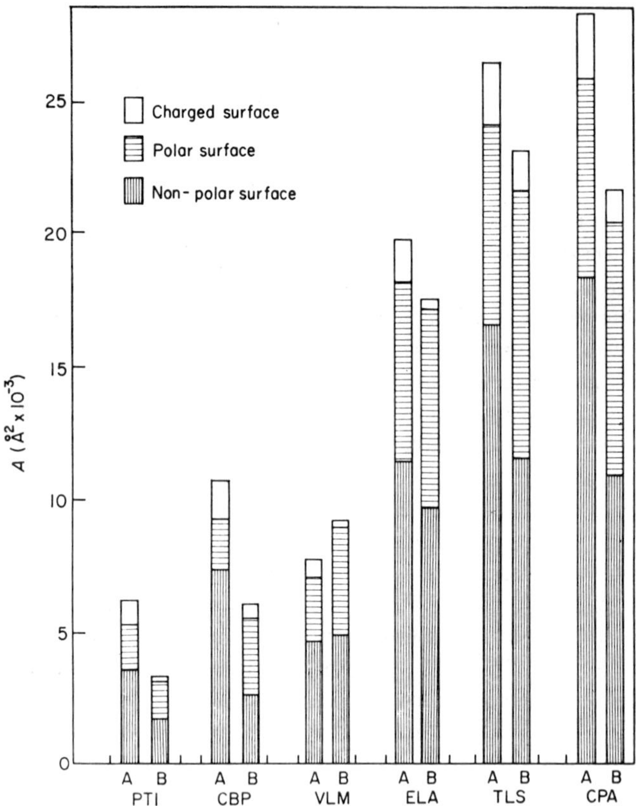

Fig. 3.15. Variations and character of accessible surface area as a result of formation of secondary structures in six proteins: pancreatic trypsin inhibitor (PTI), calcium binding protein (CBP), Bence–Jones protein REI (VIM), elastase (ELA), thermolysin (TLS), and carboxypeptidase A (CPA). (A) Surface which remains accessible; (B) surface which becomes buried. The partition, polar, charged, and nonpolar surface is indicated in all cases (according to Chothia, 1976).

formed between polar atoms of side chains which pertain to different pieces of secondary structure. More than 50% of them are on the protein surface accessible to the solvent.

The formation of tertiary structure buries a certain amount of secondary structure surfaces which varies from 32% to 60%. The proportion of the surface which is buried varies for each protein and increases with the molecular weight of the protein. In fact, it is the proportion of the nonpolar surface that becomes buried during folding, which increases with the molecular weight. From 60% for pancreatic trypsin inhibitor, it varies up to 79% in carboxypeptidase among the six proteins considered by Chothia

(1976), while the proportion of polar groups which becomes buried remains constant.

3.5. INTERACTIONS BETWEEN SOLVENT AND PROTEIN MOLECULES

The effects of solvent on conformational stability of proteins are often classified into three categories (Hagler *et al.*, 1973). The first is a nonspecific bulk effect with the solvent acting as a dielectric medium. Therefore, it influences the electrostatic interactions, mainly those which arise between groups that are located at the surface of the protein molecule. The second effect is an influence by the solvent on the interactions between atoms or groups of atoms. The hydrophobic interactions that are mainly influenced by structural changes of the solvent are in this group. These two groups include more indirect effects of the solvent. More specific interactions may arise between solvent and polar groups of proteins, such as NH and CO groups of the backbone and polar side chains. This represents the third category of the effects of the solvent. The modification of the strength of hydrogen bonds in water is an example of this effect.

The approach to describe the thermodynamics of protein solvent interactions used by Némethy and Scheraga (1962a,b,c) for the investigation of hydrophobic bonding was later extended to polar groups in aqueous solution by Gibson and Scheraga (1967). The problem of protein hydration has been extensively reviewed by Kauzmann and Kuntz (1974) and by Scheraga (1979). Different approaches were used to account for the effects of the solvent in the conformation of polypeptides. These approaches were mentioned in the beginning of Section 3.4. They were developed to evaluate the contribution of solvent–solute interactions to the conformational energy. Némethy and co-workers (1978) and Hodes and co-workers (1979a,b) distinguished in the free energy of hydration a term for specific hydration representing solute–water hydrogen bonding and a term for nonspecific hydration for the interaction of the solute with the layer of nearest neighbor water molecules. Nonspecific solvent effects were already described using the hydration shell model introduced by Gibson and Scheraga (1967) which was later modified by Hopfinger (1971). In this model, it is assumed that a characteristic sphere, which defines the hydration shell, is centered around each atom of the molecule. The size of the sphere depends on the solvent molecules and the solute atoms. The size of the sphere determines the number of molecules of solvent which can occupy the hydration shell. Furthermore, the model implies that each solvent molecule occupies an identical volume. The removal of a solvent molecule is accompanied by a

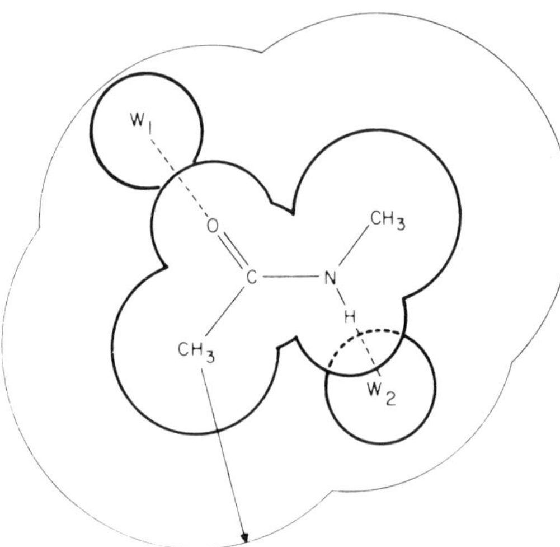

Fig. 3.16. Hydration shell model (according to Hodes *et al.*, 1979a). Hydration of *N*-methylacetamide is shown. *N*-Methylacetamide is represented by its cross section. The heavy line delineates the van der Waals contour; the thin line surrounding the molecule indicates the boundary of the hydration shell. Two specifically bound water molecules W_1 and W_2 are shown. Water molecules contained within the boundary contribute to specific and to nonspecific hydration (see text).

particular change in free energy. A linear relationship between the total free energy change and the excluded volume in the hydration shell is assumed. By this method, the energy of conformation of different polypeptides was computed in aqueous solution and also in other solvents. The earlier hydration shell models cannot account adequately for specific hydration. Modification of the model was introduced by Némethy and co-workers and Hodes and co-workers (1979a,b) to correct inadequacies of the earlier models. In the model introduced by Gibson and Scheraga (1967), as well as in the hydration shell used by Hopfinger (1971), and even in this modified model, only water molecules of the first shell around the solute are considered to interact with it. Tanaka and Scheraga (1975, 1976a,b) gave empirical parameters for the interactions between amino acids and water. The proposed model is shown in Fig. 3.16. Némethy and co-workers (1978) and Hodes and co-workers (1979a,b) used this model to include hydration terms in conformational energy computations of the terminally blocked residues of the 20 naturally occurring amino acids and also of several terminally blocked dipeptides. The data are compared with those computed in the absence of hydration. Comparison of the results clearly indicates that the relative stability of minima is altered significantly when hydration is included (Némethy *et al.*, 1978; Hodes *et al.*, 1979a,b). The effect of hydration on the probability of bend formation and stability of this structure was investigated in 23 terminally blocked dipeptides. The results were compared to those expected from X-ray structures. When dipeptides contain two polar residues, better agreement with respect to probability of bend formation was obtained when hydration was included. On the contrary, when the dipeptides were composed of two nonpolar amino acids, omission of hydration led to better agreement.

The effect of water on polypeptide conformations has been presented by Krimm and Venkatachalam (1971) for poly-L-proline. Their approach has limitations. The main limitation is that it requires excessive time for computation because of the large number of degrees of freedom, and it does not account for the interactions of all groups with water, but rather it considers only interactions due to water molecules in the vicinity of the carbonyl group.

The interactions with solvent may alter conformational preferences of amino acids and polypeptides. For this reason, besides approaches already mentioned, different attempts were and still are under investigation to account for the effect of solvent on the conformational energies of peptides and polypeptides (Hopfinger, 1971; A. Pullman and Pullman, 1974; Hagler and Moult, 1978; Owicki and Scheraga 1977a,b,c, 1978) and further developments may be expected on this problem.

3.6. TOTAL CONFORMATIONAL ENERGY OF THE POLYPEPTIDE CHAIN

The total conformational energy is the sum of the different contributions which have been described in this Chapter. These contributions must be evaluated for each peptide unit interacting with its neighbors in a polypeptide chain. The total potential energy is

$$E = E_a + E_r + E_{es} + E_l + E_\tau + E_\theta + E_h + E_{H\phi}$$

where E_a and E_r are the attractive and repulsive terms of the van der Waals contribution; E_{es}, the electrostatic potential; E_l and E_τ, the terms dependent on bond length and bond angle variations respectively; E_θ, the torsional potential; and E_h and $E_{H\phi}$, the contributions of hydrogen bonding and hydrophobic interactions respectively.

Some of these contributions are difficult to evaluate, since either the theory is uncertain or no quantitative formula is really available for the computation. The monopole approximation is preferred for the evaluation of the electrostatic potential. The computation of the energy of the hydrogen bond is most uncertain. The hydrophobic interactions, or more generally the interactions of the polypeptide chain with the solvent, remains difficult to evaluate accurately. The hydration shell model seems the most effective means for accounting for these interactions.

Even when a quantitative theory exists, different approximations leading to various formulations have been developed. As previously mentioned, there are different kinds of potential functions and different values of the parameters were proposed and used (see Section 3.2.1). Consequently, the conformational maps obtained may be significantly different according to the potential function used by the authors. Nevertheless the resulting conformational maps are in agreement with the contact map. This clearly emphasizes the important role of the repulsive term in the nonbonded interactions in determining the region of lower energy. The most significant differences between the different potential functions are in the prediction of relative stabilities of the different conformations.

The various approaches used to predict the polypeptide conformations are discussed in Chapter 4.

4

Theoretical Approach to Protein Folding

4.1. INTRODUCTION: PROBLEMS IN COMPUTATION OF PROTEIN CONFORMATION

Since it is generally accepted that all the information necessary to determine the three-dimensional structure of a protein is contained in its amino acid sequence, the knowledge of this sequence could theoretically make possible the prediction of protein conformation. Practically many problems remain unsolved and approximations are necessary to approach the problem although this area of research is in rapid progress. Conformational computations have been extensively reviewed recently. The present section is inspired principally by the analyses presented by Anfinsen and Scheraga (1975), Némethy (1974), Némethy and Scheraga (1977), B. Pullman and Pullman (1974), Ramachandran and Sasisekharan (1968), Scheraga (1968, 1971, 1973a,b, 1974), Levitt (1976), Schulz (1977), and Karplus and Mc Cammon (1980).

The goal of predicting the folding of a protein has not yet been reached. Nevertheless, interesting results have been obtained by conformational computations which are helpful for understanding protein folding. They can give reasonable estimates of the relative stabilities of different structures. Some of the earlier studies were conducted using poly-L-amino acids in order to predict their conformational preferences, and for the helical structures to explain the sense of the helix. Several analyses of dipeptides, tripeptides, and oligopeptides were done. The ability of amino acids to participate in ordered structures can be predicted by conformational computations. Many small oligopeptides possess biological activity related to the occurrence of preferred conformations. Since the theoretical approach alone does not allow the determination of the most stable conformation, it is possible to resolve a structure in some cases by comparing the theoretical predictions with experimental data mainly from IR or NMR measurements (Gibbons *et al.*, 1970). This method was successful for the determination of a low-energy structure of a cyclic decapeptide, gramicidin S (Dygert *et al.*, 1975), and more recently the structure of enkephalins, pentapeptides with a morphine-like activity (Isogai *et al.*, 1977). By combining computational and experimental data, the most likely oligopeptide structure can be found, whereas each method separately may lead to ambiguous conclusions, even for short oligopeptides. Calculations of conformational energy can be used in the refinement of atomic coordinates obtained from X-ray diffraction measurements. This aspect is discussed later.

In practice, there are several problems which make it difficult to predict the conformation of a protein from a known amino acid sequence. These problems arise mainly because of the complexity of the molecule: the great number of independent variables for generating a conformation which impose a practical limit in terms of time and cost of computations.

The main problems which remain yet unsolved at this time are at least three: (1) the multiminima problem and the difficulty in finding the global minimum, (2) the large number of interactions, and (3) the effect of the solvent. Furthermore, the potential energies calculated are not exact and imply some approximations.

4.1.1. The Multiminima Problem

The conformational hyperspace in which the energy surface of a protein can be drawn is a very complex one. To illustrate the problem, as presented by Scheraga (1973b), this hyperspace is a 500-dimensional one for a polypeptide chain of 100 amino acid residues. Indeed this polypeptide chain has 500 independent degrees of freedom, two per residue in the backbone and an average of three in the side chain, on the basis of dihedral angles of rotation. This energy surface has multiminima (Fig. 4.1) and there is no absolute method which allows one to find the global minimum. The methods of energy minimization will lead to the minimum in the same potential energy as the conformation from which it started. These methods do not allow one to surmount potential barriers and to pass from one potential energy well to another. However, mathematical approach algorithms have been employed in attempts to solve this problem. They have been successfully used for small oligopeptides and even for a decapeptide such as deca-L-alanine (Gibson and Scheraga, 1969). The large number of variables for a protein,

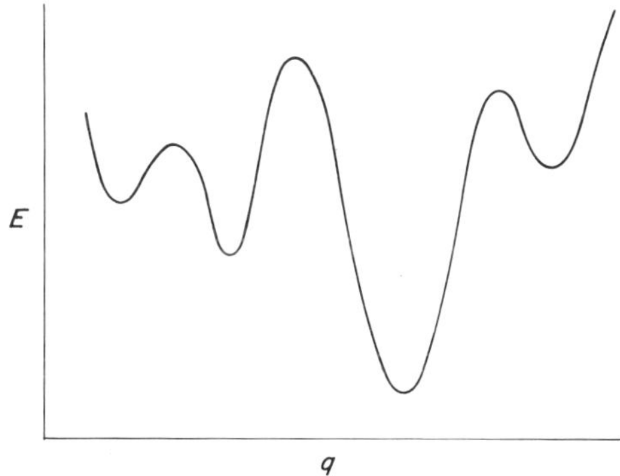

Fig. 4.1. Schematic two-dimensional representation of the energy as a function of conformation (according to Scheraga, 1973b) (see text).

imposes serious limits on time and cost of studies. Since the multiminima problem for a globular protein cannot be solved by a mathematical solution, different procedures have been developed to overcome the difficulty. Some of these procedures, the most significant and efficient, are briefly described in the following section.

4.1.2. The Large Number of Interactions

It is usually assumed for computations that all kinds of interatomic interactions can be separated and that the resulting energy is expressed as the sum of pair interactions between all atoms of the protein:

$$E = \sum_{\substack{i=1 \\ j>i}}^{n} E_{ij}(r_{ij})$$

This may not always be valid. Furthermore, the evaluation of energy requires a computation of a great number of terms; for n atoms in the molecule, there are n^2 terms in the equation. Some procedures of approximation have been proposed to reduce the time of computation.

Another problem arises from the fact that the potential energies calculated are not exact and equations for determining some of the energy contributions are empirical approximations (see following section).

4.1.3. Effect of Solvent

The correct description of the interaction with the solvent, i.e., water, is still not well developed and the evaluation of the energy of interaction is highly approximate. The interactions between atoms of the protein molecule and solvent are basically of the same kind as intramolecular interactions.

However, the solvent molecules are not fixed and their contributions can be treated only in an averaged manner, except in the case where a water molecule is bound to the protein by a strong hydrogen bond. Theories to describe solvent interactions are presented in Chapter 3 (Sections 3.4 and 3.5). As already mentioned, recent approaches use either molecular dynamics or Monte Carlo methods.

The data clearly indicate variations in conformational preferences of a polypeptide chain resulting from interactions with solvent. At this time and only in a highly approximate manner, difficulties in evaluating the effect of solvent have been partially overcome. This is a crucial problem in conformational computation and it is under investigation by several groups.

4.1.4. Importance of Short, Medium, and Long-Range Interactions

To overcome the multiminima problem alternative methods have been proposed. One of these methods assumes that the folding of a polypeptide chain is determined to a very large extent by short-range interactions. The term short-range interactions designates interactions between the side chain of a given amino acid and the two adjacent peptide groups of the backbone. This hypothesis is adopted as a first approximation in most conformational computation studies. Medium-range interactions refers to the interactions between a central residue and the neighboring residues up to four amino acids in both directions from $i-4$ to $i+4$ (Ponnuswamy *et al.*, 1973). Interactions with residues more remote along the chain are termed long-range interactions (Scheraga, 1973b; Ponnuswamy *et al.*, 1973; Anfinsen and Scheraga, 1975; Maxfield and Scheraga, 1976). In their review, Némethy and Scheraga (1977) use a slightly different definition. They term the interactions between the side chain of an amino acid and the backbone of the same amino acid as intraresidue interactions. Further, interactions of a given residue with its neighbors up to four residues in both directions are referred to as short-range interactions by Némethy and Scheraga (1977). In their terminology medium-range interactions are interactions of a residue with others within the range of 5–20 along the sequence. Those occurring beyond 20 residues are designated as long-range interactions. In some other publications, intraresidue interactions and interactions within four residues are included in the term short-range interactions (Tanaka and Scheraga, 1975, 1976a,b,c,d, 1977). In the literature, interactions are often classified into only two categories, short-range (intraresidue) interactions and long-range interactions, which include all the other kinds of interactions. Fluctuations in the adopted definitions reflect the difference in the calculation procedures and in their potentialities. In the following presentation, the first set of definitions is used, that is, short-range interactions (intraresidue), medium-range interactions (within four neighboring amino acids along the sequence), and long-range interactions (beyond four residues).

It is commonly accepted that, protein folding is determined, mainly but not exclusively, by short-range interactions and that the conformation is stabilized by medium- and long-range interactions (see Anfinsen and Scheraga, 1975; Nemethy and Scheraga, 1977). The basis of many studies was the consideration of the predominance of short-range interactions (Kotelchuck and Scheraga, 1969; Lewis *et al.*, 1970; Ptitsyn and Finkelstein, 1970; Chou and Fasman, 1974a,b; Burgess *et al.*, 1974). This originates from the behavior of homopolymers. Under given conditions, referred to as the θ point, the conformation of a homopolymer chain from random coil is governed only by short-range interactions. It is the case when $\langle \bar{r}^2 \rangle^{1/2}$, the root-mean-square (rms), end-to-end distance of the chain, varies with the root-square of the molecular weight which corresponds to ideal conditions. Medium- and short-range interactions, when they operate, tend to increase $\langle \bar{r}^2 \rangle^{1/2}$. Under appropriate conditions (solvent, ionic strength), $\langle \bar{r}^2 \rangle^{1/2}$ can reach its ideal value which defines the θ point (Flory, 1953, 1969). Under these conditions, interactions with solvent and long-range interactions neutralize each other. A protein in solution cannot be at the θ point. Nevertheless, the predominance of short-range interactions appears as a reasonable and efficient statement which makes possible conformational computations with proteins.

However, the importance of long-range interactions was emphasized and through the 1970s attempts were made to take them into account in the determination of the conformation of proteins (Schiffer and Edmundson, 1967). Progressively, models that included medium-range interactions were developed (Ptitsyn *et al.*, 1972; Ponnuswamy *et al.*, 1973; Robson and Pain, 1971, 1974a,b; Robson and Suzuki, 1976; Lim, 1974a,b). Models that include long-range interactions have been recently introduced (Ptitsyn and Rashin, 1975; Tanaka and Scheraga, 1975; Honig *et al.*, 1976).

4.1.5. Principles for Computation of Protein Conformation

Some of the different theoretical approaches, prediction methods, and simulation of protein folding are briefly introduced here. The methods for evaluating, first the conformational preferences of a single peptide unit, and second the conformational preferences of a polypeptide chain, are discussed. Models that use short- and medium-range interactions are briefly analyzed. The results of these different predictive methods of secondary structures are compared. Actually the refinements of these predictive methods are under study in the main laboratories of this field of study. It is evidently of great importance for further simulation of protein folding; however, it is not the intent of this discussion to describe and compare the validity of all the refinements in computation, but rather the intent is only to give the general principles of the different methods. Long-range interactions models are under study for predicting interactions between segments of secondary structure. To simulate protein folding, long-range interactions must be included; several models are examined. The information obtained from these approaches regarding the dynamics of protein folding are considered and discussed in following sections. Conformational fluctuations that occur in proteins under thermal motion are also considered.

4.2. METHODS FOR EVALUATION OF CONFORMATIONAL PREFERENCES OF A SINGLE PEPTIDE UNIT

4.2.1. Terminally Blocked Amino Acid

An amino acid side chain and the planes of two adjacent peptide bonds were previously incorrectly called a dipeptide and are now referred to as a terminally blocked amino acid (Fig. 4.2). In this model, only the intraresidue interactions are taken into account [i.e., those associated with rotations of the backbone (ϕ_i, ψ_i) and with rotations of the side chain (χ_i^{jk})]; all the other interactions are neglected. This model considers that the conformation of a given amino acid does not depend on the nature of its neighbors. The hypothesis that short-range interactions are dominant is a significant first approximation. Its validity was supported by different researchers, such as Kotelchuck and Scheraga (1968, 1969) who calculated that the conformation of the lowest energy for a given amino acid in lysozyme is independent of the nature of its neighbor. Many empirical computations and the quantum mechanical ones have been performed on terminally blocked amino acids. Energy contour maps have been obtained by this procedure.

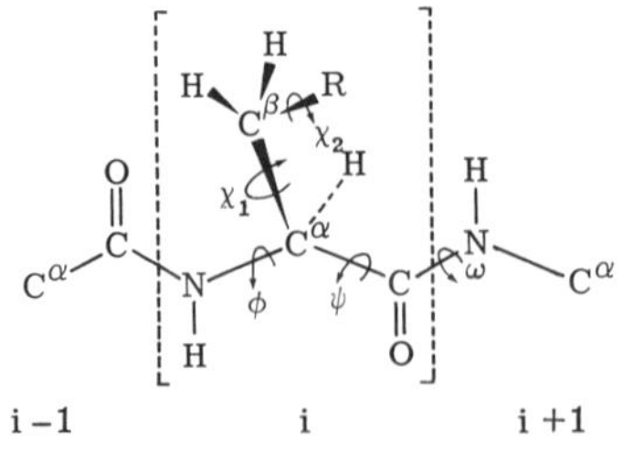

Fig. 4.2. The terminally blocked amino acid.

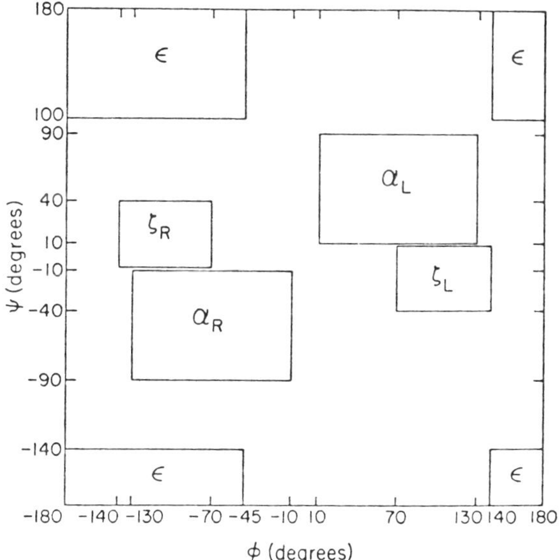

Fig. 4.3. Localization of the different conformations in the Ramachandran diagram (according to Burgess *et al.*, 1974; from Némethy and Scheraga, 1977).

According to Burgess and co-workers (1974), six conformational states for amino acids can be localized in the conformational map. They are denoted α_R, α_L for right- and left-handed helical states, ε for extended states, ζ_R and ζ_L for right- and left-handed bridge conformations. The rest of the map corresponds to the coil states (Fig. 4.3). The conformational energy minima computed for the *N*-methylamides of the 20 naturally occurring amino acids are confined in limited regions of the map as illustrated in Fig. 4.3 (Lewis *et al.*, 1973b; Zimmerman *et al.*, 1977). The data obtained from the terminally blocked amino acid model can be used as starting point for studies on protein conformation.

4.2.2. Empirical Methods

The empirical methods of computation of conformational energy are called empirical because they use functions which are not derived from a basic theory and some parameters are evaluated from experimental data obtained from model compounds. The hard sphere approximation was introduced by the Madras group who originated the conformational computation studies on proteins. From this simplest approximation, refinements have been progressively developed. These refinements consist mainly of the different contributions included in the potential energy functions and of the values of some of the parameters. The most complete expressions are the sum of (1) nonbonded interactions (with different approximations of the repulsive term), (2) barriers to internal rotations, (3) hydrogen bonding, and (4) energies from bond angle or bond length variations and interactions with solvent. Depending on the structure to be resolved, the researchers take into account all or part of these contributions. At this point, one should note a statement by Hopfinger (1973): "A general rule which seems to apply is that the level of sophistication used in the energy calculations is inversely proportional to the size of the molecule."

One must mention that some researchers in their computations use the parameters of nonbonded interactions from one published set of data and the parameters for electrostatic interactions from another, introducing some heterogeneity and therefore some source of inaccuracy. It is recommended that, for the different kinds of interactions, the parameters determined by the same researcher be used even when another researcher reports more refined parameters for a given type of interactions.

Efforts have been made to improve the physical basis of empirical methods. To refine the parameters of empirical energy functions, molecular critical calculations have been carried out for homologous compounds (Momany *et al.*, 1968; Scheraga, 1971). Parameters of empirical energy functions were calculated for the crystal structure of models and amino acids (Momany *et al.*, 1974a,b). They were refined to adjust rotational barriers close to experimental values (Momany *et al.*, 1975). Different programs were developed (Dunfield *et al.*, 1978; Snirr *et al.*, 1978a,b; Nemenoff *et al.*, 1978a,b). These evaluations include hydration effects.

4.2.3. Semiempirical Quantum Mechanical Methods

The principle of quantum mechanical methods consist basically to evaluate the total energy of a molecule by solving an N-electron Schrödinger equation

$$H(N)\psi(N) = E(N)\psi(N) \qquad \text{with } N > 1$$

where $\psi(N)$ is a nonspecific wave function; $E(N)$, the total energy of the molecule; $H(N)$, the N electrons Hamiltonian which contains nonseparable $1/r_{ij}$ terms corresponding to the relative interaction position of electrons i and j. It is indeed a complex calculation; however, development of molecular orbital methods with suitable approximations allow computation using molecules as complex as amino acids. All procedures are derived from the method of molecular orbitals which expresses the wave function of polyelectronic systems as a combination of individual one-electron wave functions [usually a linear combination as in Linear Combination of Atomic Orbitals (LCAO)].

The principal methods used for predicting the preferred conformations of amino acids are the Extended Huckel Theory (EHT), the second version of Complete Neglect of Differential Overlap (CNDO/2) and Perturbative Configuration Interaction over Localized Orbitals (PCILO). [For the details of these methods, see Hopfinger (1973) and B. Pullman and Pullman (1974).] These methods represent different suitable approximations which allow energy computation. The use of one method as opposed to another seems to have mainly operational justification. The EHT was employed by several groups to evaluate the conformational preference of amino acids, particularly by Hoffman and Imamura (1969), Momany and co-workers (1971); CNDO/2 was used by the group of Scheraga (Momany *et al.*, 1971). The use of PCILO method was introduced by the Pullman group (Maigret *et al.*, 1970; Maigret, 1971; Perahia, 1971; see also B. Pullman and Pullman, 1974). Figure 4.4

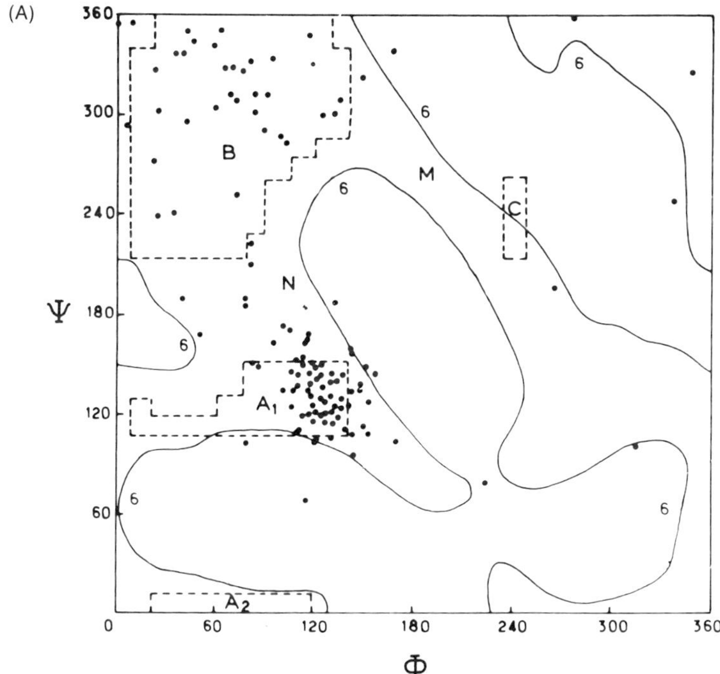

Fig. 4.4. Conformational maps for the alanyl residue obtained by different methods: (A) PCILO computations within 6 kcal above the global minimum; allowed limits of hard sphere approximation are plotted in the map. (B) Partitioned potential function (according to Brant *et al.*, 1967). (C) Partitioned potential function (according to Ponnuswamy as cited by B. Pullman and Pullman (1974). (D) Partitioned potential function, (according to Popov *et al.*, 1968a,b). (E) Extended Hückel Theory (according to Hoffmann and Imamura, 1969). (F) CNDO/2 (according to Momany *et al.*, 1971; from B. Pullman and Pullman, 1974) (*Continued*, see pp. 186–188).

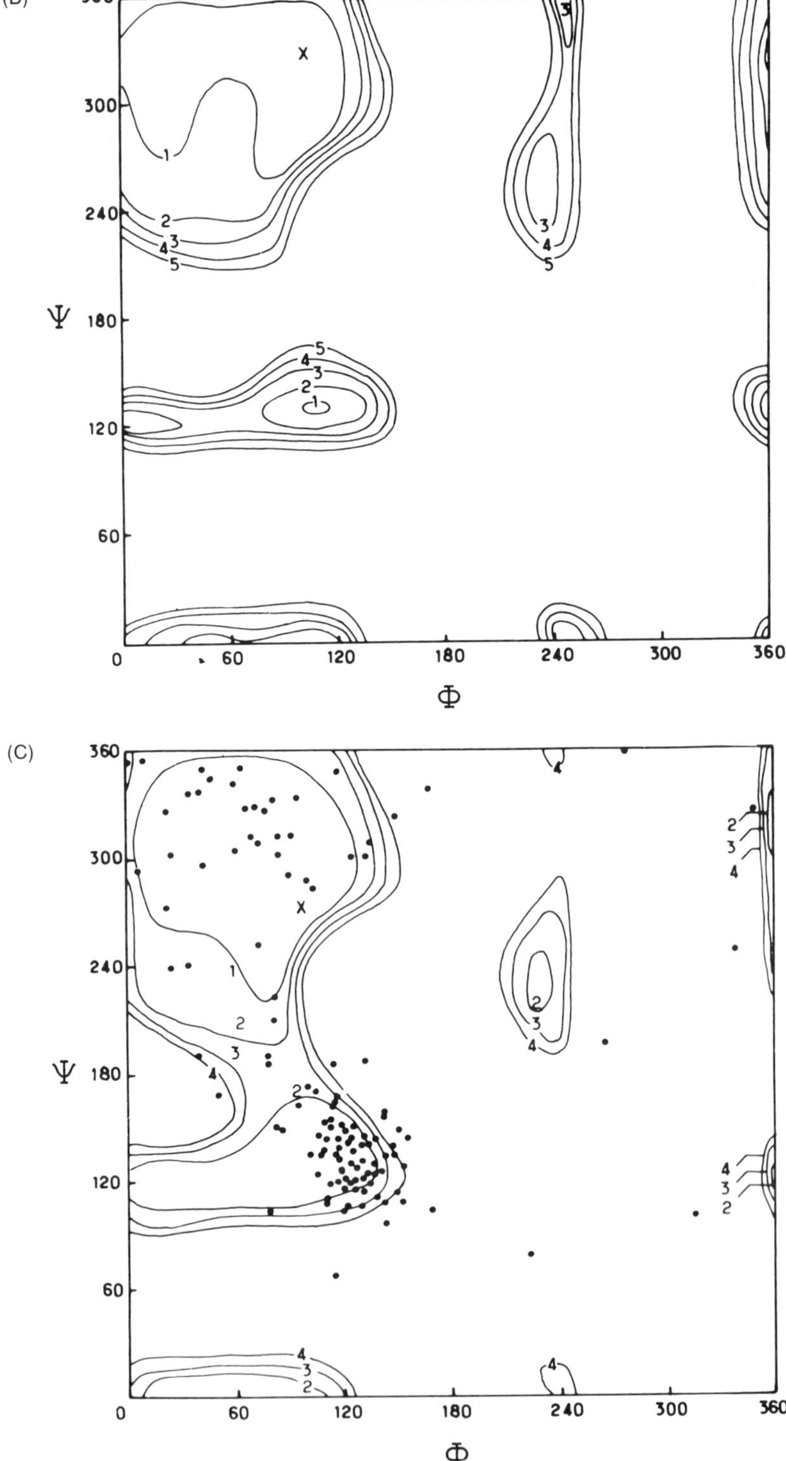

Fig. 4.4 (*continued*), see p. 185 for legend.

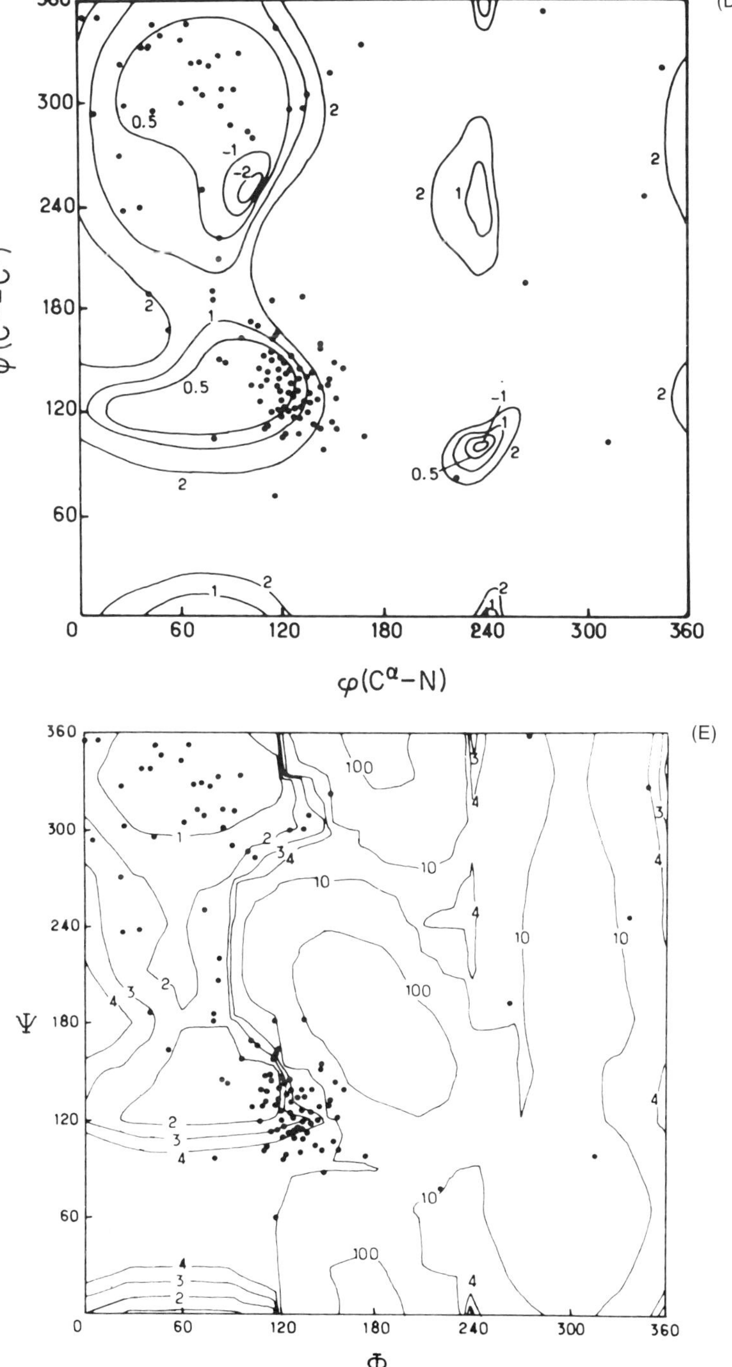

Fig. 4.4 (*continued*), see p. 185 for legend.

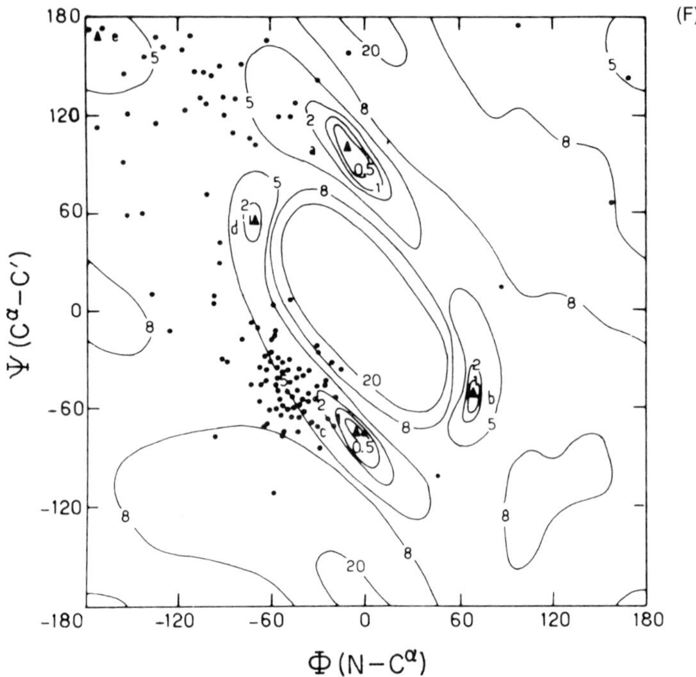

Fig. 4.4 (*continued*), see p. 185 for legend.

shows the energy contour maps for alanyl residue obtained by different procedures which include the empirical methods. In a first approximation, the results of these different methods are in rather good agreement. However, some discrepancies appear in detail. For instance, for the alanyl group, the PCILO method indicates a global minimum located in a disallowed region for both EHT and the empirical methods ($\phi = 90°$; $\psi = -60°$). No experimental point was found in this region. For aspartyl residue, the PCILO method does not give satisfactory agreement with other methods and with the experimental data.

We are not specialists of conformational computations and we have not enough competence to compare and to discuss in detail the accuracy of the different procedures; we let the specialists argue on the accuracy of the method they use. Nevertheless, we may emphasize that quantum mechanical procedures as well as empirical methods contain approximations, and the employment of one method rather than another can be justified by the degree of information needed, by the computation time required, and last but not least, by the agreement between theoretical predictions and experimental data.

4.3. METHODS FOR EVALUATION OF CONFORMATIONAL PREFERENCES OF A POLYPEPTIDE CHAIN

This field has been extensively reviewed by Némethy and Scheraga (1977). The most important aspects are summarized here. To overcome the difficulties encountered in calculations for larger molecules caused by the great number of parameters and by the problem of finding the global minimum, some procedures were progressively introduced which reduce the computation time and, thus, make possible studies of globular proteins.

The determination of conformational energy of polypeptides requires several steps. The first involves as a starting point, a good representation of the geometry of the molecule (i.e., bond angles, bond length, and a set of dihedral angles). At this level some simplified representations of the polypeptide chain have been introduced. In the second step, there is the choice of the method to use for calculation of potential energy (empirical with the different contributions possibly included, or quantum mechanical). The third step is the energy minimization for the search of a stable conformation. The problem at this level is to find the right energy well of the conformational hyperspace before energy minimization which requires a good starting point in the first step. Some procedures can help in the choice of the starting point.

4.3.1. Simplified Representation of the Polypeptide Chain

4.3.1.1. Virtual Bonds

This simplification describes the backbone of a polypeptide chain by a sequence of virtual bonds extending from one C^α to the next one. Assuming that all peptide bonds are in a planar trans form, the distance between each C^α atom is constant and can be considered as a virtual bond. In this model a torsion angle α about each virtual bond replaces the two torsion angles ϕ and ψ; only one degree of freedom replaces two degrees of freedom per residue. This procedure introduced by Brant and Flory (1965b) was utilized by different researchers. It was employed by Nishikawa and co-workers (1974) for the analysis of bend conformation in dipeptides. It is also the basis of the simplified model of the polypeptide chain used by Levitt in computing simulations of the folding of bovine pancreatic trypsin inhibitor (BPTI)

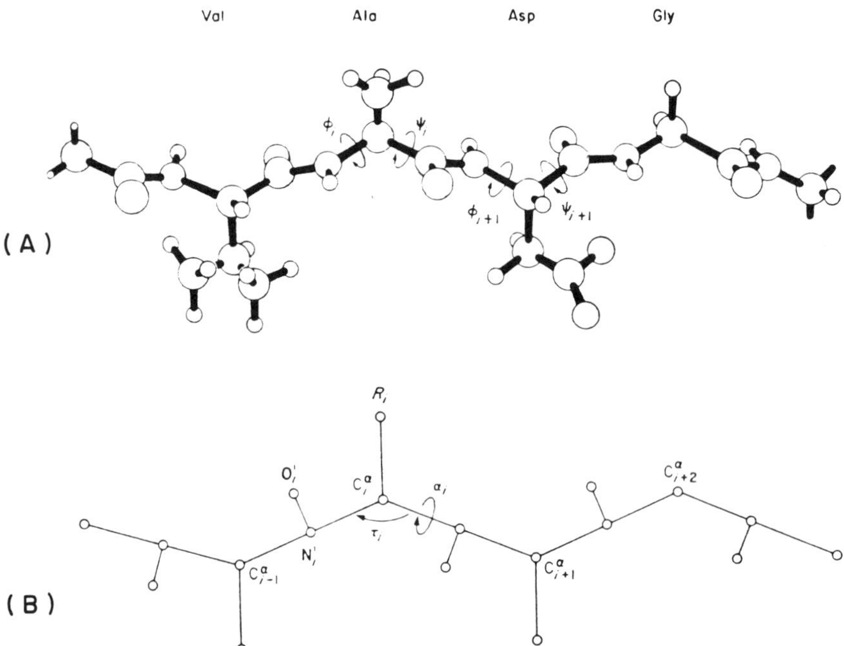

Fig. 4.5. Simplified representation of the polypeptide chain (according to Levitt, 1976). (A) The real structure of the tetrapeptide: Val–Ala–Asp–Gly. (B) Simplified model, virtual bonds are represented by the distances between C^{α}; the only dihedral angle of the backbone is α_i (see text).

(Levitt and Warshel, 1975; Levitt, 1976), which is discussed in Section 5.2 (Fig. 4.5).

4.3.1.2. United Atom Approximation

The united atom approximation reduces a group of atoms, for example CH_3, CH_2, to a unique atom considered as an effective and single center of interaction. This procedure has the advantage of reducing the time of computation.

4.3.1.3. United Residue Approximation

Side chain residues can be reduced to their center of interaction. The approximation is valid only when two interacting residues are at distances larger than 10–15 Å (Pincus and Scheraga, 1977). Each residue is represented by only two centers, the centroid of the side chain and the C residue (Fig. 4.6). Each side chain is represented by a sphere whose radius is equal to its average radius of gyration, and a uniform distribution within the sphere is assumed. The simplified model of Levitt and Warshel also used a united residue

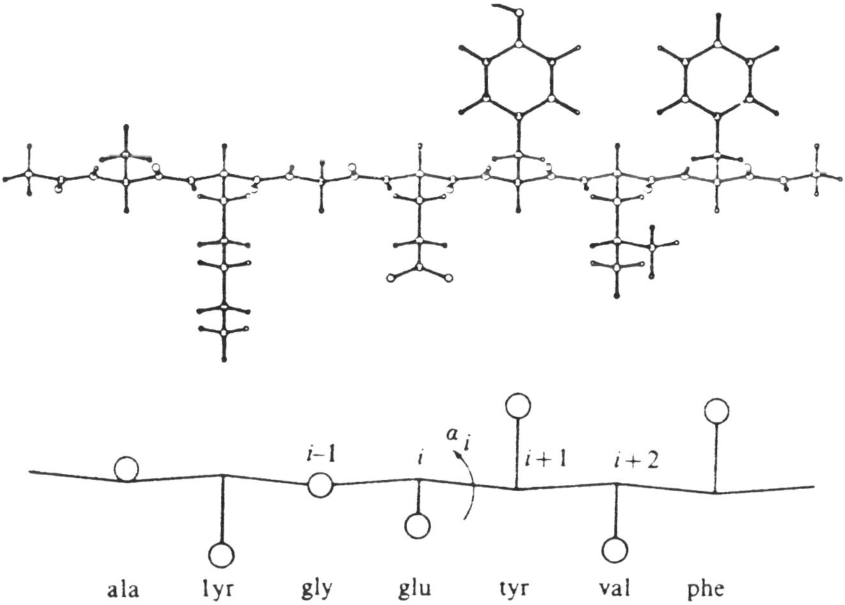

Fig. 4.6. Simplified representation of the polypeptide chain exemplifying united residue approximation (according to Levitt and Warshel, 1975). Polypeptide sequence: Ala–Lys–Gly–Glu–Tyr–Val–Phe. Each residue is represented by two centers, that of the C^α and the center of the side chain. All the side chains of a given type have the same simplified geometry.

approximation, but these researchers used it even for distances smaller than 10–15 Å.

4.3.1.4. Spherical Representations

To evaluate contact regions between residues, Tanaka and Scheraga (1975, 1976a,b,c,d) represented backbone and side chains as interacting spheres. A similar representation was used by Kuntz and co-workers (1976).

4.3.2. Energy Minimization Methods

The problem of finding the conformation of lowest energy is rather simple for a terminally blocked amino acid (such as a glycyl or alanyl residue) since it is a problem including only two variables (ψ, ϕ); however, the location of the minimum becomes more difficult to determine even for a terminally blocked amino acid when rotations χ are also included. *A fortiori*, when the molecule contains more than one amino acid, difficulties in finding the deepest minimum arise. As previously discussed (Section 4.1.1), the multidimensional energy surface has multiple minima. Since there is no procedure to surmount the potential barrier, energy minimization is made toward the minimum in the same energy well as the conformation from which it was started. The minimization procedures usually employed attain the minimum by convergence toward this minimum where the first derivative of the potential energy function is null ($\partial E/\partial \alpha_i = 0$, for $i = 1, 2, 3 \ldots n$). These methods are known as gradient methods. They are local minimization procedures since the convergence would take place only toward the minimum of the local energy well.

There are many minimization procedures. Their relative accuracy and efficiency for the computation of proteins have been investigated and discussed (see Gibson and Scheraga, 1967). This aspect is not expanded here.

4.3.3. Conformational Entropy

Once the energy minimum associated with a given conformation is found, then a question may be asked: What is the probability of finding the molecule in such a conformation? This question may be answered by the computation of the conformational entropy associated with each energy minimum, if such a computation is possible. There are in fact two sources of entropy, one contribution from solvation, the other from libration. Computation of the entropy of a system requires the knowledge of the system in all points, which is difficult for a protein, even for a small oligopeptide. Different approximate procedures exist to evaluate conformational entropy. The Scheraga group uses the matrix of the second derivatives of the conformational energy for the evaluation of librational entropy (Gō and Scheraga, 1969, 1976; Gō *et al.*, 1974).

4.3.4. Choice of Initial Conformations

The problem of finding the right starting point of the conformational hyperspace before minimization is certainly not yet solved. However, different procedures can reduce the number of possibilities in some cases.

4.3.4.1. Use of Terminally Blocked Amino Acid Conformation

Computation of the energy of terminally blocked amino acids indicates that only limited regions of the (ϕ, ψ) Ramachandran map are allowed conformations. The knowledge of these conformations narrows the choice of starting points. However, the number of combinations of these energy minima also increases with the number of amino acids in the polypeptide chain and the problem becomes rapidly insoluble, even for small peptides. Some additional criteria might be employed to reduce the number of possibilities.

4.3.4.2. *Use of Statistical Predictive Schemes*

Conformational preferences of amino acids have been derived from statistical analysis of their observed conformations in proteins whose three-dimensional structure is known. The application of the data to a known protein sequence allows one to determine probable conformational states which can be used as starting conformations for energy minimization.

4.3.4.3. *Refinements of X-Ray Structures*

The observed three-dimensional structures as obtained from X-ray analyses at about 2 Å resolution can be refined by energy minimization. This method has been utilized for some proteins particularly, lysozyme, rubredoxin, actinomycin D, and α-chymotrypsin (see Scheraga, 1973b).

4.3.4.4. *Use of Homology*

Close similarity in polypeptide sequences can generate proteins of similar three-dimensional structure. Therefore the known refined X-ray structure of a protein can be used as a starting point in conformational computations of proteins with homologous sequences. As an example, lactalbumin structure was computed by using the known X-ray structure of lysozyme as starting point (Warme *et al.*, 1974). The amino acid sequences of the two proteins are very similar; the homologies were utilized by equating the conformation of amino acid residues at corresponding positions in both sequences.

4.4. PREDICTION OF ORDERED STRUCTURES IN PROTEINS

Predictions of protein conformation use short-range and medium-range interaction models. They can be divided into two main categories, those using experimentally deduced predictive rules, which are arbitrary, and those involving more rigorous methods to deduce the probability of the different conformations. Since this problem has been extensively reviewed, particularly by Scheraga (1973b), Burgess and co-workers (1974), Anfinsen and Scheraga (1975), Némethy and Scheraga (1977), a complete survey of the subject is not given here, but only the most significant predictive approaches.

4.4.1. Models Using Empirically Deduced Predictive Rules

In this category of methods the frequency of occurrence of each amino acid in a defined conformational state is deduced from the observation of X-ray structures of many proteins. Several algorithms for the prediction of protein folding based on short- and medium-range interactions have been developed during the 1970s. Many of the earlier studies were based on only a few proteins, the ones which were available at the moment (Guzzo, 1965; Prothero, 1966; Cook, 1967; Schiffer and Edmundson, 1967; Kotelchuck and Scheraga, 1968, 1969; Low *et al.*, 1968; Ptitsyn, 1969; Leberman, 1971; Robson and Pain, 1971). Only predictions derived from more detailed methods are described.

Chou and Fasman (1974a,b) developed a more complete analysis based on the consideration of 15 proteins containing 2473 amino acid residues. They calculated the frequency of occurrence of each amino acid in the f_α, f_β, and f_c (coil) conformations and determined their conformational parameters P_α, $P_{\alpha i}$, P_β, and P_c which are respectively the parameters for the helix ($P_\alpha = f_\alpha/\langle f_\alpha \rangle$), the inner helix in which the three helical residues on both the N- and C-terminal of a helical segment are omitted ($P_{\alpha_i} = f_{\alpha_i}/\langle f_{\alpha_i} \rangle$), the β regions ($P_\beta = f_\beta/\langle f_\beta \rangle$), and the coil regions ($P_c = f_c/\langle f_c \rangle$). The $\langle f \rangle$ are the respective average fractions of residues in the different conformations. The percentage of residues in the considered proteins found in helical, β structure and coil regions were respectively 36%, 17%, and 47%. The normalization procedure gave P_α and P_{α_i} values resembling the Zimm–Bragg helix growth parameters (see further discussion). In their study Chou and Fasman introduced a detailed analysis of the helix and β sheet boundary residues which indicate the amino acid frequencies of the N- and C-terminal ends allowing one to delineate helical and β regions. Thus, charged residues are found with a great frequency at both helical ends, with a preference of the negatively charged residues (Asp, Glu) for the N-terminal end and positively charged residues (His, Lys, Arg) at the C-terminal end. This was previously noted by Cook (1967) and Ptitsyn (1969) but on fewer proteins. Charged residues were found to be mostly absent in β sheet regions. These rules were successful in correctly predicting 80% of helical and 86% of β sheet residues in 19 proteins. The conformational parameters used in this study were reevaluated on a large number of proteins (Fasman *et al.*, 1976; Chou and Fasman, 1978).

The procedure developed by Kabat and Wu (1973a,b) and Wu and Kabat (1971, 1973) used tripeptide sequence and was based on the influence of neighboring amino acids $(n - 1)$, $(n + 1)$ on the tendency of the nth amino acid to be in an ordered (helical or extended) or in a nonordered conformation. In fact, this method rules out certain conformations; applying the tripeptide analysis from proteins of known tridimensional structure to a protein of unknown structure, it predicts which central residue n can distrupt either a helix or an extended structure according to the nature of the nearest neighbors. The portion of amino acid sequence between helix-breaking residues is permissive if helix is long enough; likewise for β structures. Their predictions were based on few proteins. The second procedure of Wu and Kabat (1973) always used the tripeptide consideration; furthermore it is based on sequence homologies and states that all homologous sequences have the same backbone structure. They applied this procedure to the family of cytochrome *c*.

Lim (1974a,b) proposed algorithms for predicting regions in α helices and β structures in proteins based on a stereochemical theory of globular protein secondary structure, with particular attention to the polar or hydrophobic character of residues according to their location. Lim presented some structural principles of the organization of the protein molecules which emphasize the shielding of the nonpolar side groups from water, the solvation of polar groups, and the packing of the nonpolar side groups in the core of the protein. A theory for formation of α helical and β structures was presented. Three types of β structures were considered: (1) the internal type in which structures are immersed into the protein, (2) the surface type with the plane of peptide groups located tangentially to the surface, and (3) the semisurface type with the plane of the peptide groups located near the periphery of the molecule and perpendicularly oriented to the surface. In globular proteins, helices can also be divided into two different types, internal and surface types, but separate consideration was not required in this case. A set of rules for α helix and β structure formation were deduced from the consideration of pairs of residues $(i, i+1)$, $(i, i+2)$, and triplets $(i, i+1, i+4)$ according to the hydrophobic or hydrophilic, or both character of the side groups. This *a priori* theory, which is not based as the preceding on the examination of known structure but rather on the stereochemistry of regular structure deduced from space filing models, allows one to predict α helical and β structural regions with an accuracy of 81% and 85% respectively in 25 proteins (i.e., 4149 residues) of known three-dimensional structure (this accuracy is expressed by the percentage of residues whose state is predicted correctly). The state of 70% of all the residues with a simultaneous identification of all the types of secondary structures was correctly identified.

The empirical predictive schemes based on the analysis of X-ray structures are arbitrary since there are no precise rules for characterizing secondary structures in proteins. In more recent studies ϕ and ψ angles calculated from atomic coordinates have been used to assign the different local conformations for each residue (Burgess *et al*., 1974; Robson and Pain, 1974a,b; Tanaka and Scheraga, 1976a,b,c,d; Maxfield and Scheraga, 1976). Five distinct local conformations have been thus defined by Burgess and co-workers (1974), right- and left-handed helical (α_R and α_L), extended (ε) and the right and left handed turns (ζ_R and ζ_L); their location on the Ramachandran map is shown in Fig. 4.3. The distance between pairs of C^α was used to identify reverse turns (Lewis *et al*., 1971; Kuntz, 1972; Crawford *et al*., 1973). Some researchers prefer to use C^α distances and α torsion angles rather than ϕ and ψ angles to identify secondary structure. Crystallographers pay more attention to hydrogen bonds than to the ϕ and ψ angles for the assignment of regular structure in proteins. As already mentioned (Chapter 2), Levitt and Greer (1977) proposed the use of patterns of peptide hydrogen bonds, the distances between C^{α} and α torsion angles to find the secondary structure of proteins. Levitt and Greer (1977) developed a computer program to analyze automatically the atomic coordinates of a large number of proteins. These data allow analysis of the tendencies of the different amino acid residues to occur in the different types of structure and can also be used for predicting the packing of regular structures to form the globular protein (see Chapter 2).

4.4.2. Methods Using Statistical Mechanical Theory

The use of statistical mechanical methods was introduced initially by Lewis and co-workers (1970), Lewis and Scheraga (1971a,b), and Gō and co-workers (1971). Assuming the predominance of short-range interactions, these investigators used the nearest neighbor one-dimensional Ising model to determine quantitatively the conformational preferences of amino acids. For the first time a quantitative theory of helix-making and helix-breaking character of amino acids was presented. The models used are extension of the helix–coil transition models developed by Zimm and Bragg (1959) and by Lifson and Roig (1961).

The elementary processes of nucleation and growth of helices are defined by the parameter σ and s respectively, which represent equilibrium constants. The equilibrium for the elongation of the helix by one unit (i.e., the growth process) can be represented as follows:

$$\cdots hhc \cdots \overset{s}{\rightleftharpoons} \cdots hhh \cdots$$

h and c represent the amino acid in helical and coil conformation respectively. The nucleation process (i.e., the formation of a helix in a coil region) requires the participation of three consecutive residues that allow the formation of only one hydrogen bond, therefore the equilibrium constant is σs^3:

$$\cdots ccc \cdots \overset{\sigma s^3}{\rightleftharpoons} \cdots hhh \cdots$$

The parameters σ and s can be determined experimentally. Theoretically, the simplest way to obtain this information is the study of the thermally induced transition of a homopolymer of each amino acid. However, none of the 20 naturally occurring amino acid homopolymers satisfies the necessary conditions, i.e., (1) solubility in water, (2) helical conformation in water at low temperature, and (3) coil transition as the temperature increases. Therefore, copolymers of two amino acids are investigated for this purpose. The σ and s parameters are usually determined for guest residues from the effect of increasing amounts of guest residues on the helix–coil transition of the homopolymer host amino acid. Based on this principle, a matrix method is applied to evaluate a partition function and to compute the various probabilities of residues to be in a given conformation. Initial developments were proposed to determine quantitatively the tendency of amino acids to form helices (Lewis *et al.*, 1970; Lewis and Scheraga, 1971a,b; Gō *et al.*, 1971). They classified the residues as helix breakers, helix makers, and helix indifferent.

Using the same principles, a more extensive approach including the five ordered conformational states was developed by the same research group (Tanaka and Scheraga, 1976a,b,c,d, 1977). Their treatment includes right- and left-handed helices, extended structure, coil, chain reversal, and left- and right-handed bridged regions and coil.

In this approach, the conformational probability that n residues beginning by residue i will be in a given conformational state ρ, is given by

$$P(i|n|\{\rho\}) = Z^{-1}\mathrm{e}_1\left[\prod_{j=1}^{i} \mathrm{W}_j\right]\left[\prod_{k=i+1}^{i+n-1} \frac{\partial \mathrm{W}_k}{\partial \ln(m_{k;\eta_{k-1}\eta k})}\right]_{\{\rho\}}\left[\prod_{l=i+n}^{N} \mathrm{W}_l\right]\mathrm{e}_N^*$$

where N is the number of residues in the polypeptide chain; ρ is given by

$$\{\rho\} = \eta_i\eta_{i+1}\cdots\eta_{i+n-1}$$

in which $\eta_i(1 \leqslant i \leqslant N)$ is one of the conformational state; Z is the partition function

$$Z = \mathrm{e}_1\left[\prod_{i=1}^{N} \mathrm{W}_i\right]\mathrm{e}_N^*,$$

with

$$\mathrm{e}_1 = [1\ \ 0\ \ 0\ \ 0\ \ 0]$$

and

$$\mathrm{e}_N^* = \begin{bmatrix} 1 \\ 1 \\ 1 \\ 1 \\ 1 \end{bmatrix}$$

W_i is the matrix which correlates the state of residue i with residue $i+1$.

Using also a statistical mechanical treatment for the different structural states in proteins, Finkelstein and Ptitsyn (1976) concluded that only α helices can initiate self-organization of the tertiary structure of proteins since they preexist in the unfolded chain as *fluctuating embryos* while β bends and other irregular structures do not preexist in the unfolded state.

4.4.3. Models Using Information Theory

Robson and Pain (1971, 1974a,b) and more recently Robson and Suzuki (1976) applied information theory to determine the conformational preferences of amino acid residues in proteins. The procedure was used to evaluate the role of each amino acid in determining its conformation and that of its neighbors. The effects of intraresidue information and of directional information on conformation were evaluated. Directional information is defined as the information a residue carries about the conformation of another residue irrespective of the nature of the side chain of this other residue. The directional information between near neighbor was evaluated within 8 residues away in the N-terminal direction and 8 residues away in the C-terminal direction.

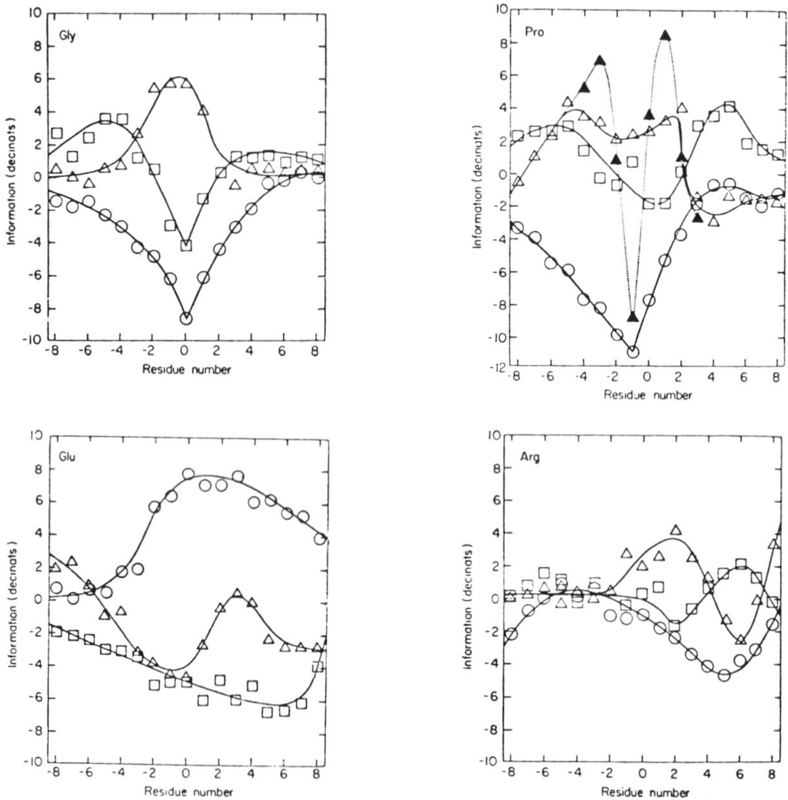

Fig. 4.7. Directional and intraresidue information for forming α helical (○), extended structure (□), and turn (△) in proteins. The type of residue is at zero. The information about the conformation of neighboring residues is plotted versus the distance from the residue along the sequence. The helix information at −4 is the information carried by the residue about the conformation of any residue located 4 away in the terminal direction. The data are given for Gly, Pro, Glu, and Arg (from Robson and Susuki, 1976).

The information measure was estimated from statistical analysis of known structures. Robson and Suzuki (1976) considered 25 proteins (i.e., about 4500 residues). Directional information was represented graphically for the different residues. In Fig. 4.7 examples are given for four amino acids taken in the classification of Lewis and co-workers (1971) (i.e., Gly and Pro as helix breakers, Glu as helix former, and Arg as helix indifferent). In Fig. 4.7, the residue considered is taken as zero. Information at -4 is the information which the named residue transmits to any residue, 4 residues away along the sequence in the N-terminal direction independently of the nature of this residue. The information for an α helix, extended structure, and turn was evaluated.

Information theory was also used by Maxfield and Scheraga (1976) who developed a prediction procedure derived also from Bayesien methods (from mathematician Bayes, 1763) from the analysis of 20 proteins of known structure. In the principle the procedure resembles that of Robson and Pain (1971, 1974a,b). However, it contains some important differences, in the procedure proposed by Maxfield and Scheraga (1976) an estimation is made of the statistical error due to the size of the data set. A greater emphasis is placed on weighting the different terms used for prediction, according to their statistical significance. This treatment which includes medium-range interactions has a reliability of 56% (i.e., 56% average of the residues were assigned correctly to one of the five conformational states defined by Maxfield and Scheraga). Therefore, it can be considered as a good base for future predictions with the inclusion of many proteins in the data set to increase the reliability of the method.

4.4.4. Tendency of Amino Acids to Form Ordered Structures; Comparison of Different Prediction Methods

Some of the methods used for prediction of ordered regions in proteins based on short- and medium-range interactions have been described. This rapid and certainly incomplete survey indicates the variety of the procedures. Burgess and co-workers (1974) emphasized the necessity of having some criteria by which one can evaluate the ability of any predictive algorithm for accurate assignment of the conformational state of each residue. Otherwise the algorithm cannot be used to predict the native conformation of a protein from the sequence of the polypeptide chain. The authors proposed the use of an index of prediction P defined as follows:

$$P = N_s Q - 1 \text{ for } 0 \leqslant Q \leqslant 1/N_s$$

$$P = (N_s Q - 1)/(N_s - 1) \text{ for } 1/N_s \leqslant Q \leqslant 1$$

where Q is the fraction of correctly predicted states (i.e., the number of residue states assigned correctly divided by the number of residues in chain); N_s,

TABLE

Evaluation of Different Models for

Protein	Values of prediction index P^b						
	α Helix						
	Low *et al.* (1968)	Guzzo (1965)	Prothero (1966)	Schiffer *et al.* (1967)	Kotelchuck *et al.* (1968) (1969)	Leberman (1971)	Lewis *et al.* (1970)
Lysozyme	0.05	−0.1	0.7	0.5	0.6	0.6	0.5
Ribonuclease	0.7	−0.2	0.5	0.5	0.6	0.8	0.2
Myoglobin	−0.5	−0.1	0.6	0.4	0.4	0.0	0.2
α-Chymotrypsin							
B chain	0.6	−0.7	0.0	0.4	0.8	0.9	0.3
C chain	0.4	−0.5	0.1	0.0	0.3	0.6	0.7
Subtilisin	—	—	—	—	0.5	0.5	0.4
Carboxypeptidase	—	—	—	—	—	—	—
Staphylococcal nuclease	—	—	—	—	—	0.5	0.5
Cytochrome *c*	0.6	—	—	0.1	—	0.4	0.2
Carp myogen	—	—	—	—	—	—	—
Cytochrome *b*	—	—	—	—	—	—	—
Rubredoxin	—	—	—	—	—	—	—
Papain	—	—	—	—	—	—	—
Concanavalin A							

[a]According to Burgess *et al.* (1974).

[b]The index of prediction P is evaluated for the different ordered structures: (a) calculated from both

the number of possible states for each residue; and P is equal to -1 when no states are predicted correctly. If all states are correctly assigned, then P is equal to 1. If the fraction of residues assigned is the same as that expected from a random distribution of states (i.e., $1/N_s$ for a N_s states model), then P is null. The value of predictive methods based on a consideration of the frequency of occurrence of each type of amino acid in a given conformational state is a question of statistics and depends on the number of proteins used for statistical evaluation. Table 4.1 according to Burgess and co-workers (1974) gives the values of the prediction index evaluated for the different analyses. It is an indication of the reliability of the procedure employed by the investigators. Predictive methods are validated by the results and therefore must be compared with the experimental data. The results of different predictive methods applied to the determination of secondary structure of adenylate kinase according to Schulz and co-workers (1974) are presented in Fig. 4.8. Adenylate kinase, a small protein of MW 21,700, was used by Schulz and co-workers as a test object to compare the results of different prediction methods with the experimentally determined (by X-ray crystal-

4.1

Prediction of Protein Secondary Structure[a]

Values of prediction index P^b

				Bends			Extended or β structure		Four state
Robson *et al.* (1971)	Robson *et al.* (1971)	Ptitsyn *et al.* (1970)	Burgess *et al.* (1974)	Lewis *et al.* (1971)	Kuntz (1972)	Burgess *et al.* (1974)	Ptitsyn *et al.* (1970)	Burgess *et al.* (1974)	Burgess *et al.* (1974)
0.7	0.5	0.5	0.6	0.5	0.1	0.6	0.4	0.6	0.3
0.9	0.6	0.9	0.6	0.3	0.4	0.4	—	0.5	0.2
0.7	0.3	0.6	0.2	—	0.2	0.8	1.0	0.9	0.4
1.0	1.0	0.7	0.8	0.5	0.4	0.6	0.4	0.4	0.3
0.7	0.7	0.3	0.5	0.5	0.2	0.5	0.4	0.4	0.2
0.8	0.1	0.6	0.7	0.0	—	0.5	0.5	0.5	0.2
—	—	—	0.4	0.2	—	0.4	—	0.4	0.1
—	—	—	0.5	0.4	—	0.6	—	0.5	0.3
—	—	—	0.5	0.4	—	0.5	—	1.0	0.4
—	—	—	0.3	—	—	0.6	—	0.9	0.2
—	—	—	0.4	—	—	0.5	—	0.8	0.4
—	—	—	0.8	—	—	0.6	—	0.2	0.1
—	—	—	0.7	—	—	—	—	0.4	—
			0.8	—	—	—	—	0.2	—

single and double residue information; (b) calculated only from single residue information.

lography) secondary structure of a protein. Information obtained by adding prediction of different methods was also reported and the histograms are shown in Fig. 4.8. These histograms allow one to assign most of the secondary structure of the protein; however, it must be remarked that adenylate kinase is a protein which contains a great number of secondary structures and thus, such a prediction for other kinds of proteins might be less consistent. The molecule contains one domain composed of three helices (between residues 41 and 84). The consistent predictions provided by most of the methods indicate that secondary structures, at least in this region, are dominated by short-range interactions. Similar estimations have been made for phage T_4 lysozyme (Matthews, 1975). However these comparisons are based on data obtained with a single protein.

Lenstra (1977) used data from 33 different proteins to compare four different methods of prediction, those of Argos and co-workers (1976), the stereochemical method of Lim (1974a,b) which is based on the relative positions of hydrophobic and hydrophilic residues, the statistical–mechanical method of Tanaka and Scheraga (1976a,b,c,d), and the statistical method of Nagano

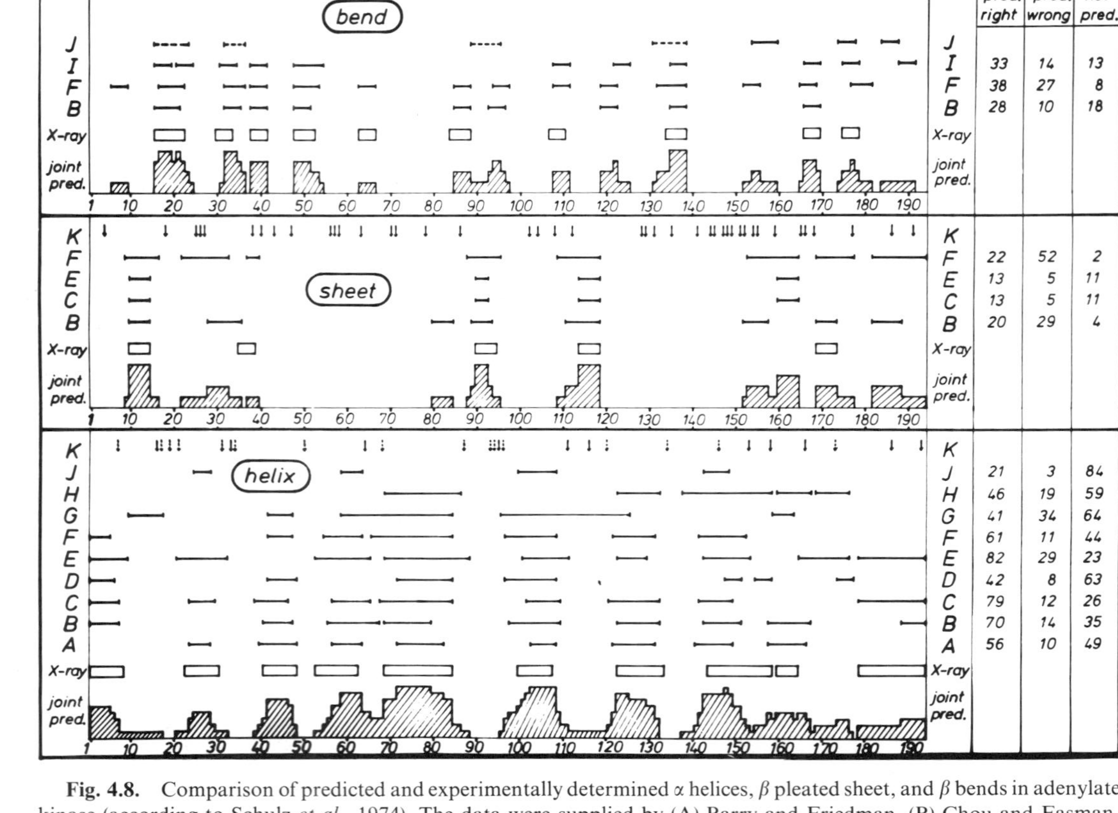

Fig. 4.8. Comparison of predicted and experimentally determined α helices, β pleated sheet, and β bends in adenylate kinase (according to Schulz *et al.*, 1974). The data were supplied by (A) Barry and Friedman. (B) Chou and Fasman. (C) Finkelstein and Ptitsyn. (D) Levitt and Robson. (E) Lim. (F) Nagano. (G–J) by Burgess and Scheraga. Data obtained by different procedures. (K) Kabat and Wu (courtesy of Schulz).

(1974). The best prediction of β structure was obtained by the method of Nagano (1974). Helix predictions were not significantly different by the method of Argos and co-workers (1976) and the method of Nagano (1974). Both methods were more accurate than the prediction according to Lim (1974a,b). The method of Tanaka and Scheraga (1976a,b,c,d) which is based on the statistical weights of individual amino acids yields the least reliable results. The success obtained by the method of Nagano (1974) is attributable to the consideration of the influence of the residues at sites $(i + 1)$ and $(i + 4)$ in a helix and at sites $(i + 2)$ and $(i + 4)$ in a β structure. That emphasizes the importance of interactions between these pairs of residues in determining the structure. Predictive methods based only on short-range interactions cannot provide 100% reliability even when statistical weights are optimized because medium-range and maybe long-range interactions are omitted. It must be noted that similar sequences, at least short sequences, may lead to entirely different conformations, for example the sequence 27–32 in trypsin is identical to the sequence 3–9 in subtilisin BPN′ (Val–Pro–Tyr–Glu–Val–Ser), but their conformations are entirely different.

The importance of having very reliable predictive methods becomes more evident when one considers that the results may be used in the simulation of the folding of the polypeptide chain. For this purpose, methods of prediction are continuously improved.

4.5. PREDICTION OF SUPERSECONDARY STRUCTURES

The occurrence of supersecondary structures with common patterns in proteins has suggested the existence of rules for the assembly of secondary structures (Chothia, 1973; Chothia *et al.*, 1977; Levitt and Chothia, 1976; Nagano, 1977a,b; Sternberg and Thornton, 1976, 1977a,b; Richardson, 1977; Finkelstein, 1977; Ptitsyn and Finkelstein, 1978). Most of the structural characteristics of the different folding units, formed from interactions between segments of secondary structures, have already been presented in Chapter 2. Their occurrence in proteins has been examined there. In the present section consideration is limited to the attempts to find predictive methods for the formation of supersecondary structures. Such methods when available with a good reliability may very well represent a new step that will prove very useful in simulation of protein folding.

Sternberg and Thornton (1976, 1977a,b, 1978) Cohen and co-workers (1980a,b), Cohen and Sternberg (1980), Cohen *et al.* (1981) have attempted to rationalize several empirical observations about the structure of the β sheets and of the different folding units (e.g., β-strand–α-helix–β–strand, β-strand–β-sheet–β-strand, β-strand–coil–β-strand). These investigators have evaluated the probability of right-handedness of these different connections. They have determined some rules concerning their relative orientations and the proximity of the strands, their length, the order of their arrangement. The data still present some serious limitations (for example no account has been taken of the influence of disulfide bridges on the connectivities) to the determination of accurate parameters for predicting protein conformation. However, they can be used to test the feasibility of a structure generated by computation procedures as used by Levitt and Warshel (1975) or by Tanaka and Scheraga (1975). The typical ($\beta\alpha\beta$) unit was analyzed by Nagano (1977a,b) on the basis of statistical predictibility. Nagano suggested that the requirement for an adjacent ($\beta\alpha\beta$) unit is "a well balanced and fairly strong potential of the respective β, α, and β segments." Some predictive rules were proposed and tentatively applied to 36 proteins. It was used for prediction of tertiary structure and applied to α/β proteins (Nagano, 1980). An important property that determines supersecondary structures is the right-handedness of these folds. This problem was reinvestigated by different authors and explanations were proposed to explain the stability of such supersecondary structures. The handedness of supersecondary structure, as for secondary structures, seems to be caused by the chirality of naturally occurring amino acids. Possible implications for protein folding were deduced; it was suggested that the right-handedness of the ($\beta\alpha\beta$) units can determine the relative positions of substructures and the size of domains in proteins. The problem of the handedness of supersecondary structures has already been discussed in detail in Chapter 2, Section 2.5.

Tentative rules for the prediction of the packing of helices have been proposed and applied to predict the structure of myoglobin (Ptitsyn and Rashin, 1975; Richmond and Richards, 1978; Cohen *et al*; 1979; Richards *et al.*, 1980).

As previously reported (Chapter 2, Section 2.5), Rackovsky and Scheraga (1978) have examined different types of model molecules containing a coil section and one or two sections of ordered backbone structure to locate their low-energy conformations in water. They have suggested, from their data, that the folding of proteins which contain preformed sections of ordered structure, probably proceeds through interactions between ordered and coil segments rather than between two ordered segments. However in this study the nature of individual amino acid side chains was not considered. A very interesting approach was used by Lifson and Sander (1979, 1980a,b). These researchers tried to detect specificity of recognition between pairs of amino acid residues in β sheets by determining residue–residue pair frequencies in the β sheets of 30 proteins of known three-dimensional structure. They suggested that strongly correlated pairs [such as Ser–Thr and Ala–Ile in β_A (antiparallel), Ile–Leu in β_P (parallel)] play a special role in the formation of β sheets. The interest of this approach is the introduction of specificity in the formation of folding units and then in the folding process, and the research of the factors which are involved in the specificity of recognition.

The analysis of supersecondary structures is one of the most interesting and promising approaches for the near future. It could provide quantitative data allowing the definition of folding parameters in a way similar to those obtained for the prediction of secondary structures. Thus, secondary structures may be reasonably predicted from the sequence of a polypeptide chain, with a rather good reliability when medium-range interactions are taken into account. When quantitative rules allow a reliable prediction of a higher level of folding (i.e., formation of supersecondary structures which include long-range interactions) a new step is reached in the prediction of tertiary structure of proteins from a given polypeptide sequence. Simulation of protein folding may be obtained by a stepwise process, including the formation of secondary structures, then the formation of supersecondary structures, and finally the assembly of these pieces. However, most of the studies that try to determine the rules of prediction for supersecondary structures are simply frequency analyses and have not yet reached the determination of rules of prediction. Serious limitations still remain. For example, the set of data is not large enough to allow good statistics.

In the near future further progress might reasonably be expected from this approach which started in the late 1970s. When this new and more complex level of structure becomes accessible to predictive computations, it could be an important step in simulating the formation of the overall conformation of a protein.

4.6. SIMULATION OF PROTEIN FOLDING

Some models including long-range interactions have been proposed for simulating protein folding. The evaluation of long-range interactions requires a simplified description of the polypeptide chain and rapid methods for computing the interaction energy. Different methods of simulation for the folding of a polypeptide chain have been reported. The Russian group used a simplified mechanical model-building approach. Energy minimization methods have been employed by several investigators, and Monte Carlo procedures have been introduced for simulation of protein folding.

4.6.1. Approaches Using Mechanical Models

The methods essentially developed by the Russian group used manual folding of simplified mechanical models. Ptitsyn and Rashin (1975) have applied this approach for the self-organization of the apomyoglobin molecule. In the mechanical model, helices were simply represented by cylinders.

Ptitsyn and Rashin proposed a scheme for the folding pathway of an apomyoglobin molecule which consists of a sequential process of several steps: (1) the formation of the α helices found in native myoglobin; (2) then two or three helices which are neighbors along the polypeptide sequence fold against each other and form centers of crystallization; (3) the attachment of neighboring helices provokes the growth of these centers, and (4) several centers fold against each other forming the tight packing of the molecule. They have even assigned semiquantitative indices to the various intermediary and final structural states of the molecule by evaluating in each structure the number of nonpolar amino acid residues of the helices which are shielded from the solvent. This procedure to obtain the native structure of a protein which represents one of the two most favorable final structures for apomyoglobin by a simple sequence of four steps, has however some important inherent limitations. It applies only to proteins that have a high content of helices and in this method, once regions folded, they cannot unfold. Furthermore, this method which represents an interesting attempt, has not been proven.

More recently, the Russian group proposed a slightly different model of protein folding (Lim, 1978), in which the first stage of protein folding is assumed to be the formation of α helices, then in the second stage the packing of these *s* helices into a globule, and at the third stage, some intramolecular rearrangements occur with the transformation of helical structures into β structures or irregular ones. Thus, in this model, the final secondary structure takes place after formation of the tertiary one. Ptitsyn and Finkelstein (1978) proposed an alternative pathway for all β proteins such that the β structure is formed at the earlier stage of folding. In fact, they consider folding pathways to be dependent on the structural classes of proteins as defined by Levitt and Chothia (1976). In their model, all α protein folding starts by the formation of a long initiating helical hairpin, whereas β hairpin, resulting from two long β segments adjacent along the chain, initiates formation of β structures. (These pathways of folding obtained by manual simulation with mechanical models are also discussed in Chapter 2, Section 2.5.)

4.6.2. Approaches Using Energy Minimization Methods

Levinthal (1966) and Katz and Levinthal (1972) described protein molecules in a united atom approximation and used a molecular modeling system in conjunction with an interactive computer graphic program. More recently (Hönig *et al.*, 1976), the same group developed a procedure based on the same principles, but with some modification to minimize rapidly the energy of rigid polypeptide segments of a protein connected by flexible joints, such small variations in the orientation of the joints can change the coordinates significantly. Calculations were used to analyze subtilisin BPN. Eight bend regions of the molecule were tested successively. Three were very rigid, but the other five were flexible and allowed movement of the structural fragments. The results of the computation indicated the possibility of predicting the formation of tertiary structure from smaller prefolded fragments, and suggested the preferred pathway for the last steps of protein folding, (i.e., the assembly of structured fragments directed by long-range interactions). This kind of approach analyzed only the last steps of protein folding.

Levitt and Warshel (1975) and Levitt (1976) developed another type of approach. To reduce the number of variables and the number of minima in the conformational energy hyperspace they introduced a simplified representation of the polypeptide chain. Residues are represented by spheres, C^{α} atoms are linked by virtual bonds, and the torsion angle α around each virtual bond replaces the ϕ and ψ angles as the only degree of freedom of the backbone. (This representation is discussed in Section 4.3.1.1. and shown in Figs. 4.5 and 4.6.) The energy function is the sum of four contributions, van der Waals interactions, side-chain–solvent interactions, peptide–hydrogen bonds interactions, and torsional potential as a function of α. Energy was minimized and when a minimum was reached the conformation was perturbed by a method termed normal mode thermalization that allowed motion to another minimum. This method is in fact a simulation of thermal motion. The procedure, under certain conditions, yielded refolded bovine pancreatic trypsin inhibitor (BPTI) into a conformation similar to that of the native one (Fig. 4.9). Two interesting aspects of the method must be emphasized: a simplified representation of the polypeptide chain geometry can be used to fold a protein, and normal mode thermalization seems to be a useful method. However, the procedure employed by Levitt (1976) was seriously criticized by Némethy and Scheraga (1977) and by Hagler and Hönig (1978), this procedure which uses arbitrary pulling and pushing functions seems ambiguous. Indeed the main criticism concerns the torsional potential function used which favors the extended chain for most residues except Gly, Asp, Asn. For these three amino acids the favored structure is the reverse turn. Since Levitt (1976) has assigned a glycine-like potential to all residues which occur in turn in BPTI, and started from an extended structure having a helix at one extremity, the correct structure is already given in the starting conformation. Hagler and Hönig (1978) obtained a structure similar to that of BPTI by using a sequence of only alanine and glycine and placing the glycine in the same position as the

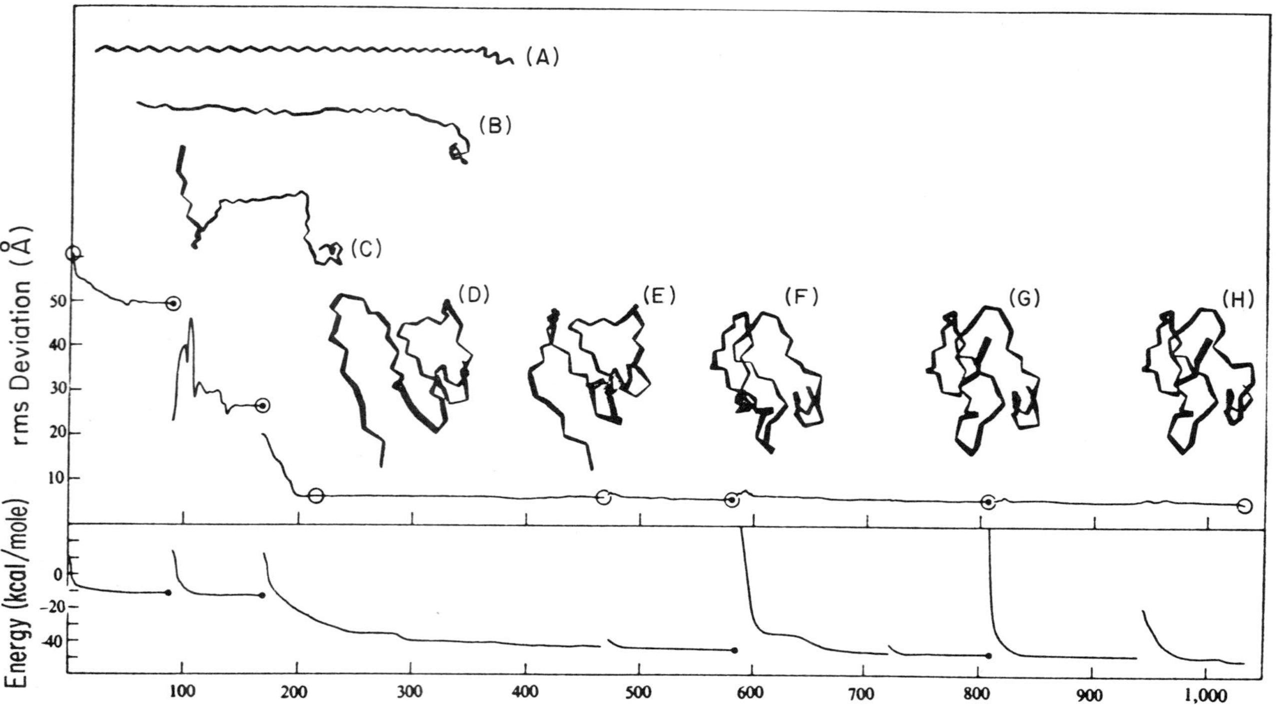

Fig. 4.9. Simulation of the folding of BPTI from the extended polypeptide chain using a simplified representation of the protein according to Levitt (1976). The conformation was perturbed by normal mode thermalization after each minimization (see text). Upper drawing represents, in ribbon diagram, the conformations obtained as a function of the number of cycles of computation; lower curves give the associated energies. The last five ribbon diagrams (upper drawing) correspond to the conformations which have progressively lower energies, each being a little closer to that of the native protein ($E = -43.4, -45.7, -46.0, -46.9, -48.9$).

glycine-like residues in the study of Levitt and Warshel (1975). They obtained a conformation which had 6.2 Å rms deviation from native BPTI. Hagler and Hönig (1978) concluded that the similarity with the native structure of BPTI obtained by Levitt and Warshell (1975) is mainly due to specific features of the protein that were included in the model. Several of the approximations and also the problem of reproducibility were argued by Robson (1975). The author also emphasized that a protein which contains a great part of helix cannot be correctly predicted by this method which underestimates the stability of helices.

Ponnuswamy and co-workers (1973) and Burgess and Scheraga (1975b) in a previous approach introduced medium-range interactions to predict the conformation of a protein and computations were made by minimizing the energies of a nonapeptide. The minimization was performed with respect to the central residue of the nonapeptide. Then an overlapping nonapeptide was selected with a shift of one residue toward the C-terminal. Energy minimization was performed for all the nonapeptide segments of the chain. The procedure is shown schematically in Fig. 4.10. As pointed out by these researchers, the lack of correlation between calculated and observed conformation states of the side chain indicates their dependency on long-range interactions (Fig. 4.11). Each single amino acid may exist with different side chain conformations which differ little in energy (Zimmerman *et al.*, 1977). The preference for a given conformation is dictated by long-range interactions.

4.6.3. Approaches Using Monte Carlo Procedure

To introduce long-range interactions into the search of folded conformation of BPTI, Tanaka and Scheraga (1975) used a Monte Carlo procedure. The procedure described the folding of a globular protein in a given medium by a three-step mechanism, all three of which may proceed simultaneously. In step A, ordered backbone structures, α-helical, extended, and chain reversal conformations are formed; the system is at equilibrium above the denaturation temperature. In step B the equilibrium is shifted by changing physical conditions (temperature, solvent) and small contact regions arising from long-range interactions between residues which approach each other are nucleated. These contact regions may be formed from interactions between amino acid residues both in ordered and in unordered structures. They are stabilized by solvent. The ordered structures formed in step A may be rearranged. Various types of random conformational deformations are generated in step B, local deformations of the chain and important

Step	1	2	3	4	5	6	7	8	9	10	11	12	13 ....
1	β	β	α	α	α	α	α	β	β				
2	β	β	α	α	$\phi = -52°$ $\psi = -48°$	α	α	β	β				
3		β	α	α	$\phi = -52°$ $\psi = -48°$	α	α	β	β	β			
4		β	α	α	$\phi = -52°$ $\psi = -48°$	$\phi = -51°$ $\psi = -49°$	α	β	β	β			
5			α	α	$\phi = -52°$ $\psi = -48°$	$\phi = -51°$ $\psi = -49°$	α	β	β	β	β		
6			α	α	$\phi = -52°$ $\psi = -48°$	$\phi = -51°$ $\psi = -49°$	$\phi = -52°$ $\psi = -49°$	β	β	β	β		

Fig. 4.10. Schematic illustration of successive steps in the computation of dihedral angles of a protein by minimizing energies of a nonapeptide (according to Scheraga, 1974). In step 1, one conformational state is assigned to each residue of the nonapeptide inside the protein. In step 2, minimization of the energy is carried out with respect to the dihedral angles of residue 5. In step 3 a shift of one residue toward the C-terminal is made and a conformational state is assigned to residue 10; in step 4, energy is minimized with respect to dihedral angles of residue 6, etc.

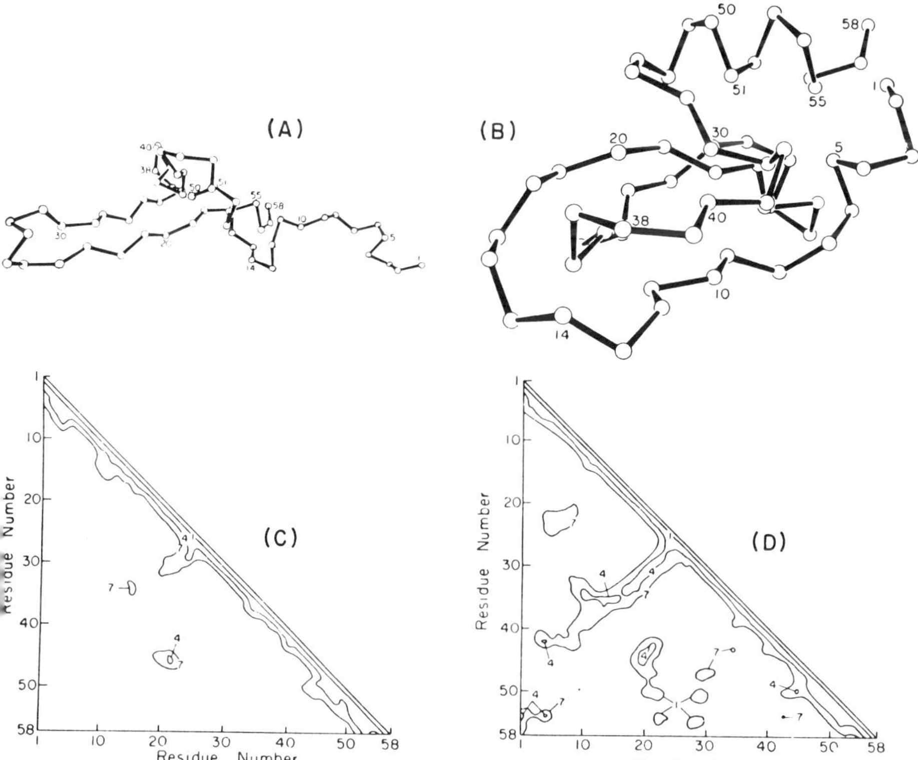

Fig. 4.11. (A) Spatial structure of BPTI computed by the procedure described in Fig. 4.10 taking into account short- and medium-range interactions (according to Burgess and Scheraga, 1975b). (B) Spatial structure as determined by X-ray crystallography. (C and D) Distance contour maps between C^{α} related to (A) and (B) structures, respectively (according to Ooi and Nishikawai 1973).

changes in the overall conformation; only the first ones are effective. Conformational changes are made by randomly selecting the number and position of residues to vary. In step C, the small contact regions associate with possible rearrangements of the intermediate structures formed in A and in B. In the Monte Carlo simulation used by Tanaka and Scheraga, chain conformations were generated randomly until one is obtained with no hard sphere overlaps between atoms. Then the free energy of all pair contacts is computed. The conformational change can give a conformation of larger free energy, therefore unfavorable and it is discarded. Contrarely, it can generate a conformation of lower energy which is then retained as starting conformation for further changes.

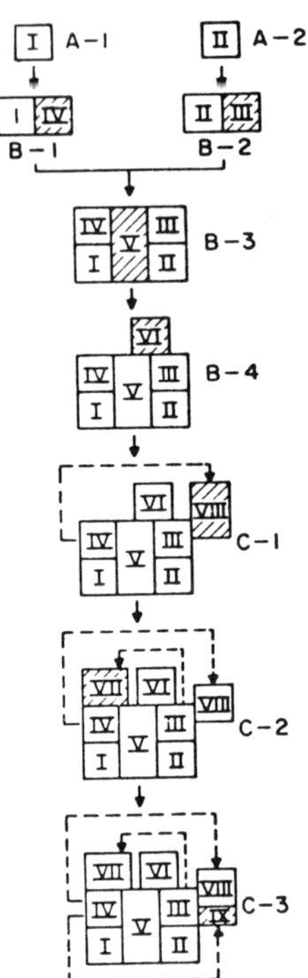

Fig. 4.12. Folding of rubredoxin (from Tanaka and Scheraga, 1977) (see text).

This method was applied to BPTI. The final structure obtained in step C had many conformational features analogous to that of the native protein. This procedure may be regarded as a three-dimensional model in which residues are allowed to move from one conformational state to another. In this respect it differs from the other long-range interaction models. More recently, the authors verified that the actual structures of some known proteins are consistent with this mechanism of folding; it was checked for lysozyme, rubredoxin, and ferricytochrome *c* (Tanaka and Scheraga, 1977). As an example, Fig. 4.12 shows the postulated mechanism of folding of rubredoxin. In this representation, the folding takes place by association of contact regions. At each stage, a new one (shadowed square) takes place. In step A, the helical and extended regions are formed; their formation is governed by short- and medium-range (below ± 4 residues) interactions. Furthermore contact region I is formed through nucleation of helical segments 15–17 which involves contact residues 13–19. Region II is formed through nucleation, by the helical segment 30–32 making contact between residues 28–34. In step B, ordered structures and contact regions formed in step A associate. Contact regions III, IV, and V are formed by association of other regions. Region IV is formed by association of region III and V. Several chain reversals are formed at various stages of step B. In step C, further associations are formed governed by long-range interactions (beyond 20 residues). Contact regions VIII, VII, and IX take place. Thus the native conformation of the protein is obtained by a well defined sequence of steps of folding.

A similar pathway was deduced by Némethy and Scheraga (1979) for ribonuclease, not from Monte Carlo simulation, but rather from consideration of the contact map. The obtained pathway agreed well with experimental information.

A two-dimensional very simplified lattice model of proteins was used by Gō and co-workers (Taketomi *et al.*, 1975; Gō, 1976; Gō and Taketomi, 1978, 1979a,b, Gō *et al.*, 1980). The time course of conformational changes in the lattice was followed by Monte Carlo simulation. Results of simulation were compared for different relative weights of short- and long-range interactions (Gō and Taketomi, 1978). The investigators concluded that long-range interactions are essential for very cooperative stabilization of the native proteins. However folding and unfolding transitions are accelerated by short-range interactions. A three-dimensional lattice model of lysozyme was introduced and studied by the Monte Carlo simulation method. With this model complete refolding from the completely unfolded state was not possible to observe; native conformation could be attained only when starting from partially folded structures (Ueda et al., 1978).

These simulations of protein folding are based on different models. They all involve different stages in the folding pathway. Some of them accept that the substructures formed at each stage persist. Others assume more or less drastic conformational rearrangements during the course of the folding.

4.6.4. Diffusion–Collision Model of Protein Folding

Another interesting, but quite different approach, was reported by Karplus and Weaver (1976) who proposed an explanation of the dynamics of protein folding by a diffusion–collision model. The model considers the protein to be divided into several microdomains; each of them is short enough that it is able to move rapidly through all possible conformations, but its native secondary structure is not stable enough in itself because of the random fluctuations. Two or several structured microdomains have to diffuse and collide allowing the formation of a stable coalescence, this coalescence only occurs for the units which have the native structure when they collide. The time, τ, required for such a coalescence is given by the following expression:

$$\tau = (1/\beta)(l\Delta V/DA)$$

corresponding to radial diffusion of spherical units, where ΔV is the volume of the diffusion space (the space between two concentric shells); A, the area of the inner shell; l, the length which is the average of the inner and outer shell radii, this last one corresponding to the maximum distance of the two units; D, the diffusion coefficient; and β accounts for the fact that only a fraction of each microdomain has the native structure. It can be evaluated by using an equilibrium model:

$$\beta = K_1K_2/(1 + K_1)(1 + K_2)$$

K_1 and K_2 being the (native/unfolded) equilibrium constants for each of the two microdomains.

Using $D \cong 1.10^{-6}$ cm^2/sec and $A = 20$ Å, Karplus and Weaver evaluated $\tau \cong \beta^{-1}$ ($10^{-7} - 10^{-8}$ sec). Assuming K_1 and K_2 equal to 10^{-3}, 10^{-4} respectively, which is the order of magnitude for helix nucleation, the folding time of a pair of units is of the order of 0.01–10 sec. The parameter β, which determines the fraction of the time that the secondary structure exists, is very important in the effective time of diffusion–collision mechanism. If several units have to collide together at the same time to form the stable entity, each with a K_1 value less than 10^{-3}, the folding time would be very long. A stepwise process seems more likely: the collision of two microdomains to form a stable entity which in turn coalesces with a third, etc. The rate of folding is thus expected to increase as the folding advances, as for a cooperative transition.

According to the authors, the occurrence of incorrectly folded intermediates, as detected for BPTI (Creighton, 1974b,c), might be expected if the diffusion–collision mechanism plays an important role. Incorrectly folded species could occur from segments in a structure differing from the native one which collide and form relatively unstable intermediates because of the illicit interactions.

4.7. ANALYSIS OF PREDICTION AND SIMULATION RESULTS

Criteria have been proposed to define the reliability of simulation results as well as to evaluate the validity of prediction methods. The estimation of rms deviations between the native and the simulated structure of a protein is not a sufficient criterium for such an evaluation. It provides a single number for the goodness of fit of many pieces of numerical information which describes the conformation. Different methods may be used to determine the success of simulation procedure; they allow one to compare either the overall conformation, or local structures, or both. The method used by

Rossmann and Argos (1976, 1977) to compare the folding of the three-dimensional structure of proteins can be used to estimate the validity of simulation. It allows for a comparison of the overall conformation by a method of superposition of residues and structural elements. Its general principle and the main concepts are reported in Chapter 2, Section 2, since this method was used to detect the evolutionary relationship between proteins from their three-dimensional structure.

For comparison of local structures, Rackovsky and Scheraga (1978) used differential geometry.

The best method to estimate the validity of a simulation is the construction of C^α—C^α distance map. It allows one to analyze local structures as well as overall conformation. However this method of analysis is rather crude because it gives for each pair of residue one all-or-none answer depending on the allowed C^α—C^α distance. Crippen and Kuntz (1977) have introduced a novel representation of the backbone conformation referred to as direction matrices. In such a representation, a matrix is calculated from C^α atom cartesian X-ray coordinates where the *ij* elements of the matrix is the cosine of the angle between the direction of the chain at residue *i* and the residue of the chain at residue *j*. This representation gives distinctive pattern for the most important structural features.

Despite the limitation of the contact map, it is always necessary to use such a representation to estimate the validity of a simulation.

4.8. CONFORMATIONAL FLUCTUATIONS OF PROTEINS

Crystallographic studies provide a rather rigid view of protein conformation which is presented as a well defined compactly packed structure. Certainly some degrees of freedom which actually exist in protein in solution are partly frozen in the crystal. Nevertheless, even crystallographic data indirectly provide information about flexibility of proteins. In some cases conformational changes as a result of ligand binding are revealed by X ray crystallography. Hemoglobin and carboxypeptidase are classic examples. As already mentioned in Chapter 1, the recent refinements of crystallographic analysis have provided detailed information about the thermal motion and intramolecular vibrations. Small motions have been observed in hen egg white and human lysozymes (Sternberg *et al.*, 1979; Artymiuk *et al.*, 1979). The greatest displacements were observed in the part of the active site cleft that undergoes conformational changes induced by ligand binding. Motions of a similar magnitude were reported and analyzed in crystalline myoglobin (Frauenfelder *et al.*, 1979). Other experimental data and also theoretical considerations indicate that there is considerable motion in proteins at ordinary temperature. Moreover, the fluidity of a protein which allows displacements of residues is at the origin of the activity of proteins; catalytic activity as well as allosteric properties. The problem of motion and conformational fluctuations in proteins was discussed by Careri (1974), Careri and co-workers (1975, 1979), Weber (1973a,b, 1975), and Gurd and Rothgeb (1979). The fluctuations on different time scales were analyzed. Sizeable conformational fluctuations in proteins may cover a time range from nanoseconds or less to minutes or even hours. These fluctuations can originate enzymatic activity for instance which might arise from some active fluctuations as was proposed by Careri (1974). During this event which may last only a very short time, the increase of free energy at the active site is available for the chemical reaction. According to the concept of fluctuating enzyme developed by Careri (1974), this active fluctuation is the bottleneck for catalysis. Moreover, Careri (1974) suggested that the active conformational fluctuation corresponds to the part of the frequency spectrum around 10^{-8} sec. Some other oscillations recognized as breathing motions correspond to low frequency. Vibrational frequency centered about 30 cm^{-1} was observed by Raman laser spectroscopy in chymotrypsin (Brown *et al.*, 1972). All the experimental evidence for conformational fluctuations of native proteins is presented in Part II and a special discussion of this aspect and its biological meaning is introduced in Chapter 12 of this volume.

Cooper (1976) reported an application of the theory of thermodynamic fluctuations to proteins. Other approaches were proposed to quantify the flexibility of proteins (Gelin and Karplus, 1975; Karplus and Weaver, 1976; McCammon *et al.*, 1977; see also Robson, 1977; Karplus and McCammon, 1980).

McCammon and co-workers (1977) studied the molecular dynamics of a folded globular protein, bovine pancreatic trypsin inhibitor. The theoretical equations of motion of all the atoms of the molecule were solved with empirical energy functions. The method allowed them to investigate the motions of a protein in the neighborhood of its equilibrium configuration. Bovine pancreatic trypsin inhibitor was selected because it contains only 58 amino acid residues. The classical equations of motion for all the atoms of the molecule were solved simultaneously for a chosen time period and the resulting atomic trajectories were analyzed. The potential energy function was an empirical function represented by the sum of terms associated with bond length, bond angles, dihedral angles, hydrogen bonds, and nonbonded interactions. Iteration of the equation of motion was performed with time steps of 9.78×10^{-16} sec. The time averaged structure thus obtained was near the X-ray structure, but not identical to it. The larger variations are located at the two ends of the molecule and at the external loop (residues

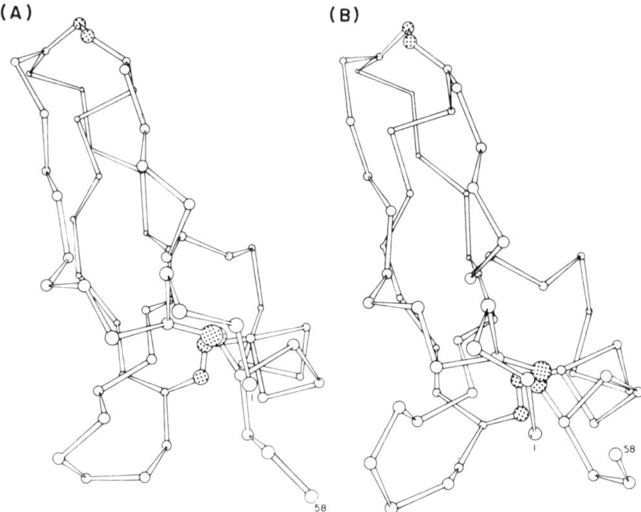

Fig. 4.13. (A) The peptide backbone and disulfide bonds of BPTI as determined by X-ray crystallography. (B) Time-evolved structure after 3.2 psec of dynamic simulation (according to McCammon *et al.*, 1977).

25–28). These parts also have particularly large fluctuations. Figure 4.13 shown time evolution of a structure after 3.2 psec of dynamic simulation. It is clear that the three residues at the C-terminal end have important fluctuations. The same is observed for residues 25–28 of the loop connecting the two strands of a β sheet, indicating an intrinsic softness of this region. The results suggest that the protein interior is fluidlike where internal motion has a diffusional character governed by collisions between neighboring nonbonded atoms. Moreover, McCammon and co-workers (1977) emphasized the occurrence, at the atomic level at ordinary temperatures, of "a rich variety of motional phenomena."

This method has been recently applied to a larger molecule, ferrocytochrome *c* which has 104 amino acids (Northrup *et al.*, 1980).

4.9. COMMENTS ON THEORETICAL APPROACHES TO PROTEIN FOLDING

The theoretical studies have provided a large body of information. It is worthwhile to examine the extent to which they have helped the understanding of protein folding. This field has progressed tremendously these recent years, as a result of the increasing data set for three-dimensional structure of proteins and the development of refinement methods. This progress which even allows one to account for long-range interactions, has not reached the level which could permit one to predict unambiguously the tertiary structure of a protein from the one-dimensional amino acid sequence of the polypeptide chain. The final structure obtained by simulation of protein folding is generally not the native one. The failure to obtain the native structure by simulation of protein folding (and that in a relatively long time compared to the time needed for the biosynthesis), indicates that researchers are still far from reaching the goal of a correct prediction.

Nevertheless, conformational computations have produced important pieces of information. They have indicated the importance of the different energy contributions in determining protein folding. If they have not proposed a definite pathway of protein folding, they have at least suggested some plausible steps. In fact, different models are proposed in computations and nothing for the moment allows one to make a decision between them.

The initial stage of the pathway of folding, the nucleation of some limited regions of the polypeptide chain was suggested and supported mainly by theoretical studies. The most significant analysis was probably reported by Tanaka and Scheraga (1975). Agreement on the nature of the nucleus is not yet realized. Several researchers have proposed ordered structures and even more, helical structures only, as possible candidates for nucleation centers. Finkelstein and Ptitsyn (1976) concluded that only α helices can initiate further folding of a polypeptide chain because they preexist in the unfolded protein; moreover, they are formed very rapidly compared to the other structures. The formation time was estimated to be as small as 10^{-8} sec for growth of helices (Zana, 1975). Therefore, they can be formed at the early stages of protein folding. Formation of β structures requires at least 10^{-2} sec. Finkelstein and Ptitsyn (1976) suggested that β bends cannot serve as initiators for the tertiary structure, but rather that they are formed at later stages of self-organization. However, in 1978, they proposed for all β proteins that β structures can be formed at the first stage of folding, since they are the only ones capable of competing energetically with the α helix to a certain extent. Anfinsen and Scheraga (1975) had suggested that nucleation sites may be regular or nonregular, but specifically folded forms, as postulated for ribonuclease. Matheson and Scheraga (1978) suggested they are formed by hydrophobic interactions. The role of hydrophobic interactions is certainly important in the early stages of protein folding.

It is generally accepted that the formation of ordered backbone structures is dominated by short- and medium-range interactions. It is supported by the rather good agreement between experimental data and predictions based only on intraresidue interactions. That strongly suggests the importance of these structures in early stages of protein folding before stabilization by long-range interactions. Furthermore these data indicate that pieces of secondary structures are persisting in the three-dimensional structure of proteins. They are not greatly modified during the structural refinements which are dominated by long-range interactions. However, as previously mentioned there is not a general agreement on this point. The Russian group, for example, assumes that final secondary structures are established through the formation of tertiary structure (i.e., dominated, rather than only stabilized by long-range interactions). Different pathways involving a sequential process of folding have been proposed. Further steps following the formation of ordered structures are determined by long-range interactions. In some models, the formation of secondary structure is followed by the formation of small contact regions which associate in a next step, association being driven by long-range interactions. A process involving (1) the interaction of structured fragments to generate some crystallization centers and then (2) the association of other structured fragments to give the compact packing of the molecule, is a common feature of several models. These models are compatible with the existence of domains which form independently and associate to give the native whole molecule.

In several models, rearrangements in the conformation of structured fragments and contact regions or crystallization centers may be induced by further interactions and association. In some others, fragments or regions, once folded, are unable to unfold or to undergo further conformational changes.

Theoretical studies associated with the crystallographic data might allow one to determine some rules for the association of structured fragments (i.e., for the formation of substructures such as supersecondary structures, and the assembly of these building blocks to form domains, and then the association of domains to form the globular compact structure of a protein).

Perhaps one of the most promising aspects is the prediction of supersecondary structures in proteins which will allow one to reach a new level of prediction in protein structure. A further step would be to determine the rules for the formation and then the association of domains in a protein.

The diffusion–collision model proposed by Karplus and Weaver (1976) as well as the approaches mentioned previously assumes a folding by steps, based on the formation of several microdomains which associate to form larger structures which on turn associate and so on. The formation of such microstructures would permit one to understand how nature had worked during the early stages of evolution for building a protein from elementary pieces with a given function and also what kind of assembly of these building blocks were necessary to differentiate.

These studies concerning flexibility of proteins are of great importance since the biological activity originates from internal motion. They probably mark the starting point of a new period of investigation: this aspect is detailed and discussed in Chapter 12.

We can briefly summarize the main points emphasized in this first part.

A. The precision of crystallographic data obtained on a great number of proteins allows one to outline several invariants in the organization of protein structure which are:

1. Proteins contain a great proportion of segments of ordered structure, α helices, β strands, and β turns, alternating with nonordered structure.

2. Polypeptide chains fold in a limited number of well-defined structural patterns.

3. Segments of ordered structure assemble to give supersecondary structures according to precise rules that are not yet very well elucidated; hydrophobic interactions play a fundamental role in this assembly and in the general packing of a protein.

4. Proteins are formed of building blocks at different levels of organization and complexity.

5. The atoms pack closely at the interior of proteins; however, there are cavities inside the molecules allowing motion.

6. Parts of the polypeptide chain are flexible and there are fluctuations in protein structure.

B. High resolution X-ray structure allows one to refine empirical methods for prediction of protein structure:

1. Segments of ordered structures are predicted with a satisfying accuracy.

2. Research of rules for predicting formation of supersecondary structure are now under study but do not yet allow prediction.

C. Various methods of simulation of protein folding are under study; however, none of them gives unambiguously the native structure.

II

Experimental Approaches

As emphasized in the introduction, experimental developments in protein folding historically reached three maxima located around 1935, 1955, and 1975. Each of them followed some important technical progress and consequently some fundamental advance in the knowledge of protein structure.

During the first period, it was accepted that denaturation is a monomolecular transformation of a protein molecule and can be a reversible process under well-defined conditions (Northrop, 1932; Anson and Mirsky, 1934a,b; Mirsky and Anson, 1935; Mirsky and Pauling, 1936). The idea was supported by a great number of studies. The origin of the thermodynamic hypothesis was the observation of the spontaneous refolding of proteins.

From these earlier studies, it was thought more or less explicitly that the formation of the tertiary structure of native proteins is a thermodynamically controlled process. The conformational studies which used synthetic polypeptides strongly supported this. As an alternative to the thermodynamic hypothesis, the kinetic control of protein folding appeared when Levinthal (1968a,b) on the basis of time consideration discarded the possibility of a random search of the most stable conformation even for a small protein. Now, these two hypotheses do not appear mutually exclusive: the final state can be under thermodynamic control and intermediary steps under kinetic control.

For several years attention was focused on another thermodynamic aspect: the-all-or-none theory whose quantitative formulation known as two-state theory is based on the cooperativity of the folded–unfolded transition. Many equilibrium experiments have been directed to verify the two-state model. In most studies, the two-state model in a first approximation reasonably accounted for the data.

The first kinetic studies carried out by slow temperature jump experiments (Pohl, 1968a,b, 1969, 1972a,b, 1976), followed by kinetic analyses reported by Baldwin and co-workers (see Baldwin, 1975) originated the interpretation of the folding process in terms of sequential steps of reactions. It was proposed that the first one, nucleation, is dominated by entropy. Progressively, it was accepted that protein folding is a multistep process and the identification of intermediates became an important goal in the study of the formation of protein structure.

The first studies were necessarily whole approaches, the lack of knowledge regarding the details of protein structure prevented researchers from obtaining quantitative data. It was only during the second period (marked by the determination of many protein sequences) and more from the third period (marked by the resolution of the three-dimensional structure) that the experimental studies of protein folding were based on the description of invariants and local organization of the structure. Experimental approaches were developed to examine the local situation at each point of a molecule during the refolding of the polypeptide chain. These approaches used various methods such as nuclear magnetic resonance, chemical, enzymatic, or immunochemical methods.

For a long time, the problem of protein folding was the field of interest of physical chemists and also protein chemists. Progressively with the increasing knowledge in the field of gene expression, the biological interest of protein folding was more emphasized and protein folding was considered as a critical posttranslational process.

Most of the main concepts regarding the mechanisms of protein folding accepted today originated from both theoretical conformational computations and the determination of structures at atomic resolution. The amount of experimental data is still insufficient to allow a high degree of generalization. The number of known proteins for which a detailed and complete study of the folding process is available, remains indeed very small. Well documented systems such as ribonuclease, staphylococcal nuclease, BPTI, lysozyme, serine proteases, and few other proteins are used frequently as examples in the discussion that follows.

5

Simulation of Protein Folding: Studies of *in Vitro* Denaturation–Renaturation

A quantitative study of protein folding under biological conditions (i.e., during the elongation of a polypeptide chain on the ribosome) was and still is a challenge. An operational, perhaps not perfectly adequate, model system which mimics the folding of a nascent polypeptide chain, is the refolding process after previous and complete unfolding of a protein. Certainly the conditions are very different; local pH, ionic strength, and ion concentrations on the ribosome are certainly determining conditions for directing the folding process. The *in vitro* refolding of a protein is a process very sensitive to these experimental conditions. However, the validity of the *in vitro* model may be questioned: do proteins fold in the test tube and cells via an analogous pathway? As stated by Tanford (1972), one can accept, at least as a working hypothesis, that the "two processes are sufficiently similar to allow consideration of the renaturation reaction as a model for the *in vivo* process." A necessary, but not sufficient, condition for using such a model is obviously the reversibility of the unfolding–folding process. The reversibility of the denaturation–renaturation under well-defined conditions has been demonstrated experimentally for at least three decades. More recent studies with well-known proteins, either with or without disulfide bridges, are more detailed, better controlled, lead to more significant interpretations.

Very early in the beginning of the studies on protein denaturation, the reversibility of the process was observed. The first proteins known to undergo the return to the native state were proteases such as trypsin and chymotrypsin and their corresponding zymogens (Northrop, 1932; Anson and Mirsky, 1934a,b; Mirsky and Anson, 1935; Eisenberg and Schwert, 1951). Very poor criteria were available for evaluating denaturation and renaturation. Aside from biological activity, the only ones available were solubility and ability to crystallize; both criteria are very crude probes of the overall conformation of a protein. Several results on the reversibility of denaturation were also obtained with hemoglobin (Anson, 1945; Steinhardt and Zaiser, 1951). In a review, Anson (1945) emphasized this aspect of reversible denaturation:

> "Hemoglobin which has been denatured in a variety of ways can be converted back into native protein which has the same solubility as the original protein, is crystallizable, has the same characteristic spectrum of native hemoglobin, can combine loosely with oxygen, has the same relative affinities for carbon monoxide and is not readily digested by trypsin."

After more than 30 years of additional study, it is interesting to note that some important and still valid ideas concerning protein folding are contained in Anson's (1945) review. Not only did he present the concept of reversibility, but also included the concept of the thermodynamic stability of the native structure, i.e., the return to the "same native structure being achieved whatever was the way of denaturation, and perhaps the denatured state." Later, many experiments were directed toward the demonstration of the reversibility of the denaturation process and the search for the optimal conditions for complete reversibility.

For many years, it was believed that reversibility of the denaturation process could be attributed to the presence of covalent disulfide bonds in proteins. It was proposed that only proteins stabilized by disulfide linkages are able to undergo reversible denaturation (Neurath and Bull, 1938). As long as the S—S bridges are not disrupted, the return to the native state may be expected; irreversibility occurs as a result of the disruption of disulfide bonds. Kauzmann (1955) used the same model to explain the denaturation of serum albumin by urea.

With the first experiments using ribonuclease that had been previously reduced in the presence of denaturing reagents, it was demonstrated that even when the disulfide bridges are broken a protein is able to refold spontaneously (Sela *et al.*, 1957; Harrington and Sela, 1959; White, 1961; Anfinsen *et al.*, 1961). The refolding of a protein represents the reaction of interest only if the product of denaturation has been shown to be completely unfolded (i.e., a random polypeptide chain without residual cross-linking covalent structures such as S—S bridges and without any structured segment). The characterization of the denatured state is a prerequisite for studying the refolding process. In earlier studies, there were no means by which to check whether a denatured protein was structureless. The only criteria available were measures such as loss of activity or decrease of solubility. These did not give sufficient information. Even in more recent studies, the examination of spectroscopic properties (ORD and CD spectra) is not always sufficient to determine easily and unambiguously if the protein is totally unfolded. Some other criteria such as the accessibility of side chain groups located at different places along the sequence, accessibility of particular peptide bonds to proteolytic enzymes, and rates of hydrogen deuterium (or tritium) exchange must be used to complement other data. The complete unfolding of proteins denatured by guanidine hydrochloride was reported by Tanford and co-workers (1966a, 1967), Tanford (1968), and Salahuddin and Tanford (1970). Nevertheless it must be demonstrated for each studied protein. The demonstration of the reversibility of the unfolding–folding process of a protein requires the knowledge of the native structure of the protein and a clear demonstration that by denaturation a completely unfolded randomly coiled polypeptide chain is obtained. The spontaneous refolding of different categories of proteins is detailed in following sections.

5.1. PROTEIN UNFOLDING INDUCED BY PHYSICOCHEMICAL PERTURBATIONS

Since the unfolding of proteins induced by physicochemical perturbations was extensively reviewed and analyzed by Tanford (1968) and Yon (1969), it is only briefly discussed in this section. Some aspects are more detailed as necessary.

5.1.1. Protein Unfolding Induced by Temperature

Thermal denaturation has been recognized for several years. The loss of enzymatic activity and solubility following the increase of temperature was described from the early period of denaturation studies. Thermal denaturation may be reversible or irreversible.

Thermodynamic parameters can be defined only for a reversible process. The midpoint corresponding to a thermally reversible folding–unfolding transition of a protein is defined by T_m, the temperature of transition, (i.e., the temperature at which the apparent transition constant K is equal to 1, and ΔG, the variation of free energy, is nul under conditions where a two-state approximation is valuable). Under these conditions T_m is equal to $\Delta H/\Delta S$ and ΔG is equal to $\Delta H(1 - T/T_m)$. The different parameters are defined in Chapter 6; however, it is noted here that Privalov and Khechinaschvili (1974) have shown that ΔH is a linear function of temperature for a number of proteins that undergo thermal denaturation in acidic conditions:

$$\Delta H = a + bT$$

with $b = \Delta C_P$, the change in heat capacity which accompanies the unfolding of the protein; therefore

$$\Delta H = \Delta H_m + \Delta C_p(T - T_m)$$
$$\Delta S = \Delta S_m + \Delta C_p \text{ In } (T/T_m)$$

ΔH_m and ΔS_m, being, respectively, the enthalpy and entropy variation at the temperature of transition. Transitions from the folded state to the completely unfolded state (random coil) cannot necessarily be reversed by a simple procedure. The conditions for reversibility may be restricted to a small range of the different external parameters (e.g., pH and ionic strength) and under slightly different conditions the process may be irreversible.

For proteins with disulfide bridges, denaturation is usually reversible at low pH because the reactivity of disulfide bonds and thiol is very low and therefore side reactions cannot take place. The reversible transition of ribonuclease, for example, is a highly cooperative process (Brandts and Hunt, 1967). The first goal of the study of the unfolding–folding process is the delineation of the conditions of reversibility. This is true for any procedure leading to denaturation. Irreversibility is frequently attributable to the formation of aggregates and is then accelerated by an increase in protein concentration. The formation of aggregates presumably results from intermolecular disulfide cross-links or intermolecular noncovalent associations with a probable predominance of hydrophobic interactions. Irreversibly denatured protein is not always formed from previously totally or partially unfolded molecules. Molecules without significant unfolding can form aggregates (Foster, 1960). This was also shown for myoglobin under certain conditions (Acampora and Hermans, 1967).

Thermal unfolding occurs at temperatures which vary greatly according to the protein. The temperature range in which a protein is stable depends on the balance between hydrogen bonds and hydrophobic interactions which stabilize the native structure. A relation between thermal stability and hydrophobicity index has been reported (Bull and Breese, 1973; Stellwagen and Wilgius, 1978). At low pH thermal unfolding occurs at lower temperatures than at neutral pH.

In some cases the products of thermal denaturation are not completely unfolded and retain structured regions. The addition of guanidine hydrochloride often induces another transition. This has been described for ribonuclease, chymotrypsinogen, and lysozyme (Tanford, 1968).

Taking into account the degree of cooperativity of thermal transition in ribonuclease and chymotrypsinogen, Aune and co-workers (1967) estimated that about 25% of the nativelike structure remains in thermally denatured proteins. Biltonen and co-workers (1965) have shown that chymotrypsinogen is not as extensively unfolded after thermal denaturation as it is after urea denaturation. This was also shown for ribonuclease (Brandts and Hunt, 1967). Many examples of thermally induced denaturation are examined in Chapters 6 and 7.

5.1.2. Protein Unfolding Induced by pH

Two kinds of effects must be considered: (1) global effects, resulting from a large variation in the electric charges, and (2) local effects induced by the ionization of residues involved in critical interactions.

5.1.2.1. Effects Induced by Variation of the Global Charge

Proteins are generally denatured at extreme pH values. The unfolding may be induced by a global effect resulting from the ionization of many side chain residues that for the most part are accessible to the solvent. At low as well as high pH, electrostatic repulsions attributable to the high density of charges can destabilize the native conformation and the swelling of the molecule can be followed by the unfolding of the protein.

The electrostatic effects alone are insufficient to account for the denaturation of proteins that occurs at extreme pH values. The alteration induced either by acid or by alkaline pH may be more or less drastic according to the proteins. For some proteins there are similarities between denaturation at high and low pH; for others, the alkaline denaturation is different from the acid one. Generally, proteins retain less residual structures when exposed to high pH values than to acid denaturation because buried residues, such as tyrosines, tend to become accessible to the solvent.

Schematically, at low pH

$$P_{\text{folded}} + n\text{H}^+ \longrightarrow P^*$$

P^* being an unstable form; at high pH

$$P_{\text{folded}} \longrightarrow P^* + n\text{H}^+$$

and in both cases

$$P^* \longrightarrow P_{\text{unfolded}}$$

An attempt to quantitate the effect of electrostatic repulsions on protein unfolding was proposed by Linderstrøm-Lang (1942). The rate of denaturation was expressed as a function of electrostatic repulsions:

$$k_z = k_0 \exp(-XZ^2) \qquad (5.1)$$

k_0 being the rate of denaturation when the global charge of the protein is null; Z is the global charge of the protein and X is given by the following equation:

$$X = w\,\Delta b/(w + \Delta b)$$

b is the radius of the protein considered as a sphere; Δb is the increase of this radius as a result of electrostatic repulsion; w is the Debye coefficient. This electrostatic interaction coefficient is given by

$$w = [\varepsilon^2/2DkT][(1/b) - (\chi/(1 + \chi a))]$$

ε is the charge of the electron; k, Boltzmann's constant; D, the dielectric constant; T, the absolute temperature; a, the exclusion radius; and χ a constant which depends on the ionic strength which is equal to $0.38 \times 10^8\ \mu^{1/2}$, $\mu^{1/2}$ being the ionic strength.

5.1.2.2. Local Effects Induced by Ionization of Critical Residues

Some proteins undergo transitions attributable to the ionization of a very limited number of residues buried in the native structure. Such transitions, which involve ionization of specific residues, were analyzed by Tanford (1961) and Tanford and Taggart (1961). The relationship between transconformation and ionization were quantitated by these investigations.

The molecular transconformation of β-lactoglobulin which depends on the ionization of two buried carboxylate groups was analyzed assuming two conformational states of the molecule in equilibrium with their protonated forms:

$$\begin{array}{ccc} N & \overset{K_1}{\rightleftharpoons} & D \\ K_N \updownarrow & & \updownarrow K_D \\ NH & \underset{}{\overset{K_0}{\rightleftharpoons}} & DH \end{array}$$

The apparent ionization constant depends on transconformation equilibrium constants:

$$K_{\text{app}} = K_0 \cdot K_D(1 + K_1)/K_1(1 + K_0) = K_N(1 + K_1)/(1 + K_0) \qquad (5.2)$$

If NH is less stable than N, and D is less stable than DH, then $K_0 = (DH)/(NH) \gg 1$, and $K_1 = (D)/(N) \ll 1$.

Reciprocally, the apparent transconformation equilibrium is depending on the ionization constant:

$$K_{\text{tran app}} = K_1[1 + (H^+/K_D)]/[1 + (H^+/K_N)] = K_0(K_D + H^+)/(K_N + H^+) \qquad (5.3)$$

There are many examples of transitions induced by the ionization of a limited number of buried groups in proteins. The conformational change of serine proteases attributable to the modification of the ionization of either the N-terminal amino group or the carboxylate of Asp 194 (numeration is referred to chymotrypsin) is also a very specific conformational transition induced by pH variations. Figure 5.1 shows this transition for δ-chymotrypsin according to Ghélis (unpublished data). In these cases, the molecules do not undergo complete unfolding. To summarize, the effects of pH on protein conformation may be a result of either electrostatic effects or the titration of specific buried groups. In the first case, it is possible to observe normal pK values; they are abnormal in the second situation. The following expression gives the apparent constant:

$$K_{\text{trans app}} = K_0 \prod_{j=1}^{n} (1 + (K_{\text{a,j,D}}/\text{H}^+)) \Big/ \prod_{j=1}^{n} (1 + (K_{\text{a,j,N}}/\text{H}^+)) \qquad (5.4)$$

K_0 is the value of the equilibrium constant in the fully protonated state of the protein; $K_{\text{a,j,N}}$ and $K_{\text{a,j,D}}$ are the dissociation constants of the jth group in the native and denatured form respectively. In many instances, the differences between $K_{\text{a,j,N}}$ and $K_{\text{a,j,D}}$ are attributable to differences in coulombic interactions in the native and in the unfolded states. If one group is not titratable in a conformational state (buried group) it may be accounted by setting $K_{\text{a,j}} \ll \text{H}^+$ if the group is permanently in its acid form and $K_{\text{a,j}} \gg \text{H}^+$ if it is permanently in its basic form.

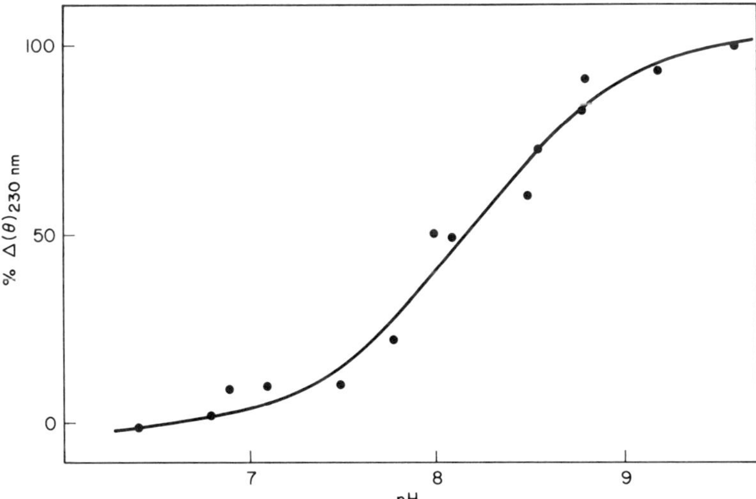

Fig. 5.1. Conformational transition of δ-chymotrypsin (10 μM) at 25°C induced by alkaline pH (from Ghélis, unpublished data).

5.1.3. Protein Unfolding Induced by Organic Solvents

Organic solvents are known to denature proteins. Generally, the products are not completely unfolded and possess an ordered conformation which differs from that of the native state. The action of organic solvents on proteins was reviewed by Singer (1962). The observed results arise from different contributions of the effects on hydrogen bonds, on hydrophobic interactions, and on electrostatic interactions.

The action of solvents on proteins is a function of the proton donor or proton acceptor character. Indeed the energy of hydrogen bonds depends on the solvent and the competition between peptide hydrogen bonds and hydrogen bonds with solvent molecules *S*,

$$\mathrm{NH}\cdots S + \mathrm{C{=}O}\cdots S \rightleftharpoons \mathrm{NH}\cdots \mathrm{O{=}C} + S\cdots S$$

which depends on the strength of hydrogen bonds between solvent molecules. When solvent molecules are linked by strong hydrogen bonds the equilibrium is shifted toward the right (i.e., the stabilization of peptide hydrogen bonds). Solvents such as dioxan, acetonitrile, dimethylformamide, pyridine, and dimethylsulfoxide, which are good proton acceptors but weak proton donors, have a very weak tendency to disrupt peptide hydrogen bonds (Singer, 1962).

By favoring peptide hydrogen bonds, some organic solvents may induce ordered conformation in globular proteins. For example, 2-chloroethanol has a tendency to induce helical conformation (Urnes and Doty, 1961; Doty, 1959). Some solvents favor the α helix $\rightarrow \beta$ structure transition in globular proteins (Blout and Asadourian, 1956). Studying the conformation of the synthetic polypeptides such as poly-γ-benzyl-L-glutamate in various solvents, Doty and co-workers (1956), Yang and Doty (1957), Doty and Yang (1956) distinguished four categories of solvents: (1) solvents in which the polypeptide chain is unsoluble (e.g., formic, acetic, propionic acids, formamide, ethylene diamine, and CCl_4); (2) solvents which lead to the aggregation of the polypeptide (e.g., benzene, ethylene, dioxan, trichlorethylene, and dichlorethylene); (3) solvents which favor helical conformation (e.g., *N*, *N*-dimethylformamide, formamide, *m*-cresol, dioxan, $CHCl_3$, pyridine, dichlorethylene, and 2-chloroethanol); (4) solvents which tend to desorganize the structure (e.g., dichloroacetic acid and trifluoroacetic acid). These organic acids are powerful solvents for synthetic polypeptides and powerful denaturants (Yang and Doty, 1957): they induce the loss of ordered structure since they form hydrogen bonds with some groups of the polypeptide. Nuclear magnetic resonance measurements indicate molecular association between the peptide groups of the polymer and the organic acid (Stewart *et at*., 1967).

Solvents which have only a very weak tendency to form hydrogen bonds with peptide groups generally induce helical conformation. Table 5.1 indi-

TABLE 5.1

Parameters of the Moffitt–Yang Equation for Several Proteins in Alcohol or Dioxan Solution[a]

Protein	Solvent	a_0	b_0	Reference
Ribonuclease	Native (globular)	−415	−95	Herskovits and Mescanti (1965)
	2-Chloroethanol	−100	−385	Herskovits and Mescanti (1965)
Lysozyme	Native (globular)	−275	−145	Herskovits and Mescanti (1965)
	80% aq. 2-Chloroethanol	—	−280	Hamaguchi and Kurono (1963)
	2-Chloroethanol	+10	−350	Herskovits and Mescanti (1965)
β-Lactoglobulin	Native (globular)	−150	−70	Tanford *et al.* (1960)
	70% aq. Ethanol	−100	−380	Tanford *et al.* (1960)
	70% aq. *n*-Propanol	−100	−370	Tanford *et al.* (1960)
	70% aq. 2-Chloroethanol	−80	−340	Tanford *et al.* (1960)
	2-Chloroethanol	+25	−455	Herskovits and Mescanti (1965)
	Methanol + 0.01 *M* HCl	+10	−505	Herskovits and Mescanti (1965)
	75% aq. Dioxan	−70	−390	Tanford *et al.* (1960)
α-Casein	Native (random coil)	−530	−45	Herskovits and Mescanti (1965)
	2-Chloroethanol	+5	−300	Herskovits and Mescanti (1965)
	Methanol + 0.01 *M* HCl	−55	−355	Herskovits and Mescanti (1965)
Tropomyosin	Native (helical rod)	—	−650	Imahori and Doty (1957)
	2-Chloroethanol	—	−680	Imahori and Doty (1957)

[a] According to Tanford (1968).

cates the effect of various solvents on the Moffitt–Yang parameters for some proteins. The alcohols (2-chloroethanol, ethanol, *n*-propanol and to a larger extent methanol + 0.01 *M* HCl mixture) decrease the b_0 value indicating an increase in helical structure. This effect was observed for all studied proteins. The introduction of water which forms hydrogen bonds with peptide groups strongly reduces the tendency of these solvents to favor helical conformation. Organic solvents also have an action on hydrophobic interactions, since nonpolar groups are more soluble in organic solvents than in water.

Polyhydric alcohols (e.g., glycerol, ethylene glycol, and propylene glycol) and sucrose which are weakly protic are less effective denaturants. Native proteins even in the presence of high concentrations of these reagents remain stable; conformational transitions occur only for concentrations of ethylene glycol higher than 50%. Therefore, these reagents are suitable as conformational probes in the method of solvent perturbation. The effectiveness of polyhydric alcohols as denaturing agents might increase in the following order: glycerol, ethylene glycol, propylene glycol, which corresponds to their increasing ability to dissolve the nonpolar groups.

Glycerol is commonly used to stabilize the native structure of proteins. A study of the mechanisms of protein stabilization by glycerol was reported by Gekko and Timasheff (1981a,b). In glycerol–water mixtures (between 10 and 40 vol % glycerol), all the seven studied proteins were found to be preferentially hydrated. The thermodynamically unfavorable interaction, protein–glycerol, tends to exclude glycerol from the surface of the protein (Gekko and Timasheff 1981a). The stabilizing effect of glycerol on the thermal denaturation of chymotrypsinogen and RNase is reflected by an increase in the T value, which results primarily from a decrease in the entropy change (Gekko and Timasheff 1981b).

The action of solvents on electrostatic interactions is a function of their dielectric constant. There are two categories of electrostatic interactions in proteins: (1) the attractive interactions arising from opposite charges which lead to the formation of salt linkages, and (2) the repulsive interactions caused by the net charge of the protein which at the extreme pH values destabilizes the native conformation. The solvent effects on these interactions are very difficult to evaluate quantitatively.

5.1.4. Protein Unfolding Induced by Organic Solutes

Organic solutes such as urea or guanidine hydrochloride (GuHCl) are powerful denaturing agents for proteins. Native globular proteins usually undergo a marked transition in the presence of GuHCl. Generally the transition is complete at a concentration of about 6–8 *M* GuHCl at room temperature except for some exceptionally stable proteins. When the transition is complete, proteins are found to be randomly coiled without any residual ordered structures. They are either linear random coils when disulfide linkages (if any) are disrupted, or cross-linked random coils if they contain intact disulfide bridges. As emphasized by Tanford (1968), the dependence of intrinsic viscosity on molecular weight is the most effective diagnostic tool for detecting linear random coils. A linear relationship is observed on a logarithmic plot of $|\eta|\, M_0$ versus the chain length, in a good agreement with the general equation:

$$|\eta|\, M_0 = An^b \tag{5.5}$$

with $A = 77.3$; $b = 0.666$; $|\eta|$ is the intrinsic viscosity; M_0 the mean residue weight, and n the number of residues. Figure 5.2 shows this linearity (Tanford, 1968). The presence of disulfide cross-links decreases the viscosity as the result of the physical restrictions imposed on the polypeptide chain (see Table 5.2). On the basis of optical rotation results, this decrease is not attributable to the presence of ordered regions in the molecule but rather it represents a purely physical effect which depends not only on the number of disulfide bridges, but also on their location in the polypeptide chain.

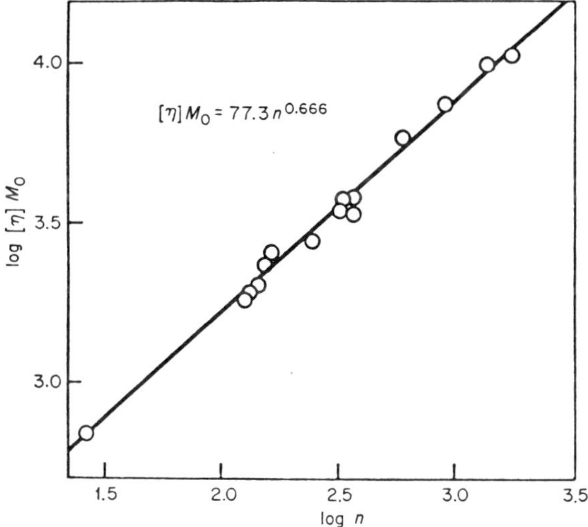

Fig. 5.2. Linear variation of intrinsic viscosity as a function of chain length for polypeptide chains in 5–7.5 *M* GuHCl at 25°C, in agreement with Eq. 5.5 (from Tanford, 1968).

TABLE 5.2

Effect of Intrachain Disulfide Bonds on Intrinsic Viscosities of Proteins in Concentrated Guanidine Hydrochloride (GuHCl) Solutions[a]

		$[\eta]$ (cm^3/g) in GuHCl Solution		
Protein	Disulfide Bonds per 100 Residues	Disulfide Bonds Broken	Disulfide Bonds Intact	Percent Decrease
Ribonuclease	3.2	16.3	9.4	42
Lysozyme	3.1	17.1	6.5	62
Serum albumin	2.8	52.2	22.9	56
Chymotrypsinogen	2.0	26.8	11.0	59
β-Lactoglobulin	1.2	22.8	19.1	16
Pepsinogen	0.8	31.5	27.2	14

[a] According to Tanford (1968).

The results shown in Table 5.3 obtained from optical rotatory measurements clearly demonstrate that proteins in concentrated GuHCl solutions are unfolded (randomly coiled) with their disulfide bonds intact as well as after reduction of disulfide bridges. Since GuHCl is a strong electrolyte, the influence of pH on unfolded chains in a high concentration of GuHCl is negligible.

TABLE 5.3

Optical Rotatory Dispersion in Concentrated Guanidine Hydrochloride: Parameters of the Moffitt–Yang Equation[a]

	Disulfide bonds reduced		Disulfide bonds intact	
Protein	*a*	*b*	*a*	*b*
Lysozyme	−449	0	−494	−22
Insulin	−470	+4	−563	−4
Pepsinogen	−507	+6	−525	+1
Ribonuclease	−521	−4	−571	−7
Chymotrypsinogen	−556	+2	−576	−7
IgG (Fab fragment)	−574	+6		
Serum albumin	−587	+8	−628	+6
Aldolase	−615	+12		
β-Lactoglobulin	−650	+6	−648	−7

[a] According to Tanford (1968).

Although urea is also a strong denaturing reagent, many proteins do not undergo complete unfolding at room temperature in urea. Examples are CO-hemoglobin, lysozyme, and immunoglobulins. Complete unfolding was shown for β-lactoglobulin (Pace and Tanford, 1968); ribonuclease (Nelson and Hummel, 1962) undergoes complete transition in 6 *M* urea at pH 4.8, and quite, but not totally, complete transition in 8 *M* urea at pH 7.3. Proteins stabilized by disulfide bonds do not always undergo complete transition even in 8 *M* urea. The completeness of unfolding is often achieved by disrupting disulfide bridges. That explains the observed differences in intrinsic viscosities of proteins with S—S bonds intact in 8 *M* urea and in 6 *M* GuHCl (Table 5.4).

The same linear dependence as given by Eq. (5.5) is also observed in the logarithmic plot of viscosity versus the polypeptide chain length for proteins with S—S bonds broken in 8 *M* urea, with $A = 76$ and $b = 0.655$. The difference in parameters between urea and GuHCl is not very significant and may be attributed to solvent effect.

The effect of temperature on the optical rotation of unfolded proteins in urea and in GuHCl was analyzed by Schellman (1958), Brandts (1969), and Tanford (1968). Proteins in urea are sensitive to the effect of pH and ionic strength because of the electrostatic interactions between the charged groups of the proteins. This has not been observed with high concentrations of GuHCl which provide a high ionic strength and therefore eliminate these effects.

The tendency of urea to decompose with the formation of cyanate ion at high temperature may introduce some irreversibility in denaturation. Cya-

TABLE 5.4

Intrinsic Viscosities[a] in 8 M Urea at 25°C

Protein	$\lvert\eta\rvert$ (cm^3/g)			
	Disulfide bonds broken		Disulfide bonds intact	
	8 M Urea	6 M GuHCl	8 M Urea	6 M GuHCl
Ribonuclease	15.6	16.3	7.6	9.4
Lysozyme	16.0	17.1	—	—
β-Lactoglobulin	21.6	22.8	16.2	19.1
Chymotrypsinogen	22.6	26.8	10.8	11.0
Takaamylase A	40.5	40.5	—	—
Serum albumin	43.2	52.2	16.6	22.9

[a] According to Tanford (1968).

nate ion reacts with free amino groups to produce carbamyl derivatives (Stark *et al.*, 1960). However, this process is relatively slow and is negligible at low pH.

The action of GuHCl and urea on protein conformation can be explained on the basis of free energy effects localized at hydrophobic interactions and peptide groups. It may be analyzed in terms of the binding of denaturant to some sites of the protein. However, it is not a strong binding since transitions occur at high concentration of ligand. This makes binding measurements rather imprecise. Tanford (1970) has analyzed the denaturation quantitatively both in terms of the binding of GuHCl and in terms of the binding of GuH^+ ions.

In the first attempt, it was assumed that the ligand is neutral GuHCl, which involves a concerted binding of both GuH^+ and Cl^- ions. Although unlikely, and without evidence that it is a possible mechanism, this situation was considered since it is easy to treat mathematically. The dependence of the equilibrium constant for the transition folded → unfolded forms was expressed as a function of the denaturant activity:

$$K_{obs} = K_0(1 + K_1 a_{GuHCl})^{\Delta n_1}(1 + K_2 a_{GuHCl})^{\Delta n_2} \cdots \tag{5.6}$$

the simplest equation being

$$K_{obs} = K_0(1 + K a_{GuHCl})^{\Delta n} \tag{5.7}$$

For the denaturation of hen egg white lysozyme by GuHCl, the best fit was obtained by using Eq. 5.7 and taking the best values of the parameters (i.e., $\Delta n = 7.84$ and $K = 3$) (Tanford, 1970). The small value of Δn which is usually found for the GuHCl induced transitions disagrees with the assumption that peptide bonds are the main binding sites for this denaturant. Aromatic side chains were also considered as possible binding sites.

The other analysis presented by Tanford (1970) implies that GuH^+ and anions play a role in the denaturation process. The simplest equation proposed to account for these effects was

$$K = K_0(1 + K_1 a_{GuH^+})^{\Delta n_1}(1 + K_2 a_{A^-})^{\Delta n_2} \tag{5.8}$$

A^- being the anion. However, since GuHCl is less effective than other salts of GuH^+, the anion term can be neglected. The activity of guanidium ion has not been estimated; Tanford (1970) used the mean ion activity of the salts ($a_\pm = a_{GuHCl}$) as a reasonable measure of it. With these assumption and approximation, the best fit was obtained for $\Delta n_1 = 21.5$, $K_1 = 1.2$, and $\log K_0 = -10.45$.

These relationships can be considered as approximations. Both of them can fit with experimental data with accurate values of the parameters. These values may be more or less plausible. However, the transition induced by GuHCl occurs in a very narrow range of denaturant concentration. Therefore, one cannot expect to obtain accurate support for a mechanism only from this analysis.

For an accurate analysis, one must take into account that in GuHCl or urea denaturation, solute occupies half of the volume of the solution and a rigorous multiple binding theory is misleading for very concentrated solutions where several ligands, even in random distribution, are into contact with the protein. Thus, Schellman (1978) recommended the use of a general thermodynamic treatment to describe solvent denaturation; a stoichiometric binding model is only a particular case. Such a treatment was applied by Dobster and Hess (1981) to study the GuHCl and urea induced unfolding–folding of rabbit muscle pyruvate kinase.

The possibility of a competition between GuHCl and water was also considered. However, it cannot account for all the denaturing action of GuHCl. A direct action on the water solvent can be also discarded since GuHCl has no effect on the activity of water.

A phenomenological treatment was also used by Tanford (1968) and by Elwell and Schellman (1975) and Elwell (1976). Assuming that the unfolding reaction $N \rightleftarrows D$ (with $K_0 = (D)/(N)$ is followed by the formation of a stoichiometric complex DG_n, such as

$$D + nG \rightleftarrows DG_n$$

with a global constant for denaturant binding

$$K_b = (DG_n)/(D)(G)^n$$

the observed denaturation constant is given by

$$K_{D\,\mathrm{obs}} = K(G)^{\bar{n}} \tag{5.9}$$

with $K = K_0 K_b$ and

$$\log K_{D\,\mathrm{obs}} = \log K + \bar{n} \log G \tag{5.10}$$

In that treatment, which represents an idealized situation, the $\bar{n}$ coefficient has a phenomenological meaning of an index of cooperativity, as the Hill number in the cooperative binding of ligand

$$\bar{n} = d \ln K_{D\,\mathrm{obs}} / d \ln G \tag{5.11}$$

This coefficient is related to c_m, the concentration of denaturant which corresponds to the half-transition by the following relation:

$$\log c_\mathrm{m} = -(1/n) \log K_0 K_b \tag{5.12}$$

Table 5.5 gives the $\bar{n}$ and c_m values for the denaturation of several proteins either by urea or by GuHCl. For GuHCl a value very close to (or higher than) 16 was found for very cooperative transitions. Similar values were obtained for proteins with intact disulfide bonds and for proteins without

TABLE 5.5

Parameters Obtained for Unfolding–Folding Transitions Induced by Organic Solutes

Protein	Denaturant	pH	c_m	$\bar{n}$
Phage T_4 lysozyme[a]	GuHCl	5.0	2.70	15.9 ± 2.6
		2.5	1.70	10.7 ± 1.2
Elastase[c]	GuHCl	5.4	1.6	8.2 ± 1.0
		8.0	2.6	9.5 ± 0.8
	Urea	5.4	3.8	14.4 ± 0.9
		8.0	5.9	10.8 ± 0.9
DIP–Elastase[c]	GuHCl	5.4	2.3	8.6 ± 0.7
		8.0	3.25	14.9 ± 1.6
Hen egg lysozyme[b]	GuHCl	5.5		15.85
β-Lactoglobulin[b]	GuHCl			19
	Urea			17
Myoglobin[b]	GuHCl			16
DIP–Chymotrypsin[b]	Urea	4		15
		7		9.5

[a] From Elwell and Schellman (1975).
[b] Reported by Tanford (1970).
[c] Obtained by Ghélis and Zilber (1982).

disulfide bridges clearly indicating that disulfide linkages have no influence on the cooperativity of the folding–unfolding process.

Denaturation by GuHCl and urea probably proceeds by different mechanisms. Hibbard and Tulinsky (1978) succeeded in equilibrating crystals of α-chymotrypsin either with $2\,M$ GuHCl or with $3\,M$ urea and obtained electron density maps. According to their results, GuHCl induced changes exclusively on the surface of the molecule; its denaturing power appears to be attributable to its ability to make water a better solvent for nonpolar chains (see also Gordon, 1972). By contrast, urea produced changes not only on the surface of the protein, but also in the hydrophobic interior of the molecule. On the surface and in the interior of the molecule urea can interact with various groups and denature the protein by making nonpolar groups more soluble because of their interaction with urea, and also by making the aqueous solvent more nonpolar (see Roseman and Jencks, 1975). The effect of urea is more complex than the effect of GuHCl.

5.1.5. Protein Denaturation Induced by Detergents

Detergents are very strong denaturants and produce very cooperative transitions at very low concentrations. Sodium dodecyl sulfate (SDS) is able to induce complete denaturation of proteins at concentration of 3 mM for β-lactoglobulin (Solomons, cited in Tanford, 1968) and at 8 mM for ribonuclease (Bigelow and Sonenberg, 1962). These values are very close to the critical micelle concentration (cmc). For β-lactoglobulin the transition is complete below the cmc, for ribonuclease it is just above. Lower concentrations are required for detergent with longer hydrophobic moities. For example, complete denaturation of serum albumin is obtained with 10 μM dodecyl benzene sulfonate.

The action of detergents can be explained by the occurrence of strong binding forces between the detergent and the protein molecule in its unfolded conformation but not in the native one, so that the equilibrium is shifted toward the denatured state; in most cases this effect is irreversible. Generally if an added reagent induces the denaturation through its binding to the protein, the equilibrium constant is a function of the concentration of the reagent:

$$\begin{aligned} K &= \text{Denatured/Native} \\ &= \sum_{j=0}^{n_{\mathrm{D}}} (DA_j)/(NA_j) \\ &= K_0\left[1 + \sum_{j=1}^{n_{\mathrm{D}}} (1 + K_{\mathrm{A,j,D}}(A))\right] \Bigg/ \left[1 + \sum_{j=1}^{n_{\mathrm{N}}} (1 + K_{\mathrm{A,j,N}}(A))\right] \end{aligned} \tag{5.13}$$

(A) is the concentration of denaturant; $K_{A,j,D}$ and $K_{A,j,N}$ are the dissociation constants of the denaturant for the denatured and the native forms of the protein, respectively; n_D and n_N are the total number of binding sites in denatured and native forms, respectively.

It was reported that several molecules of detergent can be associated to form a micelle at a single binding site of a protein (see Tanford, 1968).

Conformational changes induced by detergent do not lead to complete unfolding since the detergent does not destroy the helical structure of serum albumin and for several proteins (e.g., chymotrypsinogen, β-lactoglobulin) the presence of detergent increases the content of α helix (Jirgensons, 1967). It has been suggested that the hydrophobic part of the detergent might interact with ordered structure to form micellar regions. The strong binding of detergent on the unfolded protein has a consequence: the irreversibility of the process.

5.1.6. Effects of Inorganic Salts

The effects of salts on protein stability have been reviewed by several researchers (Tanford, 1968, 1970; Yon, 1969; von Hippel and Schleich, 1969; Record *et al.*, 1978).

Inorganic salts can induce conformational transitions in proteins. For example, LiBr, $CaCl_2$, KSCN, NaI, and NaBr are strong denaturants. Transition of ribonuclease was observed for a salt concentration of ca. 2.5 *M* at 30°C with $CaCl_2$ and KSCN, ca. 4 *M* for LiBr and NaI, and at 7 *M* for NaBr (Bigelow and Geschwind, 1961). Von Hippel and Wong (1964, 1965) extensively analyzed the effect of various electrolytes on ribonuclease stability. They have examined the added effect of various salts and temperature. They found that LiCl is less effective than NaBr. Further, NaCl and KCl did not induce transition. The transition induced by salts did not lead to completely unfolded protein. They noted that T_m increased or decreased approximately linearly with the concentration of neutral salts [see Fig. 5.3 which represents the variations of transition temperatures, T_m as a function of concentration of the added salt (a); the effect of guanidium salts and urea on T_m is compared (b)]. At 20°C salt denatured ribonuclease was not completely unfolded; the residual ordered structure was disrupted by increasing temperature.

The total effect of a given salt on T_m is the sum of the effect of its constituent ions. It is independent of ionic charge. The effectiveness of the various ions closely follows the Hoffmeister series. For the anions, it corresponds to the following series: CNS^-, I^-, Br^-, NO_3^-, Cl^-, CH_3COO^-, SO_4^{2-}, CO_3^{2-}. A similar behavior was observed for thermodenaturation of β-lactoglobulin (Dupont and Yon, 1961). Von Hippel and Wong (1964, 1965), also found a linear variation of T_m versus salt concentration for guanidinium salts and

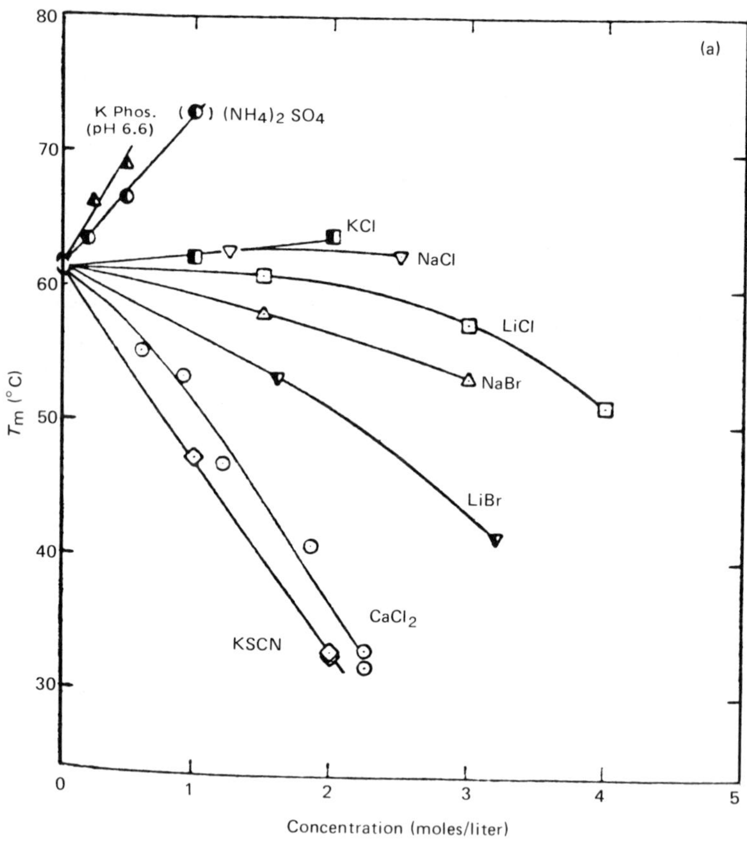

Fig. 5.3. Transition temperature of ribonuclease: (a) as a function of concentration of various added salts; (b) as a function of concentration of various denaturants, urea and guanidium salts at pH 7.0, all solutions containing 0.15 *M* KCl and 0.013 *M* sodium cacodylate; protein concentration 0.385 m*M* (from von Hippel and Wong, 1965).

tetralkylammonium salts. They observed that GuHSCN is a much more effective denaturant than GuHCl and by contrast, $(GuH)_2SO_4$ appears to protect the native protein against denaturation. This was explained to be a result of a stabilizing effect of SO_4^{2-} which is large enough to overcome the denaturing effect of GuH^+. A protection against denaturation by neutral salts on native proteins was also reported in earlier studies (Yon, 1955a,b, 1958a,b). In a recent study, Creighton (1980c) observed the same effects of salts according to the Hoffmeister series in the refolding of reduced BPTI.

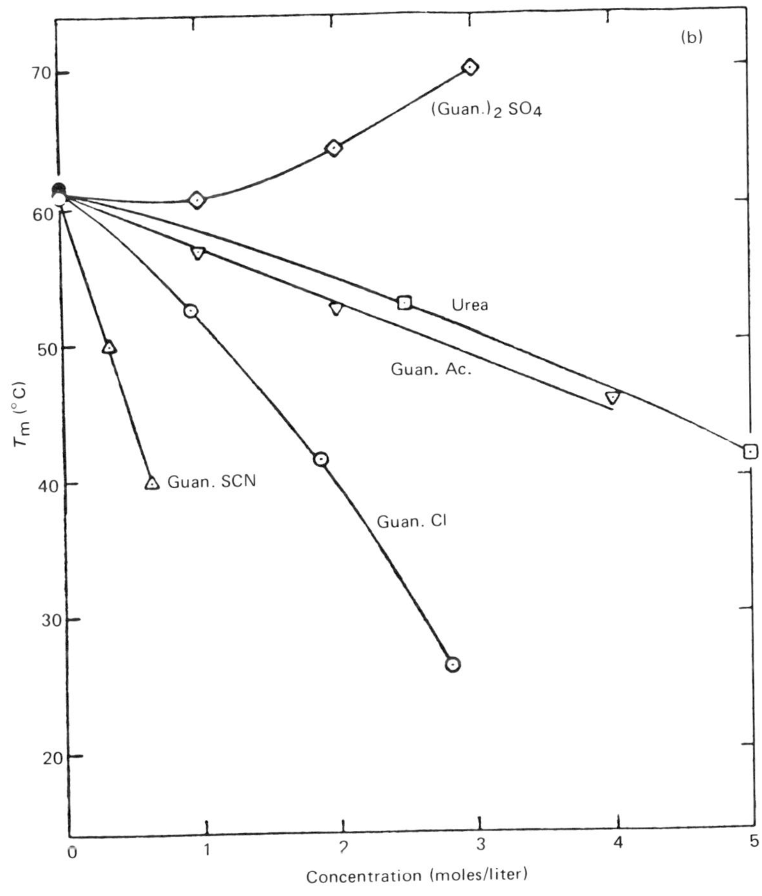

Fig. 5.3b. (*continued*)

5.1.7. Effects of High Pressures and Other Procedures

Other procedures, using radiation and high pressure, can induce denaturation, but they are not commonly used as denaturation procedures in the study of protein folding. Surface denaturation has also been described, however there are very few systematic studies on this particular subject (see Graham and Phillips, 1979). These different treatments proceed by rather complex mechanisms that are not well elucidated.

The effects of high pressures, in the range of 10^{-3} to 12 kbar have been under investigation for several years. It was recently reinvestigated on monomeric and oligomeric proteins (Weber, 1979; Schmid *et al.*, 1975, 1978; Schade *et al.*, 1980; Chryssomalis *et al.*, 1981; Jaenicke, 1981). The effects of high pressures on protein structure are complex and often difficult to interpret. They involve the breaking and forming of hydrogen bonds and hydrophobic interactions, the variation in ionization of side chains, and the modification in the structure of water. Reversible conformational changes are observed. Assuming that denaturation can be considered as a two-state process, a variation in the volume that these two states of the protein occupy in solution, ΔV^0, may be calculated. For monomeric proteins, the values of ΔV^0 are small, of the order of 30–100 ml/mole, i.e., less than 1% of the total volume of the protein. The commonly accepted explanation of the instability of proteins under high pressures refers to the oil droplet model of proteins, with the nonpolar residues packed in the interior as in an homogeneous fluid. Unfolding induces a volume decrease resulting from a higher compressibility of nonpolar side chains in contact with water. Weber (1979) adopted a different view and proposed that the volume occupied by the residues in the protein interior is strictly determined by the "incompressible" covalent bond cage made up by the folding of the peptide backbone. Unfolding changes the volume by permitting the packing of the solvent around the side chains and by the disappearance of the "dead spaces" of the native structure. The very low compressibility of proteins (with values of the order of 0.03 kbar^{-1}) suggests that the interior of a protein molecule resembles a crystal lattice rather than an oil droplet in water. That agrees with the packing density determined by Richards (1974).

The conformational changes induced by high pressures are generally detected by variations in the absorption or the emission of the aromatic chromophores or of extrinsic probes such as anilino naphtalene sulfonate (ANS) bound to the protein. Recent developments of the techniques allow detection of fluorescence polarization at pressures up to 11 kbar (Chryssomalis *et al.*, 1981). Detectable spectroscopic changes occur at pressures higher than 4 kbar for monomeric proteins. The results obtained with lysozyme indicated a reversible increase of 60% in the apparent hydrodynamic volume between 6 and 10 kbar. A simple interpretation in terms of solvent penetration was proposed. The data obtained with chymotrypsinogen are more difficult to interpret since aggregation occurred (Chryssomalis *et al.*, 1981).

An effect of hydrostatic pressure on oligomeric proteins is the dissociation into monomers which decreases the volume. Dissociation into monomers occurs at lower hydrostatic pressure (between 0.1 and 2 kbar). It was observed in such systems as actin or tubulin. The effect of hydrostatic pressure on dissociation was investigated in detail with enolase (Weber 1979, Paladini and Weber, 1981) and with lactate dehydrogenase (Schmid *et al.*, 1975, 1978;

Schade *et al.*, 1980). From the results, it was deduced that the dissociation is dictated by the existence of dead spaces between subunits (Paladini and Weber, 1981). The change in the dissociation constant K with pressure is given by:

$$-RT\,d\ln K/dp = \Delta V_{\mathrm{p}}^{0}$$

$\Delta V_{\mathrm{p}}^{0}$ is the standard volume change at pressure p. For a dimeric molecule:

$$K = 4C_0\left(\frac{\alpha^2}{1-\alpha}\right)$$

α is the degree of dissociation and C_0 the concentration. Therefore:

$$\frac{2-\alpha}{\alpha(1-\alpha)}\cdot\frac{d\alpha}{dp} = \frac{d\ln K}{dp} = \frac{\Delta V_{\mathrm{p}}^{0}}{RT}$$

With enolase, a midpoint dissociation of 1.5 kbar was observed. In the dissociation reaction, the standard volume change $\Delta V_{\mathrm{p}}^{0}$ was equal to -65 ml mol^{-1}. The reversibility of the process was better than 95% (Paladini and Weber 1981).

The effect of high pressures on proteins can be evaluated by NMR from variations in the rotational motions of internal aromatic rings (tyrosine and tryptophan) (see Chapter 8, Section 8.2). The dependence of the aromatic ring rotations in the interior of BPTI on hydrostatic pressures in the range of 1–12 kbar was recently determined by Wagner (1980) and by Wüthrich *et al.* (1980). A collisional interpretation by Karplus and McCammon (1981) indicates that this dependence results from the interactions between the ring and the protein matrix atoms. It is the pressure dependence of small packing defects that play a role in initiating the ring rotations at the interior of proteins.

5.2. CHARACTERIZATION OF THE UNFOLDED STATE

An important problem in the study of protein folding is the control of the initial state (i.e., obtaining as starting material a completely unfolded protein). Therefore a very careful characterization of the denatured protein is the first step of the experimental work; the second step is finding good conditions for reversibility. Different methods can be used to characterize the unfolded proteins: hydrodynamic methods such as viscosity and sedimentation coefficient determination; spectroscopic methods (absorption and fluorescence), fluorescence depolarization, CD, ORD and optical rotation, and NMR. Chemical methods provide a more direct tool for measuring the accessibility of side chain residues localized in different parts of the molecule by the evaluation of their chemical reactivity. The determination of the accessibility of specific peptide bonds to proteases also gives local information. Hydrogen–deuterium or hydrogen–tritium exchange indicates the accessibility of all exchangeable protons (Hvidt and Nielsen, 1966) (see Chapter 8 for details).

The different methods are not discussed in this section; they are presented later with the experimental data. However, the importance of a precise characterization of the denatured state should be noted. Methods which refer to the global conformation are often very insufficient to detect the occurrence of residual structured regions. Local methods are more indicative since the conditions allow one to use probes localized at various sites of the polypeptide chain. In any case the use of several carefully chosen methods is required.

5.3. SPONTANEOUS REFOLDING OF PROTEINS WITHOUT DISULFIDE BRIDGES

The refolding of different proteins without disulfide bridges has been investigated. Very little information is available on monomeric proteins. The data corresponding to oligomeric proteins are reported and analyzed in Section 5.9. In the present section consideration is given only to the refolding of single polypeptide chains. The best documented and most complete study is certainly that performed with staphylococcal nuclease. The unfolding–folding process of proteins such as phage T_4 lysozyme, carbonic anhydrase, and cytochrome *c* was also studied, but less extensively. Little data concerning phosphoglycerate kinase and bacterial amylase have also been reported.

The three-dimensional structure is known for most of these proteins. The different parts of secondary structure in these proteins are indicated in Table 5.6. The data concerning folding were not always obtained with the protein of the same origin as indicated in this table (for cytochrome *c* and carbonic anhydrase for example); nevertheless it was for proteins very close to those reported in Table 5.6 and the main characteristics of the structure are probably very similar. These proteins pertain to the two classes of $\alpha + \beta$ and α/β proteins as defined by Levitt and Chothia (1976).

Despite the scarcity of the data, for several of these proteins, common features appear in the unfolding–folding process, particularly the reversibility of the process. The existence of intermediates was deduced for only few of them. The ability of a polypeptide chain to refold after the removal of a short peptide was studied extensively only for staphylococcal nuclease.

TABLE 5.6

Observed Structure in Proteins without Disulfide Bonds

Protein	Number of residues	α Helix	β Sheet	Class of protein
Phage T_4 lysozyme	164	3–10, 39–49, 60–79, 82–90, 95–106 115–122, 126–134, 137–141, 143–155	14–16, 15–34, 57–58	$\alpha + \beta$
Nuclease (*Staphylococcus aureus*)	149	54–63, 64–69, 100–106, 121–135	8–19, 21–27, 30–36, 38–41, 71–76 81–83, 87–95, 97–99, 109–112	$\alpha + \beta$
Ferrocytochrome *c* (tuna)	104	1–15, 50–55, 60–68, 71–75, 87–101	19–21, 31–33, 69–70, 83–85	$\alpha + \beta$
Phosphoglycerate kinase (yeast)	373	21–36, 52–63, 73–77, 91–102, 124–129, 147–163, 177–178, 195–202, 248–262, 307–317, 348–363	2–4, 10–12, 14–19, 40–49, 54–72, 103–108, 111–113, 117–119, 131–139, 167–171, 187–191, 230–236, 237–239, 265–270, 272–274, 297–303, 318–319, 320–326, 364–371	α/β
Carbonic anhydrase (human)	264	10–16, 127–133, 151–155, 156–161, 176–181 215–224	1–3, 6–9, 24–28, 34–36, 46–48, 52–58, 61–67, 72–74, 83–93, 102–105, 112–121, 136–146, 168–172, 186–195, 199–209, 210–214, 226–228, 234–236, 243–245, 252–255	α/β

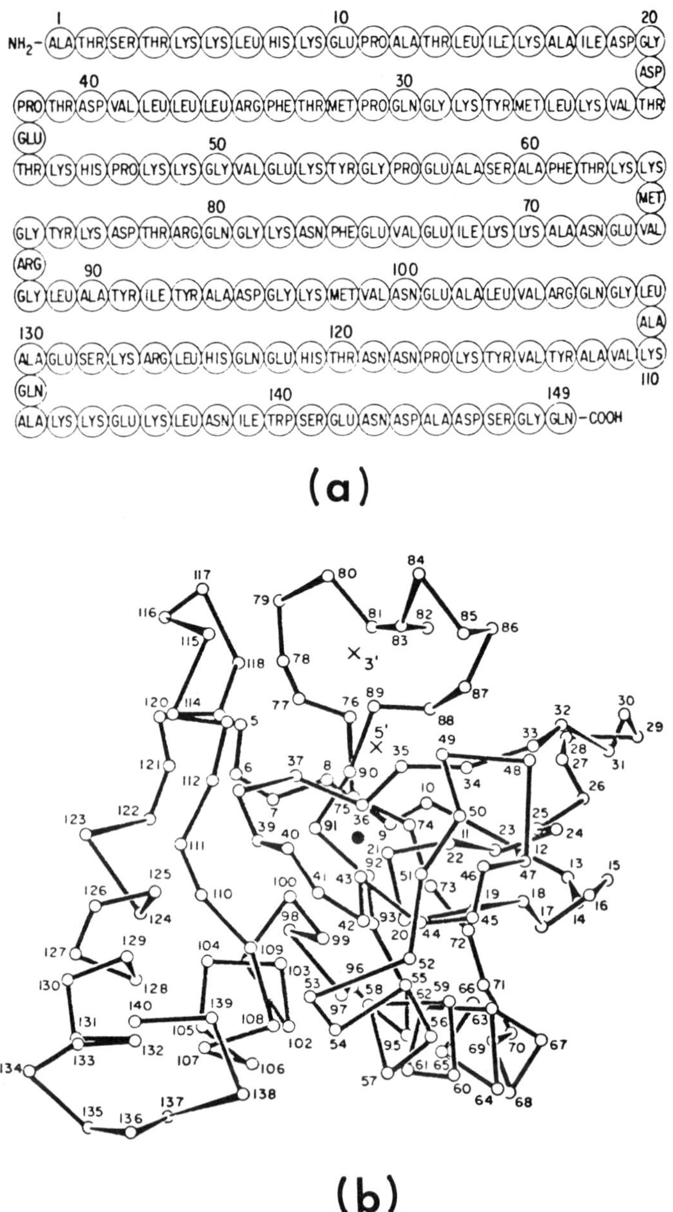

Fig. 5.4. Structure of staphylococcal nuclease: (a) amino acid sequence: (b) drawing of the backbone conformation indicating the location of the carbon atoms along the polypeptide chain (from Cotton and Hazen, 1971). Position of 3′- and 5′-phosphate of pdTp is indicated by X; Ca^{2+} by ●.

5.3.1. Reversibility of the Unfolding–Folding Process

Evidence for reversibility of the unfolding–folding process was obtained for all the studied proteins: the failure to demonstrate the reversibility of the process in the case of subtilisin (Ralston, 1974) might be attributable to difficulties in finding the right conditions for refolding.

Staphylococcal nuclease is a rather small protein composed of a single polypeptide chain of 149 amino acids and has a molecular weight of 16,800. Both sequence (Cone *et al.*, 1971) and three-dimensional structure (Arnone *et al.*, 1971) are known. Figure 5.4 shows the primary and tertiary structures. There is a single tryptophan in the sequence in position 140 (Trp 140) located very close to the C-terminus and buried in the native structure. There are also seven tyrosyl residues; Tyr 91 and Tyr 93 are clearly buried in the hydrophobic core of the protein. Three helical segments are present in the native molecule.

Important structural transitions occur in extreme pH values, i.e., below pH 4.0 and above pH 10.5. The reversibility of the transition which occurs between pH 3 and 4 was demonstrated by following different parameters (e.g., fluorescence of the unique tryptophan, fluorescence of tyrosine, tyrosine absorption, molar ellipticity, and reduced viscosity). The same transition curve is observed with all parameters, with a midpoint corresponding to pH 3.9 (Fig. 5.5).

Phage T_4 lysozyme is a small protein of MW 18,700 and is devoid of disulfide bridges (Fig. 5.6A). For this protein, reversibility of unfolding–folding was reported under different conditions, either after thermal unfolding or after denaturation by guanidine hydrochloride (Elwell and Schellmann, 1975, 1977; Elwell, 1976; Desmadril and Yon, 1981). Reversibility was checked by different techniques which included fluorescence emission, CD at 223 nm and in aromatic region, and some enzymatic assays.

Bovine carbonic anhydrase is a metalloenzyme which contains one atom of zinc bound per molecule of MW 30,000. The presence of zinc is required for the enzymatic activity. Moreover, the metal seems to play a role in the conformational stability of the protein. Bovine carbonic anhydrase may be easily denatured by guanidine hydrochloride and refolds spontaneously by removal of the denaturing reagent to regain quite full enzymatic activity (more than 95%). Reversibility of the transition was checked by using CD and UV spectroscopy and also enzymatic activity on *p*-nitrophenyl acetate (Wong and Tanford, 1970, 1973; Wong *et al.*, 1972; Yazgan and Henkens, 1972; Carlsson *et al.*, 1973, Ikai *et al.*, 1979; McCoy *et al.*, 1980; McCoy and Wong, 1981). The unfolding–folding transition of apoenzyme is also thermodynamically reversible. The midpoint is however lightly shifted (1.0 *M* in the absence of metal to 1.5 *M* in the presence of zinc). Under well-

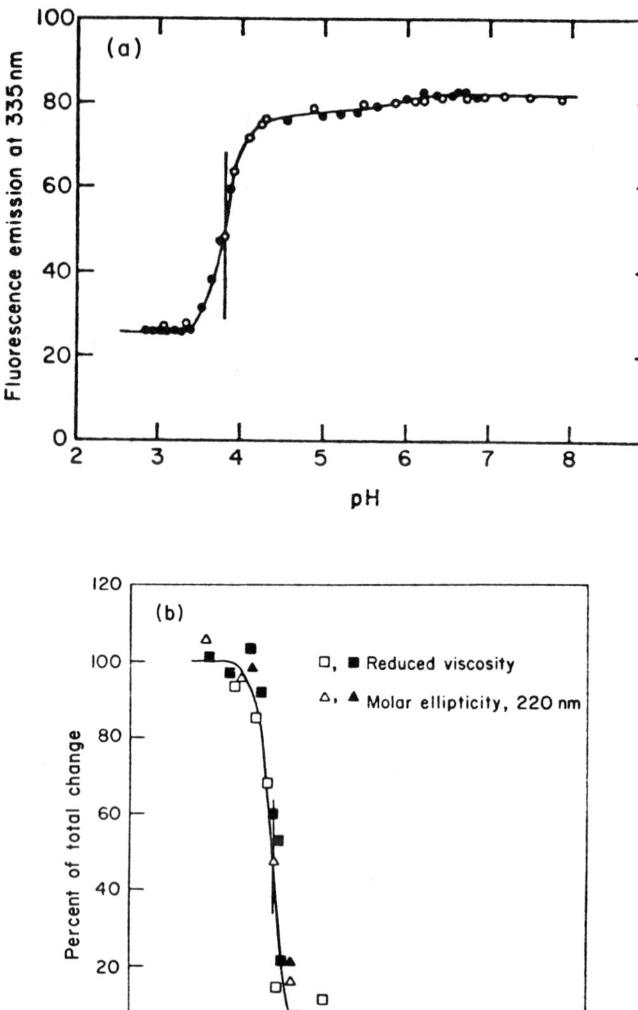

Fig. 5.5. Reversible acid induced transition of staphylococcal nuclease observed by different methods: (a) emission fluorescence of Trp 140 (according to Epstein *et al.*, 1971a); (b) changes in reduced viscosity (□, ■) and molar ellipticity (△, ▲) at 220 nm; (□, △) measurements made on decreasing pH; (■, ▲), measurements made upon increasing pH (from Anfinsen *et al.*, 1972); (c) emission fluorescence of Trp 140 (□), tyrosine (■), and tyrosine absorption at 287 nm (●) (from Anfinsen *et al.*, 1972); (d) areas of imidazole C_2 proton resonances of each of the four histidine H-1, H-2, H-3, H-4 (from Anfinsen *et al.*, 1972).

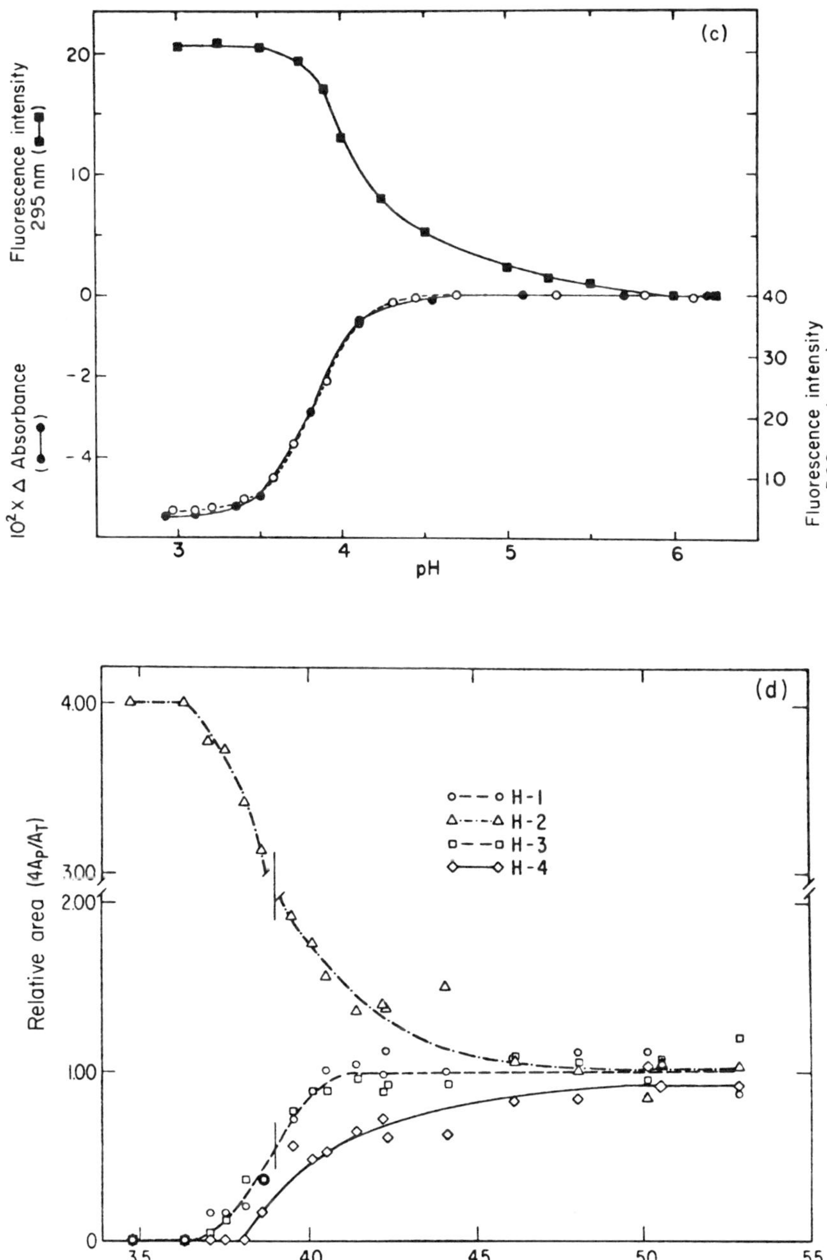

Fig. 5.5. (*continued*)

(a)
Gln 123
Arg 125
Trp 126
Tyr 88
90
Gln 122
120
130
Thr 151
80
Arg 76
Val 75
Ala 74
Ala 73
10
Gln 69
Ile 3
Phe 4
160
Trp 158
Asn 2
Met 1
Tyr 139
Trp 138
110
Asn 140
Gln 141
Gln 105
Thr 21
Asp 20
Glu 22
Gly 23
Tyr 24
Tyr 25
30
Thr 34
Lys 35
Ser 36
Pro 37
Ser 38
Leu 39
Asn 40
50
60
(b)
MET 80
HIS 18
PRO 30
103
TRP 59
1

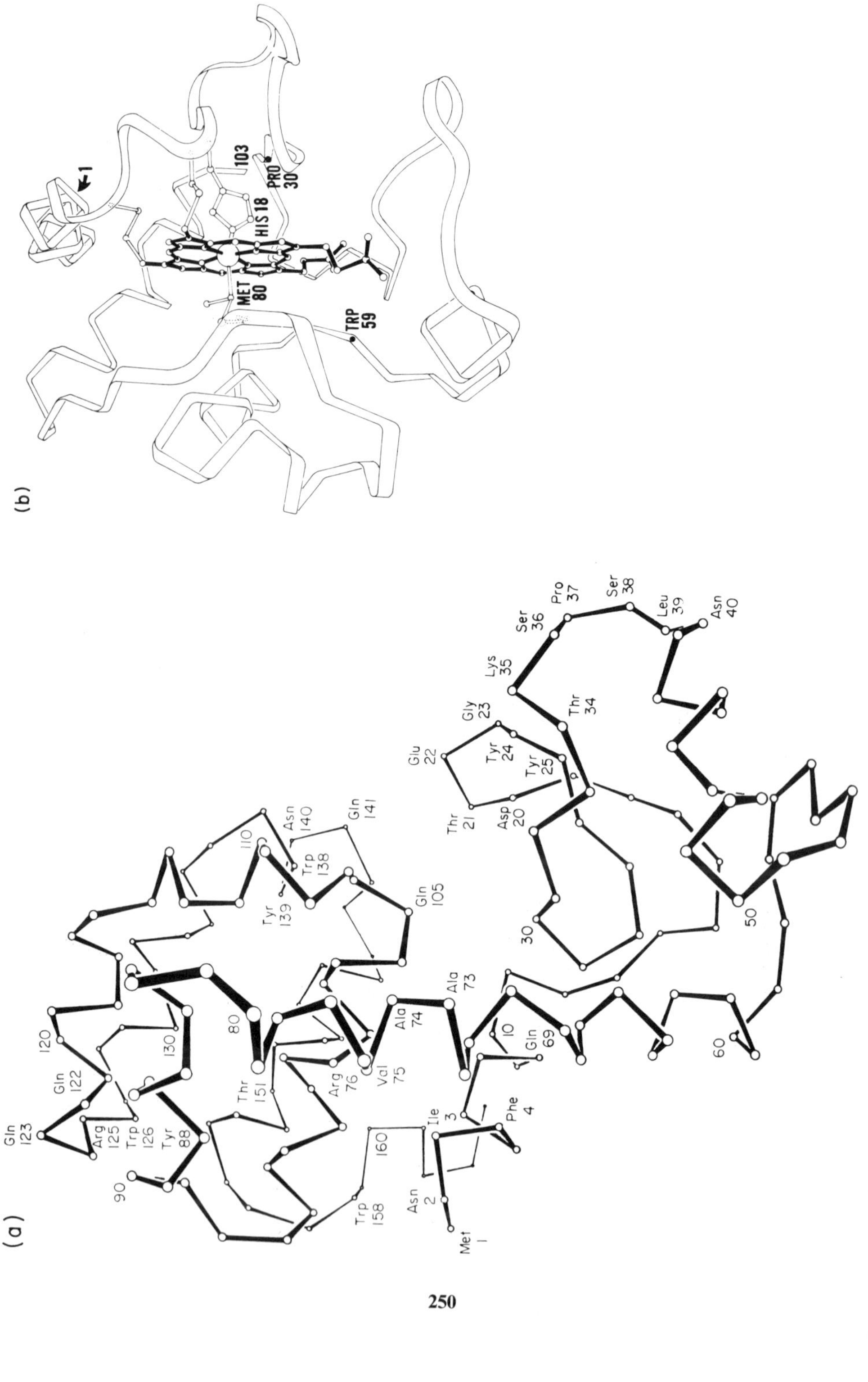

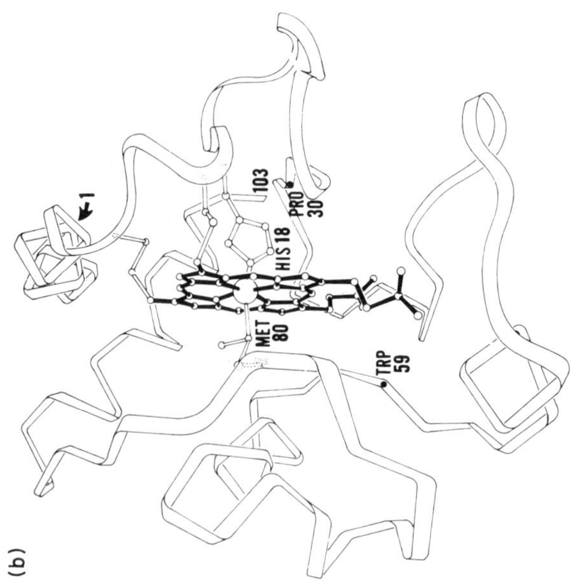

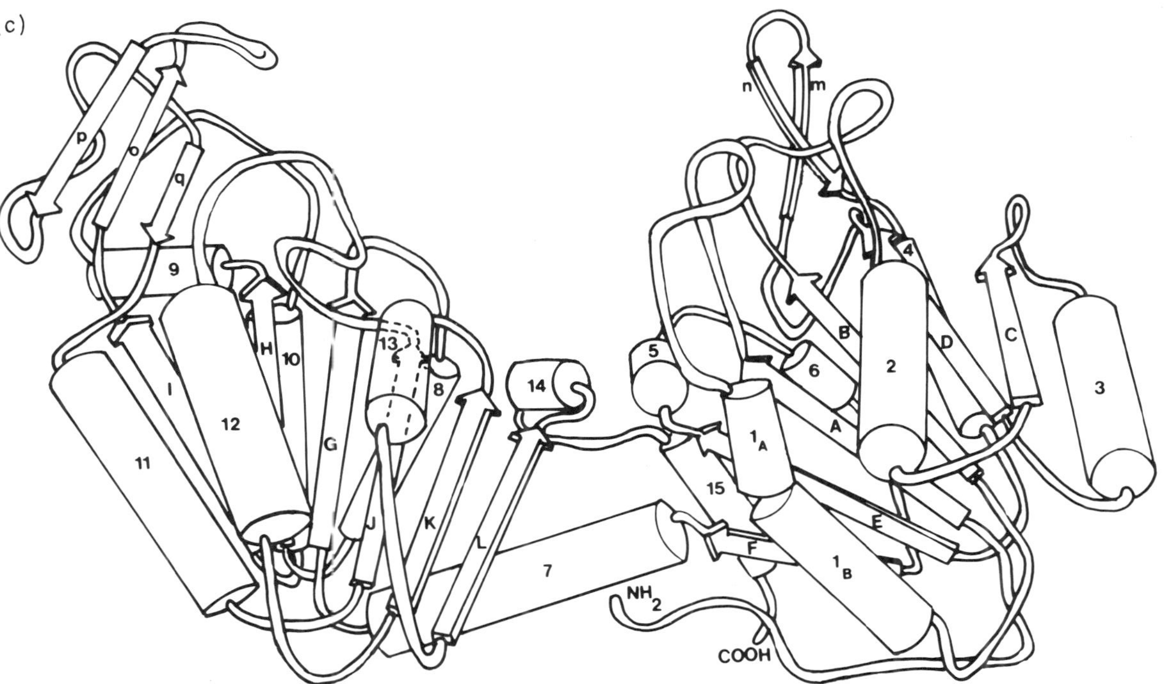

Fig. 5.6. Three-dimensional structure of (a) phage T_4 lysozyme (from Remington *et al.*, 1978, courtesy of Matthews); (b) cytochrome *c* (from Dickerson *et al.*, 1971; Takano *et al.*, 1972); and (c) horse phosphoglycerate kinase (from Banks *et al.*, 1979, courtesy of Blake).

defined conditions, when refolding is monitored by dilution from 4 *M* to 1 *M* GuHCl (pH 6 and 25°C) the presence of zinc in early stages favors refolding. When zinc is added later, it decreases the rate of refolding. The presence of a cofactor, metal or else, may introduce constraints during the refolding process (Wong and Hamlin, 1975).

Another example of a protein devoid of disulfide bonds is cytochrome *c* which consists of a single polypeptide chain of 104 amino acids and contains a prosthetic group covalently bound to the apoprotein through thioester bonds with Cys 14 and Cys 17 (Fig. 5.6b). The ferric ion is liganded with His 18 and Met 80 (Dickerson *et al.*, 1971); (see Chapter 2, Fig. 2.6). Guanidine hydrochloride induced unfolding was studied by several groups, Babul and Stellwagen (1972), Ikai and co-workers (1973), Fisher and co-workers (1973), and Tsong (1974, 1975). The conformational state was characterized by absorption spectroscopy either at 290 nm or at 405 nm in the Soret band of the protein and by optical rotation. The fluorescence of Trp 59 was also used as a conformational probe. The unfolding–folding process was shown to be thermodynamically reversible. According to Ikai and co-workers (1973) 6 *M* GuHCl induces complete unfolding of the polypeptide chain, although the heme group is covalently bound to the protein and could act, even in GuHCl solution, as a nucleus for folding. However, the protein retains certain residual structures after the major transition was achieved in 4 *M* GuHCl; Trp 59 fluorescence represents less than 50% the intensity of free tryptophan. These residual structures continue to unfold at higher GuHCl concentrations. Spontaneous refolding of cytochrome *c* denatured by acid pH was also reported (Ikai and Tanford, 1971).

The thermal unfolding of phosphoglycerate kinase from yeast and from a thermophilic bacteria, *Thermus thermophilus*, in the presence of GuHCl was shown to be completely reversible. The reversibility of the transition was studied by measuring CD and fluorescence intensity (Nojima *et al.*, 1977).

Bacillus subtilis α-amylase which contains no disulfide linkage has been denatured in 8 *M* urea under acidic conditions. The reversibility of the process after removal of the denaturant was shown by the return of the optical properties (ORD and UV absorption) and also by the return of the enzymatic activity (Fukushi *et al.*, 1963).

5.3.2. Occurrence of Intermediates during the Folding Process

The regaining of activity by a protein without disulfide bridges can be a rather rapid process. For staphylococcal nuclease, the overall rate of refolding is rather rapid. The halftime for folding is 0.1–0.2 sec. Cytochrome *c* also recovers its native conformation in a short time (several seconds). The kinetic aspect is detailed in Section 5.3.

Despite the rapidity of refolding observed for several proteins under well determined experimental conditions, a set of experimental data indicates the occurrence of intermediates during the course of refolding of different proteins. The acidic transition of staphylococcal nuclease, for example, was followed by different methods previously mentioned. It was also followed by 220 MHz proton magnetic resonance spectroscopy, by evaluation of the chemical shift of the C-2 proton resonance of the four histidine residues, 8, 46, 121, 124, (Epstein *et al.*, 1971b). Three of the histidyl groups follow a single sharp transition with a midpoint of 3.9. The fourth residue follows a biphasic transition indicating the presence of intermediates. This is in agreement with the fact that the refolding process as followed by tryptophan fluorescence may be described by two first-order reactions (Fig. 5.5).

Evidence for intermediates during the refolding of bovine carbonic anhydrase was also provided by both equilibrium and kinetic studies (Wong and Tanford, 1973). Three conformational intermediates have been proposed to account for the unfolding of cytochrome *c* in methanol and acid (Drew and Dickerson, 1978).

For bacterial amylase, the return of the different properties after denaturation by 8 *M* urea under acidic conditions does not follow the same kinetics. Physical parameters such as optical properties reach their final values immediately after removal of the denaturant whereas the return of the activity follows complex kinetics. Activity rapidly recovers to 20–40% and then gradually increases to 60–90% within 6 hr (Fukushi *et al.*, 1963). Fukushi and co-workers (1963) have interpreted their results by assuming that, after removal of the denaturant, the unfolded protein can rapidly reach a conformation very close to that of the native one. This immediate refolding is followed by a slower intramolecular rearrangement which generates the biologically active protein.

Spontaneous refolding of monomeric proteins without disulfide cross-links is illustrated by these few examples; only a small number of detailed studies are available. Nevertheless from these few examples, one can tentatively deduce some features:

(1) The reversibility of the unfolding–folding process has been observed in all the mentioned examples.

(2) The occurrence of intermediates during the folding process has been detected for staphylococcal nuclease and has been deduced for bacterial amylase (this point is detailed in Chapter 8).

(3) The presence of a cofactor or a prosthetic group in protein structure, by helping the folding process and stabilizing the structure, represents a particular case (this point is discussed in detail in Chapter 11).

5.4. SPONTANEOUS REFOLDING OF PROTEINS WITH DISULFIDE BRIDGES

To understand the way by which reduced and unfolded proteins refold and reoxidize, one must first consider the structural organization of the studied proteins and the location of disulfide bridges in the structures. This could be helpful in analyzing how the noncovalent interactions can direct the reformation of disulfide bridges. The refolding of unfolded and reduced proteins has been extensively studied only for a small number of proteins (e.g., ribonuclease, lysozyme, bovine pancreatic trypsin inhibitor, serine proteases). Figure 5.7 shows the structure of these proteins. Their structural

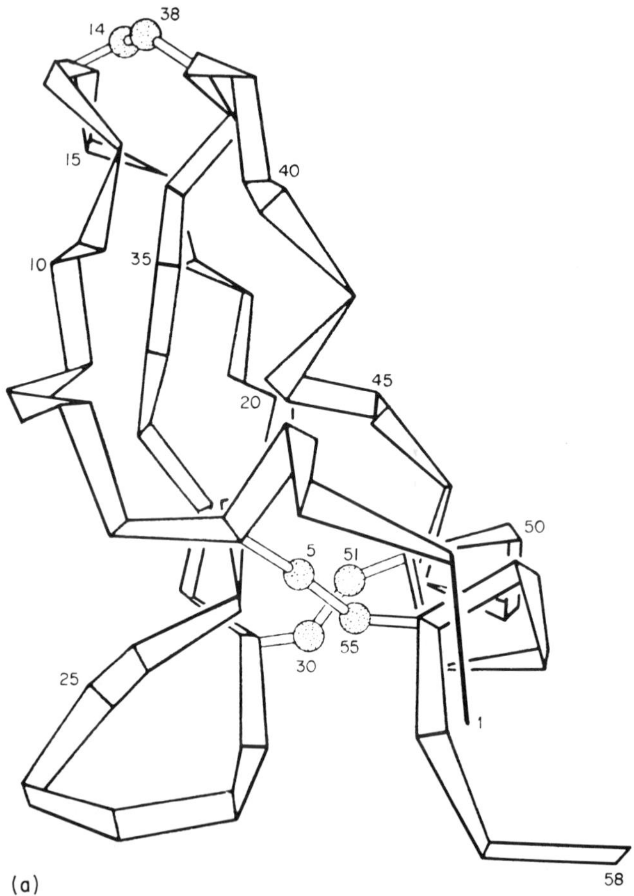

Fig. 5.7. (pp. 254–258). Three-dimensional structure of (a) BPTI (from Creighton, 1975a); (b) ribonuclease A (from Dickerson and Geis, 1969); (c) hen egg white lysozyme (from Imoto *et al.*, 1972); (d) trypsin (from Hartley *et al.*, 1972); (e) α-chymotrypsin (from Hartley *et al.*, 1972); and (f) elastase (from Hartley *et al.*, 1972).

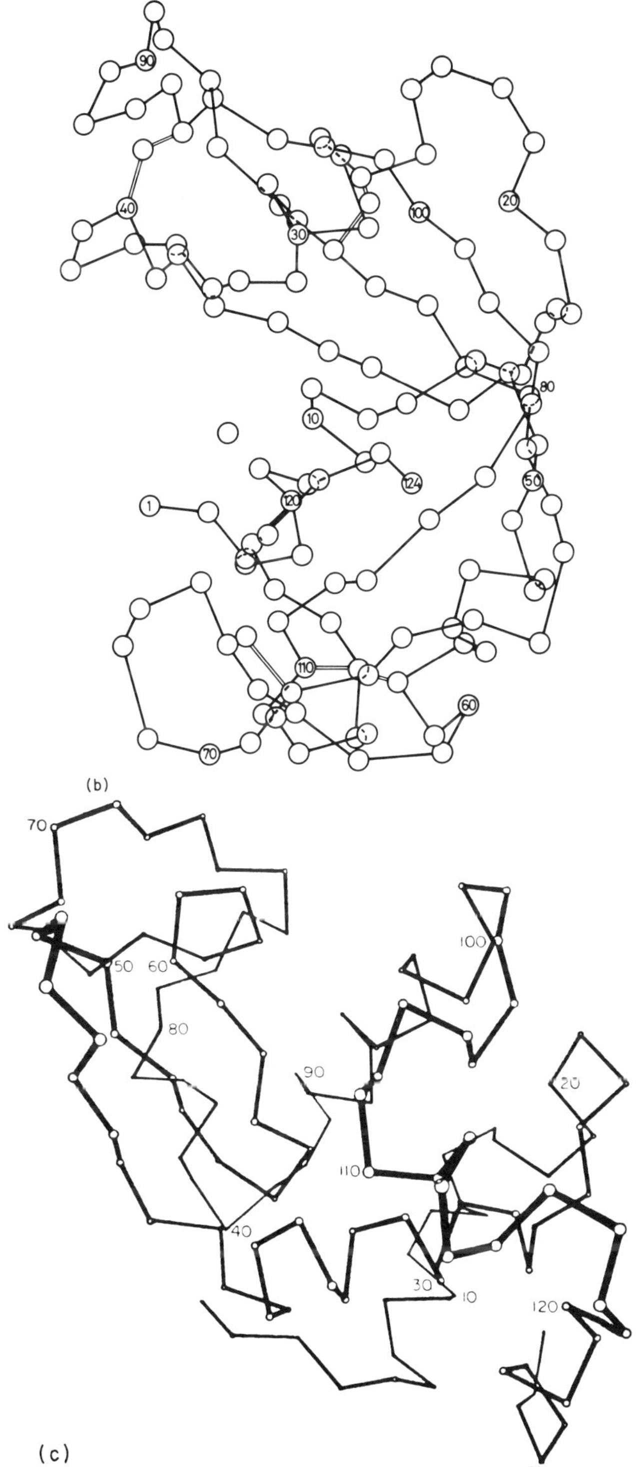

Fig. 5.7b, c. (*continued*), see p. 254 for legend.

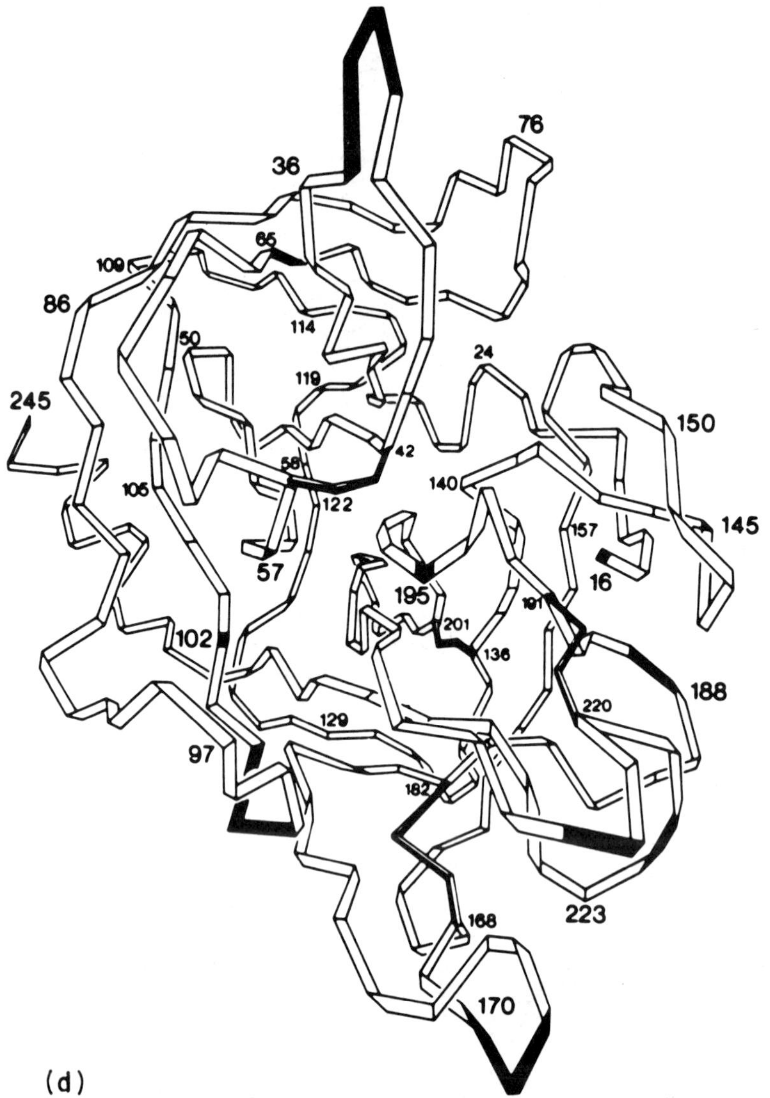

Fig. 5.7d. (*continued*), see p. 254 for legend.

characteristics, and the location of disulfide bridges are indicated in Table 5.7. These proteins are either of the class of $\alpha + \beta$ proteins or of the class of all-β proteins. It appears in a first approximation that most of all the α proteins and α/β proteins have no disulfide bridges (probably this kind of structure does not require further stabilization) whereas most of the all β proteins need to be stabilized by disulfide bridges.

The number of disulfide bridges with respect to the size of the molecule is variable. The smallest protein, BPTI, has 58 amino acids and three disulfide bonds whereas elastase has only 4 disulfide bridges for 240 amino

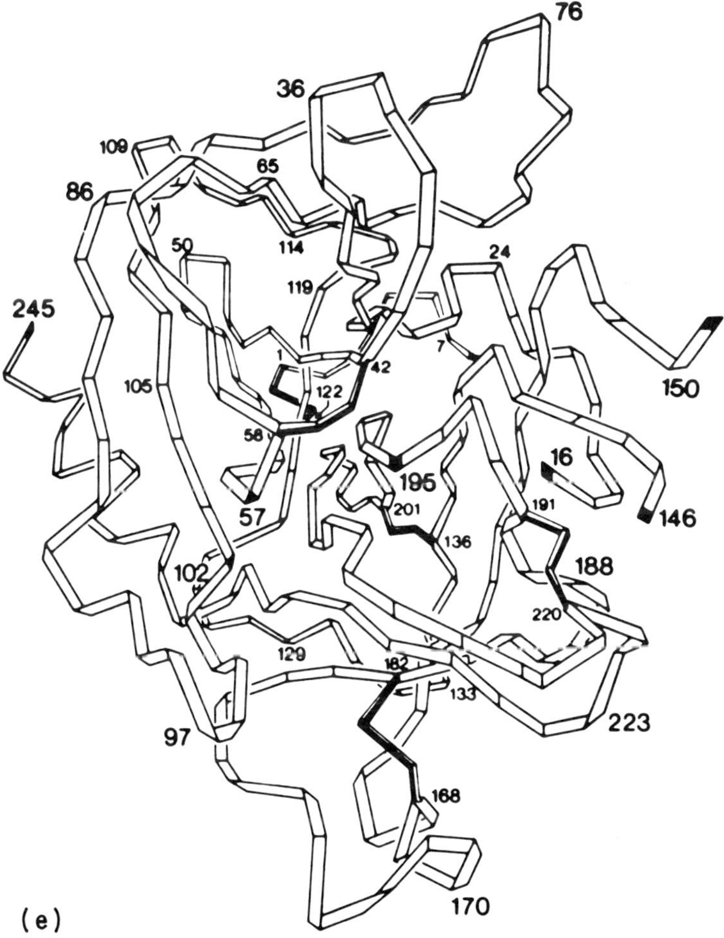

Fig. 5.7e. (*continued*), see p. 254 for legend.

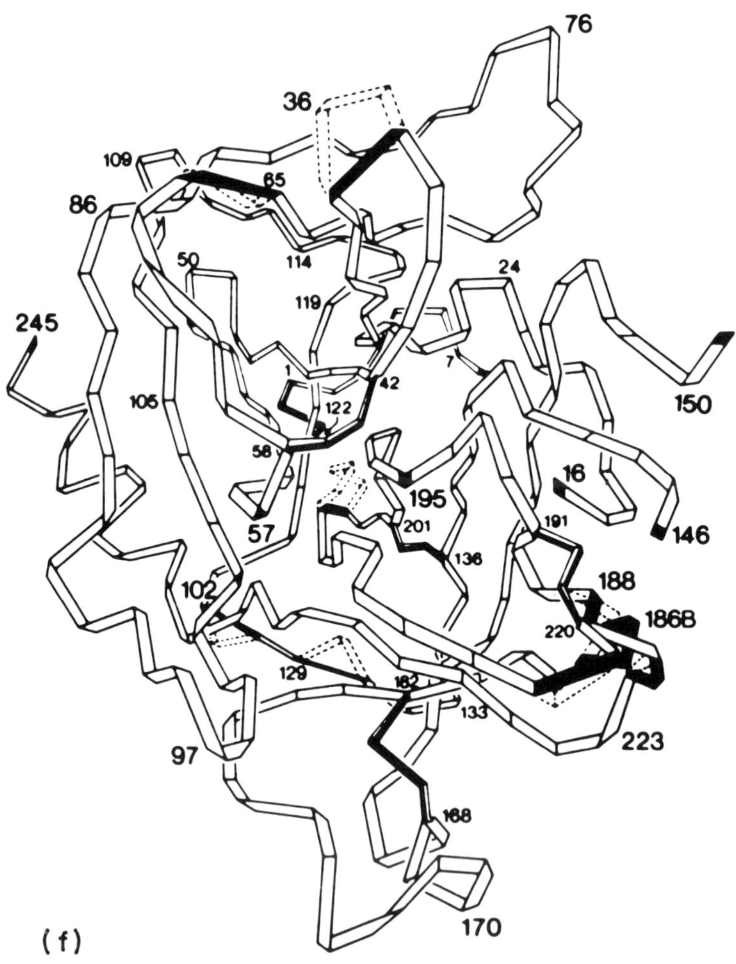

Fig. 5.7f. (*continued*), see p. 254 for legend.

acids. Disulfide bridges do exist between segments of secondary structures, either between two helices, or two β segments, or between a segment of ordered (α or β) structure and a segment of coil structure, or also between two segments of coil structure.

For all these proteins the spontaneous refolding of reduced and unfolded molecules was obtained under well-defined conditions. However, for disulfide bridged proteins different kinds of experiments were performed. The refolding process was studied either for unfolded proteins with disulfide bridges kept intact or for unfolded and reduced proteins. These different types of data are presented in following sections.

TABLE 5.7

Ordered Structures in Disulfide-Linked Proteins

Protein	Number of residues	Disulfide bond number	Disulfide bond position	α Helix	β Sheet	Class of protein
Trypsin inhibitor (bovine)	58	3	30–51 5–55 14–38	2–7, 47–55	14–25, 28–37, 43–46	$\alpha + \beta$
Ribonuclease A (bovine)	124	4	26–84 40–95 58–110 65–72	4–12, 24–33, 50–58	13–14, 34–36, 43–49, 60–64, 71–76, 77–87, 96–112, 118–123	$\alpha + \beta$
Lysozyme (chicken)	129	4	6–127 76–94 64–80 30–115	4–15, 24–35, 79–85, 88–101, 108–115, 120–125	1–3, 38–40, 42–47, 49–54, 58–61, 74–76	$\alpha + \beta$
α-Lactalbumin	123	4	6–120 28–111 61–77 73–91			$\alpha + \beta$
Elastase (porcine)	240	4	30–46 127–194 158–174 184–214	44–48 154–160 227–238	1–3, 5–7, 14–22, 25–35, 38–43, 52–59, 68–79, 94–100, 123–132, 145–153,	all β

(*continued*)

TABLE 5.7 *(continued)*

Ordered Structures in Disulfide-Linked Proteins

Protein	Number of residues	Disulfide bond number	Disulfide bond position	α Helix	β Sheet	Class of protein
Chymotrypsinogen A (bovine)	245	5	1–122 42–58 136–201 168–182 191–220	11–16 164–172 232–243	29–35, 39–47, 50–55 64–69, 80–91, 93–96 99–109, 121–123, 132–141 155–163, 180–186, 197–203 206–217, 223–231	all β
α-Chymotrypsin A (MRC) (bovine)	241	5	1–122 42–58 136–201 168–182 191–220	164–170 234–243	16–18, 20–22, 29–35, 39–47, 50–55, 64–69, 80–91, 93–96, 99–109, 121–123, 132–144, 150–152, 155–163, 180–185, 188–193, 195–203, 206–216, 224–231	all β
Trypsin–DIP (human)	223	5	13–143 31–47 115–216 122–189 179–204	151–159 219–228	8–14, 21–27, 29–37, 40–45, 53–59, 68–80, 83–85, 88–90, 92–99, 119–131, 137–139, 142–150, 167–173, 177–182, 184–190, 193–201, 208–216	all β

5.4.1. Spontaneous Refolding of Denatured Proteins without Reduction of Disulfide Bridges

Most thermodynamic and kinetic studies of lysozyme, chymotrypsinogen, and ribonuclease were performed using proteins denatured by urea or GuHCl, with their disulfide bonds intact. A polypeptide chain without any residual noncovalent interaction can be obtained, but with restrictions because of the disulfide linkages which decrease the conformation entropy of the polypeptide chain. Complete unfolding has not been demonstrated in all cases. Tanford and co-workers (1966a,b, 1967, 1973), and Aune and Tanford (1969a,b) have shown full reversibility of the unfolding–folding process for lysozyme, the unfolding being induced by GuHCl, or pH, or temperature. In the different cases, the randomly coiled structure of the polypeptide chain on denaturation was checked by ORD, or hydrogen titration, or by difference spectrophotometry. The complete reversibility of the unfolding–folding process with disulfide bonds intact was also obtained for RNase and chymotrypsinogen. The refolding of ribonuclease without reduction of disulfide bonds has been the subject of a great number of investigations. Tanford has presented the evidence that small globular proteins such as RNase as well as lysozyme are thoroughly unfolded in 6 *M* GuHCl solutions; however, it is probable that pH or temperature induces transitions which lead to incompletely unfolded products. The results of Nall and co-workers (1978) have confirmed the suggestion by Tanford that thermal unfolding transitions are more complex than denaturation induced by GuHCl. The transitions are complicated by additional species whose nature is unknown.

The goal of most of these studies was either to check the validity of the two-state theory or to determine the presence of intermediates during the refolding process. Since Chapters 6 and 7 detail the kinetic or the thermodynamic studies, it is only emphasized here that most of these studies have revealed a multistep (at least two-step) behavior. For proteins with intact disulfide bonds, even the slow step is very rapid (a few seconds) compared to the time needed for the return of the activity of a protein when disulfide bridges have been previously reduced (see following discussion).

This is clearly shown in Fig. 5.8 by the case of elastase (Ghélis *et al.*, 1982); the return of the activity of the protein unfolded with and without disruption of disulfide linkages are shown on the same graph. When disulfide bonds are not cleaved, the activity is instantaneously recovered, whereas for the enzyme with disrupted S—S bonds, this process is slow.

The presence of S—S bonds severely limits the possible conformations accessible to the randomly coiled polypeptide chain and favors the refolding process by limiting the formation of intermediates other than those leading more directly to the native state. The process is therefore more rapid.

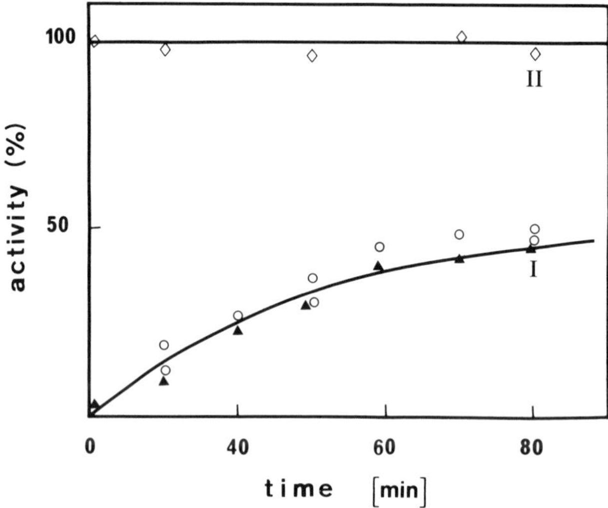

Fig. 5.8. Recovery of enzymatic (—) and antigenic activities (○, ▲) of reduced and denatured elastase (by 6 *M* GuHCl) curve I, of denatured (by 6 *M* GuHCl) and unreduced elastase (curve II) (from Ghélis *et al.*, 1982).

5.4.2. Probability of Disulfide Bond Formation in Proteins

The statistics of the process of disulfide bond pairing from SH groups in a polypeptide chain has been analyzed by Kauzmann (1959b), and Sela and Lifson (1959). The number of ways to form j disulfide bonds from 2 n SH groups in a random pairing is given by the following expression:

$$N^{j}_{2n} = (2n)!/[2^{j}(2n - 2j)!(j)!]$$

Thus when n increases, the number of possible combinations becomes tremendously large. For two disulfide bridges there are three possible combinations and for four disulfide bridges the number of combinations reaches 105. Table 5.8 is an illustration of the situation.

Kauzmann calculated the relative probability of the different arrangements. Thus, for two disulfide bridges, among the three possible geometries, the first arrangement (Fig. 5.9) which corresponds to the pairing of two adjacent sulfhydril groups is the most probable. The small chance of the correct pairing indicates that a random search for disulfide formation is unlikely; noncovalent interactions, at least for a large contribution, direct the formation of disulfide bridges.

In a recent study, Thornton (1981) analyzed in detail the distribution and topology of disulfide bridges in proteins of known sequence and structure (28 proteins) and in proteins of known structure and disulfide connectivities

TABLE 5.8

Ways in Which 2*n* Sulfhydryl Groups Can Combine to Form *j* Disulfide Bonds[a]

Number of bonds	Number of combinations
1	1
2	3
3	15
4	105
5	945
6	10395
7	135135
8	2027025
9	34459425
10	654729075
11	13749310575
12	316234143225
13	7905853580625
14	213458046676875
15	6190283353629375
16	191898783962510625
17	6332859870762850625
18	221643095476699771875
19	82200794532637891559375
20	319830986772877770815625
21	131113070457687988603440625
22	563862029680583509947946875
23	25373791335626257947657609375
24	1192568192774434123539907640625
25	58435841445947272053455474390625

[a] According to Kauzmann (1959b), Sela and Lifson (1959); from Anfinsen and Scheraga (1975).

(26 proteins). Several general patterns which dictate disulfide formations were suggested. (1) Local disulfide bridges are preferred where possible. (2) Disulfide bridges between N and C terminus of the polypeptide chain are also favorable. (3) Secondary structures such as α helix and β sheet often prevent formation of local disulfide bridges. (4) The loop connecting half-cystines generally includes a tetrapeptide with positive β turn potential. (5) There is a strong preference for shorter connections; 49% of the analyzed disulfide-bridged half-cystine are separated by less than 24 residues; the most frequent separation is 10 to 14 residues. (6) Right- and left-handed disulfide bridges occur equally. (7) Cystines are generally well conserved in contrast to cysteines; if a disulfide is not conserved, both cystines are mutated.

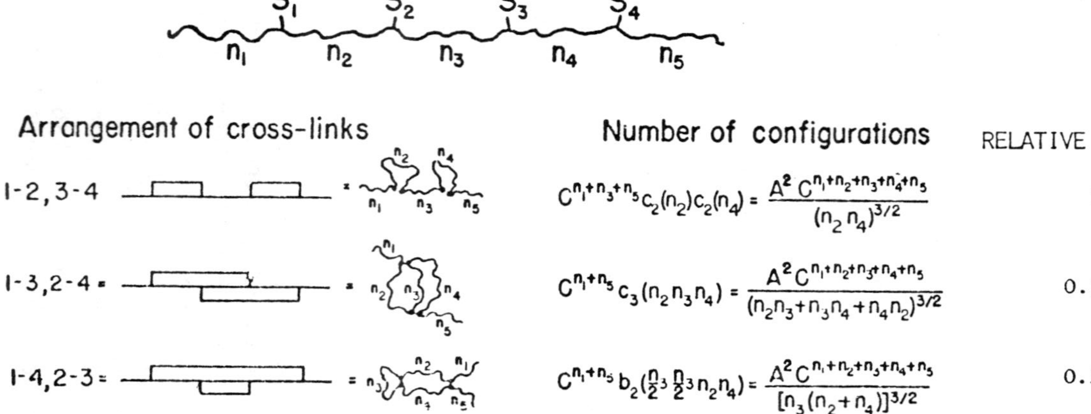

Fig. 5.9. Various arrangements of disulfide bonds and their relative probability (adapted from Kauzmann, 1959a,b).

5.4.3. Formation of Incorrect Disulfide Bridges and Reshuffling Processes during Protein Folding

5.4.3.1. Evidence for the Occurrence of Incorrect Disulfide Bridges

The refolding of denatured and reduced proteins was achieved in many cases; however, for several proteins the presence of small amounts of reducing agents is required for the refolding. This indicates the occurrence of incorrect disulfide bonds during the early stages of the refolding process which have to be disrupted to allow the correct pairing.

Direct and indirect evidence exists for the formation of incorrect disulfide bonds in early stage of refolding of reduced and denatured proteins:

(1) The delay between the disappearance of free SH and the return of activity has often been interpretated as an argument for the early formation of incorrect disulfide linkages; nevertheless it is not a proof since, even for proteins without disulfide bonds, delay between reappearance of optical signals for example and return of activity can be observed.

(2) The acceleration of the reactivation process in suitable conditions of reagents (either trace of reducing reagents or redox mixtures) allowing the interchange of incorrectly formed disulfide bonds is a better argument.

(3) The results of chemical analysis directly indicate the sulfhydryl groups which are paired.

The first well documented studies of the formation of the tertiary structure of proteins from the linear polypeptide chain were initially performed using as a model system the refolding of denatured and reduced ribonuclease (see Fig. 5.10). Bovine pancreatic ribonuclease is a small $\alpha + \beta$ protein (MW 13,700) with 37% β structure and 23% α helix, stabilized by four disulfide bridges (see Fig. 5.6 and Table 5.7). Reformation of tertiary structure was achieved from a protein unfolded in 8 *M* urea and reduced by β-mercaptoethanol with a yield approaching 100% (Anfinsen and Haber, 1961). It was confirmed that the unfolded reduced protein was devoid of any residual folded structure (Young and Potts, 1963). In earlier studies reduction was performed with thioglycolate. With this reducing agent, thiolation of amino groups as a result of the presence of thiol ester impurities explained the low yield of recovery of the native structure (20%) (White, 1961; Anfinsen, 1966). With β-mercaptoethanol this artfact was prevented. When the enzyme was removed from the denaturing conditions and diluted in buffer solutions at pH ranging from 7.0 to 8.5, the activity was recovered. Figure 5.11 illustrates the return of the different properties of the molecule during its refolding. A marked lag is observed before the appearance of the enzymatic activity as measured against either urydilic-2′3′-cyclic phosphate or ribonucleic acid.

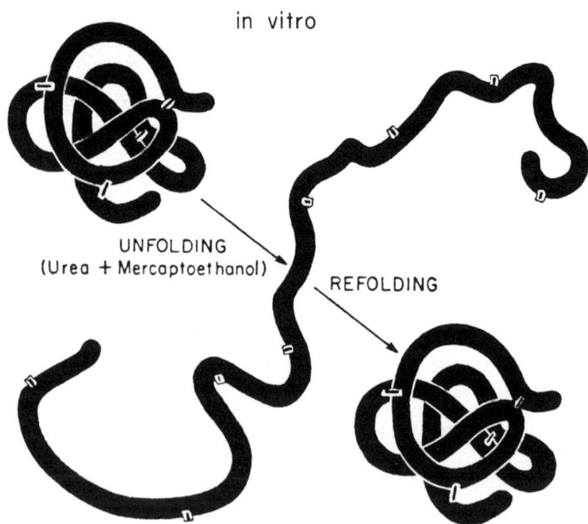

Fig. 5.10. Schematic drawing of the unfolding–folding process of RNase as a model system of the folding of the nascent chain (according to Epstein *et al.*, 1963).

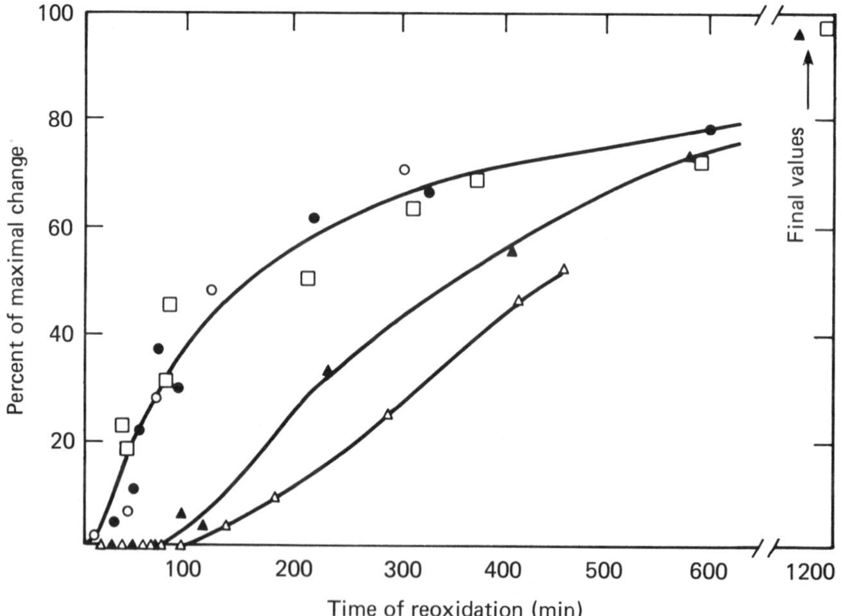

Fig. 5.11. Return of the different properties of RNase after reoxidation (according to Anfinsen *et al.*, 1961). Sulfhydryl titration with *p*-chloromercuribenzoate (●), with radioactive iodocetate (○), change in optical rotation (□) enzymatic activity measured on RNA (▲) and on uridylic-2′-3′ cyclic phosphate (△).

The oxidation of SH groups precedes the return of the enzymatic activity suggesting an incorrect pairing of sulfhydryl groups in the earlier steps of the process.

Haber and Anfinsen (1962) have shown that reoxidation of RNase in 8 *M* urea or 4 *M* GuHCl gives a product which contains a great number of species with incorrectly formed S—S bonds. This scrambled ribonuclease is able to regain its native structure by exposure of the material to a small amount of reducing reagent and removal of urea to allow disulfide interchange (Fig. 5.12).

Hantgan and co-workers (1974) were able to demonstrate (by peptide mapping) the initial formation of a large number of incorrect pairs of cysteine residues followed by a slow rearrangement to the native structure of ribonuclease. Kinetics of refolding of reduced RNase were reinvestigated by Creighton (1977a, 1979b) using the method he developed to follow the appearance and disappearance of molecules linked by one, two, three, and four intramolecular disulfide bridges. Creighton observed the formation of incorrectly paired species and the return of the native structure by slow intramolecular interconversion of molecules with three or four disulfide bridges. The method used by Creighton to follow kinetics of refolding of reduced proteins (ribonuclease and BPTI) is detailed in Chapter 8. Hen egg white lysozyme can be denatured and reduced by treatment with 0.6 *M* β-mercaptoethanol in 8 *M* urea or 6 *M* GuHCl. As for RNase, the return of the native conformation is also accelerated by the presence of β-mercaptoethanol (Epstein and Goldberger, 1963).

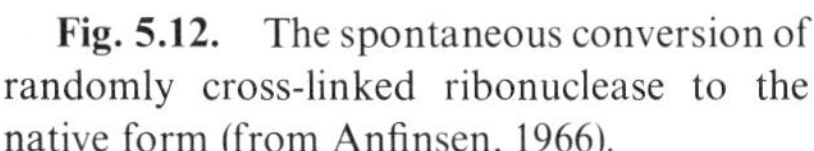

Fig. 5.12. The spontaneous conversion of randomly cross-linked ribonuclease to the native form (from Anfinsen, 1966).

In lysozyme (Ristow and Wetlaufer, 1973; Anderson and Wetlaufer, 1976) only a very limited number of structures are formed in the early stages of the reoxidation process. Acharya and Taniuchi (1976) studying air reoxidation of lysozyme at pH 8.0 and 37°C found an inactive form which contains at least 24% of the population with incorrect disulfide bonds involving Cys 6, Cys 30, Cys 115, and Cys 127. In the case of bovine pancreatic trypsin inhibitor, the transient intermediates were detected by trapping free sulfhydryl groups by carboxymethylation and they were identified by chemical analysis (Creighton, 1974a,b,c, 1975a,b,c, 1977a,b,c). The results indicate also a limited number of incorrect disulfide bridges formed in the early stages of reoxidation (see following discussion).

5.4.3.2 Rearrangement of Disulfide Bonds during the Refolding Process

The rearrangement of disulfide bonds during the refolding of a protein can be monitored under different conditions. Several possible methods for reshuffling have been described (1) oxidation by air, (2) oxidation by air catalyzed by the presence of trace metal ions, (3) the presence of reducing reagents, (4) the presence of a mixture of reduced and oxidized thiol compounds, and (5) a particular enzyme.

In the initial research of Anfinsen and co-workers (1961), on ribonuclease as in most studies of the same period, disulfide bonds reformation were produced by air reoxidation. In air reoxidation regeneration, correction of incorrect disulfide bonds occurs through disulfide–sulfhydryl interchange, either by an intramolecular process such as

$$\underset{\diagdown SH}{\overset{S-S}{\bigvee_{P}}} \rightleftharpoons \underset{\bigvee_{S}}{\overset{SH\diagdown}{P-S}}$$

or by an intermolecular process such as

$$\underset{\diagdown SH}{\overset{S-S}{\bigvee_{P}}} + \overset{SH}{\underset{|}{P}}\genfrac{}{}{0pt}{}{\diagup SH}{\diagdown SH} \rightleftharpoons \genfrac{}{}{0pt}{}{HS\diagdown}{HS\diagup}P\diagup S-S-P\genfrac{}{}{0pt}{}{\diagup SH}{\diagdown SH}$$

$$\rightleftharpoons \; HS\diagdown P\genfrac{}{}{0pt}{}{\diagup S}{\diagdown S}\Big| + \overset{SH}{\overset{|}{P}}\genfrac{}{}{0pt}{}{-SH}{\diagdown SH}$$

(Saxena and Wetlaufer, 1970).

Regeneration of reduced RNase by air oxidation was shown to depend strongly on the temperature (Ahmed *et al.*, 1975). For RNase, the process is considerably slower than for lysozyme which undergoes a one-half regeneration by air oxidation in 15–20 min. Air reoxidation of reduced polyalanyl-chymotrypsinogen and reduced chymotrypsinogen as well occurred very slowly even at pH 10 and at low concentrations of protein. The formation of protein aggregates interferes with the renaturation, but it can be limited by the presence of moderate concentrations of urea or GuHCl during the reoxidation (Orsini *et al.*, 1975).

The reduced BPTI slowly regains activity at approximately neutral pH by air reoxidation. Complete activity is recovered after 1 day (Creighton, 1974a).

Denatured and reduced pig pancreatic elastase regains activity by air reoxidation after removal of the denaturant and reducing reagent either by dialysis or by dilution (Ghélis, 1980).

Air oxidation can be catalyzed by trace metals such as Cu^{2+}, Co^{2+}. The catalytic role of metals in reoxidation of a sulfhydryl was described by Cecil and Mc Phee (1959). Takagi and Isemura (1963, 1964) and Yutani and co-workers (1968) reoxidized lysozyme in the presence of 5 μM Cu^{2+} at pH 8.0 and room temperature. When metal ions are precluded from participation by addition of EDTA no significant reoxidation is observed after 30 min (Saxena and Wetlaufer, 1970). The catalytic effect of various metal ions was studied by Saxena and Wetlaufer (1970). Of the six metal ions tested (Cu^{2+}, Co^{2+}, Mn^{2+}, Fe^{3+}, Zn^{2+}, Ni^{2+}), only Cu^{2+} enhanced the rate of regeneration of RNase activity. However, both Co^{2+} and Cu^{2+} catalyzed air oxidation of the enzyme, since both can accelerate thiol oxidation. The efficiency of these metals has been noted for concentrations ranging between 0.1 and 10 μM.

The rearrangement of disulfide bonds during the refolding of a protein can be accelerated by facilitating the thiol–disulfide exchange either with a trace of reducing or oxidizing reagent or with a regeneration mixture of reduced and oxidized thiols. The basic reaction of thiol–disulfide exchange is

$$R_1S^- + R_2SSR_2 \rightleftharpoons R_2S^- + R_1SSR_2$$

It involves the ionized form of the thiol and therefore is pH dependent. The reoxidation of sulfhydryl in proteins can be monitored either with a linear disulfide reagent and proceeds as follows:

$$P\begin{matrix} \diagup SH \\ \diagdown SH \end{matrix} + RSSR \rightleftharpoons P\begin{matrix} \diagup SS{-}R \\ \diagdown SH \end{matrix} + RSH$$

$$P\begin{matrix} \diagup SSR \\ \diagdown SH \end{matrix} \rightleftharpoons P\begin{matrix} \diagup S \\ \diagdown S \end{matrix}\Big| + RSH$$

or with a cyclic reagent such as

$$\mathrm{P}\begin{matrix}\diagup \mathrm{SH}\\ \diagdown \mathrm{SH}\end{matrix} + \mathrm{R}\begin{matrix}\diagup \mathrm{S}\\ \ \ |\\ \diagdown \mathrm{S}\end{matrix} \rightleftharpoons \mathrm{P}\begin{matrix}\diagup \mathrm{SSRSH}\\ \diagdown \mathrm{SH}\end{matrix}$$

$$\mathrm{P}\begin{matrix}\diagup \mathrm{SSRSH}\\ \diagdown \mathrm{SH}\end{matrix} \rightleftharpoons \mathrm{P}\begin{matrix}\diagup \mathrm{S}\\ \ \ |\\ \diagdown \mathrm{S}\end{matrix} + \mathrm{R}\begin{matrix}\diagup \mathrm{SH}\\ \diagdown \mathrm{SH}\end{matrix}$$

The thiol disulfide reaction is very rapid (Eldjarn and Pihl, 1957) under the pH conditions that lead to ionization of thiols.

The presence of trace of β-mercaptoethanol accelerated the renaturation process of RNase and lysozyme (Haber and Anfinsen, 1962; Epstein *et al.*, 1962; Epstein and Goldberger, 1963). In the case of RNase, β-mercaptoethanol catalyzed the disulfide interchange in incorrectly reoxidized enzyme. The effect of β-mercaptoethanol on the reactivation of lysozyme is shown in Fig. 5.13. It prevents aggregation of the protein and is required to obtain the maximal rate of reactivation of lysozyme, in contrast with RNase where dilution alone is able to overcome the rate limiting effect of aggregation.

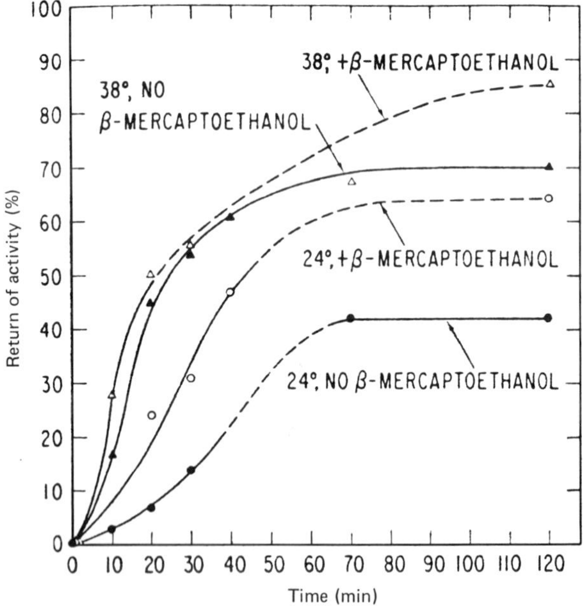

Fig. 5.13. Effect of β-mercaptoethanol on the reactivation of reduced lysozyme at two temperatures 24°C and 38°C (according to Epstein and Goldberger, 1963).

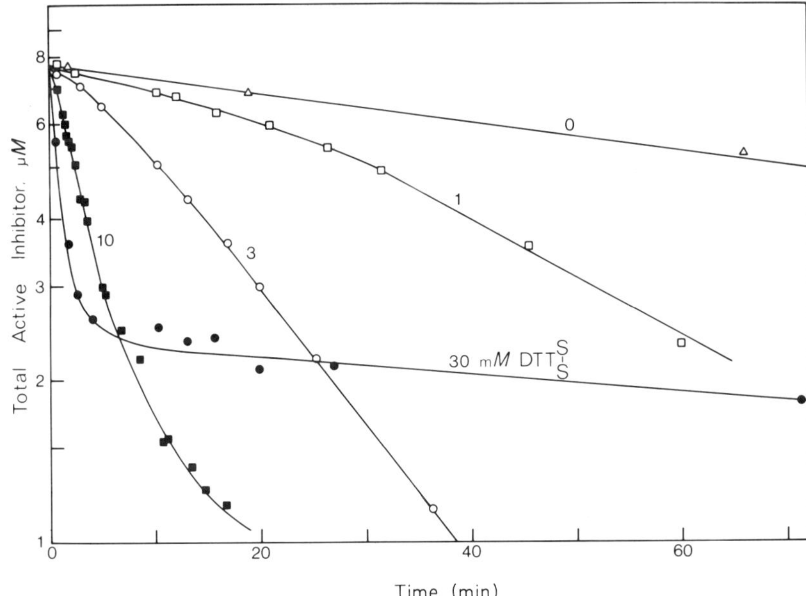

Fig. 5.14. Effect of oxidized DTT on renaturation of reduced BPTI at pH 8.5 (courtesy of Creighton, 1974a); concentrations of oxidized DTT are indicated (in m*M*) in the graph. The renaturation proceeds more rapidly in the presence of disulfide reagent. In each case, the reaction follows a pseudo-first-order kinetics with respect to reduced BPTI, except at higher concentrations of disulfide reagent when the reaction becomes biphasic (see the curve for 30 m*M* DTT).

Mixed disulfide derivatives of proteins and cystine for example can be reactivated by a reducing agent (Bradshaw *et al.*, 1967).

Reoxidation of reduced and denatured BPTI proceeded much more rapidly in the presence of disulfide reagents (Creighton, 1974a). Hydroxyethyl disulfide, oxidized dithiothreitol (DTT), glutathione, and lipoic acid were tested. The acceleration of the reaction depends on the nature of the disulfide reagent; oxidized glutathione and hydroxyethyl disulfide react at similar rates and approximatively ten times faster than oxidized DTT. The effect of oxidized DTT on renaturation of reduced BPTI is shown in Fig. 5.14. The addition of low concentrations of reduced DTT (0.4 m*M*) to the renaturation mixture (which contained 10 m*M* oxidized DTT) inhibited the rate but not the extent of the renaturation of the inhibitor. Higher concentrations of reduced DTT completely prevented renaturation (Creighton, 1974a) (see Fig. 5.15).

To facilitate the thiol–disulfide exchange Saxena and Wetlaufer (1970) used a mixture of reduced (GSH) and oxidized (GSSG) glutathione in the

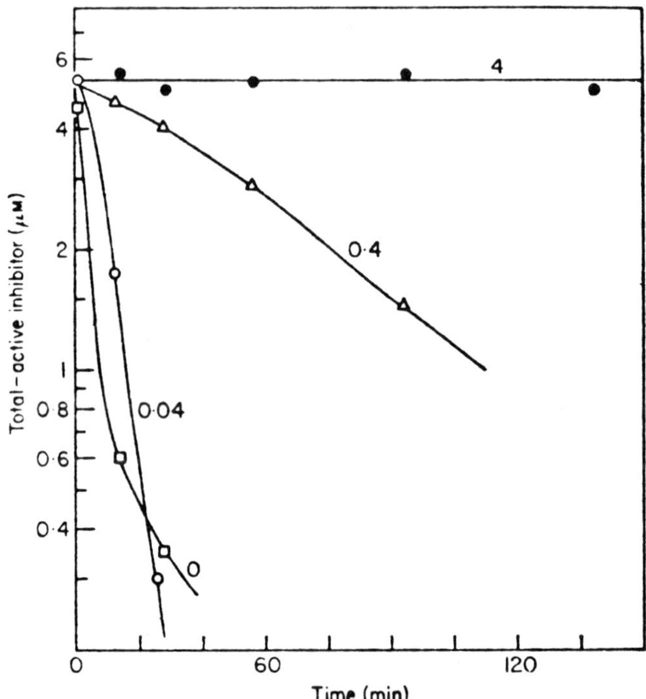

Fig. 5.15. Inhibition of renaturation of reduced BPTI, by reduced DTT. Reduced inhibitor (0.35 m*M* in 6.0 *M* GuHCl, 1 m*M* EDTA, 1 m*M* HCl) was diluted 1/50 into 0.1 *M* NH_4HCO_3 1 m*M* EDTA, 10 m*M* oxidized DTT at pH 8.5°; m*M* concentrations of reduced DTT are indicated in the graph. Remaining inactive inhibitor is plotted as a function of time. After 1 day all mixtures have regained full activity except that containing 4 m*M* reduced DTT (courtesy of Creighton, 1974a).

absence of air. Saxena and Wetlaufer carefully investigated the most satisfactory conditions for the refolding of reduced and denatured hen egg white lysozyme. They obtained maximal rates of reactivation for a mixture of 5 m*M* GSH and 0.5 m*M* GSSG (Fig. 5.16). Under these conditions 50% of the activity was recovered in less than 5 min and under maximal conditions (i.e., 37°C and pH 7.4–7.85), ca. 60–85% was recovered after 30 min. No effect on the efficiency of the redox regeneration system was observed when EDTA was used (Fig. 5.17). Regeneration systems containing either cystine and cysteine, or cystamine and cysteamine, result in yields and rates similar to those obtained with oxidized and reduced glutathione at the same level.

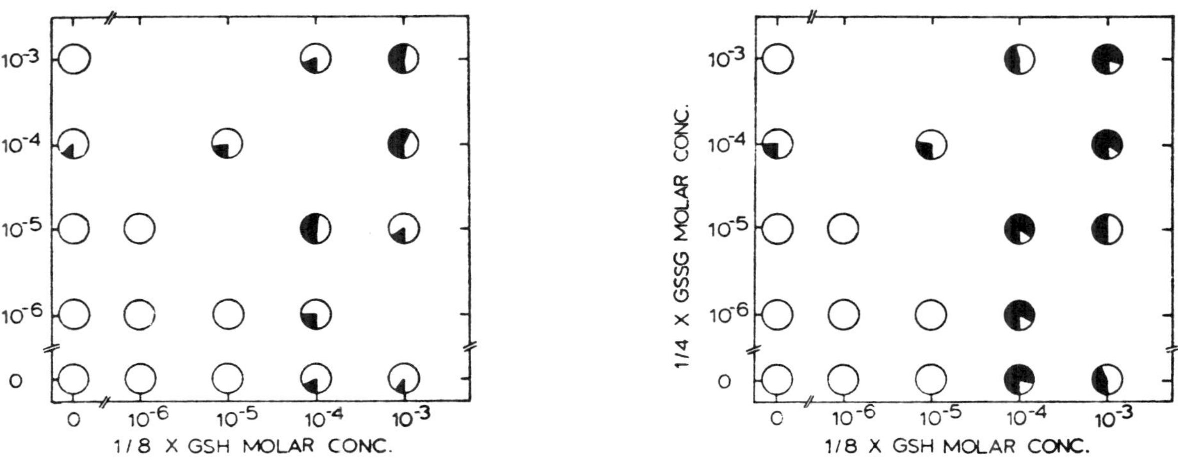

Fig. 5.16. Effect of different initial concentrations and ratio of GSSG and GSH on the extent of recovery of lysozyme activity (according to Saxena and Wetlaufer, 1970): (a) after 5 min; (b) after 30 min. Regeneration conditions were 37°C, 0.08 *M* Tris buffer at pH 8.0. Initial lysozyme concentration was 1 μM. Enzymatic activity was measured on aliquots quenched at pH 5.0.

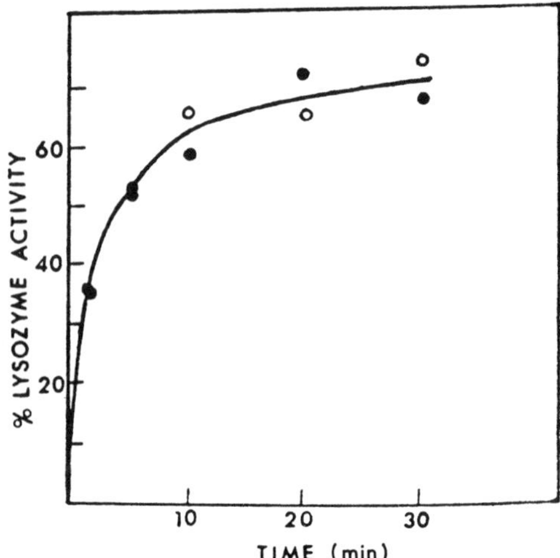

Fig. 5.17. Effect of EDTA on lysozyme renaturation in regeneration system containing 6 m*M* GSH and 0.6 m*M* GSSG (open circles 0.1 m*M* EDTA). The rate of reactivation is the same in the absence and in the presence of EDTA (0.1 m*M*) (from Saxena and Wetlaufer, 1970).

TABLE 5.9

Regeneration Systems Used for Renaturation of Reduced Chymotrypsinogen[a]

Reagent concentration	Renaturation yield (% of control)
$CuSO_4$ 0.1–10 μM	2
Oxidized dithiothreitol 1–10 m*M*	2
Oxidized dithiothreitol 5 m*M* and reduced dithiothreitol 0.5 m*M*	25
Oxidized dithiothreitol 10 m*M* and reduced dithiothreitol 0.5 m*M*	35
Oxidized glutathione 0.5 m*M* and reduced glutathione 5 m*M*	45–50
Oxidized glutathione 5 m*M* and reduced glutathione 50 m*M*	6

[a] Chymotrypsinogen was reduced with 10 m*M* or 100 m*M* reduced dithiothreitol in 6 *M* GuHCl at pH 8.5. The resulting protein showed a titer of 9.9–10.2 SH/mole. The reduced protein was diluted to 20–50 μg/ml in 2 *M* GuHCl solutions at pH 8.5 in the presence of the indicated concentrations of reagent and dialysed overnight against 20 m*M* Tris HCl 0.1 *M* KCl. From Orsini *et al.* (1975).

The efficiency of various regeneration systems was investigated for reactivation of reduced chymotrypsinogen (Orsini *et al.*, 1975). The redox system of 0.5 m*M* reduced dithiothreitol and 5 m*M* oxidized dithiothreitol was the most effective, allowing a renaturation yield of 45–50% at pH 8.5 (see Table 5.9).

Renaturation of reduced and denatured Sepharose–trypsinogen requires continual disulfide interchange probably to correct incorrect pairing of sulfhydryl groups during refolding. Different regeneration systems were tested (Fig. 5.18). The regeneration curves shown in Fig. 5.18 indicate the requirement of both a reducing and an oxidizing reagent. The most effective among the systems tested was the mixture of 4 m*M* GSH and 0.4 m*M* GSSG (Light *et al.*, 1974). The role of specific disulfide bonds has been examined (Odorzynski and Light, 1979; Light and Odorzynski, 1979).

Hantgan and co-workers (1974) used equimolar concentration of reduced and oxidized glutathione (5 m*M*) and found, as did Haber and Anfinsen (1962), that the initial formation of a large number of incorrect disulfide bonds is followed by a slower rearrangement to the native structure. Figure 5.19 shows the disappearance of SH groups with time, the variations of the different optical properties, and the return of the enzymatic activity. The presence of the reshuffling enzyme extracted from beef liver microsomes accelerates the interchange by a factor of about 5–10. The existence of an enzyme found in different tissues, which is associated with the membrane of endoplasmic reticulum and able to accelerate the formation of disulfide bridges, was discovered independently by two research groups: in rat and beef liver by Anfinsen and co-workers (Goldberger *et al.*, 1963; de Lorenzo

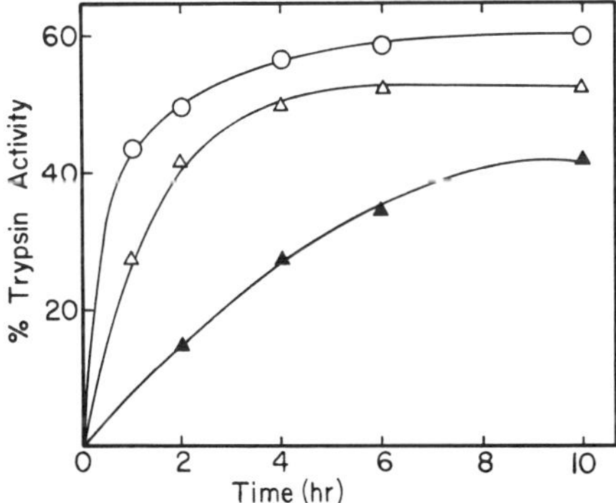

Fig. 5.18. Refolding of reduced Sepharose trypsinogen in different regeneration systems (from Sinha and Light, 1975): (▲) 0.5 m*M* β-mercaptoethanol; (△) 0.5 m*M* β-mercaptoethanol + 1 m*M* dehydroascorbic acid; (○) 4 m*M* GSH + 0.4 m*M* GSSG. In all cases Tris 0.05 *M* $CaCl_2$ buffer at pH 8.5, 35°C was used. The best regeneration mixture is the redox one containing 4 m*M* GSH and 0.4 m*M* GSSG.

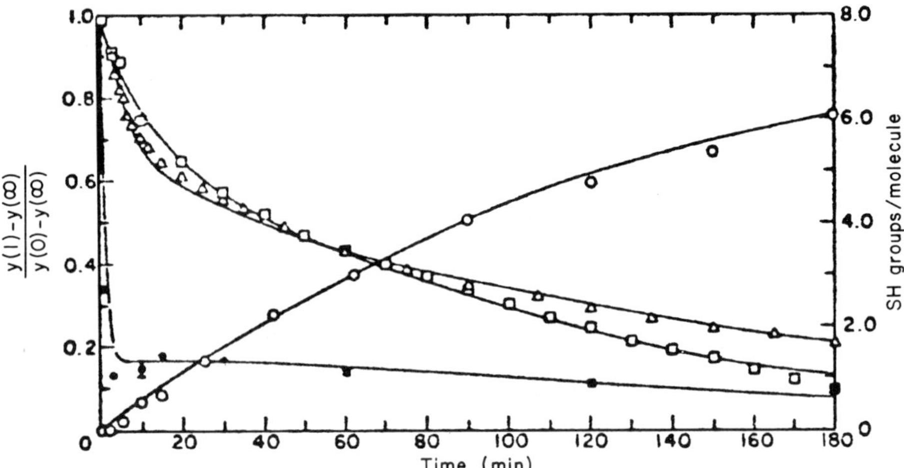

Fig. 5.19. Time dependence of the reoxidation of ribonuclease in the presence of oxidized and reduced glutathione (from Hantgan *et al.*, 1974). Key: (○), fraction of activity; (□), absorbance; (△), fluorescence; (■), number of SH groups per protein molecule. SH groups disappear very rapidly prior to the variations of the other signals. The slower rate corresponds to the return of enzymatic activity.

et al., 1966) and in pancreas of rat, chicken, and pigeon by Straub and coworkers (Venetianer and Straub, 1963, 1964). This enzyme has been purified and well characterized. It is of MW 42,000 and contains a sulfhydryl group essential for the enzymatic activity (de Lorenzo *et al.*, 1966; Fuchs *et al.*, 1967). This enzyme accelerates the rate of formation of disulfide bonds in proteins. It was assayed for lysozyme, ribonuclease (Goldberger *et al.*, 1964), soybean trypsin inhibitor (Steiner *et al.*, 1965), and for serum albumin (Teale and Benjamin, 1976a,b). This enzyme probably works as a reshuffling enzyme. catalyzing disulfide–sulfhydryl interchange. Presumably it is able to correct the incorrect pairing of bridges if some are formed in the first steps of the refolding process (Givol *et al.*, 1964, 1965). By its function and by its location, this enzyme could play a role *in vivo* in the folding process; however, its physiological function remains unclear. Other enzymes acting as protein disulfide oxidoreductases have been described (Tietze, 1969; Carmichael *et al.*, 1977, 1979). At present no data permit one to make a decision about their implication for correcting the incorrect disulfide bridges of the nascent polypeptide chain in ribosomes.

Aside from the possible role of the reshuffling enzyme which is not elucidated, the biological significance of the requirement for reducing conditions was emphasized by Saxena and Wetlaufer (1970). They proposed that protein biosynthesis occurs under substantially reducing conditions. Even the *in vitro* cell-free biosynthesizing systems seem to function optimally with mercaptoethanol or reduced DTT (Matthei and Nirenberg, 1961). However, the conditions for regeneration are different for each protein. The same glutathione system which regenerates lysozyme rapidly ($t = 5$ min) regenerates reduced RNase at a 15-fold slower rate. Trypsin inhibitor regeneration occurs more rapidly in the presence of oxidized reagent and at a slower rate in the presence of reduced thiol. The efficiency of oxidized DTT or of hydroxyethyl disulfide in regeneration of BPTI indicates that the formation of S—S bridges follows a direct pathway. It is not necessary to add a reducing reagent to correct the incorrect disulfide bridging. In fact, S—S bridge (Cys 30–Cys 51) is formed initially and represents 70% of the one-disulfide intermediates. Baldwin (1975) suggested that the concentrations of the possible one-disulfide and two-disulfide intermediates are governed by their thermodynamic stabilities. The incorrect disulfide intermediates disappear when the second bridge is formed because they are unstable. The particular stability of the Cys 30–Cys 51 bridge and its rapid formation may be correlated to its location between a β-structured segment (28–37) and a short helix (47–55). Furthermore, incorrect pairing in the intermediates occurs between Cys 5 and Cys 30; Cys 5 being included in the helix (2–7) or Cys 5–Cys 14 between a α helix and a β structure. When the correct Cys 14–Cys 38 bridge is formed in a two-disulfide intermediate, it must be disrupted until the last step of the folding.

Ribonuclease and lysozyme refolding from the reduced protein requires disulfide interchange. In the early stages highly scrambled species of RNase are formed. If refolding of BPTI and RNase are similar in that they do not occur by a simple stepwise reformation of disulfide bonds, then they are significantly different concerning the pathway and the intermediates formed in early stages.

5.4.4. Spontaneous Refolding of Denatured and Reduced Proteins

When suitable conditions are identified to rearrange the incorrect disulfide pairing and to prevent aggregation (and autolysis in particular cases of proteases), then the unfolding–folding process is also obtained for disulfide linked proteins. As mentioned previously, it has been clearly demonstrated for ribonuclease, for which a yield of renaturation of 95–100% was obtained.

The reversibility of the process under defined conditions was also established for reduced and denatured lysozyme (White, 1961, 1976; Epstein and Goldberger, 1963; Yutani *et al.*, 1968). In this case, the best conditions for regeneration were carefully determined by Saxena and Wetlaufer (1970) and a suitable mixture of oxidized and reduced thiols allowed them to obtain

60–85% regeneration in periods of 30 min. For lysozyme as for RNase, the reactivation rate was inversely proportional to the protein concentration; concentrations smaller than 0.1 mg/ml must be used to avoid aggregation (Epstein and Goldberger, 1963). Larger concentrations can be used for BPTI since complete regeneration is obtained with concentrations 30 times larger than those permissible for RNase (Creighton, 1974a,b,c, 1975a,b,c, 1977a,b,c,d; Anfinsen, 1967) and lysozyme (Saxena and Wetlaufer, 1970). To overcome this difficulty and also to avoid possible autolysis (as in the case of some proteases and zymogens such as trypsin, trypsinogen, and chymotrypsinogen) renaturation was achieved either by immobilizing the enzyme on a solid matrix (Epstein and Anfinsen, 1962a; Sinha and Light, 1975; Light and Sinha, 1976 Mozhaev and Martinek, 1981), or by polyalanylation (Epstein and Anfinsen, 1962b; Cook *et al.*, 1963; Orsini *et al.*, 1975). Insoluble carboxymethylcellulose trypsin after complete reduction of disulfide bonds in 8 *M* urea and β-mercaptoethanol recovers 4% of its original activity; however, this weak reactivation is significant, since it is much greater than would be expected for a random reoxidation process (0.01%). Carboxymethylcellulose RNase is able to recover up to 40% of the original activity under the same conditions (Epstein and Anfinsen, 1962a). Poly-DL-Alanyl trypsin reduced with β-mercaptoethanol in 8 *M* urea recovered 8% of the starting activity after removal of the reducing and denaturing reagents; this activity is also weak but significant.

Sinha and Light (1975) obtained better yield by using trypsinogen and trypsin immobilized on Agarose beads. With 0.2–0.6 mg of protein bound per ml of gel up to 60–70% regeneration yield was obtained in 24 h. Furthermore, the incorrectly folded structures continued to refold when placed in a mixture of reduced and oxidized glutathione to allow disulfide interchange.

The regeneration of denatured and reduced chymotrypsinogen covalently attached to a solid matrix (porous succinyl glass beads), was also reported (Brown *et al.*, 1972; Brown and Horton, 1972); the recovered material obtained with a yield of 53% had the same properties as the original one (i.e., esterolytic activity of the activated zymogen). In a detailed study of the renaturation of reduced and denatured chymotrypsinogen in solution Orsini and co-workers (1975) obtained a regeneration yield of 35–50% under suitable conditions (0.5 m*M* oxidized–5 m*M* reduced glutathione) (pH 8.5) (Table 5.9). Formation of aggregates interfered with the renaturation process. In fact, reduced chymotrypsinogen requires for its renaturation more exacting conditions than lysozyme or ribonuclease.

Another serine protease, elastase, when denatured and reduced, is able to refold spontaneously after removal of denaturant and reducing reagent with a yield of 65% under optimal conditions. This has been shown by using several probes of the native conformation. The recovery of the overall con-

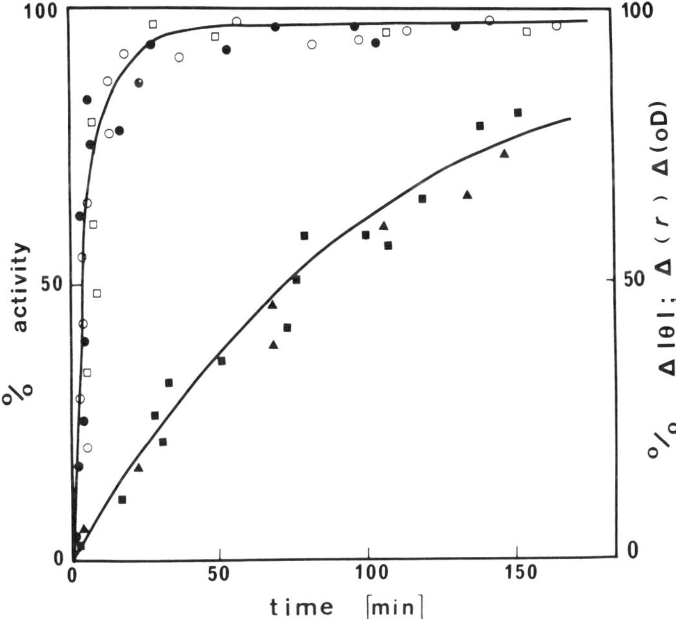

Fig. 5.20. Return of different properties of elastase upon renaturation of denatured and reduced protein: (▲), enzyme activity; (■), antigenic activity; (□), differential absorption at 282 nm; (●), circular dichroic signal at 235 nm; (○), chemical reactivity of Met 180 with [^{14}C]iodoacetamide, of Val 16 with [^{14}C]acetic anhydride, and of Tyr 234 with [^{14}C]dimethyl sulfate (from Ghélis, 1980).

formation followed by the optical signals or by the decrease of the chemical reactivity of several amino acid side chains buried in the native protein precedes the return of both antigenic and enzymatic activity (Ghélis, 1980) (Fig. 5.20).

Proinsulin, which is a single polypeptide chain, is able to refold spontaneously after reduction of disulfide bridges (Steiner and Clark, 1968), although A and B chains of insulin show only a very weak tendency to refold and associate correctly. The long connecting peptide, which represents 33 amino acids of the 54 of the entire polypeptide chain, certainly plays a determining role in ensuring a correct conformation of the insulin molecule.

For all these proteins the spontaneous refolding of the reduced and denatured molecule (i.e., of the linear polypeptide chain) was obtained. However, more or less exacting conditions are required for the different proteins indicating more or less restricting conditions in their folding pathway. Table 5.10 summarizes the data obtained for several proteins and gives the yield of regeneration.

TABLE 5.10

Refolding of Denatured and Reduced Proteins[a]

Protein	Disulfide bonds	Recovery of activity	Random recovery
Ribonuclease (bovine pancreatic)	4	95–100%	ca. 1%
Lysozyme (egg white)	4	50–80%	ca. 1%
Takaamylase A (*Asp. oryzae*)	4	48%	ca. 0.3%
Trypsin (bovine)	6		ca. 0.01%
a. CM-cellulose trypsin (insoluble)		4%	
b. Poly-DL-alanyl trypsin (soluble)		8%	
Insulin (bovine)	3	5–10%	ca. 6.7%
Alkaline phosphatase (*E. coli*)	2	80%	ca. 33%
Pepsinogen (swine)	3	50%	ca. 6.7%
Serum albumin (human)	17	50%	
Chymotrypsinogen	5	35–50%	ca. 0.1%
Elastase	4	65%	ca. 1%

[a] Adapted from Epstein *et al.* (1963). Recovery of activity was measured experimentally; it is the return of enzymatic activity of a protein previously reduced. Random recovery was estimated, taking into account the probability of disulfide bond formation according to Sela and Lifson (1959) and Kauzmann (1959b).

For BPTI, as for ribonuclease, the data emphasize the importance of the noncovalent interactions in directing the correct disulfide pairing. In fact, the formation of disulfide linkages, which involves the proximity of the two sulfhydryl groups, occurs for a protein which is partially or totally refolded. Two sets of experimental data are particularly significant in this regard for BPTI. First, the formation of disulfide was followed in the presence of 8 *M* urea or 6 *M* GuHCl (Creighton, 1977c) as it was for RNase (Haber and Anfinsen, 1962). A completely random formation of the species was observed. The data are quantitatively consistent with the formation of all possible disulfide bonds as expected on a statistical basis. The great difference between the species encountered in the presence and in the absence of the denaturant clearly demonstrates that noncovalent interactions guide the refolding of the protein. Second, the rearrangement of the random one and two disulfide species was followed after removal of the denaturant. The species are rapidly converted to the same intermediates as those normally observed during the refolding of the fully reduced protein, for which in contrast to RNase there exists a preferential pathway of folding.

The times needed for the refolding of the reduced proteins are not significantly slower than those observed for proteins without disulfide linkages. It depends on the proteins and on the experimental conditions. Refolding process can be accelerated by various factors, as for example anions. Schaffer and co-workers (1975) reported that K_2HPO_4 and $(NH_4)_2SO_4$ strongly accelerate the refolding of ribonuclease; total regeneration can be achieved in 10 min.

The four disulfide bonds of nine homologous short neurotoxins were cleaved by reduced dithiothreitol. The kinetics of refolding, induced by air reoxidation or in the presence of thiol–disulfide exchange reagents, indicated differences in refolding rates. Three toxins refolded 4–10 times more slowly than the six others. It was proposed from consideration of the sequence that a single additional amino acid insertion is responsible for these differences (Menez *et al.*, 1980). These data indicate the role of noncovalent interactions in the formation of disulfide bridges.

The reformation of all disulfide bridges does not appear to be a necessary requirement for the return of functional properties. Thus in BPTI, 14–38 S—S bridging is not necessary for activity. The inhibitor with the 14–38 disulfide bond broken has a nativelike conformation (Kress and Laskowski, 1967; Brunner *et al.*, 1974). Reduced and denatured Takaamylase A, can be reactivated (Isemura *et al.*, 1961, 1963); reformation of all original disulfide bonds are not essential for recovery of enzymatic activity (Yutani *et al.*, 1965). In lysozyme, it was initially reported by Yutani and co-workers (1968), that the return of the activity requires the reformation of at least two of the four disulfide bridges of the molecule. Enzymatically active intermediates during reoxidation of the reduced protein were carefully identified by Acharya and Taniuchi (1976, 1977, 1978, 1980). From their data, it appears that noncovalent interactions can maintain a functional structure of lysozyme even when one of the three native disulfide bridges between Cys 6 and Cys 127, Cys 76 and Cys 94, and Cys 64 and Cys 80 is open; however, the data also prove the essential role of bridge Cys 30–Cys 115.

In enterokinase, which is a membrane-bound enzyme, disulfide bonds holding the light and heavy chains together are not essential for the activity (Savithi and Light, 1980).

Garel (1978) detected the enzymatic activity of RNase during the early steps of the refolding under conditions where the protein was completely or partially reduced. This activity is very weak and represents 0.04% of that native enzyme. [Chavez and Scheraga (1980a,b) using immunochemical methods also found some small degree of native structure in RNase (see

Chapter 9)]. Garel's results emphasize that the refolding of a reduced protein is an intimate mixture of conformational processes and cross-link formation. The data illustrate the role of noncovalent interactions in the formation of a native and functional protein since a small fraction of reduced protein can rapidly fold into a native and active conformation. This conformation, which is probably in equilibrium with nonnative forms, must be stabilized by disulfide bond formation. According to Creighton (1977d, 1979b), two disulfide bonds are not sufficient to stabilize significantly native ribonuclease. Furthermore according to Garel's results and Creighton's data, the fraction of native conformation seems to remain constant (0.04–1.5%) until two disulfide bonds are formed. Since disulfide interchange can be a fast reaction (Eldjarn and Pihl, 1957), the aforementioned data suggest that the correct disulfide pairing occurs only on a small fraction of native molecules refolded from reduced RNase, in the total absence of disulfide formation. As for other proteins, the noncovalent conformational processes would determine the folding. These data also suggest the formation of species with illicit noncovalent interactions allowing incorrect pairing of sulfhydryl groups.

5.5. INFLUENCE OF VARIOUS FACTORS ON PROTEIN FOLDING

The influence of the external residues on protein folding has been investigated. Chemical modification of the 11 free amino groups located at the surface of the molecule does not modify the capacity of RNase to undergo completely reversible refolding of the denatured and reduced molecule. These amino groups were modified by addition of poly-DL-alanine chains containing 8 alanine residues per chain (Fig. 5.21). Likewise poly-DL-tyrosyl RNase could be reactivated after denaturation and reduction. The addition of charged groups such as succinyl or phtalyl groups or of uncharged groups such as butyryl and caproyl, if it modifies the enzymatic activity to varying extent, does not interfere with the ability of the unfolded and reduced molecule to regain its native conformation (Epstein *et al.*, 1963). These experiments, summarized in Table 5.11, suggest that the protein may accomodate mutations or chemical modifications of residues located outside. Refolding of unfolded polyalanyl chymotrypsinogen and unmodified chymotrypsinogen follows the same pattern (Orsini *et al.*, 1975).

Besides the modification of the protein itself, variations of external parameters such as ionic strength, presence of anions or cations, and pH, can modify the rate and yield of protein renaturation. Ahmed and co-workers (1975) investigated the action of some effectors on the refolding of reduced RNase in the glutathione regeneration system. Ahmed and co-workers (1975)

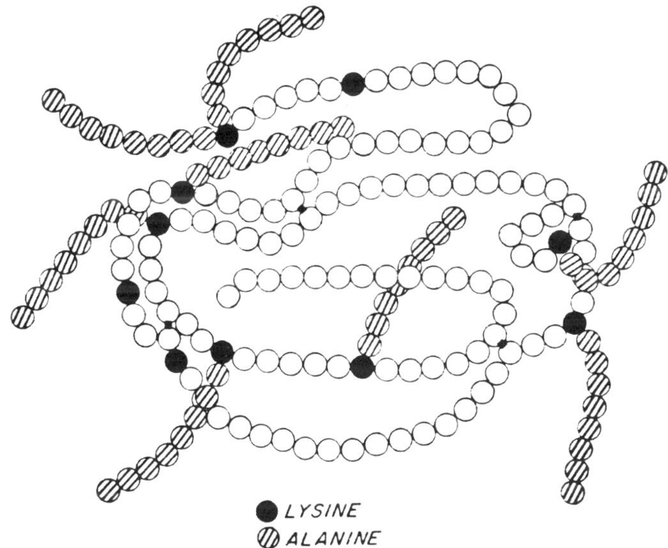

Fig. 5.21. Schematic representation of poly-DL-alanyl ribonuclease (see text) (from Cooke *et al.*, 1963; Epstein *et al.*, 1963).

reported that glutathione regeneration (3 m*M* GSH, 0.3 m*M* GSSG) was facilitated by the presence of several salts, especially phosphate and pyrophosphate which appear to act synergistically at higher salts concentration. These salts were found to stabilize native ribonuclease. With the goal of mimicking the biological environment, Ahmed and co-workers (1975) investigated the action of added lipoprotein, hen egg lecithin suspensions, and even *E. coli* ribosomes; however, no effect on ribonuclease regeneration was observed.

There are many examples of the stabilization of the native structure by specific ligands, substrates, or effectors (see Yon, 1969; Citri, 1973). However, addition of these ligands at early stages of protein folding may induce conformations that are different from the native one. For example, reoxidation of RNase in the presence of various nucleotides gives conformational isomers with altered functional properties (Gutte, 1978).

It is worthwhile to make some preliminary generalization from these data. If proteins have a three-dimensional structure which is stabilized by disulfide bridges then the fully denatured and reduced polypeptide chain is also able to refold spontaneously under favorable conditions. However, for each protein, more or less restricting conditions are required for the optimization of the process. The factors which govern the refolding process do not seem fundamentally different for proteins linked or not linked by

TABLE 5.11

Reduction and Reoxidation of Chemically Modified Enzymes

Preparation	Group covered	Number of groups covered (average)	Chain length (average)	Charge substitution	Activity of protein (% of native)	Activity after reoxidation (% of initial)
Poly-DL-alanyl ribonuclease[a]	$-NH_2$					
PAR 4	—	8 of 11	5		100	95
PAR 5	—	8 of 11	7–8		65	40
Poly-DL-tyrosyl ribonuclease[a]	—					
PTR 2	—	1 of 11	3		100	100
PTR 3	—	2 of 11	3		100	60
Poly-DL-alanyl trypsin[b]	—	12 of 15	9–10		72.5	8
Succinyl ribonuclease	—	6–7 of 11		− for +	15	62
Methylated ribonuclease	$-COOH$	7–8 of 11		0 for −	18	48
Methylated succinyl ribonuclease	$-NH_2$ $-COOH$	5–6 of 11 7–8 of 11		− for + 0 for −	2	73
DNS ribonuclease	$-NH_2$	2 of 11		0 for +	100	15[c]
Phthalyl ribonuclease	—	3–4 of 11		− for +	24	75
Butyryl ribonuclease	—	5–6 of 11		0 for +	44	77
Caproyl ribonuclease	—	2–3 of 11		0 for +	96	70

[a] Anfinsen *et al.* (1962).
[b] Epstein and Anfinsen (1962b).
[c] Unpublished data from Dr. F. H. White, Jr.

disulfide bonds, and the noncovalent interactions are decisive for the folding of the polypeptide chain. The disulfide pairing occurs when two sulfhydryl groups come to a sufficient proximity. When incorrect disulfide links are formed, they probably result from illicit noncovalent interactions, which are thus stabilized, and consequently the presence of a reducing reagent is required for reshuffling.

5.6. SPONTANEOUS REFOLDING OF OLIGOMERIC PROTEINS

The folding of oligomeric proteins is an important problem that requires a whole chapter for review and discussion. Thus, it is only discussed briefly here with regard to the occurrence of spontaneous refolding for these proteins. Emphasis is given to the particular characteristics of the refolding process in this case in Chapter 11.

5.6.1. Reversibility of the Unfolding–Folding Process

A complex process consisting of a consecutive process of folding and association steps is expected for the recovery of oligomeric proteins from their unfolded state. The correlation between the reassociation and the return of the biological function is an important but open question in many cases. Information concerning refolding of oligomeric enzymes is certainly more scattered than data on monomeric proteins and very few examples of documented studies are available. In particular, because of the scarcity of well investigated proteins whose three-dimensional structure is known at atomic resolution, it is difficult to rationalize the data.

Nevertheless many reports indicate that oligomeric proteins, when denatured, are able to refold to a nativelike conformation after removal of the denaturant. Teipel and Koshland (1971a,b) have shown the reversibility of the process for several enzymes (fumarase, enolase, aldolase, glyceraldehyde-3-phosphate dehydrogenase, lactate dehydrogenase, and malate dehydrogenase) after denaturation by GuHCl. Renaturation was achieved by lowering the concentration of denaturant either by dialysis or by dilution. Both yield and rate of reactivation differed significantly among these enzymes. Aldolase regained 72% of the activity in 120 min whereas glyceraldehyde-3-phosphate dehydrogenase (GPDH) regained only 6% during the same time. Conditions for better reactivation of GPDH were investigated by Rudolph and coworkers (1977b). Irreversible aggregation competed with the reconstitution process.

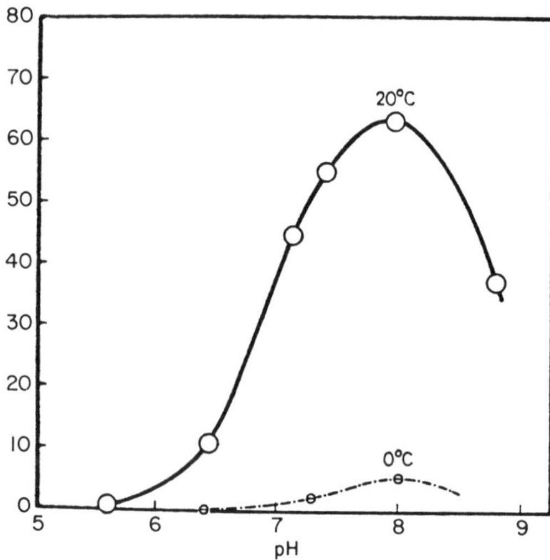

Fig. 5.22. Influence of pH and temperature on the reactivation of pig heart lactate dehydrogenase (from Jaenicke, 1974).

Lactate dehydrogenase can be reversibly dissociated and inactivated by different procedures: either by acid pH (Jaenicke, 1974), or by strong denaturants (Rudolph *et al.*, 1977a,c). The refolding process depends on pH and temperature (see Fig. 5.22 (Jaenicke, 1974). The H_4 and M_4 isoenzymes can recover up to 100% activity and the refolded molecules are indistinguishable from the native ones.

The refolding of aldolase was investigated by different groups of researchers. The enzyme can recover its native conformation after denaturation by urea, GuHCl, or acid pH. Under suitable conditions, in regeneration buffer, up to 80% of activity can be restored by acid pH (Stellwagen and Schachman, 1962; Deal *et al.*, 1963; Vimard *et al.*, 1975). Ultraviolet spectra and ORD probe indicated no residual organized structure after acid pH and GuHCl unfolding of the molecule.

The reversibility of the unfolded–folded transition of phosphofructokinase induced by GuHCl was shown by using fluorescence signals. Although a complex transition was observed (Fig. 5.23), by following the enzymatic activity, the situation appeared still more complex (Parr and Hammes, 1975).

For tryptophanase (London *et al.*, 1974), when renaturation from 8 *M* urea is performed for concentrations lower than 5 μg/ml, the recovery of the activity reaches approximately 100% depending on the conditions of renaturation and protein concentrations (see Section 5.6.2). In this case it was

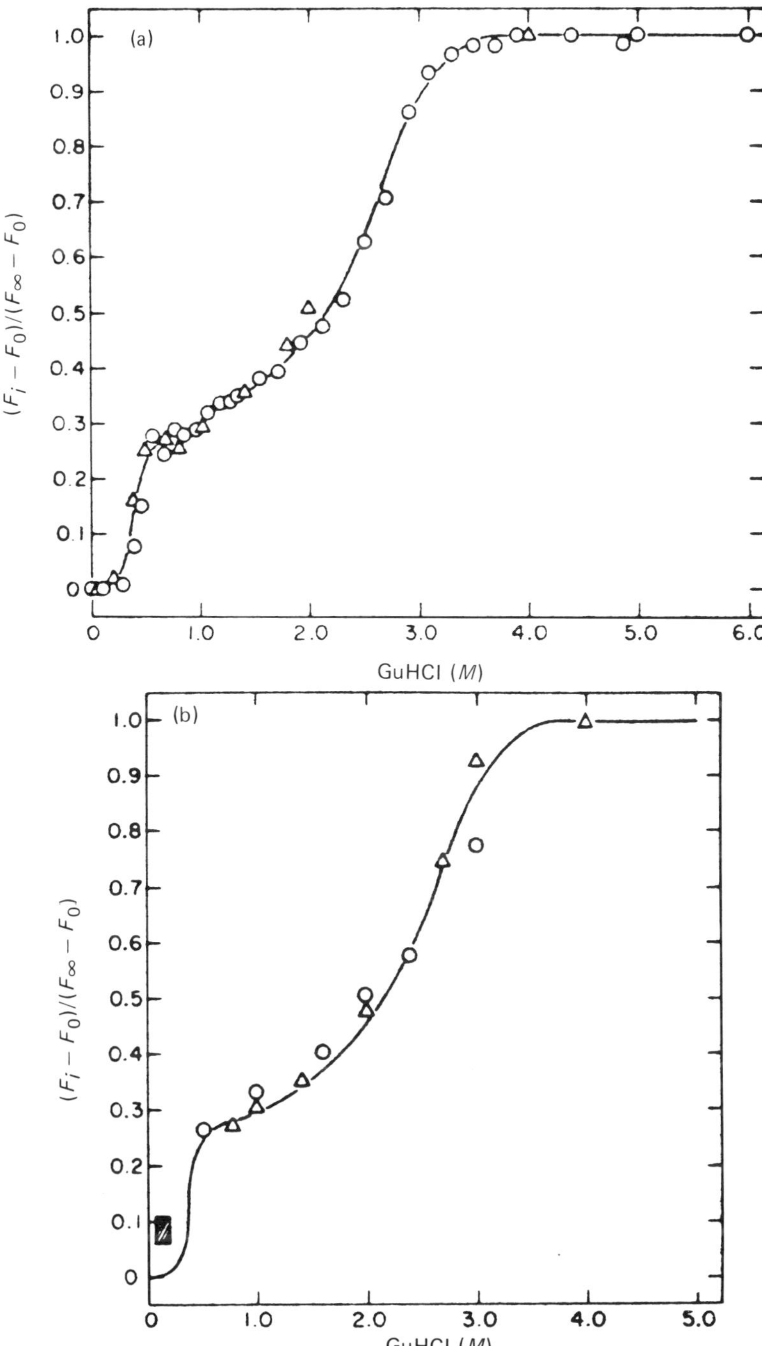

Fig. 5.23. Reversibility of the GuHCl-induced transition of phosphofructokinase followed by fluorescence emission (from Parr and Hammes, 1975): (a) starting from native protein; (b) starting from protein incubated in 4 *M* GuHCl for 1–3 hr and rapidly diluted to final GuHCl concentration.

shown that the quaternary structure is required for the activity (Raibaud and Goldberg, 1976a,b, 1977).

Reversible dissociation and inactivation of tryptophan synthetase by 4.5 *M* GuHCl at pH 2.3 can be realized. The regain of activity on removal of the denaturant by dilution can reach 90% (Groha *et al.*, 1978).

The refolding of GuHCl-denatured bifunctional enzyme aspartokinase II–homoserinedehydrogenase II was observed by following the reappearance of the two enzymatic activities (Dautry-Varsat and Garel, 1978).

The trimeric catalytic subunit assembly of aspartate transcarbamylase can be reversibly unfolded by GuHCl (Ghélis and Hervé, 1978). Up to 95% of the activity is restored under suitable conditions (i.e., 25°C, pH 7.0).

5.6.2. Characteristics of the Refolding Process

The refolding of oligomeric proteins has common characteristics: (1) a delay in the reappearance of activity compared to the conformational signals is observed; (2) the kinetics do not obey those of a simple first-order reaction; (3) specific effectors may influence the refolding process.

In several of the previously mentioned studies, the reappearance of different physical signals (e.g., fluorescence, UV spectrum, ORD, light scattering) was found to occur at a significant time before the return of enzymatic activity. The gross structural changes were complete within ca.

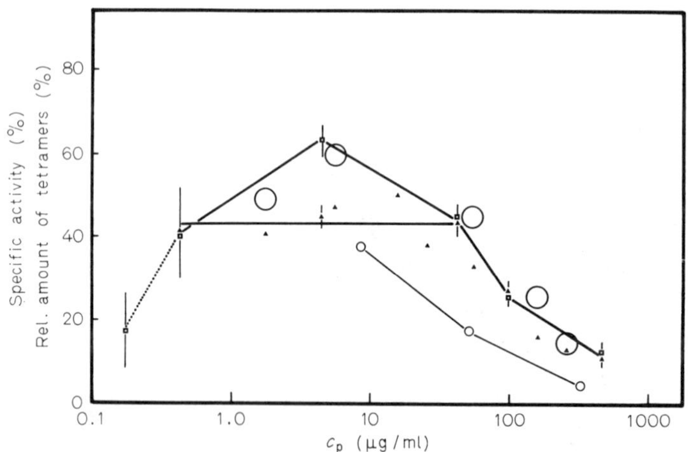

Fig. 5.24. Effect of protein concentration on the reactivation and reassociation of pig heart lactate dehydrogenase (LDH-H_4) (from Jaenicke, 1974). Enzyme was deactivated at pH 2.3 and reactivated in 0.2 *M* phosphate buffer pH 7.95 + 1 m*M* DTT + 10 m*M* EDTA at 20°C for 3 hr (▲); 20-hr (□); yields of the tetrameric species obtained from centrifugation velocity (○); (∘), reactivation after denaturation in 6 *M* GuHCl (courtesy of Jaenicke).

1 min, but very little activity was regained within this interval. The halftime for reactivation of 4 min was reported for aldolase; in this case the refolded monomer was found to be endowed with activity (Teipel, 1972). For fumarase and malate dehydrogenase, a halftime of 75 min was reported (Teipel and Koshland, 1971a). These differences in the time needed for the return of activity may be reasonably correlated with the fact that, for aldolase, refolded monomer is enzymically active whereas the activity of other enzymes requires the quaternary structure.

The rates of refolding display complex concentration dependency. They do not obey simple first- or second-order reactions, but rather follow more complex processes (i.e., unimolecular, bimolecular, or higher order reactions) according to the system (see Chapter 7). The formation of aggregates competes with the refolding process and is also concentration dependent. Figure 5.24 shows the effect of protein concentration on the refolding of lactate dehydrogenase (Jaenicke, 1974).

For oligomeric enzymes, the presence of substrate or coenzyme was found to favor both the rate and yield of reactivation. The action of specific ligands on refolding is discussed in detail in Chapter 11.

5.7. APPARENT IRREVERSIBILITY OF THE UNFOLDING–FOLDING PROCESS UNDER CRITICAL CONDITIONS

When refolding is monitored by exposure of the protein through intermediate concentrations of denaturing agent, discontinuities in the process may appear at critical concentrations. These phenomena were studied in very few cases, but seem to be independent of the existence of a quaternary structure of the enzyme and the presence of disulfide bridges in the molecule. It was observed for β-galactosidase (Goldberg, 1972), tryptophanase (London *et al.*, 1974), catalytic subunit assembly of aspartate transcarbamylase (ATCase) (Ghélis and Hervé, 1978), and elastase (Zilber, 1979). For two of these proteins, elastase and ATCase, discontinuity was observed when starting from the native conformation as well as from the unfolded one. In all cases, the phenomenon was observed when the unfolded protein was exposed to the critical concentrations of denaturing agent.

5.7.1. Apparent Irreversibility of the Refolding Process

The existence of a critical concentration of denaturant was reported for the first time in the study by Goldberg (1972) of urea denatured β-galactosidase refolding. This enzyme is completely renatured after denaturation by 8 *M*

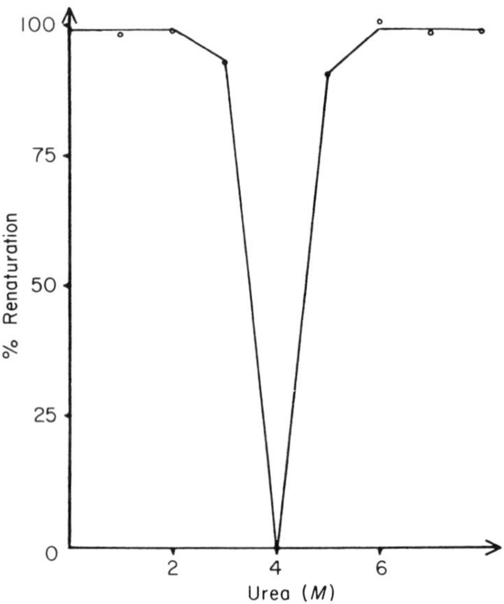

Fig. 5.25. Apparent irreversibility of the refolding process for β-galactosidase at the critical urea concentration (from Goldberg, 1972).

urea (Ullmann and Monod, 1969). However, exposure of the protein during 1 hr to the critical intermediate concentration of 4 *M* urea leads to complete irreversibility of the process. No activity was regained after complete removal of the denaturant. A very sharp minimum was observed under experimental conditions (Fig. 5.25).

The irreversibility depends on the time during which the protein is incubated in the presence of intermediary concentration of urea. For a very short time of exposure the refolding is observed.

Similar behavior was reported for tryptophanase (London *et al.*, 1974). In this case the critical concentration was 3 *M* urea. The broadness of the minimum is strongly dependent on concentration; the lower the enzyme concentration, the sharper the minimum (Fig. 5.26).

For elastase the same phenomenon was observed. The unfolding–refolding process of elastase was followed by using optical signals (UV difference spectra) and by measuring both enzymatic and immunochemical activities. Two denaturing agents were used, GuHCl and urea. On the basis of the optical signals, a perfect reversibility of the unfolding–folding process is observed (Fig. 5.27). For diisopropyl-elastase, an acyl enzyme in which the reactive Ser 195 was blocked (DIP-Ser 195-elastase), the reversibility of the process was also established on the basis of optical signals. The unmodified protein, denatured by either 6 *M* GuHCl or 8 *M* urea, without reduction of

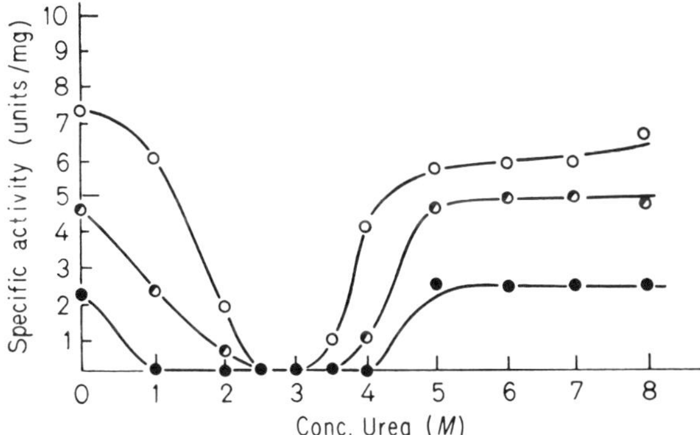

Fig. 5.26. Enzyme concentration dependence of the 3 *M* trough for tryptophanase (from London *et al.*, 1974): (○) 100 μg/ml; (◐) 200 μg/ml; (●) 500 μg/ml. This illustrates the broadening of the minimum when protein concentration increases.

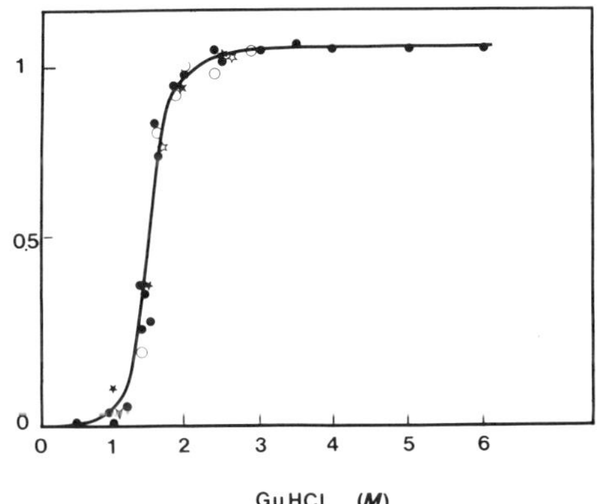

Fig. 5.27. Transition curve of elastase denatured by GuHCl at pH 5.4 (from Ghélis and Zilber, 1982); denaturation process (●,★); renaturation (○,☆).

disulfide bonds, was exposed to intermediary concentrations of denaturant before incubation in regeneration conditions, and enzymatic activity was assayed. A critical concentration of denaturant which prevents totally or partially the regain of activity was observed. The critical concentration corresponds to a value very close to the end of the transition curve (i.e., for which the protein is quite totally unfolded). The same results were found by following the return of enzymatic and antigenic activity (Fig. 5.28). The through is observed for a critical concentration of 1.8 *M* GuHCl or 4.8 *M* urea at pH 5.4 and 3.6 *M* GuHCl at pH 8.0. For DIP–Ser-195-elastase, a

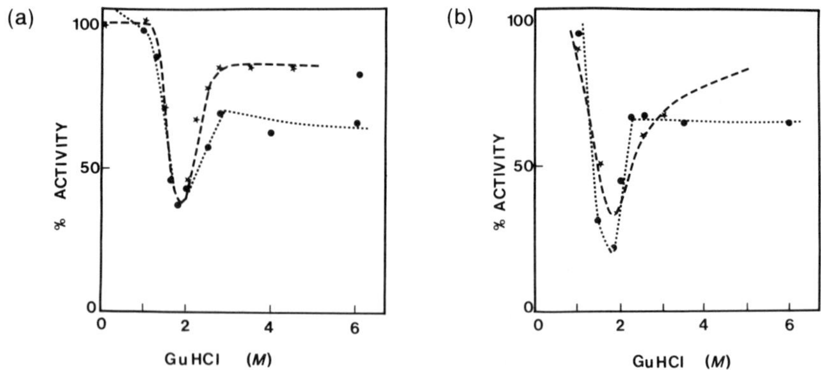

Fig. 5.28. Apparent irreversibility of denaturation–renaturation process for elastase at pH 5.4 measured by the return of antigenic activity (★), by the return of enzymatic activity (●): (a) elastase 4 μM denatured in different concentrations of GuHCl; (b) elastase 4 μM previously denatured in 6 M GuHCl and then incubated in different concentrations of GuHCl. For radial immunodiffusion measurements 0.12 nmoles of elastase were added to wells in gel. The 2 M trough at this pH value was found in all cases and by both methods (from Ghélis and Zilber, 1982).

complete reversibility was observed at all GuHCl concentrations (Fig. 5.29).

Parr and Hammes (1975) mentioned that incubation of phosphofructokinase in 0.6 M GuHCl leads to irreversibility of the refolding process.

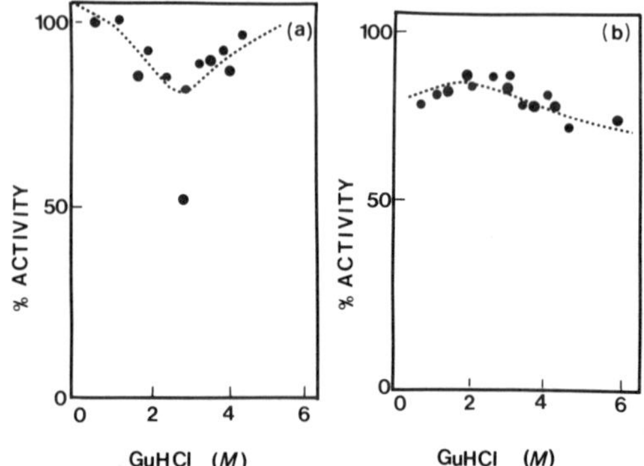

Fig. 5.29. Return of antigenic properties of DIP–Ser-195-elastase at pH 5.4, 25°C: (a) DIP–Ser-195-elastase (4 μM) incubated at different GuHCl concentrations; (b) DIP–Ser-195-elastase (4 μM) previously denatured by 6 M GuHCl and then incubated at different intermediate concentrations. Amount of DIP–Ser-195-elastase in the gel 0.12 nmoles of DIP–elastase were added to wells in gel for immunodiffusion measurements of antigen concentrations (from Ghélis and Zilber, 1982).

5.7.2. Apparent Irreversibility of the Unfolding Process

The apparent irreversibility of unfolding processes was carefully studied in the case of elastase (Ghélis *et al.*, 1982). Identical behavior was observed when starting from the native conformation and increasing progressively denaturant concentration, as well as when starting from the unfolded state and transiting in intermediate denaturant concentrations. The same critical concentration of denaturant was determined in both directions of the process, regardless of the initial state of the protein (native or unfolded), by following return of enzymatic and antigenic activities. The minimum of reversibility and the value of critical concentrations were also pH dependent. Similar apparent irreversibility was thus detected when starting from the native as well as from the unfolded state.

As indicated by transition curves, the native state of elastase is stabilized by the presence of DIP group covalently attached to the reactive serine. The unfolding process occurs for higher concentrations of denaturant; however, it is not modified.

For ATCase, which is an oligomeric protein without disulfide bridges, the same kind of phenomenon was observed (Ghélis and Hervé, 1978).

When exposed for at least 1 hr to the critical concentration of denaturant (3 *M* urea), native tryptophanase also recovers the native structure after removal of the denaturant (London *et al.*, 1974). In this case the presence of aggregates was reported and this apparent irreversibility was attributed to aggregation.

5.7.3. Formation of Aggregates

The occurrence of a critical concentration of denaturant, for which the irreversibility was observed, was interpreted to be caused by the formation of aggregates (London *et al.*, 1974). It was even proposed that aggregation arises from specific interactions between complementary association areas of two different molecules rather than within the same molecule. The dependency of the phenomenon on concentration and direct observation of aggregates was accepted as supporting this hypothesis.

The formation of aggregates during the refolding process as a competiting reaction was reported for many proteins (either oligomeric or monomeric). The refolding of a protein often results from a competition between the formation of the native structure and the irreversible occurrence of aggregated species. The best regenerating conditions are generally those that optimize the first process and minimize the formation of aggregates. The low concentration of denaturant (close to the end of the transition curve) has been reported to not favor the renaturation for β-galactosidase, tryptophanase, elastase, or ATCase. For the first two proteins it was interpreted to be a result of the trapping of the incorrectly folded aggregated species.

Other possibilities can be evoked to take these effects into account, for example, a trapping of intermediate species in a step preceeding the last one. With elastase it was confirmed that apparent irreversible species are not attributable to aggregation. Less than 10–15% aggregates were obtained on fractionation by gel permeation of a solution of 1–4 μM elastase after 6 hr incubation in 1.8 *M* GuHCl at pH 5.4 (Ghélis, unpublished data). Furthermore apparent irreversible species have a folded conformation as indicated by differential spectroscopy using unfolded protein as standard. Therefore, it was proposed that irreversible species arise from side reactions occurring from an incompletely refolded intermediate:

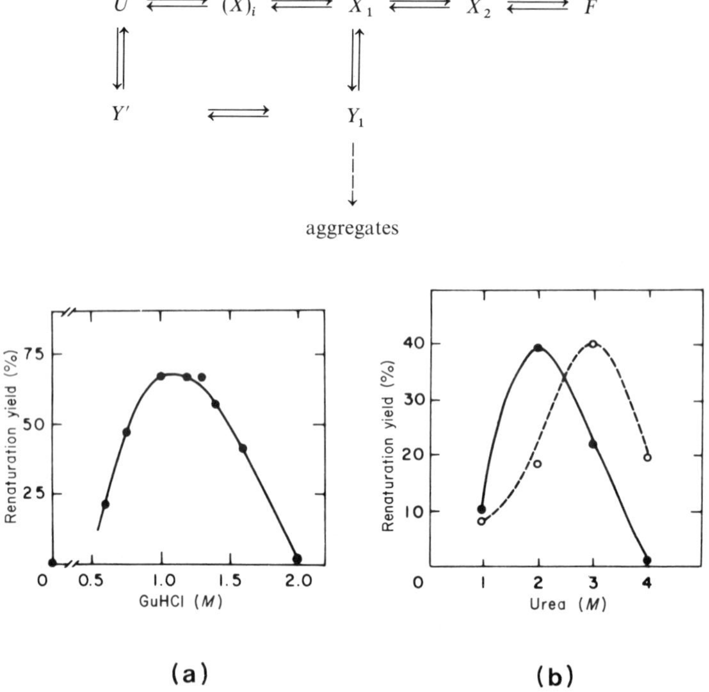

Fig. 5.30. Effect of guanidine (a) and urea (b) on the renaturation yield of reduced chymotrypsinogen (according to Orsini and Goldberg, 1978). Reduced chymotrypsinogen was renatured by dilution into the following regeneration mixture (50 m*M* Tris-HCl buffer, pH 8.2, 2 m*M* EDTA, 1 m*M* GSH, and GSSG) and varying GuHCl concentrations during 20 hr as indicated in (a) or varying concentration of urea during 5 hr as indicated in (b) in the absence (●) or in the presence (○) of 1 *M* KCl.

Conversely, low concentrations of denaturant may favor the refolding of carbonic anhydrase (Wong and Tanford, 1973) and chymotrypsinogen (Orsini and Goldberg, 1978). Figure 5.30 indicates concentrations of GuHCl and of urea for which the yield of renaturation is maximum. These values probably correspond to the early beginning of the transition curves. In this case a weak concentration of denaturant prevents or minimizes aggregation.

5.8. COMMENTS ON THE VALIDITY OF THE *IN VITRO* DENATURATION–RENATURATION MODEL

Various physicochemical perturbations induce denaturation of a protein; however, not all of them drive the protein to a completely unfolded state. A clear characterization of the unfolded state is a prerequisite to any study of protein folding, particularly to the demonstration of the reversibility of the denaturation–renaturation transition. Such a characterization requires the use of different methods, especially those which are accurate enough to detect local residual structures. Generally, the most complete denaturation is obtained at high concentrations of GuHCl. The unfolded state has not been rigorously characterized in all the research that has been examined in this chapter. Nevertheless enough information is available to conclude that the denaturation–renaturation process is reversible when conditions are found. It was established for monomeric proteins without disulfide bridges. For S—S bridged proteins, the reversibility of the transition has also been demonstrated. Starting from reduced and denatured proteins, a spontaneous recovery of the native and active conformation was observed for many proteins, however, restricted conditions for regeneration are often required, for example, the presence of a suitable redox mixture. Reversibility of the denaturation–renaturation process has also been shown with several oligomeric proteins; although the process is more complex, involving refolding and subunit reassociation.

Since oligomeric as well as monomeric proteins, with and without disulfide bonds, when totally denatured are able to spontaneously refold under suitable conditions to recover a conformation, which has the same characteristics as the native one, it can be considered that the model is valid. When reversibility does not occur with a significant yield, it is probable that good renaturation conditions have not yet been found. A kinetic competition between refolding and aggregation or illicit intramolecular interactions leading to incorrectly refolded species, often limits the reversibility of the process. The existence of a critical concentration of denaturant introducing an apparent through of reversibility reflects such a situation. In this case, the species obtained may be renatured only after a new complete denaturation.

For several proteins the conditions of reversibility are very restricted. Conditions such as pH, ionic strength, temperature, and even residual concentration of denaturant must be carefully determined in the study of the refolding of a protein. The narrower the conditions of renaturation, the fewer are the probable possible pathways to refold a protein. Nevertheless, the presence of ribosome is not a necessary requirement to refold a protein and the *in vitro* denaturation–renaturation model may be considered very significant, at least for soluble globular proteins.

For membrane proteins, the right solvent has to be found and remains an open question since the system is defined not only by the sequence of amino acids but also by the solvent. For spanning proteins, the mechanism of folding is certainly complex, since parts of the molecule interact with lipids and other parts are located outside of the membrane in an aqueous environment. Moreover, the folding process must be considered together with the process of insertion of the protein in the membrane.

To summarize, the various data presented in this chapter show that:

(*1*) *Proteins previously denatured are able to refold spontaneously when suitable conditions are found.*

(*2*) *The foregoing statement* (*1*) *is true for all proteins, monomeric as well as oligomeric, proteins with and without disulfide bonds, even with those that have been previously reduced.*

(*3*) *Refolding is directed by noncovalent interactions, even for S—S bridged proteins.*

(*4*) *Apparent irreversibility observed for various proteins is attributable to side reactions which occur with intermediates in the folding pathway and corresponds to incorrect refolding. Sometimes, but not always, this event is followed by aggregation.*

6

Physicochemical Studies of the Unfolding-Folding Equilibrium

Since denaturation–renaturation was found to be a reversible process under adequate conditions, the transition was studied for several proteins. The most extensive and detailed studies were performed on ribonuclease, hen egg lysozyme, chymotrypsinogen, myoglobin, hemoglobin, carbonic anhydrase, and cytochrome *c*. In early studies, the transition was considered to follow a two-state mechanism because of the great cooperativity of the process. This point of view was seriously challenged first by Tanford (1964) and later by Poland and Scheraga (1965) on the basis of experimental and thermodynamic data obtained with polypeptides as models and proteins. Then, the occurrence of intermediates was supported by kinetic studies (see Baldwin, 1975). The occurrence of intermediates in the pathway of protein folding has been sufficiently supported. The concentration of intermediates is certainly of considerable importance in the transition region. In fact, if the concentration of intermediates is low, the analysis of the folding–unfolding transition in terms of a two-state mechanism is straightforward. If appreciable concentration of intermediates exists, a more complex analysis is required.

In the present discussion, only the studies at equilibrium are described. From the data reported in the literature, one can discuss the validity of the analysis according to a two-state mechanism for several examples, and for other examples the deviation from a two-state behavior. The kinetic aspect of the problem is analyzed in Chapter 7. Specific methods and particular approaches to detect, trap, and identify intermediates occurring in the folding pathway of a protein are described in Chapters 8 and 9.

6.1. STUDIES OF THE UNFOLDING–FOLDING TRANSITION AT EQUILIBRIUM

6.1.1. Experimental Methods Currently Used to Study the Transition

Various experimental methods can be used to follow the conformational transition of a protein from the native to the unfolded state. In principle, each method related to the conformation of the protein may be used. However, first the sensitivity and the precision of the methods are of great importance, since the experiments are generally performed at rather low concentrations of protein to minimize aggregation; that introduces limitations in the choice of the methods. Second, each method provides a special type of information. Certain ones are suitable to describe overall conformational changes, others are related to local variations. Hydrodynamic methods are used to determine the global shape of the molecule. Although very different in principle from hydrodynamic methods, optical methods generally allow one to describe the global conformational change of a protein. Spectrophotometric measurements (absorption or fluorescence) are based on the change of optical density on variations of the environment of chromophoric groups (Tyr, Trp, Phe) when the protein unfolds. They are able to give local information only in particular cases (e.g., when the native molecule contains only one buried tryptophan). Optical rotatory dispersion and circular dichroism generally give only information concerning global transition of a protein (Jirgensons, 1973). Evaluation of the different secondary structures (helices and β structure) which unfold during the transition is matter of discussion. Optical rotary dispersion is not reliable for such a quantitative discrimination. This kind of information can be obtained from Raman spectra analysis.

More localized variations can be detected by immunochemical studies when the antibodies are purified to monospecificity. To obtain information regarding very localized parts of the polypeptide chain during the folding (or the unfolding) process, the most suitable methods are certainly those which allow one to determine the accessibility of amino acid side chains to chemical reagents in the precise position of the sequence, or the accessibility of definite peptide bonds to specific proteases. Nuclear magnetic resonance allows one to follow the behavior of individual protons when correct assignment is possible and thus provides the same type of local information as is obtained by chemical methods. The results obtained from chemical methods, from NMR measurements, and from immunochemical studies are described in detail in Chapters 8 and 9.

The results obtained from viscosity to characterize the unfolded state are presented in Chapter 5. In the present section, the data obtained by optical methods currently used in the study of the transition between native and denatured proteins are briefly discussed. Various reviews describe spectrophotometric methods and their use in the conformational studies of proteins (Wetlaufer, 1962; Hermans, 1965; Herskovits, 1967; Donovan, 1969b, 1973; Yon, 1969; Chen, 1967; Weber and Teale, 1965; Chen *et al.*, 1969; Lehrer and Fasman, 1967). Ultraviolet difference spectroscopy is the most current method used to study conformational transitions of proteins. When a protein undergoes a conformational transition from the native to the unfolded state, several chromophoric groups are transferred from the interior of the protein to the solvent. The alteration of their environment is accompanied by a small variation of the absorption spectrum, which is shifted in wavelength ($\Delta\lambda$) and in intensity ($\Delta\varepsilon$). The variation of intensity is in first approximation given by the first derivative of the absorption spectrum

$$\Delta\varepsilon = \varepsilon(\lambda + \Delta\lambda) - \varepsilon(\lambda) = -\Delta\lambda(d\varepsilon/d\lambda) + [(\Delta\lambda)^2/2!](d^2\varepsilon/d\lambda^2) - \cdots$$

the higher terms being negligible. Every change in the environment yields a difference spectrum of the chromophores commonly encountered in proteins. The effects can be bathochromic or hypsochromic. It is possible to observe a bathochromic effect and hypochromicity at the same time, or similarly hypsochromic effect and hyperchromicity; however, these effects are not necessarily associated.

The difference spectrum of unfolded versus folded proteins in general indicates several minima which reflect the variations of the environment of the different chromophores. Figure 6.1 shows an example of perturbation of amino acids by a change of their environment. Tryptophan has a contribution with a first minimum at 292 nm and a second at 284 nm; for tyrosine the main peak is at 286 nm, the other at 278 nm; for phenylalanine, there are different peaks, the main at 253 nm, the others at 260, 265, and 270 nm. Table 6.1 gives the variation in molar extinction coefficient at the wavelength of the main peak for various chromophores. There are small variations in these values for particular proteins.

Abnormalities can appear in the difference spectra. A typical abnormality is a peak near 300 nm which has been reported and analyzed. It has been attributed to the environment of a chromophore in the native protein, such as the orientation of a dipole, a charge or a polarizable group in the protein with respect to tryptophan. The same effect is produced by the respective

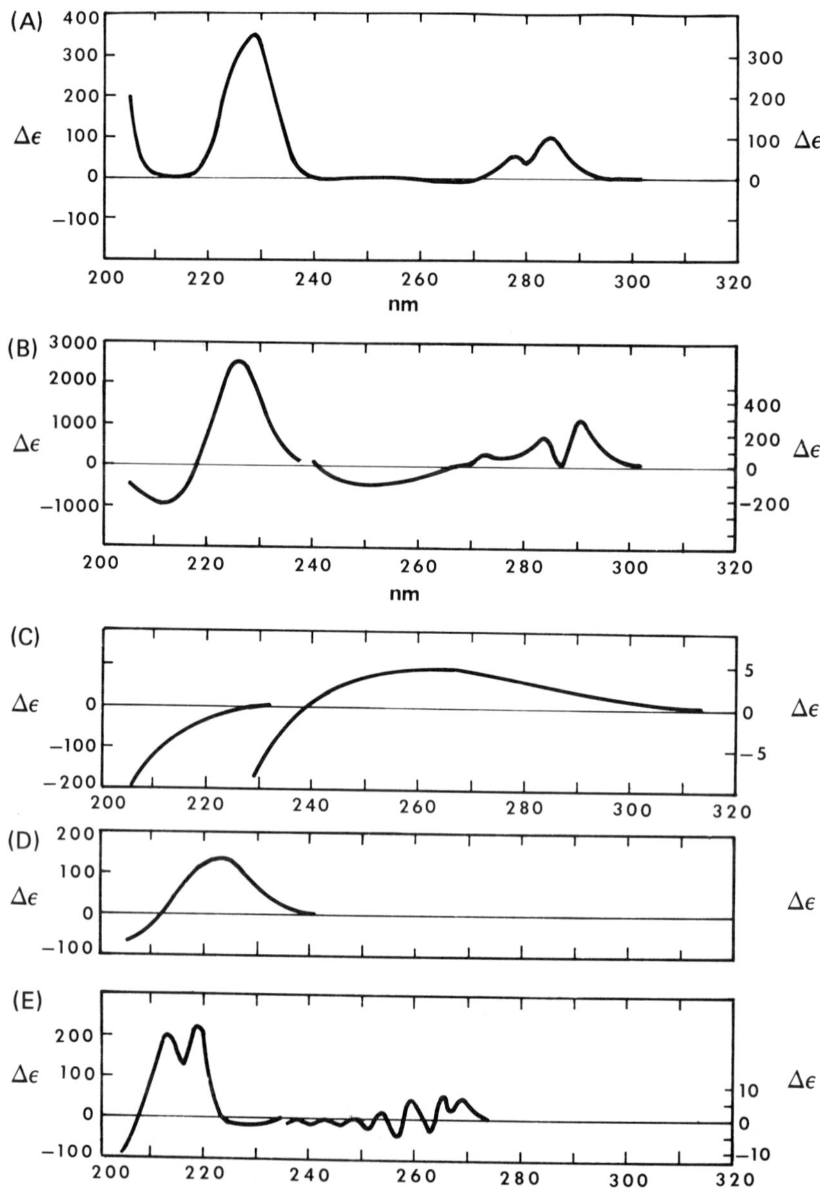

Fig. 6.1. Perturbation difference spectra of several amino acids produced by 20% (vol/vol) ethylene glycol: (A) tyrosine, (B) tryptophan, (C) cystine, (D) histidine, and (E) phenylalanine (courtesy of Donovan, 1969a).

TABLE 6.1

Variation in the Molar Extinction Coefficient of the Peak Maximum at the Indicated Wavelength (λ_{max}) on Transfer from the Interior of a Protein to Water[a]

Chromophore	λ_{max}[b] (nm)	$-\Delta\varepsilon_{max}$	λ_{max}[b] (nm)	$-\Delta\varepsilon_{max}$
Phenol	287	700	228	2000
Indole	292	1600	225	13,800
Benzene	259	36	218	1260
Disulfide	265	30	—	—
Imidazole	—	—	223	780

[a] From Donovan (1973). Calculated from the observed perturbation with 20% ethylene glycol, assuming that the protein interior is equivalent to 120% ethylene glycol.

[b] Both λ_{max} and $\Delta\varepsilon$ show changes with different proteins and solvents.

orientation of two indole groups close to each other, or by a ligand bound to the protein (Ananthanarayanan and Bigelow, 1969 a, b; Andrews and Forster, 1972). Figure 6.2 shows a typical difference spectra observed after denaturation of elastase induced by GuHCl (Fig. 6.2).

In the range where temperature does not induce conformational change, small variations in spectra are produced by variations in temperature in the

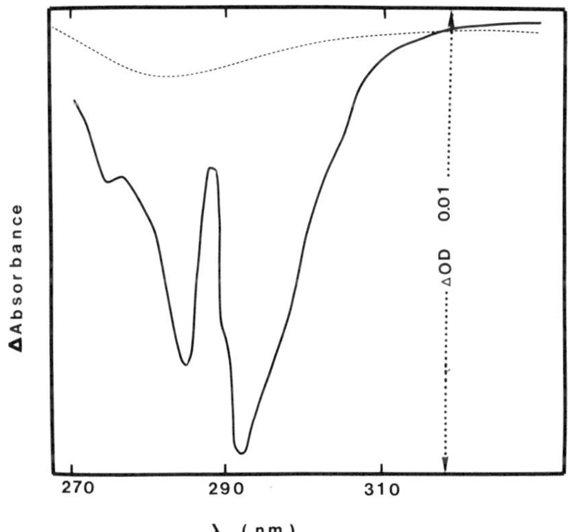

Fig. 6.2. Difference absorption spectra of denatured over native elastase (4 μM) at pH 7.5 (—); the base line (---) was obtained by differential absorption spectra of native over native elastase (4 μM) (from Ghélis and Zilber, 1982).

TABLE 6.2

Thermal Pertubation Spectra of Phenol and Indole[a]

$\Delta\varepsilon$ $[\mathrm{deg}^{-1}\ \mathrm{at}(\lambda)]$			
Phenol		Indole	
Water	Acetonitrile	Water	Acetonitrile
—	+0.9 (285)	+5.9 (290)	+1.8 (292)
+2.4 (281)	−5.1 (280)	+0.9 (282.5)	−5.5 (284)
+0.8 (274)	−5.5 (273)	—	—

[a] From Lehrer and Barker (1971). Change in absorbance per degree $\Delta\varepsilon$. A positive sign indicates positive $\Delta\varepsilon$ when the temperature of the sample solution is greater than the temperature of the reference.

native as well as in the denatured state, all chromophores being perturbated by the change of temperature. This is caused by variations in orientation and distance of the atoms induced by variations in the thermal motion. This effect is generally smaller on the chromophores in the interior of folded proteins because of the restrictions in rotation resulting from the close packing of the atoms. For chromophores exposed to water as the solvent, a red shift of the absorption spectrum caused by increasing temperature is observed (i.e., a positive difference spectrum); for buried chromophores, a blue shift with increase of temperature is expected. Temperature perturbation spectra for indole and phenol were studied in different solvents by Lehrer and Barker (1971). The perturbation produced by temperature varies with the wavelength. Table 6.2 summarizes the main variations. When measurements are performed at the longest wavelength peak of the difference spectrum, the exposed chromophores make the greatest contribution. The effect of temperature on the extinction coefficient of chymotrypsinogen is indicated in Fig. 6.3. The temperature dependence of the folded as well as unfolded state is the same at all pH values of the study. For the native state, $\Delta\varepsilon$ increases slightly with an increase of temperature from 0°C to 32°C, then decreases at higher temperatures that are below the transition. For the unfolded state $\Delta\varepsilon_D$ decreases with temperature. The characteristics of the initial and of the final state can be obtained at any point of the curve by extrapolation from the linear portions of the curve at very low as well as at very high temperature. Bello (1969) introduced the method of thermal perturbation.

The optical signals are very sensitive to the environment and they vary with any perturbant such as a chemical denaturant (urea, GuHCl), or pH.

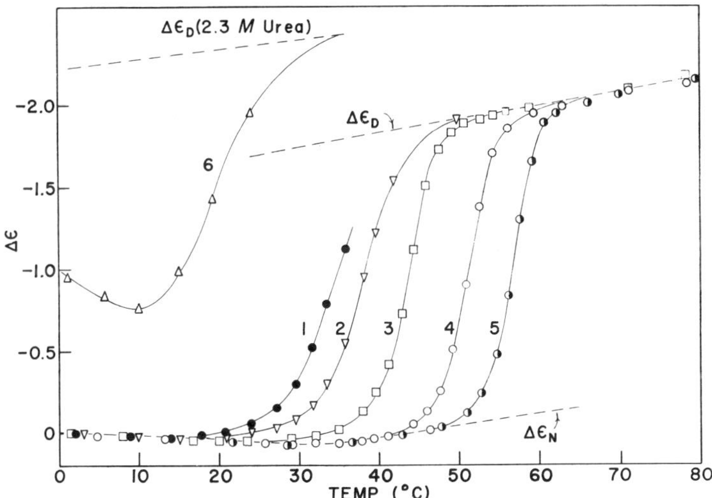

Fig. 6.3. Temperature dependence of the extinction coefficient of chymotrypsinogen at 293 nm: curve 1, pH 1.11; curve 2, pH 1.71; curve 3, pH 2.07; curve 4, pH 2.56; curve 5, pH 3.00; curve 6, 2.3 *M* urea pH 1.55. The value of $\Delta\varepsilon$ (---) varies with temperature either for the native form ($\Delta\varepsilon_N$) or for the denatured one ($\Delta\varepsilon_D$) and also the denatured one in 2.3 *M* urea ($\Delta\varepsilon_N$ 2.3 *M* urea (courtesy of Brandts, 1964a).

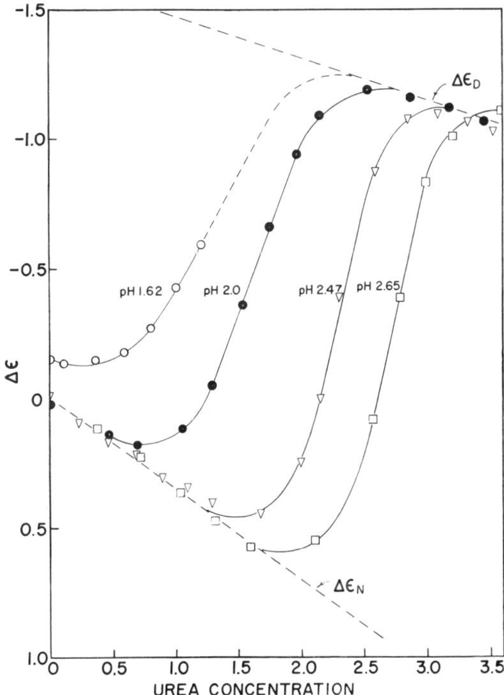

Fig. 6.4. Variation in extinction coefficient of chymotrypsinogen with urea concentration at 30.4°. Dotted lines indicate the variation in extinction coefficient with urea concentration at the same temperature for the native form ($\Delta\varepsilon_N$) and for the denatured one ($\Delta\varepsilon_D$) (courtesy of Brandts, 1964a).

Figure 6.4 displays the denaturation transition curve of chymotrypsinogen at 293 nm in urea. Variations of the extinction coefficient with urea concentration has also been observed for the native and for the unfolded states. These properties have been used to estimate the number of exposed or buried chromophores either by difference spectrophotometry or by the method of solvent perturbation (Herskovits and Laskowski, 1962, 1968; Herskovits, 1965, 1967).

Intrinsic fluorescence of a protein is also a useful method for detecting and following the conformational changes. It is based on the change of environment of chromophores, tyrosine, and tryptophan resulting from folding or unfolding of a protein. Tryptophan and tyrosine have emission bands around 350 nm and 303 nm respectively. In general, when tyrosine and tryptophan are present in a protein, only the contribution of tryptophan is detectable since tyrosine fluorescence is masked by tryptophan. The advantage of intrinsic fluorescence over absorption spectrophotometry, is its great sensitivity which permits the use of lower concentrations of proteins. In staphylococcal nuclease where only one tryptophan is present in the sequence (Trp 140), fluorescence can be used as a very local probe of conformational changes.

Despite the high sensitivity of the fluorescence technique, its use in the study of protein folding requires some caution with regard to acceptance of experimental results. Systematic errors can be introduced by the presence of the denaturant or reducing agent.

The use of an extrinsic fluorescence probe bound to a protein is not recommended for the study of the folding process since the presence of the probe may modify the pathway of folding.

Optical rotation, ORD, and CD, are also commonly used to follow conformational transitions of proteins. The advantage of CD over ORD in the Cotton effect region is that the bands are more easily resolved and assigned. Each optically electronic transition gives rise to only one CD band instead of both positive and negative signals as is the case with ORD. Brahms and Brahms (1980) developed a new approach using vacuum ultraviolet CD spectrum which extended down to 165 nm and yielded new characteristic bands in this region. Since many reviews of the analysis and interpretation of ORD and CD data are available (see Jirgensons, 1973; Adler *et al.*, 1973, and bibliography therein) it is not discussed here. Figure 6.5 shows typical variations of molar ellipticity from the native to the denatured states of elastase. The interpretation of these variations in terms of changes of the different secondary structure (α helix and β strands) is not straightforward. This problem has been analyzed by several researchers including Greenfield and co-workers (1967) and Greenfield and Fasman (1969a,b). Since parameters are generally derived from synthetic polypeptide standards, which are

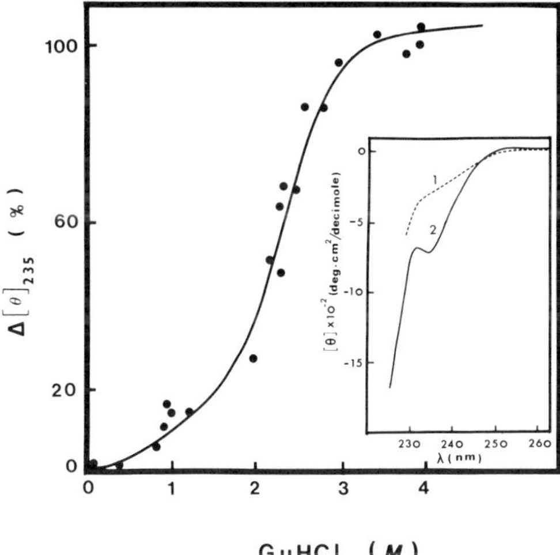

Fig. 6.5. Guanidine hydrochloride-induced transition of elastase measured by circular dichroism (10 μM elastase, pH 5.4, 25°C). Insert, typical CD spectra for (1) denatured elastase, (2) native elastase (from Ghélis and Zilber, 1982).

not sufficiently adequate models, the interpretation in terms of secondary structure should be seriously questioned. Even when the methods do not use synthetic polypeptide standards (Saxena and Wetlaufer, 1971), the agreement between the secondary structure derived from these methods and that obtained directly from the X-ray data is not satisfactory enough to make this a method for detection of the return of helical and β structures during the folding process. A new method of analysis was presented for the estimation of ordered structure in proteins from CD spectra. In this method, a CD spectrum is analyzed as a linear combination of the CD spectra of 16 proteins whose structures are known from X-ray data (Provencher and Glockner, 1981). For proteins which contain disulfide bridges, the CD spectra are modified after reduction of S—S linkages. L-Cystine displays typical CD bands in the near UV, a negative one at 250 nm and a positive CD band near 220 nm. Therefore, the presence of S—S bridges may confuse the interpretation of protein peptide contributions (Beychok, 1968). In addition to the variations of these parameters resulting from conformational transition, characteristics of the initial as of the final state vary slightly either with temperature, or with denaturant. The optical rotation of native proteins generally increases with temperature whereas that of the unfolded state decreases with temperature. Contrary effects on the initial and final states are also observed for denaturation induced by urea or GuHCl (see Fig. 6.6).

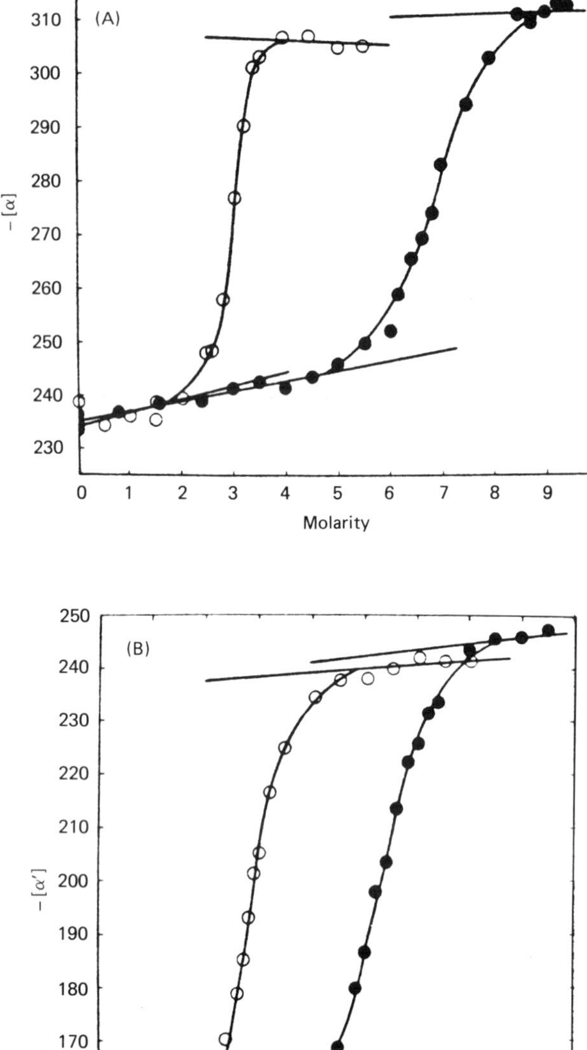

Fig. 6.6. Urea (●) and GuHCl (○) denaturation curves of ribonuclease at pH 6.6 (A) and lysozyme at pH 2.9 (B) at 25°C; |α′| is the reduced specific rotation at 365 nm (from Green and Pace, 1974).

6.1.2. Transition Curve Analysis

The classical method of analysis assumes a two-state mechanism

$$N \underset{}{\overset{K_D}{\rightleftarrows}} D$$

only the native-folded and the denatured-unfolded states are considered to exist in significant concentration. (The validity of the approximation is discussed further on.) Unfolding is followed by the variation of an observable parameter y, y_N referring to the folded protein and y_D to the denatured one. The fractions of native and denatured protein are represented by f_N and f_D respectively:

$$f_N + f_D = 1$$

values for y_N and y_D can be obtained at any point in the transition region by extrapolation of the linear variations of the parameter for the high and low concentration of denaturant.

$$y = (f_N y_N) + (f_D y_D)$$

Therefore

$$y = y_N + f_D(y_D - y_N)$$

and

$$f_D = (y - y_N)/(y_D - y_N) \tag{6.1}$$

The equilibrium constant K_D is

$$K_D = f_D/(1 - f_D) = (y_N - y)/(y - y_D) = \exp(-\Delta G_D/RT) \tag{6.2}$$

As pointed out by Tanford (1968) and by Pace (1975), the analysis in term of a two-state mechanism can be useful even when the transition is known to deviate significantly from a two-state behavior. In this case, Pace (1975) recommended to term K_{app} and ΔG_{app}, the equilibrium constant and the free energy variation of unfolding, respectively.

Tanford (1968) considered the situation where unfolding deviates from a two-state mechanism and took into account the presence of intermediary species, X_i, each being characterized by the property y_i, f_i being the fraction of intermediates present at equilibrium. Thus y becomes at any point:

$$y = y_N + f_D(y_D - y_N) + \sum_i f_i(y_i - y_N)$$

and the extent of unfolding is

$$f_{obs} = f_D + \sum_i z_i f_i \tag{6.3}$$

where

$$z_i = (y_i - y_N)/(y_D - y_N)$$

Since y_i generally falls between y_N and y_D, z_i probably ranges between 0 and 1. A relationship exists between

$$K_{app} = f_{obs}/(1 - f_{obs}) \qquad \text{and} \qquad K_D = f_D/f_N$$

$$K_{app} = K_d\left[1 + \sum_i z_i(K_i/K_D)\right] \Bigg/ \left[1 + \sum_i (1 - z_i)K_i\right] \tag{6.4}$$

where

$$K_i = f_i/f_N$$

For z_i ranging between 0.1 and 0.9, K_D and K_{app} are equal for a point very close to the midpoint of the transition ($K_{app} = 1$). Below this point, K_{app} is less than K_D, and greater above that point.

6.2. THE TWO-STATE APPROXIMATION

6.2.1. The Two-State Model

The best known example of two-state processes is the phase change solid–liquid which is characterized by the abruptness of the transition indicating a very cooperative process. The two-state theory of protein denaturation, assuming the reversible denaturation of a protein, involves only two single thermodynamic (macroscopic) states, the native one and the denatured one. In a two-state process, the microscopic properties of the molecule may be distributed in only two regions of the phase space. For temperature induced transitions, these regions are such that one is characterized by a low enthalpy and entropy, and the other is characterized by a high enthalpy and entropy. The larger the entropy difference, the more cooperative the process. The equilibrium denaturation constant K_D is the ratio of the partition function for the denatured and native molecules. The total canonical partition function is usually given by the expression

$$Q = \sum_i w_i \exp(-\varepsilon_i/kT)$$

the index i being taken over energy state, w_i is the statistical weight of the state i. The validity of the two-state hypothesis for conformational transitions in proteins was analyzed and discussed by several investigators including Poland and Scheraga (1965), Lumry and co-workers (1966), and Tanford (1964). Lumry and co-workers (1966) have analyzed different types of distribution including symmetrical, nonsymmetrical two-state, three-state, five-state, and uniform continuous state. Figure 6.7 shows the pattern of typical

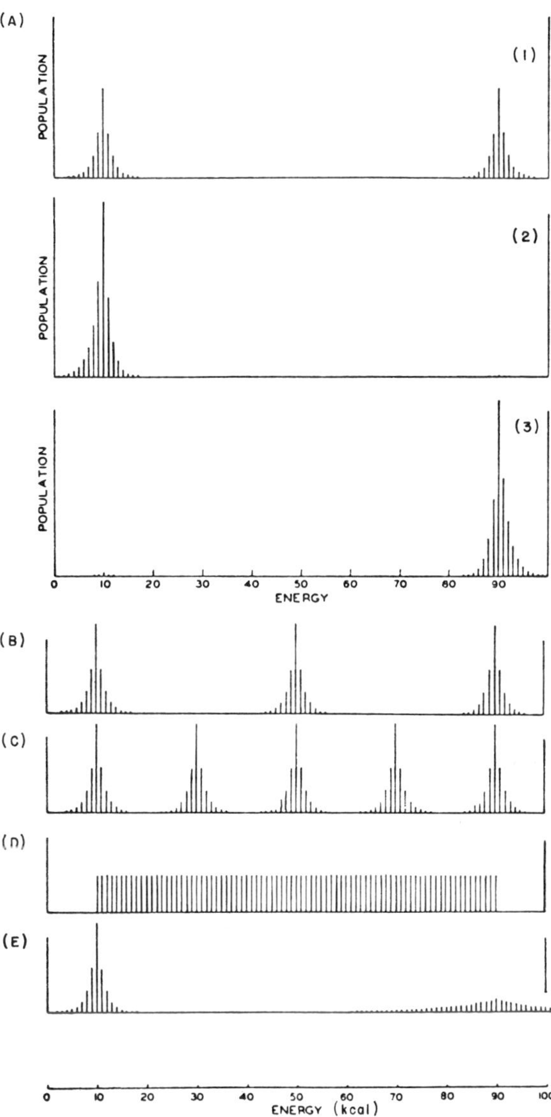

Fig. 6.7. Patterns of the different types of distribution for several transition models: (A) symmetrical two-state transition, (1) at the transition temperature T° (2) at $T^{\circ} - 20^{\circ}$, and (3) at $T^{\circ} + 20^{\circ}$; (B) three-state transition; (C) five-state transition; (D) continuous distribution; (E) nonsymmetrical two-state distribution at the transition temperature (from Lumry *et al.*, 1966).

distributions in the different cases. Lumry and co-workers also derived the variations of ΔH along the transition for the different transition models (Fig. 6.8). The value of ΔH is constant at any temperature for a symmetrical two-state process; it increases monotonically for a nonsymmetrical two-state process. A maximum at the transition temperature is observed for multistate transitions.

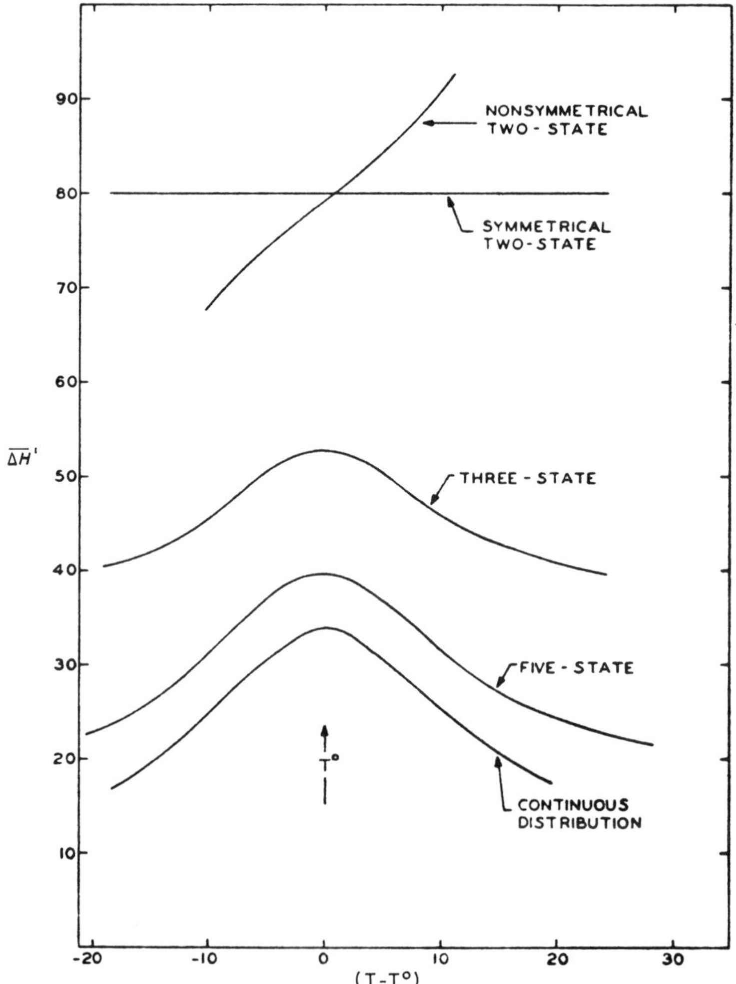

Fig. 6.8. Variations of ΔH with temperature for the different transition models (calculated values) (from Lumry *et al.*, 1966).

6.2.2. Experimental Characteristics Expected for Two-State Behavior

Several experimental consequences are expected for a two-state process.

(1) The process is characterized by a single, rather abrupt transition curve. However, a single cooperative transition curve is not a conclusive argument in itself.

(2) The equilibrium constant and consequently the free energy is independent of the observable parameter. The same transition curve is expected when different characteristics of the protein are measured. The greater the number of different methods used to study the transition, the better is the test.

(3) The enthalpy either is constant or increases monotonically with temperature along the transition when induced by temperature.

(4) The enthalpy determined from the variation of equilibrium constant with temperature is equal to the ΔH directly obtained from calorimetric measurements.

(5) The kinetics follow a monophasic first-order reaction.

In general, data from different tests are necessary to draw conclusions since each set of data is not conclusive in itself.

Conversely, the process cannot be described by a two-state approximation when: (1) the transition curve displays one, or more than one, well delineated plateau, or even a shoulder, (2) the different observables do not coincide, (3) ΔH variations display a maximum, (4) ΔH, measured from the variations of K_D with temperature, is smaller than true calorimetric ΔH, and (5) when kinetics are multiphasic. When even only one of these conditions is fulfilled, the transition cannot be described by a two-state process.

There is no doubt at the actual level of investigation that intermediates are formed during the refolding of a denatured protein. The validity of the two-state approximation depends on the concentration of the intermediates at equilibrium. It does not depend on the location of these intermediates. Indeed, in some cases, intermediary species are not present on the pathway of unfolded to folded state. For example the mechanism of refolding of cytochrome c is consistent with the scheme:

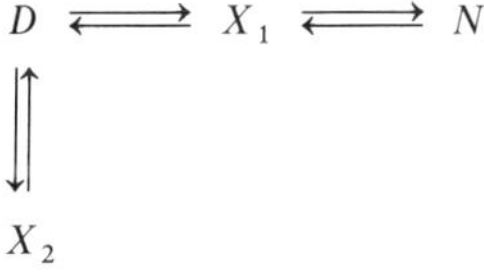

where X_2 is a partially folded species which has to unfold before undergoing the refolding to the native state. At equilibrium 40% of the species are in the state X_2 and obviously the two-state approximation does not apply (Ikai *et al.*, 1973). In several other cases intermediates are located on the refolding pathway from D to N.

It has been noted (Pace, 1975) that urea- and GuHCl-induced transitions are generally closer to a two-state mechanism than pH- or temperature-induced denaturation.

The degree of departure from a two-state mechanism varies widely with the proteins, and for a given protein varies with the denaturant. It also depends on the other experimental conditions. In many cases, only kinetic studies have revealed the occurrence of intermediates (see Chapter 7). When most of the equilibrium studies are such that conditions (1), (2), and (3), or even only (1) and (2), are satisfied, it may be reasonably concluded that the intermediate states, if they do exist, are not significantly populated, and that the equations of a two-state model may be used in a first approximation. Useful information can be obtained by such analysis. Furthermore, when conditions (3), (4), and (5) are fulfilled, the validity of a two-state mechanism is demonstrated.

6.3. ENERGETICS OF THE TRANSITION DEDUCED FROM EQUILIBRIUM STUDIES

6.3.1. Evaluation of Thermodynamic Parameters

The determination of the thermodynamic parameters from equilibrium studies is based on a two-state transition model. Free energy variation, ΔG_{D}, can be deduced directly from the transition curves:

$$K_{\mathrm{D}} = (D)/(N) = \exp(-\Delta G_{\mathrm{D}}/RT)$$

or

$$\Delta G_{\mathrm{D}} = -RT \ln K_{\mathrm{D}}$$

ΔH_{D} is obtained by a van t'Hoff plot from the variations of K_{D} with temperature. The thermodynamic relationship

$$\Delta G_{\mathrm{D}} = \Delta H_{\mathrm{D}} - T\Delta S_{\mathrm{D}}$$

allows the determination of the entropy of denaturation.

It is of interest to have a good estimate of ΔG_{D}^0, the free energy in the absence of denaturant and at physiological temperature and pH. This value is a measure of the stability of the native protein. Different procedures for estimating ΔG_{D}^0 do exist. Direct determination based on the rate of hydrogen–deuterium exchange was used by Hvidt and Nielsen (1966);

this rate is proportional to the concentration of the unfolded protein. This method is detailed in Chapter 8. The use of proteolytic cleavage was also proposed, but the interpretation is not straightforward (Matthyssens *et al.*, 1972). Calorimetric studies are sufficiently accurate to allow direct determination. However, this value is most generally determined from denaturation–renaturation studies at equilibrium by extrapolation; different extrapolation procedures have been proposed.

When denaturation is induced by denaturant such as urea or GuHCl, the value of ΔG_D^0 must be extrapolated to zero. The simplest method is linear extrapolation to zero concentration of denaturant. For lysozyme, Aune and Tanford (1969a,b) observed a rather good linearity between ΔG_D and the concentration of denaturant in a rather wide concentration range. Linear extrapolation was used by Green and Pace (1974) for denaturation of this protein by urea and GuHCl. Figure 6.9 shows their results; data were fitted according to the equation:

$$\Delta G_D = \Delta G_D^0 + m(\text{denaturant}) \tag{6.5}$$

The same value of 5.8 kcal/mole was obtained for both denaturation curves.

Aune and Tanford (1969a,b) proposed another procedure of extrapolation. They assumed that denaturation by urea or GuHCl results from a greater number of binding sites for the denaturant on the denatured molecule than on the native molecule, the sites being equivalent and independent. Under

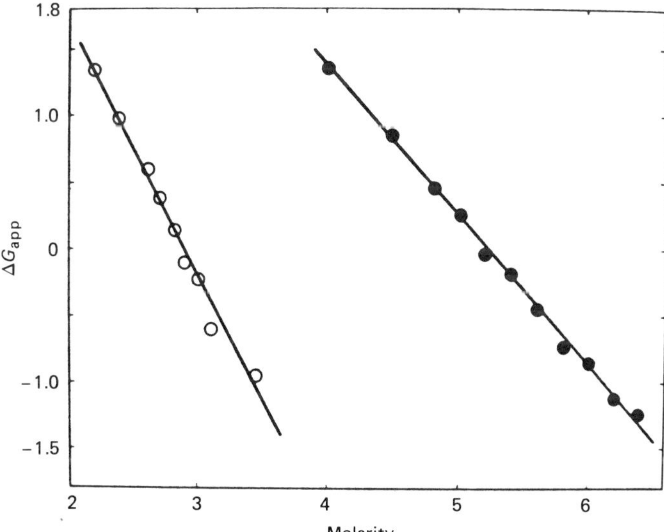

Fig. 6.9. Denaturation of hen egg lysozyme. Variation of ΔG with denaturant concentration: urea (●); GuHCl (○) (from Green and Pace, 1974).

these conditions, the following expression describes the variations of ΔG_D:

$$\Delta G_D = \Delta G_D^0 - \Delta n RT \ln(1 + Ka) \tag{6.6}$$

This expression is derived from Eq. (5.8) (Chapter 5). The best fit gives $K = 1.2$ for lysozyme (see Chapter 5); this value was adopted for similar data obtained from studies of ribonuclease, myoglobin, cytochrome *c*, and α-lactalbumin and K_D was thus estimated to be 10.44 kcal/mole. Based on data obtained from studies of several proteins, it is clear that this procedure gives a larger value than linear extrapolation.

Tanford (1964) had proposed another procedure for estimating ΔG_D^0 from denaturation by urea and GuHCl. This procedure is based on the following equilibrium:

$$\begin{array}{ccc} N \text{ in } H_2O & \rightleftharpoons & D \text{ in } H_2O \\ \Delta G_{tr,N} \updownarrow & & \updownarrow \Delta G_{tr,D} \\ N \text{ in denaturant} & \rightleftharpoons & D \text{ in denaturant} \end{array}$$

where $\Delta G_{tr,N}$ and $\Delta G_{tr,D}$ are the free energy of transfer from water to solution containing the denaturant, of the native and the denatured forms respectively. Therefore,

$$\Delta G_D - \Delta G_D^0 = \Delta G_{tr,D} - \Delta G_{tr,N} = \sum_i \alpha_i\, n_i^0\, \delta g_{tr,i}$$

this difference depends only on groups which are exposed to the solvent in denatured but not in native molecules. Thus, $\delta g_{tr,i}$ is the free energy of transfer of a group of type *i* from water to denaturant and α_i is the fractional change of the degree of exposure to solvent of groups of type *i*. The values of $\delta g_{tr,i}$ have been determined from solubility of amino acids and derivatives. These values are given in Table 6.3 (Robinson and Jencks, 1965). Introducing $\bar{\alpha}$, the average value of α_i, the equation of ΔG_D becomes

$$\Delta G_D = \Delta G_D^0 + \bar{\alpha} \sum_i n_i^0\, \delta g_{tr,i} \tag{6.7}$$

In Fig. 6.10, which represents denaturation of lysozyme by GuHCl, the curves are drawn according to Eq. (6.7) for two values of $\bar{\alpha}$ (0.35 and 0.275, respectively for curve 1 and 2). The best agreement for $\bar{\alpha}$ is obtained when not only peptide groups and hydrophobic side chains, but also polar side chains, are included in the calculation. In his review, Pace (1975) emphasized that contrary to expectation, $\bar{\alpha}$ values do not increase with increasing molecular weight. He arrived at this conclusion because for the largest protein considered he found the smallest value of $\bar{\alpha}$. To explain the observed discrepancy, the author supposed differences in the accessibility to solvent of the denatured states for the different proteins.

TABLE 6 3

Free Energies of Transfer, $-\delta g_{tr}$(cal/mole) from Water to Aqueous Urea or GuHCl of Peptide Groups and Amino Acid Side Chains

	Urea				GuHCl			
	2 *M*	4 *M*	6 *M*	8 *M*	1 *M*	2 *M*	4 *M*	6 *M*
Peptide[a]	49	86	118	130	83	134	207	245
Ala	0	−15	−10	−10	10	20	30	45
Val[b]	60	85	125	160	85	115	195	265
Leu	110	155	225	295	150	210	355	480
Ile	100	140	205	265	135	190	320	430
Met	115	225	325	415	150	245	400	535
1/2 Cys[b]	115	225	325	415	150	245	400	535
Phe	180	330	470	600	215	355	580	775
Tyr	225	395	580	735	235	385	605	770
Trp	270	505	730	920	400	630	980	1235
Pro[b]	75	105	155	200	100	140	240	320
Thr	40	60	90	115	65	90	120	125
His	100	160	205	255	180	285	385	420
Asn	135	225	330	430	200	320	490	645
Gln	80	130	190	230	135	215	315	360

[a] Based on solubility studies of two model compounds *N*-acetyltetraglycine ethyl ester and ethylacetate by Robinson and Jencks (1965).

[b] The δg_{tr} values for these side chains are estimates based on results for the other side chains and on results at a single denaturant concentration (Wetlaufer *et al.*, 1964).

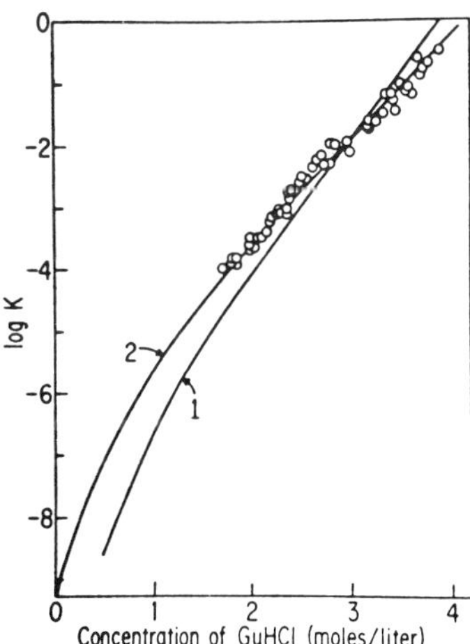

Fig. 6.10. Guanidine hydrochloride denaturation of lysozyme. Curves are drawn according to the equation $\Delta G = \Delta G_0 (1 + \alpha K)$ with $\alpha = 0.35$ for curve 1; $\alpha = 0.275$ for curve 2 (from Tanford, 1970).

However, $\bar{\alpha}$ is an average value and a more exact calculation can be made on the basis of accessible surface area when high resolution X-ray structure is available (Chothia, 1976 and J. Janin, personal communication). Such an evaluation has been made for T_4 lysozyme (Desmadril *et al.*, 1982) and compared with that obtained with $\bar{\alpha} = 0.35$ as proposed by Tanford (1970). The slopes of the curves obtained from these two kinds of evaluation are significantly different. From T the value given by Tanford, it increases to 15 in the procedure taking into account exposure of individual residues in the native molecule.

The validity of the different extrapolation procedures was discussed by Pace (1975) from results on four proteins (ribonuclease, lysozyme, α-chymotrypsin, and β-lactoglobulin) by comparison of the ΔG^0 obtained from urea and GuHCl denaturation. Table 6.4 gives the different values. It clearly appears that Tanford's model does not provide consistent values of ΔG_D^0. It is attributed to the standard used for the evaluation of δg_{tr} values for a peptide bond; these values are obtained from solubilities data of *N*-acetyltetraglycine ethyl ester and ethylacetate. Linear extrapolation leads to smaller values of of ΔG_D^0. A difference in denaturant binding in native and denatured states is probably the most reasonable approximation. However, the estimation of K (Eq. 6.6) is not rigorous. With K values of 0.1 the denaturant binding model gives values of $\Delta G_D^{H_2O}$ slightly higher than those of linear extrapolation and those using larger K values give almost double values of $\Delta G_D^{H_2O}$. When K = 1.0 for urea, a good fit of experimental data is obtained, but when one takes

TABLE 6.4

Comparison of the $\Delta G_D^{H_2O}$ (kcal/mole) Values Obtained by Various Extrapolation Procedures[a]

	Ribonuclease (pH = 6.6)		Lysozyme (pH = 2.9)		α-Chymotrypsin (pH = 4.3)		β-Lactoglobulin (pH = 3.2)	
Extrapolation procedure	Urea	GuHCl	Urea	GuHCl	Urea	GuHCl	Urea	GuHCl
Linear extrapolation	7.7	9.3	5.8	5.8	8.4	7.8	10.5	12.5
Tanford's method	12.1	14.8	6.6	9.1	8.8	10.4	12.1	18.4
Denaturant binding								
$K = 0.1$	8.8	11.9	6.9	7.1	9.9	10.7	12.3	15.8
= 1.0 (urea) or = 1.2 (GuHCl)	15.7	17.3	12.1	10.0	16.7	14.3	21.4	23.4
Average	10.0 ± 1.6		6.4 ± 0.5		9.1 ± 0.9		12.6 ± 1.2	

[a] From Pace (1975). Average values were calculated including Tanford's method and GuHCl results using denaturant binding with the largest K value.

into account the results of studies of model compounds this value is too high. With GuHCl, a value of 1.2 is correct for model compounds that GuHCl solubilizes best, but not for those which are less soluble. Thus a lower value of K should be used. The assumption that all sites are equivalent and independent is certainly oversimplified. However, since the affinity is very weak, the order of magnitude might not be very different. Whatever it is, a rather good fit might be obtained by different procedures, since experimental data extend in a very narrow range of denaturant concentrations; however, extrapolation to zero is very different for each procedure as early shown in Table 6.4. From these procedures, in any case ΔG_D^0 cannot be obtained with a better precision than ± 15–20%. The most accurate treatment is certainly the thermodynamic one used by Schellman (1978).

Another method of evaluation was proposed based on hydrogen exchange experiments (Hvidt and Nielsen, 1966). This method is detailed in Chapter 8. Amide hydrogens of the peptide groups quickly exchange if they are on the surface of the molecule accessible to solvent, and very slowly if buried at the interior. Hydrogen can be exchanged either by deuterium or by tritium (Ottesen, 1971). This method has been used by different authors for the evaluation of ΔG_D^0 (Hvidt and Nielsen, 1966; Nakanishi *et al.*, 1972; Woodward and Rosenberg, 1970, 1971; Ottesen, 1971). In many cases there are different classes of exchanging hydrogens with different rate constants, and sometimes it is difficult to resolve a distinct class. Another difficulty may be the fact that the stability of a protein can change during the course of an experiment on the exchange of hydrogen for deuterium or tritium. Table 6.5 indicates values thus determined for lysozyme, ribonuclease, and myoglobin. The ΔG_D^0 values are consistent with those obtained from extrapolation procedures, so much more that pH and temperature conditions are slightly different from those of Table 6.4. In spite of the low accuracy of the ΔG_D^0 values, whatever may be the method of determination, they indicate the weak stability of native proteins.

TABLE 6.5

Estimates of $\Delta G_D^{H_2O}$ from Hydrogen Exchange Studies[a,b]

Protein	pH	T (°C)	$\Delta G_D^{H_2O}$ (kcal/mole)
Lysozyme	5.4	20	7.4
Ribonuclease	4.7	38	6.5
Myoglobin	7	21	13.0

[a] From Pace (1975).

[b] These estimates are based on k_{ex} values from poly-DL-alanine, and on k_{obs} values for the most slowly exchanging hydrogens.

6.3.2. Evaluation of Different Energetic Contributions

As mentioned in Chapter 2, the free energy variation of the overall process (unfolded–folded) is the sum of diverse contributions arising from different forces of interaction and entropy factors:

$$\Delta G_{\mathrm{D}} = \Delta G^{0}_{\mathrm{H}_{\phi}} + \Delta G^{0}_{\mathrm{h}} + \Delta G^{0}_{\mathrm{b}} + \Delta G^{0}_{\mathrm{x}} + \Delta G^{0}_{\mathrm{elec}} + \Delta G^{0}_{\mathrm{titr}} + \Delta G_{\mathrm{binding}}$$

which are respectively, hydrophobic interactions, hydrogen bonding, backbone conformational entropy, disulfide bonds, variations of electrostatic interactions from native to denatured protein, titration of anomalous groups, and denaturant binding if any (i.e., if denaturation is induced by urea or GuHCl for example).

The diverse free energy contributions were carefully estimated in the very nice experimental study on denaturation of chymotrypsinogen by Brandts (1964a,b). This work is not recent; however, no other study has been published that is as careful and complete. A two-state process was assumed to be reasonably supported by the transition curves obtained from different observables (Brandts and Lumry, 1963). The thermodynamic parameters were estimated by analyzing the transition over the temperature range from 0° to 65°C and over the pH range from 0.5 to 3.7. The enthalpy and entropy changes are strongly temperature dependent indicating a great difference in the heat capacity of native and denatured forms. Variations of ΔG versus temperature at a given pH indicate a minimum and therefore a maximum of the stability of native protein at about 10°C (Fig. 6.11). The effect of ionic strength as well as the effect of denaturing agents, such as urea and ethanol, were also analyzed. The free energy of denaturation was evaluated by estimating the contribution from each amino acid residue involved in the transition:

$$\Delta G^{0} = p\left(N\,\overline{\Delta h}_{\mathrm{H}} - NT\,\overline{\Delta S}_{\mathrm{c}} + \sum_{i=1}^{N} \delta g_{i}^{\mathrm{tr}}\right) + \Delta G_{\mathrm{elec}} + \Delta G_{\mathrm{titr}} \qquad (6.8)$$

p is the fraction of the N residues ($N = 240$ residues) which is actually unfolded during the transition; $\overline{\Delta h}_{\mathrm{H}}$ is the average enthalpy for breaking a peptide hydrogen bond; $\overline{\Delta S}_{\mathrm{c}}$ is the average increase of conformational entropy; $\delta g_{i}^{\mathrm{tr}}$ is the free energy of transfer from its local environment in the protein to the solvent; ΔG_{elec} and ΔG_{titr} were previously defined.

The free energy of titration, ΔG_{titr} was evaluated by titration of anomalous groups:

$$\Delta G_{\mathrm{titr}} = -RT \ln \prod^{i} \left[(1 + (a_{\mathrm{H}}/K_{\mathrm{D}}^{i}))/(1 + (a_{\mathrm{H}}/K_{\mathrm{N}}^{i}))\right] \qquad (6.9)$$

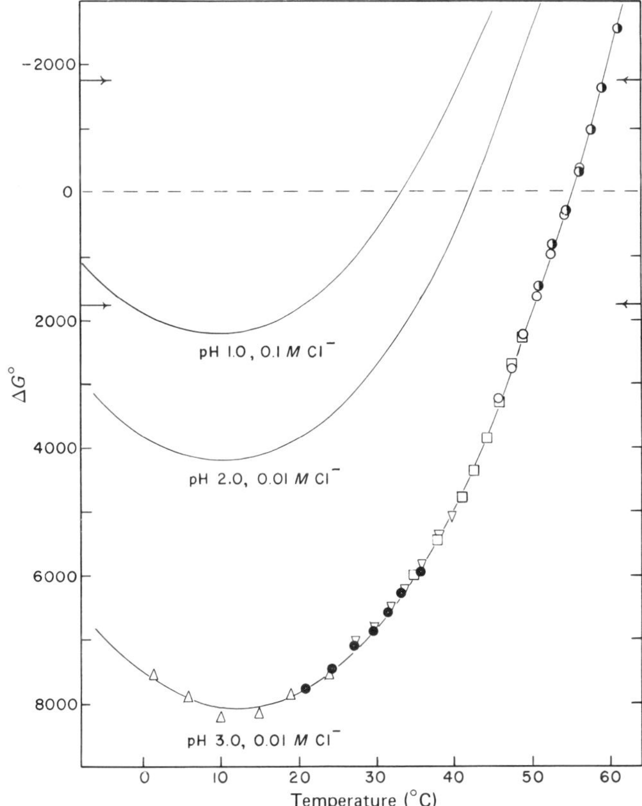

Fig. 6.11. Variations of ΔG with temperature for chymotrypsinogen under different conditions of pH and Cl^- concentration (courtesy of Brandts, 1964b).

where K_D^i and K_N^i are respectively the ionization constants of the i group in the denaturated and in the native protein. It was shown that chymotrypsinogen contains 3 (actually 3.18) anomalous carboxyl groups, with a pK of 1.3 in the native protein ($K_N = 5 \times 10^{-2}$) and a pK of 4.5 ($K_D = 3 \times 10^{-5}$) in the denatured protein. An estimation of ΔG_{elec} was made by studying the dependence of ΔG on Cl^- concentration. The value of $\sum_{i=1}^{N} \delta g_{tr}^i$ was estimated from the free energy transfer of amino acid side chains from 95% ethanol (as used for the reference solvent to mimic the interior of the protein) to water. All the commonly occurring amino acids may be grouped into six categories according to their values of δg_{tr}^i. Table 6.6 gives these values. Since the sequence of chymotrypsinogen is known, it was possible to estimate

TABLE 6.6

Free Energy of Transfer from 95% Ethanol to Water for Amino Acid Side Chains[a]

Group	Amino acid side chain	Δg^i_{tr} (cal)
I	Tyr	2800
II	Norleu, Leu, Ile, Phe, Pro, Tyr	2500
III	Val, Met, Lys (nonpolar portion)	1500
IV	Ala, Arg (nonpolar portion)	600
V	Gly, Asp (NH_2), Glu (NH_2), Ser, Thr, Cys	0
VI	Asp, Glu, His, and charged portion of Arg and Lys	<0

[a] From Brandts (1964b). The values of Δg^i_{tr} are approximate and have been estimated from solubility data.

$\sum_{\text{I–IV}} \delta g^i_{tr}$ (only categories I to IV are considered). Equation (6.8) can be written

$$\Delta G^0 = p(240\,\overline{\Delta h_H} - 240T\,\overline{\Delta S_c} - 2{,}289.2T + 18.255T^2 - 0.028128T^3) + \Delta G_{elec} + \Delta G_{titr} \tag{6.10}$$

By comparison of this expression and of the experimental curves (Fig. 6.11) a rather good fit is obtained at all temperature by choosing

$$p = 0.634 + 0.01$$
$$\overline{\Delta h_H} = 0.8 \quad \pm 0.01 \text{ kcal/mole}$$
$$\overline{\Delta S_c} = 5.1 \quad \pm 0.1 \text{ eu}$$

The values of $\overline{\Delta h_H}$ and of $\overline{\Delta S_c}$ are in agreement with other estimates found in the literature.

The total contribution to the stability of chymotrypsinogen provided by hydrogen bonding is 120 kcal/mole; that of hydrophobic interactions is 105 kcal/mole at 0°C and it rapidly increases with temperature until 75°C. These two contributions (250 kcal at 0°C) are quite balanced at each temperature by the conformational entropy which tends to destabilize the structure, so that the free energy of stabilization scarcely exceeds 10 kcal/mole. Figure 6.12 shows the different energetic contributions in a large range of temperature. One must note the rather good agreement between this first experimental evaluation and a calculated estimation of the various contributions to the free energy of unfolding presented by Finney and co-workers (1980). The native structure results from a delicate balance between stabilizing and destabilizing contributions. If the contribution of anomalous group is

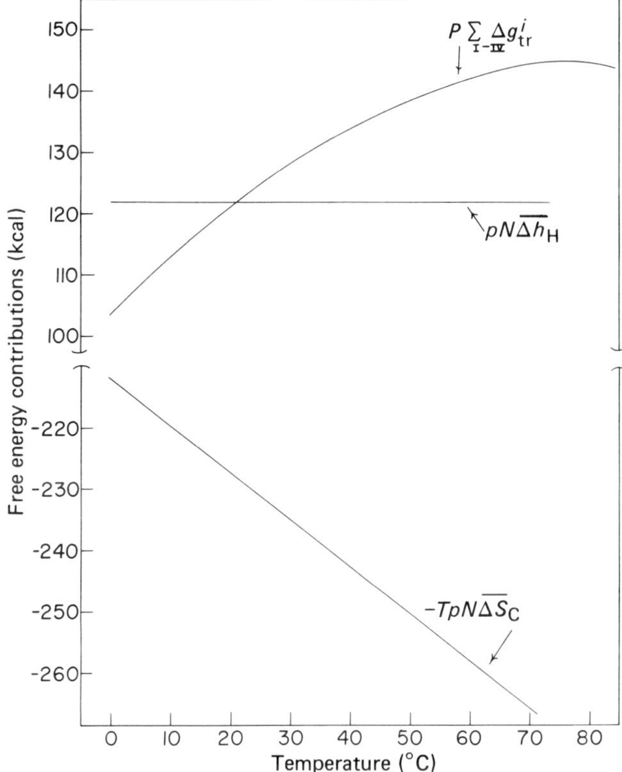

Fig. 6.12. Estimates of the different energy contributions, hydrogen bonding, hydrophobic interactions, and conformational entropy to the stability of chymotrypsinogen (courtesy of Brandts, 1964b).

negligible compared to the total stabilizing energy, it is large enough to seriously alter the equilibrium by changing the pH.

The effect of cross-links in thermodynamic stability of proteins has been evaluated by R. E. Johnson and co-workers (1978). They studied the thermal transitions of native lysozyme and of a cross-link ester derivative, the oxindole alanine lysozyme obtained by iodine oxidation (Imoto and Rupley, 1973). The change in free energy of stabilization was attributed only to the entropy decrease of the unfolded chain.

The effect of ligand binding on the melting point of a protein has been quantitatively studied by Schellman (1975). He proposed the following equation to account for the effect of ligand binding on the melting temperature of a protein:

$$T_m = (T - T_0)R/[\Delta H_D \ln(1 + K_B[S]) \qquad (6.11)$$

where T_m is the variation of the melting point in the presence and in the absence of ligand, at concentration $[S]$; K_B is the binding constant of the ligand to the proteins; T and T_0 are the T_m in the presence and in the absence of ligand. The presence of tri-*N*-acetylglucosamine stabilizes lysozyme to thermal as well as to GuHCl denaturation, in agreement with these predictions (Pace and McGrath, 1980).

The thermostability of proteins in organisms which live at high temperature is questionable. Thermodynamic studies were conducted on the thermostable proteins, phosphoglycerate kinase (PGK) (Nojima *et al.*, 1977) and cytochrome *c*-552 (Nojima *et al.*, 1978) from a thermophilic bacterium, *Thermus thermophilus*. Although the contributions of each kind of stabilizing forces were not detailed in the reports on these studies as it was for chymotrypsinogen, the different parameters were determined as a function of temperature for different concentrations of GuHCl, and the ΔG versus temperature profiles were compared for thermophilic and mesophilic proteins. Figure 6.13 shows the behavior of PGK from yeast and from *T. thermophilus*. All the coefficients of the equation,

$$\Delta G = A + BT + CT^2 + DT^3$$

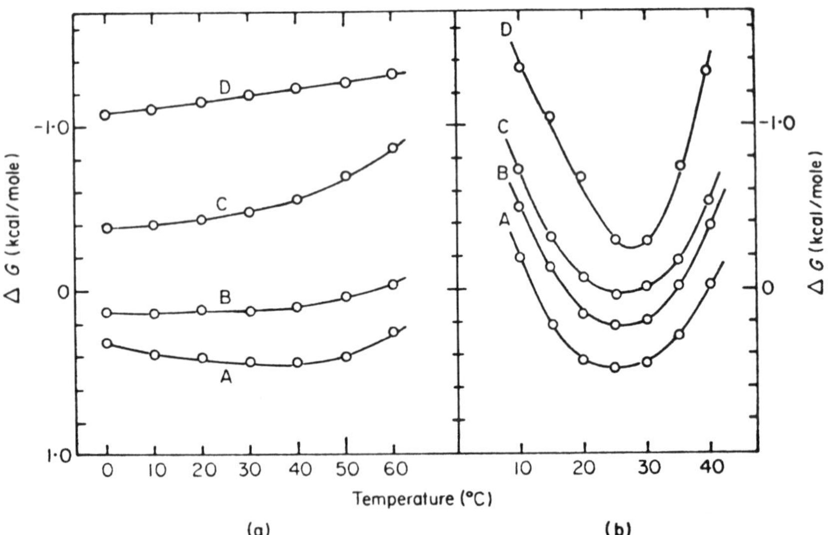

Fig. 6.13. Variations of ΔG with temperature for phosphoglycerate kinase: (a) from *T. thermophilus*, GuHCl concentration A, 2.26 *M*; B, 2.33 *M*; C, 2.45 *M*; D, 2.73 *M*, (b) from yeast, GuHCl concentration A, 0.54 *M*; B, 0.56 *M*; C, 0.60 *M*; D, 0.70 *M* (from Nojima *et al.*, 1978).

obtained for the different GuHCl concentrations, were determined. The profiles obtained are different. While yeast PGK displays a classical behavior similar to that of chymotrypsinogen, with a well defined minimum around 25°–28°C, the profile obtained for *T. thermophilus* is flattened and a minimum cannot be determined. Furthermore, for the thermophilic enzyme, the dependence of ΔH on temperature is extremely small, whereas ΔH is strongly temperature dependent for yeast PGK and varies from −27 to 31 kcal/mole when the temperature increases from 10° to 40°C. A very small ΔS of denaturation is observed for *T. thermophilus* PGK; from −3.9 eu at 0° it reaches +21 eu at 60°C. For yeast PGK it increases from −96 to 104 eu between 10° and 35°C.

A similar analysis was reported for cytochrome *c*-552 (Nojima *et al.*, 1978). The enzyme from *T. thermophilus* in this case has a ΔG versus T profile with a well defined minimum at 27° ± 1.4°C; therefore, it is quite similar to the classical profile. However, the value of the temperature of maximum stability (ΔG_{min}) is shifted to higher temperatures by comparison with that of cytochrome *c*-552 from mesophilic organisms, which are the following: bovine (12°C), horse (12°C), *C. krusei* (9°C). The temperature dependence of ΔH and ΔS in this case are similar. Nojima and co-workers (1978) proposed three possible models to explain how thermophilic protein can become thermostable: (1) by a shift of ΔG versus T profile to higher temperatures, and therefore by increasing the temperature of maximum stability (This model accounts for cytochrome *c*-552.); (2) by flattening the denaturation curve, with a very small ΔS (this mechanism applies for *T. thermophilus* PGK); (3) by lowering ΔG at the temperature of maximum stability, and therefore increasing this stability.

However, these models remain phenomenological and do not provide explanation of the thermostability in terms of interactions. Nojima and co-workers (1978) have not determined the contributions of the various interactions (e.g., hydrophobic interactions, hydrogen bonds) especially those which reinforce the stabilization energy in thermophilic versus mesophilic proteins. Crystallographic studies of thermophilic and mesophilic proteins have revealed the existence in thermophilic proteins of extra ionic bonds clamping together structural elements of the molecules. It was shown for glyceraldehyde-3-phosphate dehydrogenase (Biesecker *et al.*, 1977; Walker, 1978) and for ferredoxins (Perutz and Raidt, 1975).

The importance of the ionic bonds in the thermostabilization was emphasized by Perutz (1978). From differences in rate of denaturation in mesophilic and thermophilic molecules, Perutz (1978) has evaluated the extrastabilization energy to be no more than 2 kcal/mole in ferredoxin and 5–10 kcal/mole in glyceraldehyde-3-phosphate dehydrogenase. Proteins from thermophilic organisms offer a very good example, allowing one to understand the mechanism by which molecules adapt to various environmental conditions. Very few substitutions which do not significantly modify the folding of the polypeptide chain are needed, since only a small extra energy of stabilization is required. Substitution of a single amino acid (i.e., replacement of a glycine at position 211 with either a Glu or an Asp) does not change considerably the free energy of unfolding but increases both enthalpy and entropy changes (Matthews *et al.*, 1980).

6.4. INTERPRETATION OF EXPERIMENTAL STUDIES OF THE UNFOLDING–FOLDING TRANSITION

6.4.1. Validity of the Two-State Approximation

It appears from different research reports (Tanford, 1970; Pace, 1975) that, for many proteins, reversible transitions from the unfolded to the folded state have characteristics of a two-state process, without detectable intermediates at equilibrium, even when the existence of intermediates is supported by other experimental arguments (kinetics for example, see Chapters 7, 8 and 9). Urea and GuHCl denaturation is generally closer to a a two-state mechanism than pH or thermal unfolding.

Transition curves for several proteins with and without disulfide bonds are given in Chapter 5. For example, with staphylococcal nuclease a single transition was found (Fig. 5.5). The curves obtained from different signals (e.g., reduced viscosity, molar ellipticity at 220 nm, emission fluorescence, and absorbance at different wavelengths, and proton resonance of three of the four histidines) are superposed for the acidic transition; however, occurrence of intermediates was detected by other methods. Thermal denaturation of RNase in the acidic pH range can be reasonably analyzed according to a two-state process. A cooperative transition is observed; there is a satisfactory coincidence of the data obtained from techniques as diverse as intrinsic viscosity, optical rotation at 365 nm, and difference absorption at 287 nm (Ginsburg and Carroll, 1965). Furthermore ΔH varies monotonically according to temperature (Lumry *et al.*, 1966).

Figure 6.14 shows the two-state behavior of RNase transition. However, deviation from a two-state is observed at higher pH values.

The denaturation of hen egg lysozyme was also carefully investigated (Aune and Tanford, 1969a,b; Tanford and Aune, 1970). The equilibrium was studied as a function of denaturant (GuHCl, pH, and temperature). The transition is in all cases very close to a two-state behavior. For phage T_4 lysozyme, an abrupt transition is also observed for denaturation by GuHCl (Elwell and Schellman, 1975, 1977; Desmadril and Yon, 1981; Desmadril

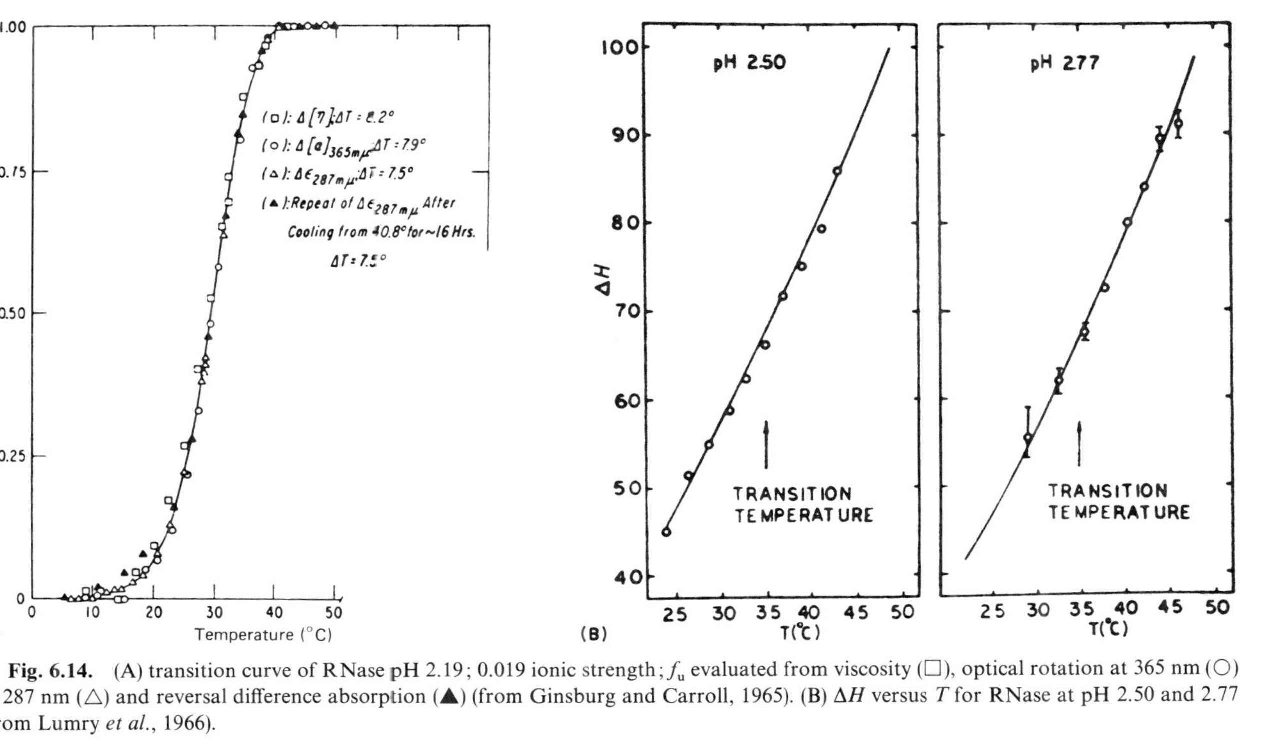

Fig. 6.14. (A) transition curve of RNase pH 2.19; 0.019 ionic strength; f_u evaluated from viscosity (□), optical rotation at 365 nm (○) at 287 nm (△) and reversal difference absorption (▲) (from Ginsburg and Carroll, 1965). (B) ΔH versus T for RNase at pH 2.50 and 2.77 (from Lumry *et al.*, 1966).

et al., 1982, in press). At pH 7.4, in spite of the great cooperativity of the transition ($n \approx 23$), T_4 lysozyme transition significantly deviates from a two-state behavior. (Desmadril and Yon, 1981, 1982; Desmadril *et al.*, 1982). Transition curves obtained for chymotrypsinogen denaturation induced by temperature and by denaturants (GuHCl or urea) (Brandts and Hunt, 1967) are reasonably interpretable according to a two-state process.

The steepness of the transition was evaluated by different authors (Tanford, 1970, Elwell and Schellman, 1975) from the $\bar{n}$ number of the Eq. (5.10). Table 5.5 (Chapter 5) gives the $\bar{n}$ values and the c_m for different proteins. For GuHCl, a value of $\bar{n}$ very close to 16, found for very cooperative transitions, is interpretable according to a two-state process. Lower values of $\bar{n}$ index might indicate the presence of intermediary species.

Several other examples of the validity of a two-state approximation in the transition curves of proteins exist in the literature, for example, GuHCl and acid denaturation of cytochrome *c* from different species (Knapp and Pace, 1974; McLendon and Smith, 1978) and myoglobins (Puett *et al.*, 1973; Puett, 1973). It is not the purpose of the present discussion to describe exhaustively all known examples. Here only the most characteristic examples are given to illustrate the validity of the approximation.

6.4.2. Deviations from Two-State Behavior

6.4.2.1. Evidence for Intermediates Deduced from Transition Curve Profile

If an appreciable concentration of intermediates exists at equilibrium, then the steepness of the transition curves is not as marked and transition curve profiles may clearly indicate the different steps. An example is provided by urea denaturation of phosphorylase *b* (Chignell *et al.*, 1972); two intermediates of folding were discerned in the denaturation profile (Fig. 6.15). It was shown that the protein is essentially monomeric at 1 *M* urea and for higher concentrations (except for the formation of an aggregated fraction at higher concentration) and that intermediates do not arise from subunit dissociation. The first transition, in the 1–2 *M* urea concentration range is associated with a large increase in the fluorescence of pyridoxamine; the resulting intermediate easily aggregates. The second transition is centered at about 4 *M* urea. Both intermediates can be observed by dichroism and difference spectra.

Other examples have been reported. Denaturation of RNase by LiCl clearly displays two transitions, when the process is followed by variations of absorbance at 287 nm. For denaturation of paramyosine by GuHCl, at pH 7.3 followed by optical rotation at 232 nm (Riddiford, 1966), different separate steps were observed. Tanford (1970) also mentioned separate stages

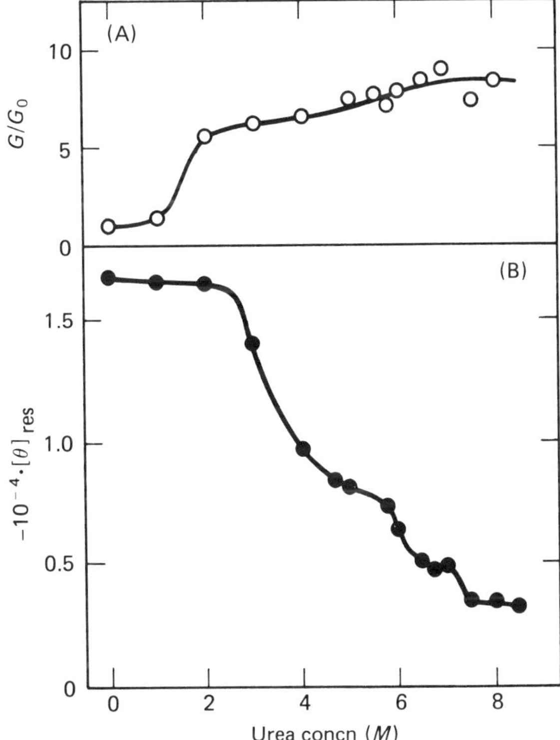

Fig. 6.15. Denaturation of phosphorylase *b* by urea: (A) fluorescence intensity at 395 nm relative to that without urea; (B) molar residue ellipticity (from Chignell *et al.*, 1972).

in the unfolding of yeast aldolase in the acid pH range detected by ORD and difference spectra.

6.4.2.2. *Noncoincidence of Transition Curves Obtained by Different Methods*

The noncoincidence of denaturation curves obtained from different methods is a clear indication of the existence of intermediates. A typical example is provided by the unfolding of carbonic anhydrase induced by GuHCl (Wong and Tanford, 1973). Denaturation was followed by CD, optical rotation, difference spectral change, and by loss of enzymatic activity. The noncoincidence of the denaturation profiles is evident in Fig. 6.16A. The transition followed by difference spectral measurements is essentially identical at different wavelengths (291.4 nm, 286.4, and 235 nm) giving a midpoint of 1.8 *M* GuHCl (Fig. 6.16B). According to circular dichroism data,

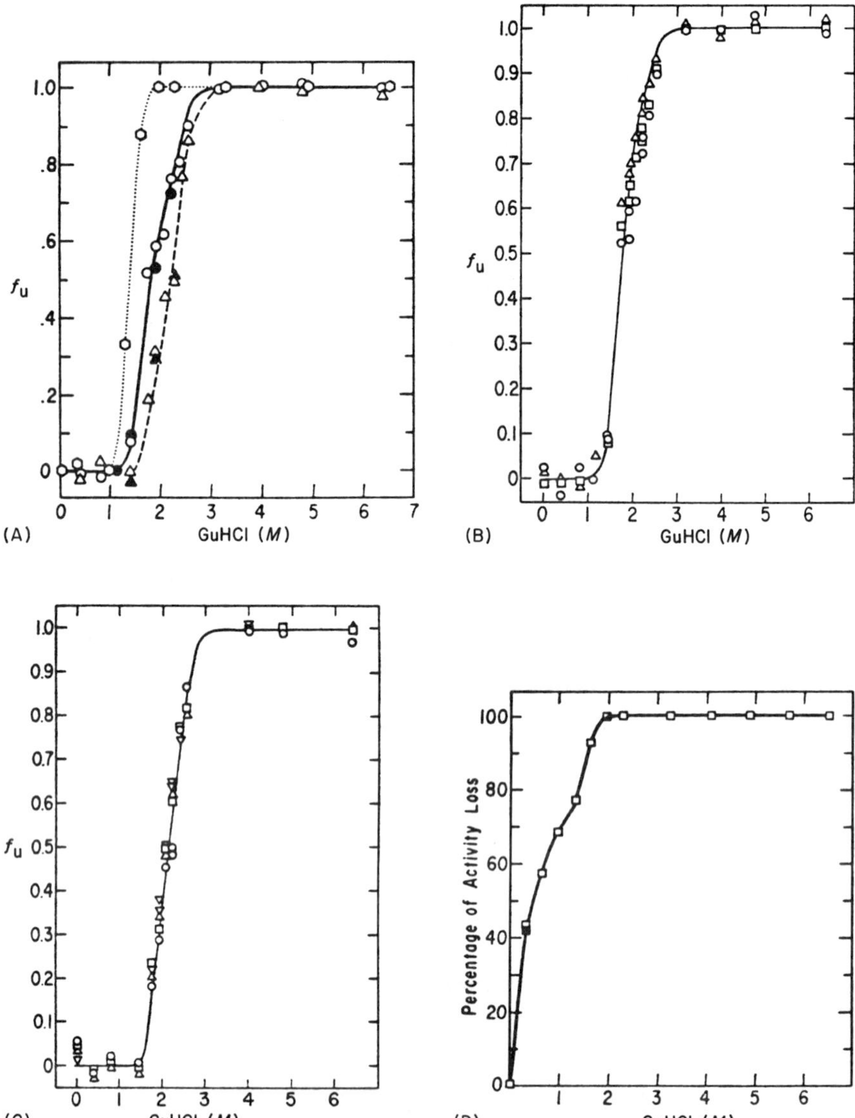

Fig. 6.16. Guanidine hydrochloride-induced denaturation of carbonic anhydrase B. The transition was followed by various methods (A) f_u was evaluated by CD measurements (··· ○ ···), by difference spectra measurements (—○—), or by optical rotation measurements (△); (B) f_u was evaluated by difference spectral change at 291.4 nm (○), 286.4 nm (△), or 235 nm (□); (C) f_u was evaluated from optical rotation changes at 400 nm (○), 350 nm (□), 300 nm (△), and 287 nm (▽); (D) f_u was evaluated by percentage of activity loss (from Wong and Tanford, 1973).

the transition takes place at a lower concentration with a midpoint at 1.4 *M* GuHCl. The optical rotation changes during the unfolding of carbonic anhydrase mainly reflects the randomization of the polypeptide backbone; the loss of the aromatic Cotton effect contributes only in a small proportion to the total change of optical rotation. Figure 6.16C shows the denaturation profile at different wavelengths. The results indicate that the unfolding of the backbone occurs at a higher GuHCl concentration than the exposure of aromatic residues. Two stages in the transition are clearly discerned by following the loss of enzymatic activity (Fig. 6.16D). The total transition is less cooperative than the transition observed for most globular proteins.

Even by study at equilibrium, the reversible unfolding of carbonic anhydrase can be separated into distinct and successive stages. The authors suggest three stages, the first being a loosening of the tight folding of the protein accompanied by the change in CD arising from aromatic groups. This may be followed by an exposure of aromatic groups to solvent. In a third stage, the randomization of the polypeptide backbone probably takes place.

In the study of GuHCl-induced transition of phage T_4 lysozyme, it was shown that transition curves obtained by CD and fluorescence (shift in λ_{max}, wave length of the maximum of fluorescence emission, and also variation of fluorescence intensity at 326 nm) do not coincide. Furthermore, transition obtained by CD can be decomposed into two separate transitions: one with a c_m of 1.95 *M* GuHCl; the other with a c_m of 2.2 *M* GuHCl and an amplitude of 20% of the total variation in ellipticity corresponds to the transition observed by fluorescence. Since all three tryptophans are localized in the C-terminal domain, these data suggest that this domain or part of this domain unfolds after the rest of the molecule (Desmadril and Yon, 1981, Desmadril *et al.*, 1982a).

Noncoincidence of transition curves was also reported for GuHCl unfolding of penicillinase from *Staphylococcus aureus* (Robson and Pain, 1973). Figure 6.17 illustrates the transition curves obtained by following changes in difference spectra, in optical rotation, and in intrinsic viscosity. The data were interpreted by the occurrence of a first transition virtually complete at 0.8 *M* GuHCl which reflects an unfolding of the molecule to a highly expanded state without modification of the helical conformation. The second transition with a midpoint at 1.3 *M* GuHCl is attributable to the helical content.

The noncoincidence of transition curves was also reported for acid denaturation of cytochrome *c* (Knapp and Pace, 1974), for denaturation of growth hormone (Holladay *et al.* 1974), and for guanidine unfolding of apomyoglobin (Balestrieri *et al.*, 1976). Reversible denaturation of α-lactalbumin has been studied (Nitta and Sugai, 1972; Sugai *et al.*, 1973; Kita

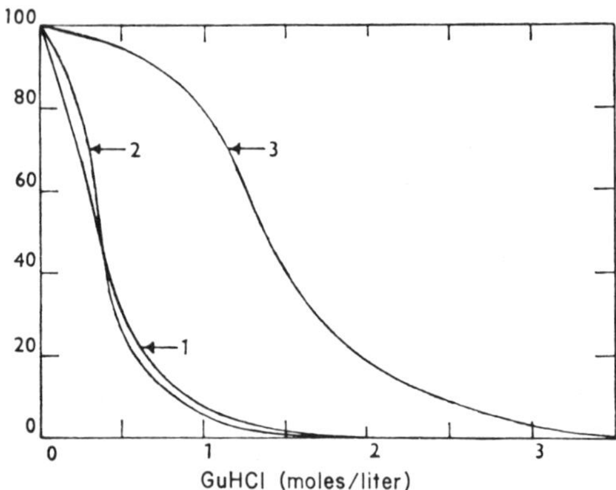

Fig. 6.17. Conformational transition of penicillinase followed by difference spectral measurements (1), viscosity (2), and circular dichroism at 234 nm (3) (reproduced from Robson and Pain, 1973).

et al., 1976; Nitta *et al.*, 1977a,b; Kuwajima *et al.*, 1976; Okuda and Sugai, 1977).From equilibrium study, a three-state mechanism was reported (Kuwajima, 1977). The unfolding was produced by GuHCl at different temperatures and pH, and studied by means of CD, difference UV absorption spectra. A preliminary kinetic approach was also reported. The data are consistent with the occurrence of an intermediate species having a structure comparable in helical content to the native one, but somewhat extended. An interpretation of the folding process is suggested involving in a first stage the formation of helical structures dictated by local interactions, in a second stage the packing of the segment of helices governed by hydrophobic interactions, and in final stage only stabilization of the structure by electrostatic interactions (Fig. 6.18). In this last step, the interactions of tyrosyl and carboxyl groups are most important, and possible mechanisms were discussed (Kuwajima *et al.*, 1981).

6.5. CALORIMETRIC STUDIES OF THE TRANSITION

Comparison of thermodynamic parameters determined from equilibrium studies, and those directly obtained from calorimetric measurements is another method to check the reliability of two-state approximation (Lumry *et al.*, 1966; Brandts, 1969; Tanford, 1968, 1970; Jackson and Brandts, 1970; Tsong *et al.*, 1970; Privalov, 1963, 1974; Privalov *et al.*, 1971, 1973; Khechinashvili *et al.*, 1973; Privalov and Khechinashvili, 1974; Tiktopulo and Privalov, 1974, 1978; Pfeil and Privalov, 1976b,c). Identity of ΔH_{cal} determined from calorimetric measurements and $\Delta \overline{H}_{vH}$ from van t'Hoff plots

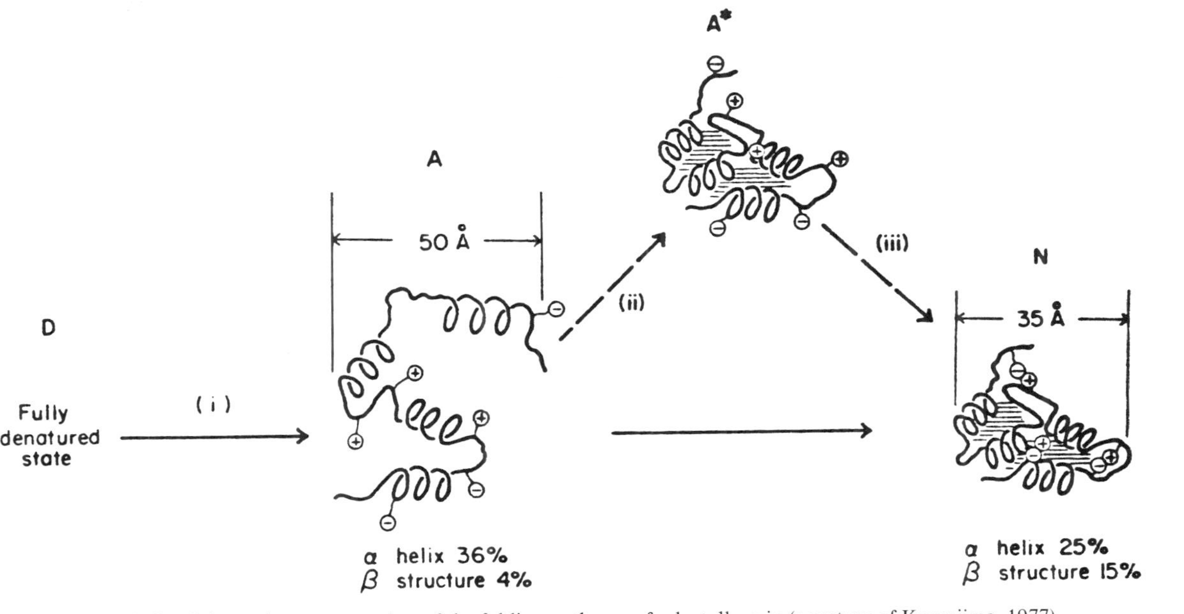

Fig. 6.18. Schematic representation of the folding pathway of α-lactalbumin (courtesy of Kuwajima, 1977).

obtained by equilibrium studies is a criterion of a two-state behavior. When significantly populated intermediary states occur between the initial and final states, $\Delta \overline{H}_{\mathrm{vH}} < \Delta H_{\mathrm{cal}}$.

6.5.1. Determination of Thermodynamic Parameters by Calorimetry

The partial molal heat capacity of a protein in native and unfolded states, related to the enthalpy of the transition by the relation,

$$\delta \Delta H / \delta T = \Delta C_{\mathrm{p}}$$

is directly determined from calorimetric measurements. This determination on dilute solutions of proteins requires a great sensitivity. The first reports of successful experiments appeared around 1970 and corresponded to the improvement of microcalorimeters. Tsong and co-workers (1970) and Jackson and Brandts (1970) used sensitive differential heat capacity calorimeters, so that the latter researchers were able to work with a moderately dilute solution of chymotrypsinogen (1.85%). The progress in scanning differential microcalorimetry and the development of precise instruments with a high base-line stability allowed researchers to work with lower concentrations. Privalov and Khechinashvili (1974) used 0.182% lysozyme solution. Such a microcalorimeteric recording is shown in Fig. 6.19; the base line with a calibration label obtained using the solvent is also indicated. The sensitivity of the calorimeter to heat capacity at heating rates from 2 to 0.5°/min was 10^{-5} cal/degree and the reproducibility of the base line at repeat fillings was not worse than 3.10^{-5} cal/degree. The deviation of the recording curve from the base line Δ_{T} gives the partial heat capacities of the protein and solvent, respectively $C_{\mathrm{p}}^{\mathrm{p}}$ and $C_{\mathrm{p}}^{\mathrm{s}}$. The mass of protein in the measuring cell is m_{p}, Δm_{s} the mass of the solvent displaced by the protein:

$$\Delta_{\mathrm{T}} = |C_{\mathrm{p}}^{\mathrm{p}}| m_{\mathrm{p}} - |C_{\mathrm{p}}^{\mathrm{s}}| m_{\mathrm{s}}$$
$$\Delta m_{\mathrm{s}} = m_{\mathrm{p}} |V|^{\mathrm{p}} / |V|^{\mathrm{s}}$$

where $|V|^{\mathrm{p}}$ and $|V|^{\mathrm{s}}$ are the corresponding partial volumes. The ratio $|V|^{\mathrm{p}}/|V|^{\mathrm{s}}$ can be considered independent of temperature. Therefore

$$C_{\mathrm{p}}^{\mathrm{p}} = |C_{\mathrm{p}}^{\mathrm{s}}|(|V|^{\mathrm{p}} / |V|^{\mathrm{s}}) - (\Delta_{\mathrm{T}}/m_{\mathrm{s}})$$
$$\Delta C_{\mathrm{p}} = C_{\mathrm{p}}^{\mathrm{d}} - C_{\mathrm{p}}^{\mathrm{n}}$$

is the difference of partial specific heat of denatured and native states. It is obtained by linear extrapolation of heat capacity before and after the process as indicated in Fig. 6.19. Difficulties in determining heat capacity of a denatured protein often arise from aggregational effects at high temperature.

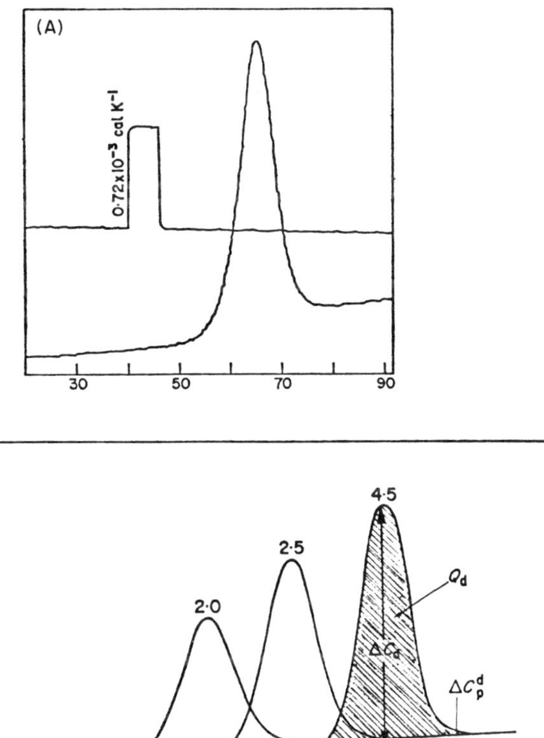

Fig. 6.19. Microcalorimetric recording indicating: (A) Heat absorption of a 0.182% lysozyme solution at pH 2.5; the base line is drawn with a calibration label obtained for the solvent. The deviation of the recorded curve from the base line gives the partial heat capacity of the protein. (B) Temperature dependence of partial heat capacity of lysozyme at pH 2.0, 2.5, and 4.5 (from Privalov and Khechinashvili, 1974).

The area of a heat absorption peak, as indicated in Fig. 6.19, gives the heat of denaturation, Q_D, per gram of protein by reference to area of the calibration label. The calorimetric enthalpy is

$$\Delta H_{cal} = Q_D M$$

M being the molecular weight. Privalov and Khechinashvili (1974) proposed a determination of the effective or van t'Hoff enthalpy from the same calorimetric curve. By assuming a two-state transition, at the middle of the transition where $T = T_d$:

$$\Delta H_{vH} = 2R^{1/2} T_d M^{1/2} \Delta C_d^{1/2}$$

The correspondance between calorimetric and van t'Hoff enthalpy is given by the expression:

$$\Delta H_{cal}/\Delta H_{vH} = M^{1/2} Q_D / 2R^{1/2} T_d \Delta C_d^{1/2}$$

When more than two states are significantly populated, the broadening of the transition leads to a reduced ΔH_{vH} and

$$\Delta H_{vH} < \Delta H_{cal}$$

except if intermediate transitions are characterized by the same equilibrium constant K and a temperature independent ΔH

$$N \underset{\Delta H}{\overset{K}{\rightleftarrows}} X_i \underset{\Delta H}{\overset{K}{\rightleftarrows}} \cdots \underset{\Delta H}{\overset{K}{\rightleftarrows}} Xn \underset{\Delta H}{\overset{K}{\rightleftarrows}} D$$

If any intermediate differing in enthalpy from N and D is significantly populated, then the inequality holds. A typical example is provided by the helix coil transition which gives very different values of ΔH_{cal} and ΔH_{vH}; the Zimm–Bragg nucleation parameter σ is related to the enthalpies ΔH_{cal} and ΔH_{vH} by the relation

$$\Delta H_{cal}/\Delta H_{vH} = 1/\sigma^{1/2}$$

generally $\sigma = 10^{-4}$.

Studying the transition of several globular proteins by differential scanning microcalorimetry, Privalov (1974) noted that: (1) the initial partial heats of

TABLE 6.7

Typical Values of Thermodynamic Parameters[a]

	ΔG (kcal/mole)	ΔH (kcal/mole)	ΔS (cal/deg/mole)	ΔC_p (cal/deg/mole)	(°
Ribonuclease thermal transition pH 2.5, 30°C	+0.9	+57	+185	+2000	−
Chymotrypsinogen thermal transition pH 3, 0.01 M Cl^-, 25°C	+7.3	+39	+105	+2600	[illegible]
Myoglobin thermal transition pH 9, 25°C	+13.6	+42	+95	+1400	<
β-Lactoglobulin 5 M urea, pH 3, 25°C	+0.6	−21	−72	+2150	[illegible]

[a] From Tanford (1970); data taken from Brandts and Hunt (1967), Brandts (1964a,b), Hermans [illegible] Acampora (1967), and Pace and Tanford (1968).

native globular proteins are very similar; (2) with increasing temperature, the specific heats of all proteins increase linearly and similarly until the temperature region at which unfolding starts; (3) the more thermostable a protein, the sharper the transition and the greater the enthalpy of denaturation.

Large positive values of ΔC_p^d are obtained from microcalorimetric measurements. According to Tanford (1970), this fact is at least qualitatively predictable since hydrophobic side chains of the protein become accessible to solvent and contact between hydrophobic groups and water are characterized by a very large anomalous capacity. Table 6.7 gives typical values of thermodynamic parameters.

6.5.2. Comparison of Calorimetric Data and of Data Deduced from Equilibrium Studies

Several well-studied proteins were used for calorimetric investigation of the native–denatured transitions.

Thermal transition of RNase was studied by calorimetry (Beck *et al.*, 1965; Tsong *et al.*, 1970). By consideration of ΔH_{cal} and ΔH_{vH} as indicated in Table 6.8 it appears that the transition follows a two-state process at pH 2, but widely deviates at higher pH. Tiktopulo and Privalov (1978) had reinvestigated the thermally induced denaturation of RNase and found a

TABLE 6.8

Thermal Transition of Ribonuclease A at Various pH Values[a]

pH	Solvent	T_m (°C)	ΔH_{cal} (T_m) (kcal/mole)	ΔH_{vH} (T_m) (kcal/mole)	ΔH_{cal} (25°)[b] (kcal/mole)	ΔC_p (kcal/deg/mole)
0.36	HCl	31.5	61.2	62	48.7	1.90
1.05	0.1 *N* NaCl, HCl	29.9	59.0	59	49.1	2.02
2.02	0.2 *N* Glycine buffer	31.2	65.6	67	53.3	1.98
2.80	0.2 *N* Glycine buffer	40.6	87.7	68	54.8	2.11
3.28	0.2 *N* Glycine buffer	45.8	105.0	67	63.4	2.00
4.04	0.2 *N* Acetate buffer	52.3	126.0	64	69.2	2.08
5.00	0.2 *N* Acetate buffer	57.8	151.0	60	77.2	2.25
6.23	0.2 *N* Acetate buffer	60.8	155.5	66	81.2	2.05
7.00	0.2 *N* NaCl	61.3	168.0	(88)	92.9	2.07
7.80	0.2 *N* NaCl	61.2	178.0	72	105.5	2.00
				Av 65		Av 2.05
				Av dev ± 3		Av dev ± 0.07

[a] From Tsong *et al.* (1970).
[b] $\Delta H_{cal}(25°) = \Delta H_{cal}(T_m) - \Delta C_p(T_m - 25)$.

TABLE 6.9

Thermodynamic Parameters for Chymotrypsinogen Denaturation[a]

		Direct calorimetric analyses		Calorimetric van t'Hoff analyses
pH	Temp (°C)	ΔH° (kcal/mole)	ΔC_p (cal/mole deg)	ΔH°(kcal/mole)
1.95	40.6	103	4200	94
2.03	42.0	102	3300	98
2.08	42.0	99	3900	98
2.59	48.0	126	3200	126
2.99	53.9	145	2800	133
3.02	54.2	135	2800	134

[a] From Jackson and Brandts (1970).

$\Delta H_{cal}/\Delta H_{vH} = 1.05 \pm 0.03$, which therefore lightly deviates from unity. This rather small deviation is interpreted to result from by the occurrence of some intermediates in very small quantity; these states appear to be unstable and the protein very rapidly passes through them.

Chymotrypsinogen was also extensively studied by microcalorimetry. In this case, satisfactory identity of ΔH_{cal} and ΔH_{vH} was found, allowing the description of this transition by a two-state process (Table 6.9) (Jackson and Brandts, 1970). Biltonen and co-workers (1971) reported calorimetric studies for chymotrypsinogen, chymotrypsin, and a chemical derivative, dimethionine sulfoxide chymotrypsin. The results supported the validity of the two-state approximation.

Privalov and Khechinashvili (1974) have investigated the thermal transition of five globular proteins: ribonuclease, lysozyme, chymotrypsin, cytochrome *c*, and myoglobin. Figure 6.20 represents the $\Delta H_{cal}/\Delta H_{vH}$ at various temperatures for all studied proteins. The average value is 1.05 ± 0.03. This deviation from a two-state process (not exceeding 5%) was attributed to the existence of highly unstable intermediates.

A $\Delta H_{cal}/\Delta H_{vH}$ ratio of 1.007 ± 0.011 indicates the validity of the two-state approximation for the thermal denaturation of lysozyme in alcohol–water mixtures (Velicelebi and Sturtevant, 1979).

For papain denaturation, an evident deviation from a two-state process was revealed by the comparison of ΔH_{cal} and ΔH_{vH} (Table 6.10) (Tiktopulo and Privalov, 1978).

These different methods of analyzing data, obtained from studies of the folding–unfolding process, have revealed in many cases the occurrence of intermediates. However, the population of intermediates is often very small

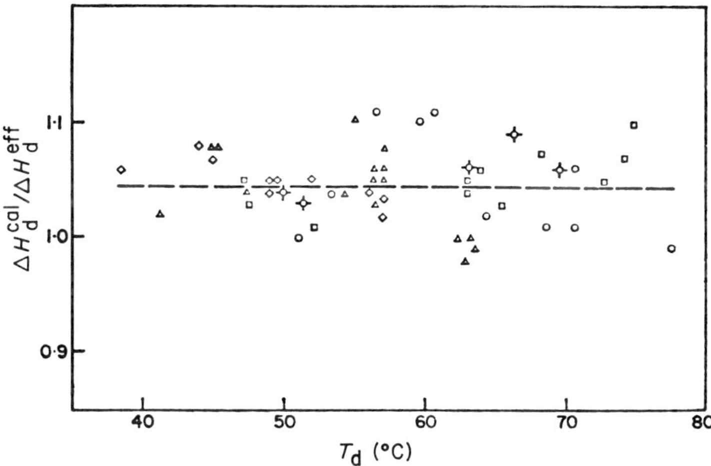

Fig. 6.20. Variations of the ratio $\Delta H_{cal}/\Delta H_{vH}$ at various temperatures for different proteins: ribonuclease (△), lysozyme (□), chymotrypsin (◇), cytochrome *c* (○), and myoglobin (-⌖-) (from Privalov and Khechinashvili, 1974).

TABLE 6.10

Thermodynamic Characteristics of Papain Denaturation[a]

pH	T_d/C	$\Delta_d H_{cal}$ (kcal/mole)	$\Delta_d H_{eff}$ (kcal/mole)	$(\Delta_d H_{cal})/(\Delta_d H_{eff})$
2.60	56.0	122.6	67.5	1.82
2.75	61.0	142.0	78.0	1.82
2.85	65.3	144.0	80.0	1.81
3.00	69.3	168.0	84.0	2.00
3.12	75.0	182.0	106.6	1.71
3.35	77.5	194.0	109.0	1.78
3.50	80.0	196.0	115.0	1.70
3.80	83.8	216.0	117.0	1.85

[a] From Tiktopulo and Privalov (1978).

when the species are unstable and rapidly transformed and the transition approaches very closely a two-state mechanism. The existence of intermediates was more evidently revealed by kinetic studies.

The data obtained from equilibrium studies emphasize the following points:

(*1*) *The protein unfolding–folding transition is generally a very cooperative process.*

(2) *In many cases but not all, protein unfolding–folding can be described by a two-state approximation.*

(3) *Deviations from a two-state behavior have been observed in several cases even in equilibrium studies:* (*a*) *either by the existence of an intermediate plateau or a shoulder in the transition curve, or* (*b*) *by noncoincidence of the transition curve obtained from different observables, or* (*c*) *by differences in* ΔH *obtained from van t' Hoff plots and directly by calorimetric measurements.*

(4) *Thermodynamic data indicate that native proteins have a relatively weak stability* (-5, -10 *kcal/mole*) *at room temperature resulting from a delicate balance between a great number of stabilizing interactions and conformational entropy which tends to destabilize the molecule.*

7

Kinetic Studies of Unfolding and Folding Processes

Even if kinetic studies gave only a phenomenological answer, they were very decisive in establishing the occurrence of more than one step in protein folding and unfolding. The early studies suggested that only a single step reaction could be observed, the amplitude of the signal being very close to the entire difference between native and unfolded states. The introduction of fast reaction techniques allowed the detection of additional faster steps for thermal-, pH-, and GuHCl-induced transitions. Biphasic kinetics were generally observed inside or below the transition zone. In some cases, biphasic kinetics could approach a two-state behavior within the transition zone when the amplitude of the fast step reaches a small value or when the time constants of the two phases become close to one another. The kinetic behavior of protein folding and unfolding is, however, not yet understood; the interpretation of the different phases remains a matter of controversy.

In the present discussion, the phenomenological aspects of the dynamics of protein folding are reported and only the kinetic studies are described. Nevertheless, the attempts to interpret the slow phase during the unfolding–folding process, evinced in many kinetic studies (Baldwin, 1975), are included. Characterization of intermediary species are detailed in Chapter 8.

7.1. KINETICS OF REVERSIBLE FIRST-ORDER TRANSITIONS

A rigorous treatment and analysis of the kinetics of protein folding has been made by Tanford's group with the hypothesis of unique N and D

states. Formalism and analysis were presented in different reports (Tanford, 1972; Tanford *et al.*, 1973; Ikai and Tanford, 1971, 1973). It is very important to detail this theoretical aspect since it is the basis of further analyses.

7.1.1. General Case: Multistep Transitions

The formalism derives from the general multistep transition occurring in a single pathway*:

$$A_1 \rightleftharpoons A_2 \rightleftharpoons A_3 \rightleftharpoons \cdots A_{n-1} \rightleftharpoons A_n$$

The reversible transition between native (N) and denatured (D) states may pass through various possible conformational states $A_1, A_2, \ldots A_n$; A_1 being the native (N) state and A_n the denatured (D) state, $A_2, A_3, \ldots A_{n-1}$ the intermediary species. Each transformation is a unimolecular process. Unstable transient species do not affect the kinetics, only those significantly accumulating do. The microscopic rate constants for transformation $A_i \rightarrow A_j$ are designated as k_{ij}.

The concentration of any species A_i as a function of time is given by the general equation:

$$C_i(t) = C_i(\text{eq}) + \sum_{s=1}^{n-1} a_{is} \exp(-\lambda_s t) \tag{7.1}$$

λ_s being a composite rate constant and a function of the microscopic rate constants k_{ij}. (They are called macroscopic rate constants by Ikai and Tanford, 1973); a_{is} are the corresponding amplitudes of the exponentials. When there are n distinct species, there are $(n-1)$ values of λ_s. The λ_s are functions of the microscopic rate constants. In the most general case, the $n-1$ values of λ_s are roots of the following equation:

$$\begin{vmatrix} \lambda - \sum k_{1j} & k_{21} & k_{31} \cdots & k_{n1} \\ k_{12} & \lambda - \sum k_{2j} & k_{32} \cdots & k_{n2} \\ k_{13} & k_{23} & \lambda - \sum k_{3j} \cdots & k_{n3} \\ \cdots & \cdots & \cdots & \cdots \\ k_{1n} & k_{2n} & k_{3n} & \lambda - \sum k_{nj} \end{vmatrix} = 0$$

The amplitude of the exponential a_{is} differs for each species; C_i (eq) is the concentration of species A_i at equilibrium. The a_{is} are solutions of the

* Solutions of the system with n first-order differential equations, which describe the concentrations of the different species as a function of time, have been provided by various investigators (see Benson, 1960; Frost and Pearson, 1953; Rodiguin and Rodiguina, 1964).

following equations:

$$a_{js}\lambda_s + \sum_{i=1}^{n} k_{ij}a_{is} - \left(\sum_{i=1}^{n} k_{ji}\right)a_{js} = 0 \tag{7.2}$$

$$\sum_{s=1}^{n-1} a_{is} = C_i(o) - C_i(eq) \tag{7.3}$$

For a unimolecular process:

$$\sum_{i=1}^{n} a_{is} = 0 \tag{7.4}$$

for each value of s. The a_{is} depend on the initial conditions and on the k_{ij}. However, a relation derived from the previous ones,

$$\sum_{i=1}^{n} (k_{ji} - k_{ij}a_{is}/a_{js}) = \lambda_s \tag{7.5}$$

indicates that the ratios a_{is}/a_{js} obtained by simultaneous solutions of all equations from Eq. (7.5) do not depend on the initial conditions.

For the reversible denaturation–renaturation process:

$$\frac{(a_{is})_D}{(a_{is})_R} = \frac{(a_{js})_D}{(a_{js})_R}$$

In the kinetic studies of conformational transitions in proteins, it is not generally possible to measure the concentration of the intermediary species. The kinetics are followed by variation of some physical property of the system which depends on the conformational state. If y is the observable, then the general expression can be written in the following form:

$$\frac{y(t) - y(eq)}{y(o) - y(eq)} = \sum_{s=1}^{n-1} P_s \exp(-\lambda_s t) \tag{7.6}$$

with the same definitions of the y as in Chapter 6; however, here y is expressed as a function of time. The coefficients of the exponential P_s are proportional to $\sum_i a_{is}, y_i$ depending on a_{is} and also on physical characteristics of species i.

$$\sum_{s=1}^{n-1} P_s = 1$$

since for $t = 0$

$$[y(t) - y(eq)]/[y(o) - y(eq)] = 1$$

The values of y_i are known only for the native and denatured states, but not for the intermediates. Therefore an auxiliary parameter α_i is defined such that

$$\alpha_i = \frac{(y_i - y_N)}{(y_D - y_N)} \tag{7.7}$$

$\alpha_i = 0$ for the native state; $\alpha_i = 1$ for the denatured state; for the intermediates, α_i varies between 0 and 1; P_s can be expressed in terms of α_i and a_{is}. The λ_s values are independent on the initial state of the transition and the same values apply in both directions, folding or unfolding. On the contrary the P_s values depend on the initial state.

There is a relation between the sum of the different intermediary species and the apparent equilibrium constant K_{app} defined as

$$K_{app} = \frac{y(eq) - y_N}{y_D - y(eq)} \tag{7.8}$$

This relation is

$$\sum_{i=1}^{n} \alpha_i C_i(eq) = \frac{K_{app}}{1 + K_{app}} C_0 \tag{7.9}$$

If P_{sF} represents the P_s for the kinetic studies in the forward direction starting from the native state and P_{sR} for the direction from unfolded to folded states, the P_{sF} and P_{sR} values generally differ.

The number of exponential terms, required to fit the experimental data in Eq. (7.6), indicates the minimum number of species that must be involved in the reaction pathway.

However, the reaction must be followed in both forward and reverse directions. Indeed, in some cases there is a good fit of the data with an exponential term in one direction, but not in the other. If a fast step is followed by a slower one in the forward direction, the two consecutive steps will be observed kinetically. However, in the reverse direction, if the slow step is followed by the fast one, it will be rate limiting for the whole reaction and the fast step will not be observed kinetically. When the number of exponential terms is not the same in both directions, it indicates that one of the terms, either P_{sF} or P_{sR} in Eq. (7.6), is equal to zero, since, according to the principle of microscopic reversibility, the same mechanism must operate in both directions.

It could also happen that two distinct λ_s have very close values and are not experimentally distinguishable.

7.1.2. Two-State Transition

When the transition is a two-state process

mechanism I $$N \underset{k_{21}}{\overset{k_{12}}{\rightleftharpoons}} D$$

kinetics follow a single exponential curve according to the relation:

$$[y(t) - y(\text{eq})]/[y(\text{o}) - y(\text{eq})] = \exp(-\lambda t) \tag{7.10}$$

It is the kinetic criterion for a two-state behavior. This variation is obtained in different conditions, as well for the reverse as for the forward reaction; no intermediary species accumulate to a significant extent during the process. There is only one value of λ, the exponential constant, equal to $k_{12} + k_{21}$. Each of these microscopic constants can depend on pH or denaturant concentration.

7.1.3. Three-State Transition

Three-state transitions involve three species, the native (N), the denatured (D) and an intermediary state (X). However, different three-state transition mechanisms were considered according to the location of species X in or out the pathway between N and D; another distinction between linear and cyclic mechanisms was made.

mechanism II $$N \underset{k_{21}}{\overset{k_{12}}{\rightleftharpoons}} X \underset{k_{32}}{\overset{k_{23}}{\rightleftharpoons}} D$$

mechanism III $$N \underset{k_{31}}{\overset{k_{13}}{\rightleftharpoons}} D \underset{k_{23}}{\overset{k_{32}}{\rightleftharpoons}} X$$

mechanism IV $$N \underset{k_{31}}{\overset{k_{13}}{\rightleftharpoons}} D, \quad N \underset{k_{21}}{\overset{k_{12}}{\rightleftharpoons}} X, \quad X \underset{k_{32}}{\overset{k_{23}}{\rightleftharpoons}} D$$

For all these mechanisms, there are two exponential terms; for mechanisms II and III, there are 4 microscopic rate constants, and 6 for mechanism IV. These three mechanisms are described by the following equation:

$$[y(t) - y(\text{eq})]/[y(\text{o}) - y(\text{eq})] = P_1 \exp(-\lambda_1 t) + P_2 \exp(-\lambda_2 t) \tag{7.11}$$

with

$$P_1 + P_2 = 1$$

The λ are the roots of the general equation:

$$\lambda^2 - \beta_1\lambda + \beta_2 = 0$$

β_1 and β_2 differing for each mechanism.

For mechanism II, where X species are located on the pathway between N and D, this equation is defined as

$$\lambda^2 - \lambda(k_{12} + k_{21} + k_{23} + k_{32}) + k_{12}k_{32} + k_{21}k_{32} + k_{12}k_{23} = 0$$

and the amplitude for reverse and forward reactions are not independent; they are given by the following equations:

$$\lambda_1 P_{1F} + \lambda_2 P_{2F} = [(1 + K_{app})/K_{app}]k_{12}\alpha$$
$$\lambda_1 P_{1R} + \lambda_2 P_{2R} = (1 + K_{app})k_{32}(1 - \alpha)$$

α being the fraction α_i of species X. Two important equations are derived for this mechanism:

$$P_{sF}/P_{sR} = -\frac{1}{K_{app}}\frac{k_{12}(k_{32} - \lambda_s)}{k_{32}(k_{12} - \lambda_s)} \tag{7.12}$$

which applies independently for $s = 1$ and $s = 2$, and

$$\lambda_1\lambda_2(K_{app}/(1 + K_{app})) = k_{12}(k_{23} + k_{32}\alpha) \tag{7.13}$$

If mechanism II and mechanism III are described by equivalent formalism, they differ from each other by the fact that, in mechanism III, X is not on the pathway between N and D. Therefore, the subscripts 2 and 3 are interchanged in the k_{ij} and a_{is}. Thus the equation which defines the λ values is

$$\lambda^2 - \lambda(k_{13} + k_{31} + k_{32} + k_{23}) + k_{13}k_{23} + k_{31}k_{23} + k_{13}k_{32} = 0$$

and the following relation between the P is obtained:

$$P_{sF}/P_{sR} = k_{13}/[K_{app}(\lambda_s - k_{13})] \tag{7.14}$$

also applying independently for $s = 1$ and $s = 2$. A relation analoguous to Eq. (7.13) was derived:

$$\lambda_1\lambda_2(K_{app}/(1 + K_{app})) = k_{13}(k_{23} + k_{32}\alpha) \tag{7.15}$$

Contrary to the previous ones, mechanism IV is not a linear mechanism but a cyclic one. Therefore, it exists a relation between the microscopic rate constants derived from the microscopic reversibility principle:

$$k_{12}k_{23}k_{31} = k_{13}k_{32}k_{21}$$

The λ values are given by the following expression:

$$\lambda^2 - \lambda(k_{12} + k_{13} + k_{21} + k_{23} + k_{32} + k_{31}) + k_{12}k_{23} + k_{12}k_{32} + k_{12}k_{31} + k_{13}k_{21} + k_{13}k_{23} + k_{13}k_{32} + k_{21}k_{32} + k_{21}k_{31} + k_{23}k_{31} = 0$$

TABLE 7.1

Inherent Ambiguities of Selected Mechanisms[a]

Mechanisms	Unknown parameters				Experimental parameters[b]				Number of undeterminable parameters
	k_{ij}	α_1	Relations	Total	K_{app}	λ_s	P_s	Total	
$A_1 \rightleftharpoons A_2$	2	0	0	2	1	1	0	2	0
$A_1 \rightleftharpoons A_2 \rightleftharpoons A_3$	4	1	0	5	1	2	2	5	0
$A_1 \rightleftharpoons A_2 \rightleftharpoons A_3$, $A_1 \rightleftharpoons A_3$ (triangle)	6	1	1	6	1	2	2	5	1
$A_1 \rightleftharpoons A_2 \rightleftharpoons A_3 \rightleftharpoons A_4$	6	2	0	8	1	3	4	8	0
$A_1 \rightleftharpoons A_2 \rightleftharpoons A_3 \rightleftharpoons A_4 \rightleftharpoons A_1$ (square)	8	2	1	9	1	3	4	8	1
$A_1 \rightleftharpoons A_2 \rightleftharpoons A_3 \rightleftharpoons A_4 \rightleftharpoons A_1$, $A_1 \rightleftharpoons A_3$	10	2	2	10	1	3	4	8	2
$A_1 \rightleftharpoons A_2 \rightleftharpoons A_3 \rightleftharpoons A_4 \rightleftharpoons A_1$, $A_1 \rightleftharpoons A_3$, $A_2 \rightleftharpoons A_4$	12	2	3	11	1	3	4	8	3
$A_1 \rightleftharpoons A_2 \rightleftharpoons A_3 \rightleftharpoons A_4 \rightleftharpoons A_5$	8	3	0	11	1	4	6	11	0

[a] From Ikai and Tanford (1973) (see text).

[b] This analysis applies only if data are available for both forward and reverse reactions, and if all theoretically possible λ_s values can be obtained from the experimental data. All strictly linear mechanisms allow one to obtain unambiguous solutions.

and the relations between amplitudes in the forward and reverse reactions are

$$P_{sF}/P_{sR} = (1/K_{app})(\gamma_2 - k_{12}\lambda_s)/(\gamma_2 - k_{32}\lambda_s) \tag{7.16}$$

and

$$\lambda_1\lambda_2(K_{app}/(1 + K_{app}) = \gamma_3 + \gamma_2\alpha \tag{7.17}$$

with

$$\gamma_1 = k_{21}k_{31} + k_{21}k_{32} + k_{23}k_{31}$$
$$\gamma_2 = k_{12}k_{31} + k_{12}k_{32} + k_{13}k_{32}$$
$$\gamma_3 = k_{12}k_{23} + k_{12}k_{21} + k_{13}k_{23}$$

All parameters are determinable from the experimental data in mechanisms II and III; in mechanism IV, there is one more unknown parameter than those determinable from experiments (Table 7.1).

7.1.4. Four-State Transition

Four-state transitions are described by three exponential terms, the values being roots of a general expression

$$\lambda^3 - \beta_1\lambda^2 + \beta_2\lambda - \beta_3 = 0$$

Different mechanisms can occur according to intermediates which are on the pathway between N and D, or which are dead-end species. Mechanism V describes the process when intermediates are on the pathway from N to D:

mechanism V $$N \underset{k_{21}}{\overset{k_{12}}{\rightleftarrows}} X_1 \underset{k_{32}}{\overset{k_{23}}{\rightleftarrows}} X_2 \underset{k_{43}}{\overset{k_{34}}{\rightleftarrows}} D$$

In mechanism VI, X_2 is not on the direct pathway, but X_1 occurs between N and D:

mechanism VI $$N \underset{k_{21}}{\overset{k_{12}}{\rightleftarrows}} X_1 \underset{k_{42}}{\overset{k_{24}}{\rightleftarrows}} D \underset{k_{34}}{\overset{k_{43}}{\rightleftarrows}} X_2$$

In mechanism VII, both X_1 and X_2 are dead-end species:

mechanism VII $$N \underset{k_{41}}{\overset{k_{14}}{\rightleftarrows}} D \underset{k_{34}}{\overset{k_{43}}{\rightleftarrows}} X_2 \underset{k_{23}}{\overset{k_{32}}{\rightleftarrows}} X_1$$

In this last case an interesting relation is obtained for the amplitudes of the forward and reverse reactions:

$$P_{sF}/P_{sR} = k_{14}/[K_{app}(\lambda_s - k_{14})] \tag{7.18}$$

Complete solutions of the equations for mechanisms V, VI, and VII have also been obtained (see Table 8.2) by Ikai and Tanford (1973). All unknown parameters can be determined if the three λ_s and corresponding amplitudes for forward and reverse reactions are known.

Mechanism VIII is a cyclic four-state mechanism. As was the case of mechanism IV, it contains one more unknown than the number of determinable experimental parameters.

mechanism VIII

$$\begin{array}{ccccc} & & X_1 & & \\ & {}^{k_{12}}\nearrow\!\!\swarrow_{k_{21}} & & {}^{k_{24}}\searrow\!\!\nwarrow_{k_{42}} & \\ N & & & & D \\ & {}_{k_{31}}\nwarrow\!\!\searrow^{k_{13}} & & {}^{k_{34}}\nearrow\!\!\swarrow_{k_{43}} & \\ & & X_2 & & \end{array}$$

We have described the main complex mechanisms analyzed by Ikai and Tanford (1973) which apply to the kinetics of protein folding and unfolding. For a mechanism involving four species, kinetics are described by three exponential values. More generally, for a mechanism involving n species, there are $n - 1$ values of λ_s, $n - 2$ independent values of P_{sF} and $n - 2$ independent values of P_{sR}, but since the P_s are related by the relation:

$$\sum_{s=1}^{n-1} P_s = 1$$

Taking into account the equilibrium constant, K_{app}, there are $3n - 4$ experimental parameters. There are also $2n - 2$ microscopic rate constants for linear mechanisms and $n - 2$ unknown values of α_i, therefore, $3n - 4$ unknown parameters. Table 7.1 summarizes the situation for several mechanisms analyzed by Ikai and Tanford (1973).

7.1.5. Independent Folded Regions

Another interesting and special mechanism, described by Ikai and Tanford (1973), refers to independent folded regions. The existence of domains in proteins is broadly analyzed in Part I. The experimental arguments for independent refolding (or unfolding) of distinct regions in protein are presented in Chapter 10. In the present chapter only the kinetic aspect of domains refolding independently will be discussed.

One can imagine a protein molecule composed of two folded continuous domains, which can unfold (and refold) by independent two-state processes. A fraction α of the total physical change is associated with one domain; the fraction $(1 - \alpha)$ with the other. Such a process has been considered by Ikai and Tanford to be a particular case of mechanism VIII, X_1 being an intermediary species, with one domain unfolded and the other not, and X_2 being an intermediate with the first domain folded and the second unfolded. Since domains fold and unfold independently, there are relations of equality for microscopic rate constants as follows:

$$k_{12} = k_{34}; k_{21} = k_{43}; k_{13} = k_{24}; k_{31} = k_{42}$$

In this case kinetics are described by only two exponential terms, and the λ are given by the expressions:

$$\lambda_1 = k_{12} + k_{21}$$
$$\lambda_2 = k_{13} + k_{31}$$

The amplitudes are expressed in terms of equilibrium constants:

$$P_{1F}/P_{1R} = K_A/K_{app} \text{ and } P_{2F}/P_{2R} = K_B/K_{app} \tag{7.19}$$

with $K_A = k_{12}/k_{21}$ and $K_B = k_{13}/k_{31}$, the equilibrium constants for the unfolding of each domain respectively.

7.2. ANALYSIS OF THE KINETICS

7.2.1. Direct Analysis: Diagnostic Rules

Figure 7.1 indicates typical theoretical kinetic plots corresponding to several of the cases described in the preceding section. From the examination and analysis of kinetic plots, it is often difficult to establish a possible mechanism. It is always necessary to use consistency checks. Diagnostic

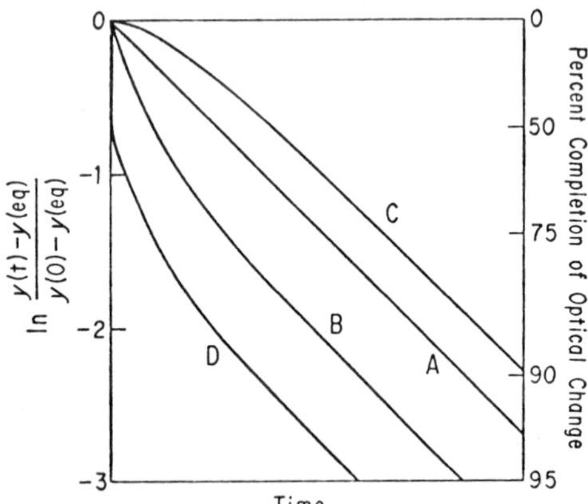

Fig. 7.1. Kinetic plots corresponding to several cases described in Section 7.1 (from Ikai and Tanford, 1973). Curve A: single exponential term. Curve B: two exponential terms, $\lambda_1 = 5\lambda_2$; $P_1 = 0.5$; $P_2 = 0.5$. Curve C: two exponential terms, $\lambda_1 = 3\lambda_2$; $P_1 = -0.5$, $P_2 = 1.5$. Curve D: three exponential terms, $\lambda_2 = 5\lambda_3$; $P_1 = 0.5$; $P_2 = 0.25$; $P_3 = 0.25$; λ_1 = too fast to measure.

tests have been theoretically deduced by Ikai and Tanford (1973) for the principal mechanisms: each mechanism is associated with obligatory relations among the set of the experimental variables.

When kinetics are described by a single exponential curve (i.e., a linear logarithmic plot) (Fig. 7.1, curve A), the mechanism follows a two-state process if this result is obtained under different conditions, in both directions with the same λ.

For a three-state process obeying a linear mechanism, a rule of diagnostic requires that

$$K_{\text{app}}(P_{1\text{F}}/P_{1\text{R}}) < 0$$

if X species are intermediates between N and D (mechanism II). Another important requirement for this mechanism is that the amplitude of the variation corresponds to the total change of the observable between initial and final state for the given conditions of the kinetic study. Since K_{app} is positive, either $P_{1\text{F}}$ or $P_{1\text{R}}$ must be negative. The λ_s are defined such as $\lambda_1 > \lambda_2 > \lambda_3$, etc. If the ratio $P_{1\text{F}}/P_{1\text{R}}$ is positive, then mechanism II can be ruled out.

If X are deadend species (mechanism III), the following relation must be obeyed:

$$K_{\text{app}}(P_{1\text{F}}/P_{1\text{R}}) > 0$$

In this case, $P_{1\text{F}}/P_{1\text{R}}$ is positive. Another diagnostic equation, entirely formulated in terms of experimental parameters can confirm this mechanism:

$$\frac{P_{1\text{F}}}{P_{1\text{R}}} = \frac{\lambda_2 + P_{1\text{F}}(\lambda_1 - \lambda_2)}{\lambda_1 + K_{\text{app}}(1 - P_{1\text{F}})(\lambda_1 - \lambda_2)}. \tag{7.20}$$

With four-state mechanisms, diagnostic equations have also been obtained. For example, when X_1 and X_2 are dead-end species (mechanism VII), a relation between experimental parameters must be expected such that

$$\frac{\lambda_1 P_{1\text{F}}}{P_{1\text{R}} + K_{\text{app}} P_{1\text{F}}} = \frac{\lambda_2 P_{2\text{F}}}{P_{2\text{R}} + K_{\text{app}} P_{2\text{F}}} = \frac{\lambda_3 P_{3\text{F}}}{P_{3\text{R}} + K_{\text{app}} P_{3\text{F}}} \tag{7.21}$$

For two regions of a protein folding and unfolding independently, an unequivocal criterion does exist: well below and well above the transition zone, the amplitudes P_1 and P_2 must be constant and their ratio must be equal to $\alpha_1/(1 - \alpha_1)$ or its reciprocal, if λ_1 is used to designate the faster macroscopic rate constant in both situations.

For cyclic mechanisms, the number of unknowns is larger than the number of experimentally determinable parameters. Characteristic diagnostic equations cannot be established and the ambiguities are inherent to the mechanisms themselves.

As usual, kinetic studies cannot rigorously demonstrate a mechanism of reaction, but the diagnostic tests, when they exist, allow one to rule out incorrect mechanisms.

For more complex mechanisms, or when the number of consecutive steps increases, kinetics may be very difficult to analyze. Computer simulation of different models of mechanisms is a helpful tool, at least in discarding the incorrect mechanisms. Mechanisms involving more than one native and one denatured states (i.e., occurring in a complex pathway) have been considered by Hagerman, 1977.

7.2.2. Location of Intermediates in the Pathway by Use of Chromophores

The use of chromophoric groups located in well defined regions of a protein can help one to identify steps involved in folding (or unfolding) of each of these regions. If one step of the process is associated with the refolding of such a region, this method could, in principle, allow one to determine if corresponding intermediates are on the pathway between N and D or dead-end species.

Such an approach was used by Baldwin and co-workers (Tsong and Baldwin, 1972a,b; Garel and Baldwin, 1975b). Kinetics of thermal unfolding of RNase have been followed by fast reaction techniques (temperature jump) in the millisecond time range and these studies have revealed the existence of intermediates; biphasic kinetics were observed. The unfolding was measured by the exposure of buried tyrosine groups at 286.5 nm (Tsong *et al.*, 1971). With the goal of determining whether the fast and slow processes are different kinetic phases of a general unfolding process or whether they correspond to separate unfolding processes, the reaction was followed with 41-dinitrophenyl ribonuclease, which was specifically labeled on Lys 41 by a dinitrophenyl (DNP) group (Tsong and Baldwin, 1972b). The kinetics of unfolding were followed by the absorbance of the dinitrophenyl group at 358 nm which is at least partially buried in the native protein. The results were compared to those obtained by following the exposure of tyrosine groups. The same slow and fast processes were observed at 358 nm and 286.5 nm and had the same time constants. The fast reaction was seen within the transition zone by both observables. Tsong and Baldwin (1972b) concluded that both DNP and tyrosine group(s) are on the same kinetic pathway of unfolding in a sequential mechanism. The details concerning kinetic studies are given in Section 7.3. The study is mentioned here only to illustrate the use of chromophore in search of folding mechanism, although in this case two different spectral probes have failed to show physical differences.

Nitrotyrosyl ribonuclease A (Garel and Baldwin, 1975b) was used to study the refolding of the protein by following the signal of nitrotyrosyl groups at 428 nm. It was confirmed that nitration of tyrosine did not modify the properties of the enzyme with regard to enzymatic activity, thermal transition, and kinetics of refolding. A physical difference was observed between the fast and slow refolding forms of the modified enzyme in the pK change of the nitrotyrosyl groups of the protein on refolding. It was estimated to be 0.046 ± 0.006 pH unit in the slow reaction, and very small (less than 0.02 pH units) in the fast reaction. These results have given additional support to the three-state mechanism previously proposed by Garel and Baldwin (1975a).

However, with nitrotyrosyl groups the pK change reflected an average effect for several tyrosine and the individual behavior of each nitrotyrosine was not known. This precluded a straightforward interpretation. For this reason, labeling of a single group was performed and fluorescein isothiocyanate was covalently attached to the amino group of ribonuclease A (Garel, 1976). The conformational changes were observed by means of the pK changes of this unique ionizable reporter group. A change in pK of the fluoresceine group by about 0.14 pH unit between native and heat-denatured RNase was found. Biphasic kinetics were observed in agreement with the previous results.

7.3. KINETIC STUDIES BY FAST REACTION TECHNIQUES

Pohl (1968a,b, 1969, 1972a,b) developed a simple method (which was in fact a slow temperature jump method) to follow reversible thermal unfolding of globular proteins. Reversible denaturation of trypsin, chymotrypsin, and chemically modified derivatives (i.e., anthranyloyl chymotrypsin and diisopropyl chymotrypsin) and ribonuclease A were studied. The thermal denaturation was followed for different pH values. From the data obtained for chymotrypsin as well as for trypsin, Pohl (1968a,b) concluded that there was a two-state process. This conclusion was supported by the following arguments:

(1) The time dependent change of the observable after a temperature jump followed a single exponential profile.

(2) The relaxation time did not depend on the parameters used to follow the kinetics, but only on the final temperature. It was found to be identical using optical rotation, absorption spectroscopy at different wavelengths, and fluorescence. Furthermore, when an extrinsic probe was used (fluorescence or absorption variations of the anthranyloyl group in anthranyloyl chymotrypsin) the same relation time was observed.

(3) The relaxation time was independent of protein concentration in a range between 0.2 and 40 μM for trypsin, thus precluding the occurrence of dimerization or aggregation reactions.

(4) The amplitude of the single exponential relaxation curve was very close to the whole difference between the equilibrium values in the initial and final states.

Therefore, kinetic criteria for a two-state process were considered to be fulfilled. However, if temperature dependence of the unfolding rate constant varied linearly indicating independence of the activation energy, anomalous temperature dependence was observed for the refolding. Figure 7.2 indicates the different temperature dependence for unfolding and folding rate constants.

On the basis of theoretical considerations, Cerf (1973) explained this anomalous behavior according to a three-state process, in which the enthalpy and entropy of each step were temperature independent:

$$N \rightleftarrows D_1 \rightleftarrows D_2$$

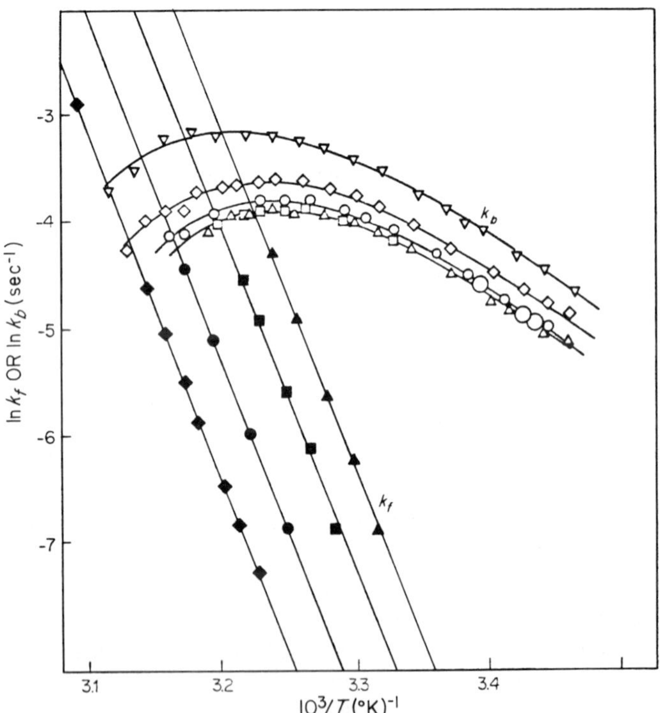

Fig. 7.2. Arrhenius plots for the rate constants for unfolding (solid symbols) and refolding (open symbols) of α-chymotrypsin at pH 1.59, 1.80, 2.01, 2.22 (according to Pohl, 1968b). Large open circles are obtained from measurements of the change in optical rotation, at pH 1.8.

with the simplest assumption that D_1 and D_2 are characterized by the same α value ($\alpha_1 = \alpha_2$), α being the observed parameter. Under these conditions, an apparently unique relaxation time is observed. Such a model also accounts for the Arrhenius' plots observed experimentally. The time scale used in these kinetic studies was of the order of seconds or tenth of seconds; it was too large to detect faster reactions.

More recently, the use of fast reaction techniques such as stopped flow and temperature jump with better resolution, has allowed the observation of more than one phase. Scott and Scheraga (1963) reported biphasic kinetics for thermal unfolding of RNase studied under certain conditions by stopped-flow techniques and followed by the change in tyrosyl absorbance at 287 nm (pH 1.3). Complex kinetics were observed in the lower one-third of the thermal transition curve, and a practically two-state behavior was described in the upper two-third of the transition curve.

Biphasic kinetics were found in thermal-induced transitions (Tsong *et al.*, 1971, 1972a,b,c; Tsong and Baldwin, 1972a,b; Tsong, 1973), as well as in pH-induced transitions (Epstein *et al.*, 1971a; Schechter *et al.*, 1970; Summers and McPhie, 1972; Shen and Hermans, 1972). Complex kinetics were observed in some cases for guanidine-induced transitions (Ikai *et al.*, 1973; Tanford *et al.*, 1973).

7.3.1. Thermal Transitions

Thermal transitions of ribonuclease, chymotrypsinogen (with disulfide bonds intact) were reinvestigated by temperature jump in a time range of observation which could extend down to few microseconds allowing detection of rapid phases, under the same conditions (pH, concentration) used by Pohl (1968a).

Reversible unfolding of chymotrypsinogen, at low pH was found to follow biphasic kinetics, a fast phase was detected in the millisecond time range in addition to the slow phase (in seconds) previously observed by Pohl. The amplitude of the fast process was small and increased at temperature near the upper end of the transition curve; the same behavior was observed for RNase. The kinetics of unfolding were followed by temperature jump and by pH jump; identical results were obtained at the same final pH and temperature (Tsong and Baldwin, 1972a,b).

Kinetics of RNase transitions were extensively and carefully studied. At pH 1.3 unfolding transition takes place between 20°C and 40°C whereas at pH 7.0 it takes place between 50°C and 70°C. A fast reaction was also observed for the thermal unfolding of this protein over the entire pH range from 1 to 7 (Tsong *et al.*, 1971; Tsong and Baldwin, 1972a,b). It occurred also in the millisecond time range, being faster by 3–4 orders of magnitude than the slow unfolding reaction. Biphasic kinetics were found both by measuring the exposure of tyrosine groups at 286.5 nm and the exposure of an extrinsic probe, a DNP group covalently attached to Lys 41 (see Section 7.2.2). The amplitude of the rapid phase was very small and depended on the progress of unfolding on the transition curve as for chymotrypsinogen. It was only detectable in the temperature range of unfolding. At pH 1.3 and 39°C, 85% of total absorbance change on unfolding occurred in the slow process, whereas at 49°C the slow contribution represents less than 25% of the total absorbance change (Summers and McPhie, 1972). The fast reactions are negligible in temperature jumps that end below or begin above the transition zone, indicating that they are on the direct pathway of unfolding.

Thus common patterns have been observed in the kinetics of thermal unfolding of small globular proteins (with disulfide bridges intact). Even when they approach a two-state behavior, they are accurately described by a biphasic process, a rapid phase in milliseconds with a small amplitude, and a major slow phase in seconds. Furthermore, for RNase, both kinetics of refolding and unfolding were studied under the same final conditions of pH and temperature (i.e., at final pH 3.9 and at various temperatures from 36°C to 65°C). Unfolding was followed by pH jump from pH 7 to pH 3.9 and refolding from pH 2.0 to pH 3.9, by the absorbance changes at two wavelengths (240 nm and 286.5 nm). First-order kinetics were observed for refolding in the upper two-third of the thermal transition zone, whereas unfolding obeyed biphasic kinetics under the same conditions (Tsong *et al.*, 1972a).

To account for all these results and also for previous conflicting observation (from equilibrium and microcalorimetric studies) (see Chapter 6), a model was proposed (Tsong *et al.*, 1972a). It was a simple sequential model of protein folding depending on a nucleation step very similar to the nucleation–propagation model used to describe helix–coil transitions (Elson, 1972a,b). This model was suggested by the thermodynamic behavior of folding and unfolding processes (i.e., the large activation energies of unfolding independent of temperature, and the activation energies of refolding temperature dependent and small at the midpoint as shown in Fig. 7.2). This behavior is characteristic of a nucleation–propagation mechanism:

$$A_0 \underset{\sigma_b}{\overset{\sigma_f}{\rightleftharpoons}} B_1 \underset{k_b}{\overset{k_f}{\rightleftharpoons}} B_2 \underset{k_b}{\overset{k_f}{\rightleftharpoons}} \cdots\cdots \underset{k_b}{\overset{k_f}{\rightleftharpoons}} B_{n-1} \underset{k_b}{\overset{k_f}{\rightleftharpoons}} B_n$$

with the nucleation constant

$$\sigma = \sigma_f/\sigma_b$$

and the propagation constant

$$s = k_f/k_b$$

Several properties characterize the model:

(1) Nucleation is the limiting rate step for refolding.

(2) There is a definite sequence of steps in the pathway of folding and unfolding. The same values of rate constants are assumed for all the other steps (different from nucleation step).

(3) Unfolding and folding are cooperative processes; the cooperativity depends on the value of the equilibrium constant for nucleation and on the number of steps between unfolded and folded steps.

(4) In principle, this model is described by a complex kinetic curve containing n exponential terms. In fact, the curve approximates a biphasic curve with a slow phase which has the main amplitude and follows an exponential time variation and a fast phase with a small amplitude.

(5) The slow reaction in folding is sensitive to intermediates and the same value of τ_1 (the slowest time constant) must be expected for any observable. Folding probably approximates a two-state process more than unfolding, since nucleation is the first step in folding.

(6) The model predicts the existence of a rapid transient phase in unfolding with a small amplitude if the process is very cooperative. The two relaxation constants must be separated by several orders of magnitude in the time scale.

The advantage of this model is the reconciliation of apparent contradiction in different data: on the one hand equilibrium studies which showed a two-state unfolding at low pH and significant deviations from a two state at pH 7, on the other hand kinetic studies of the unfolding process which approach a two-state at pH 7.0 better than at pH 1.3. Kinetic data for unfolding and refolding of ribonuclease A are consistent with this model. When temperatures are just above the transition zone, the fast phase of unfolding becomes the major kinetic phase. The slower rate in the unfolding reaction, which increases with temperature, approaches the average rate of the fast phase. However, some criticisms have been made about this model by Tsong and co-workers themselves. They concluded that the nucleation–propagation model is too simple to be physically credible despite the satisfying numerical fit of experimental data. Furthermore it fails to account for the existence of a mid-range kinetic phase which occurs at low temperature of folding and unfolding.

Studying the thermal unfolding of ferricytochrome *c*, Tsong (1973) detected three phases: a very fast one that had a relaxation time of 20 μsec, it was kinetically complex since it indicated different time constants, when the process was followed at different wavelengths that corresponded to heme absorbance. The other two phases in time scales of milliseconds and seconds respectively fall into time ranges of the unfolding of ribonuclease and chymotrypsinogen. Tsong (1973) explained the data in terms of a sequential model and ascribed the very fast reaction to the formation of local structures preceding nucleation. It must be noted that for unfolding of chymotrypsinogen A (Tsong and Baldwin, 1972a) an additional very fast relaxation in the microsecond time range was also reported at acidic pH, but with very small amplitude. For ribonuclease A at higher temperature a new, much faster kinetic phase was also observed and reported by Hagerman and Baldwin (1976).

This first model of nucleation–propagation was forsaken after new studies on refolding of heat-unfolded ribonuclease showed that the fast reaction yields native enzyme as well as the slow one does (Garel and Baldwin 1973, 1975a,b; Hagerman and Baldwin, 1976; Nall and Baldwin, 1977). A new model was established after consideration of the following facts:

(1) Both fast and slow refolding species yield native enzyme possessing both ability to bind 2′CMP and catalytic activity (hydrolysis of CpA). The binding constant of RNase formed by fast refolding process for 2′CMP is identical to that of native RNase. Further, the presence of 2′CMP does not affect the kinetics of refolding as followed by the exposure of tyrosine groups.

(2) Both species have nearly the same optical properties as measured by the exposure of tyrosine groups.

To account for these new facts, the following model involving two unfolded forms U_1 and U_2 was proposed:

$$U_1 \underset{\text{slow}}{\rightleftarrows} U_2 \underset{\text{fast}}{\rightleftarrows} N$$

in this model fast and slow reactions originate from fast and slow refolding species.

It was shown that fast refolding species exist in a pH-dependent equilibrium with slow refolding species. When equilibrium was changed by variation of initial conditions, it readjusted slowly. If only one species was present initially, its concentration could not be varied by changing initial pH. Figure 7.3 shows the pH dependence of α_2, fraction of fast refolding species; values of τ are pH independent.

In and above the transition zone, fast and slow species exist at equilibrium; they are forms of the heat-unfolded ribonuclease. The ratio of their concentrations (U_2^0/U_1^0) does not depend on temperature.

Possible other three-state mechanisms have been discussed. Two forms of native enzyme such as

$$U \underset{\text{fast}}{\rightleftarrows} N_1 \underset{\text{slow}}{\rightleftarrows} N_2$$

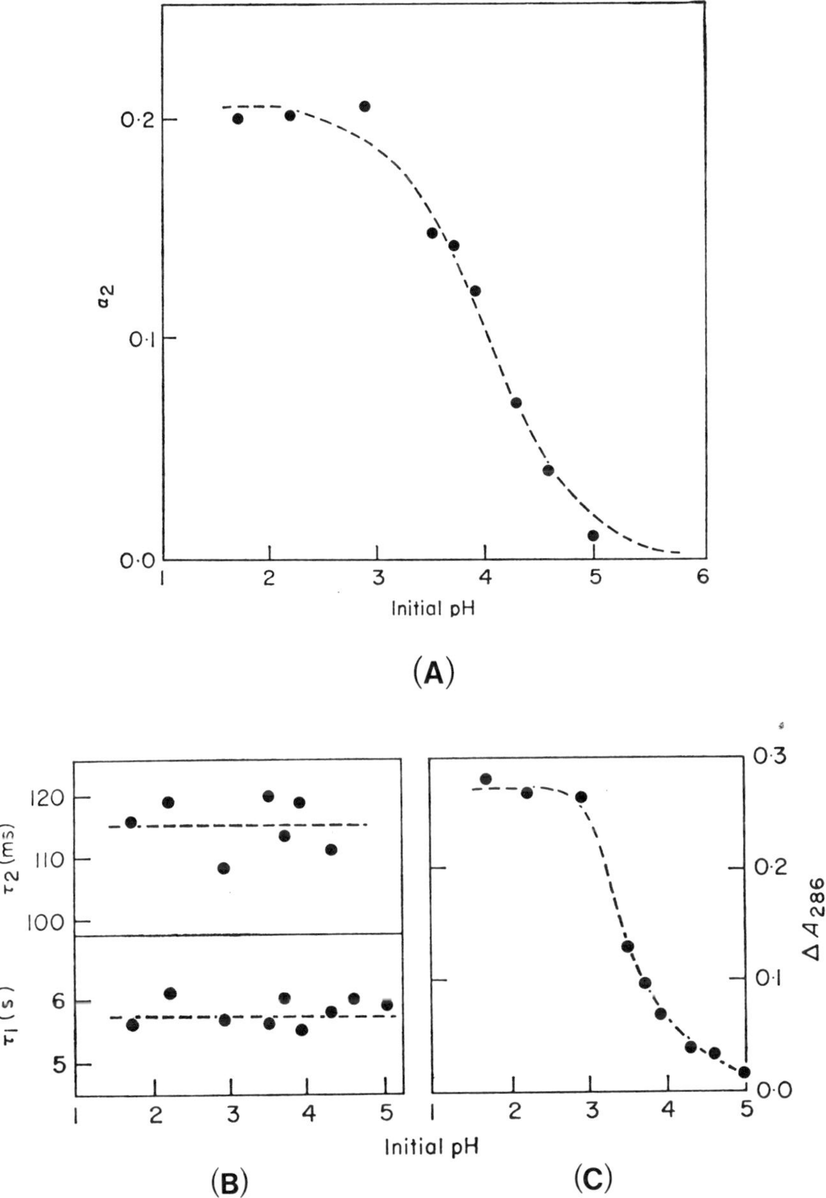

Fig. 7.3. Dependence on initial pH of amplitude and time constants for the refolding of RNase at pH 7 and 50°C (final conditions): (A) variations with pH of α_2, the amplitude of the fast phase; (B) independence on pH of the fast (τ_2) and the slow (τ_1) refolding time constant; and (C) dependence on pH of the total change in absorbance (0.15 mM RNase) (from Garel and Baldwin, 1975a).

were ruled out because such a mechanism implies only a single form of unfolded enzyme; experimental data fit with an equilibrium between two unfolded forms existing in initial conditions. Evidence was provided later that U_1 and U_2 are completely unfolded (Garel *et al.*, 1976).

A cyclic mechanism such as

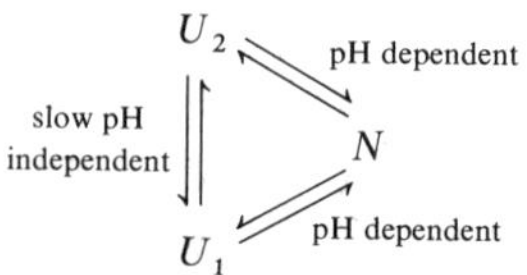

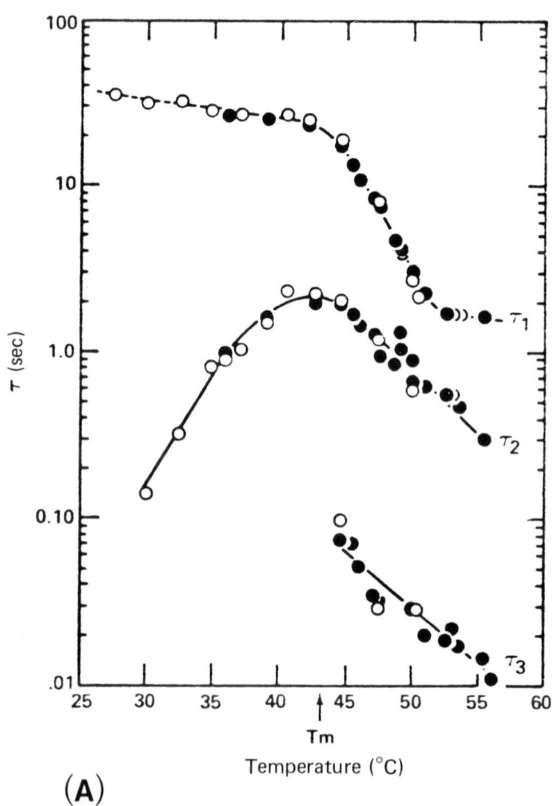

Fig. 7.4 (pp. 358–359). Temperature dependence of the kinetic parameters for unfolding (●) and refolding (○) of RNase (final pH 3.0) determined by pH jump experiments; (●) for unfolding pH 7.0 → 3.0; (○) for refolding pH 2.0 → 3.0; RNase 0.5–1 mg/ml. Kinetics were followed by optical changes at 287 nm: (A) kinetic constants τ_1, τ_2, τ_3; (B) corresponding amplitudes (from Hagerman and Baldwin, 1976).

was also discarded: indeed, the fraction of fast refolding species changes with pH only because the rate constants change with pH (Garel and Baldwin, 1975a).

A careful kinetic analysis was reported by Hagerman and Baldwin (1976). It allows one to rule out another mechanism:

$$U_1 \rightleftarrows N \rightleftarrows U_2$$

which predicts that α_2 at T_m should increase as τ_2 approaches τ_1. Furthermore, three relaxation times were obtained (Fig. 7.4); their variations with

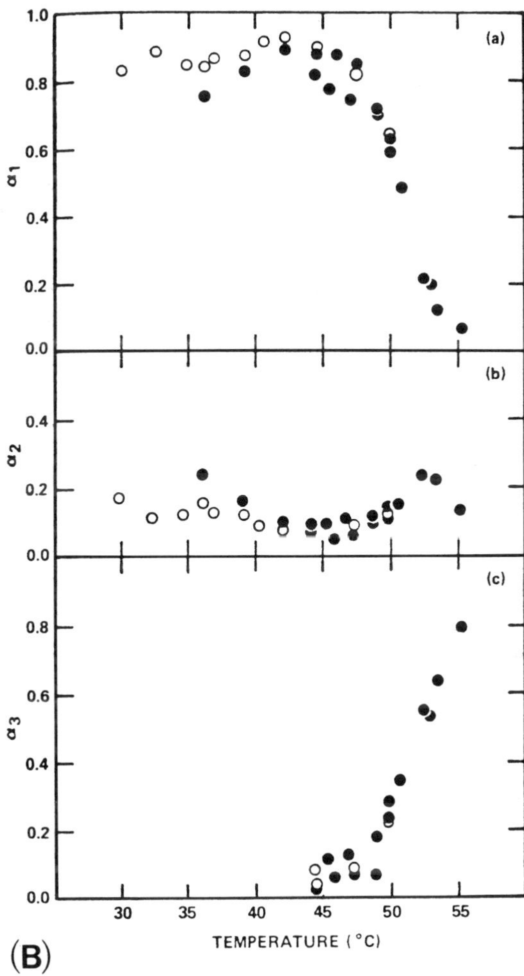

Fig. 7.4. *(continued)*

temperature are indicated in Fig. 7.4, as are the respective amplitude of these three phases. According to their analysis, Hagerman and Baldwin posed the following mechanism:

$$U_1 \rightleftarrows U_2 \rightleftarrows I \rightleftarrows N$$

All successive and possible interpretations in terms of characterization of the species are discussed in detail in Section 7.5. However, it should be noted here that when the existence of fast and slow refolding species was established, several properties of equilibrium surprised the researchers: the fact that this equilibrium between the two species is slow, pH dependent, and temperature independent.

7.3.2. pH-Induced Transitions

Staphylococcal nuclease undergoes a reversible transition between pH 3 and 4. Kinetics of refolding of the acidified enzyme was studied by a stopped-flow method by measuring the increase of fluorescence of Trp 140 (Schechter *et al.*, 1970; Epstein *et al.*, 1971a). Biphasic kinetics were observed as shown in Fig. 7.5. Two first-order processes correctly describe the kinetics with rate constants of 12.4 and 1.9 sec^{-1}, respectively for fast and slow processes. The two processes are not so broadly separated in the time scale as is the

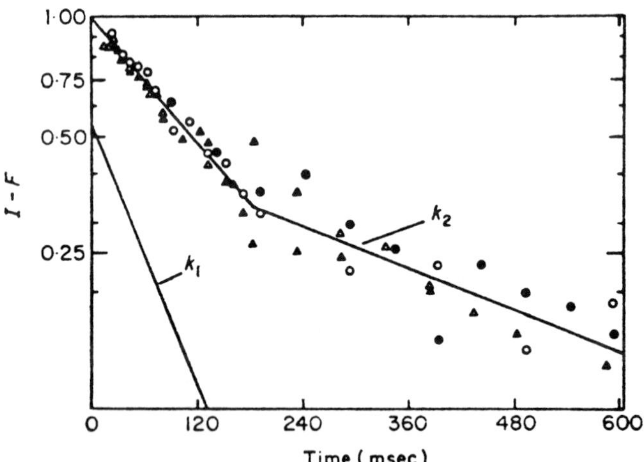

Fig. 7.5. Biphasic kinetics for refolding of staphylococcal nuclease denatured at low pH, observed by variation of fluorescence of Trp 140. Initial pH values (●) 3.27; (○) 3.45; (▲) 3.65; (△) 3.80. Final pH was 6.7 in all experiments (from Epstein *et al.*, 1971a).

case for thermal unfolding–refolding. The faster rate does not change significantly with temperature, but the slower rate decreases when temperature increases (Fig. 7.6). After consideration of their results, Epstein and co-workers (1971a) proposed a mechanism as follows:

$$D \xrightarrow{k_{DX}} X \xrightarrow{k_{XN}} N$$

with $k_{DX} = 12\ \text{sec}^{-1}$ and $k_{XN} = 1.9\ \text{sec}^{-1}$. A good fit was obtained by analyzing the data according to this model. However, the data on kinetics of unfolding and data on equilibrium are very incomplete compared to those obtained with RNase, so that an analysis of the amplitudes under different conditions is missing.

The acidic transition of horse heart metmyoglobin becomes also a biphasic process when pH decreases (Summers and McPhie, 1972). For sperm whale myoglobin, pH-induced unfolding and refolding follow complex kinetics (Shen and Hermans, 1972). Unfolding takes place in a three-step process at low pH and apparently in a two-step process in the transition range. The changes between the two types of kinetics are continuous. For refolding, a single rate was observed in the transition range, but at higher pH two steps were detected with apparent first-order rate constant values of $0.2\ \text{sec}^{-1}$ and $0.08\ \text{sec}^{-1}$ between pH 5 and 6.

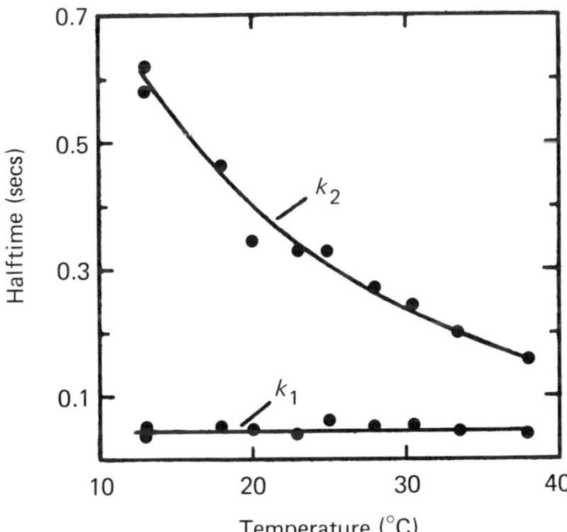

Fig. 7.6. Variations with temperature of the two rate constants observed for refolding of staphylococcal nuclease. Activation entropy for the slower process was 33 eu, and for the faster one 58 eu (from Epstein *et al.*, 1971a).

7.3.3. Guanidine Hydrochloride- and Urea-Induced Transitions

Detailed kinetic studies of protein unfolding induced by GuHCl and the refolding process from denaturant, have been reported by Tanford's group, mainly from a study of lysozyme with intact disulfide bridges (Tanford, 1972; Tanford *et al.*, 1966b, 1973), of cytochrome *c* (Ikai *et al.*, 1973), and of immunoglobulin light chain (Rowe and Tanford, 1973). Stopped-flow methods with detection of spectral change at 301 nm was used to follow the kinetics of denatured lysozyme refolding at pH 2.6 and 25°C at low and nondenaturing concentrations of GuHCl. The transition occurs between 3 *M* and 5 *M* GuHCl and is complete near 5 *M* GuHCl. Kinetics of refolding can be fitted with a two-exponential equation. The unfolding process was correctly described by first-order kinetics until a concentration of 6.5 *M* GuHCl was attained. Kinetic data were analyzed according to the equations given in Section 7.1. The simplest mechanism for lysozyme denaturation was pro-

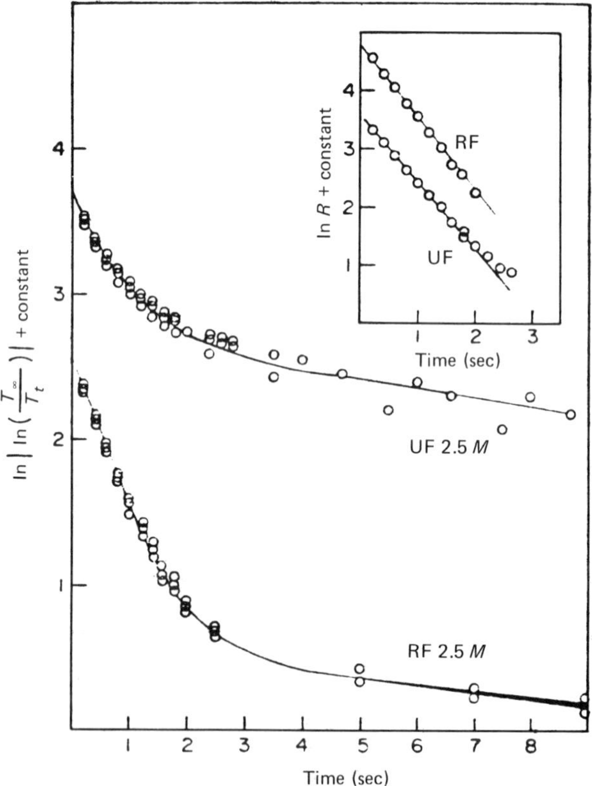

Fig. 7.7. Biphasic kinetics for unfolding and refolding of cytochrome *c* at 2.5 *M* GuHCl pH 6.5; protein concentration 0.07 mg/ml (from Ikai *et al.*, 1973). The insert indicates logarithmic plots of the residue *R* after substraction of the second exponential term.

posed, analogous to mechanism VIII (Section 7.1):

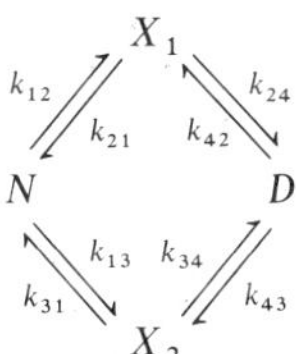

X_1 and X_2 being two species in the pathways between native and unfolded states. Under most conditions, the unfolding and refolding of lysozyme can be described by a two-state process. Intermediates accumulate to a sufficient extent to affect the kinetics only at low and very high concentrations of GuHCl. The existence of X_1 was deduced from data at very low concentrations of GuHCl and evidence for the existence of X_2 was obtained from data at very high concentrations of GuHCl. These intermediates appear to be on the direct pathways between N and D. The individual rate constants of the scheme were evaluated.

Kinetics of unfolding and refolding of cytochrome *c* have also been described in terms of only two exponential decays. Figure 7.7 shows the biphasic kinetics and Fig. 7.8 the variations of the λ with concentration of GuHCl. It should be noted that the transition occurs between 2 *M* and 3 *M* GuHCl. A careful analysis of all possible mechanisms has been performed by studying forward and reverse reactions starting from different GuHCl concentrations and by analyzing the corresponding amplitudes and λ_n values. A mechanism involving three species, with an intermediate between N and D (mechanism II) is excluded because P_{1F}/P_{1R} was positive (see Section 7.1). These conditions are required for mechanism III. However, this last mechanism is not satisfactory; in fact the values of λ_1 differ by about 50% for folding and unfolding at 2.28 *M* GuHCl. A more adequate description requires the introduction of a third exponential term. The mechanism involving independent folded regions (see Section 7.2.5) is also excluded, since it requires that P_{1F}/P_{2F} at high GuHCl concentrations and P_{1R}/P_{2R} at low GuHCl concentrations be equal to each other or one the reciprocal of the other; neither of these conditions are fulfilled. To account for the experimental data the following mechanism is suggested:

$$N \rightleftarrows X_1 \rightleftarrows D_2 \rightleftarrows X_2$$

(mechanism VI), with formation of an incorrectly folded form on a dead-end pathway.

A mechanism involving independent unfolding of domains was reported by Rowe and Tanford (1973) for the human immunoglobulin light chain giving further support for the domain hypothesis. Furthermore, it was demonstrated that the two domains have similar intrinsic stabilities (i.e., 5.5 kcal/mole at pH 7.0 and 25°C). Kinetics of the complete unfolding by 4 *M*

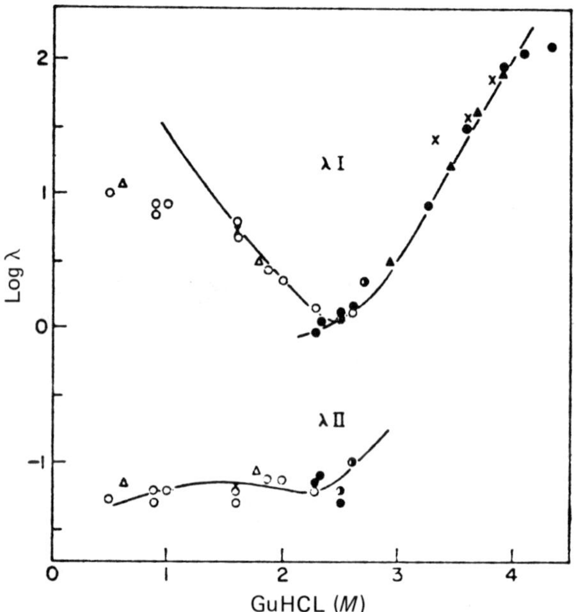

Fig. 7.8. Variations of λ_1 and λ_2 as a function of GuCHl concentration for cytochrome *c*: (●) denaturation starting from native state; (▲) denaturation starting from equilibrium mixture at 2.5 *M* GuHCl; (○) renaturation starting from denatured state; (△) renaturation starting from equilibrium mixture at 2.5 *M* GuHCl. All data were obtained at 400, 405, or 425 nm for protein concentration ranging from 0.04 to 0.13 mg/ml except *X* obtained at 290 nm with a concentration of 0.4 mg/ml of protein (from Ikai *et al.*, 1973).

GuHCl and refolding of the $C_\gamma 3$ domain of human IgC_1 (peptic fragment F'_c) were reported (Isenman *et al.*, 1979). For the refolding three first-order processes were observed with roughly equal amplitude; the unfolding process was biphasic. A scheme involving three nucleated reoxidized, but unfolded forms was proposed, only one of these forms yields the native domain, as follows:

$$\begin{array}{lll} & U_1 & \\ \nearrow & \updownarrow & \\ R \longrightarrow & U_2 & \\ \searrow & \updownarrow & \\ & U_3 \longrightarrow N & \end{array}$$

R being the reduced and unfolded form. The slow rate constants observed were attributed to complex processes involved in the formation of β sheet arrangement in immunoglobulin domains.

Complex kinetics were also found for the refolding of carbonic anhydrase, where intermediate species seem to accumulate during the reaction (Yazgan and Henkens, 1972).

The experiments with RNase were reinvestigated by denaturing the protein by GuHCl or urea at high concentrations of denaturant, 6 *M* and 8.5 *M* respectively, with the goal of suppressing any elements of residual structures in the unfolded state (Garel *et al.*, 1976). Indeed many proteins do not show any detectable elements of organized structure in GuHCl at high concentrations (see Chapter 5). Well separated fast and slow refolding phases were found, as previously obtained for the heat-unfolded RNase. At pH 2 the fraction of fast refolding species is the same for the denaturant unfolded RNase and for the heat-unfolded protein (20% and 80% of fast and slow refolding species respectively). The same proportion was found from pH 2 to pH 6. The results indicated that fast refolding species are present before refolding is initiated. Thus, these experiments have confirmed the previous results and supported the mechanism previously proposed. Recent data obtained for pH and GuHCl unfolding of RNase (Nall *et al.*, 1978; Schmid and Baldwin, 1978a,b; Hagerman *et al.*, 1979) indicate the presence of intermediates and suggest more complex pathway for the folding and unfolding of the protein:

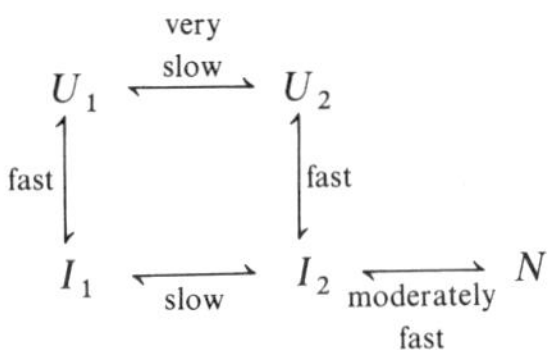

$$\begin{array}{ccccc} U_1 & \overset{\text{very slow}}{\rightleftarrows} & U_2 & & \\ \text{fast}\updownarrow & & \updownarrow\text{fast} & & \\ I_1 & \underset{\text{slow}}{\rightleftarrows} & I_2 & \underset{\text{moderately fast}}{\rightleftarrows} & N \end{array}$$

$U_1 \rightleftarrows U_2$ corresponds to the proline isomerization; I_1 is a species partially folded from U_1 in which the proline is in the incorrect configuration; I_2 is also a partially folded species, but it arises from U_2, the form in which proline is in the native conformation.

A similar pathway was proposed by Matthews and Crisanti (1981) to account for urea-induced unfolding–folding of the β subunit of tryptophan synthase; such a model is consistent with equilibrium and kinetic results.

This rapid survey of kinetics studies reveals that unfolding and refolding of globular proteins are more complex than a two-state process; in some cases they could approach a two-state process, especially when the unfolded structure is still constrained by the presence of disulfide bonds. The difference in behavior between lysozyme (which approaches a two-state process) and cytochrome *c* (which remains complex) was interpreted by Tanford to be the result of the presence of disulfide bonds in the former protein which severely limits the number of possibilities of refolding.

Kinetic studies have demonstrated the occurrence of several steps in the folding pathway of proteins, whatever might be the procedure of unfolding (i.e., temperature, pH, or GuHCl and urea). The presentation of the different mechanisms, progressively assigned to the different phases, and (or) progressively discarded, illustrates the difficulty involved in demonstrating a mechanism from kinetic studies. Even at this degree of refinement in the methods, only phenomenological informations could be obtained and other approaches are required for characterizing the intermediary species. Different situations were reported concerning the location of intermediates on the folding pathway. Although the characterization of intermediates is discussed in Chapter 8, here it is discussed briefly in regard to the attempts to make assignments of the different steps.

7.4. OCCURRENCE OF ABORTIVE INTERMEDIATES

In cytochrome *c* dead-end species were observed (mechanism VI). These abortive intermediates were interpreted as incorrectly folded states (Ikai and Tanford, 1971, 1973) of the polypeptide chain that cannot progress by direct folding to the native state, but must necessarily unfold to the state D (completely unfolded polypeptide chain) before refolding. Probably different incorrectly folded states could be formed from an unfolded polypeptide chain, and folding is the result of a kinetic competition between the correct kinetic pathway(s) which drive(s) the protein to the native form and the incorrect pathways which lead to incorrectly folded species. The occurrence of abortive intermediates was also reported for β-lactoglobulin (Tanford, 1972) where both X_1 and X_2 intermediates were found on a dead-end pathway. In mechanisms III, VI, and VII, it was assumed that the intermediates X_n are never populated at equilibrium outside the transition zone.

In the last mechanism proposed for RNase, both U_2 and U_1 are present at equilibrium above the transition zone (Baldwin, 1975).

Another type of abortive intermediate was emphasized by Hijazi and Laidler (1972). It was an abortive intermediate in unfolding, and not in refolding:

$$U \rightleftarrows N \rightleftarrows X$$

7.5. NATURE OF FAST AND SLOW REFOLDING REACTIONS

Various interpretations of fast and slow phases in refolding of RNase A were proposed from the beginning of the studies, but this problem still remains a matter of controversy.

With the model of nucleation–propagation proposed in the first studies, it was assumed the first slow limiting step in refolding was the formation of nucleated species. The formation of nucleus at the initial stage of a sequential folding as a limiting process was considered as a plausible hypothesis; the polypeptide chain first has to cross an entropy barrier to achieve a relatively stable conformation in which a series of bond formation can take place to drive the protein to the native structure (Tsong, 1973). However, it seems unlikely that one can kinetically detect the formation of nucleus even by fast reaction techniques, they might have a very short halftime, if one refers to the conclusion of theoretical approaches such as those of Karplus and Weaver (1976).

When the existence of an equilibrium between two unfolded species under initial conditions was shown (Garel and Baldwin, 1975a,b), U_2 was considered as a species nucleated for refolding, the alternative being that U_1 was incorrectly folded, preventing its rapid refolding to the native form. Another possible explanation for the existence of the two unfolded forms was proposed by Brandts and co-workers (1975). Brandts and co-workers (1975) assumed that the slow phase arises from the cis–trans isomerization of proline residues in the unfolded polypeptide chain. They emphasized the ability of proline to isomerize in proline-containing peptides and the existence of the well-known transition between polyproline I (all prolines in the cis form) to polyproline II (all prolines in the trans form). Simple proline dipeptides contain nearly 40% cis configuration at neutral pH. From different data on model compounds, Brandts and co-workers (1975) estimated the cis character of proline residues to be about 10–30% in a random polypeptide chain. In native proteins, each proline is found either in a cis or trans configuration. The results on refolding of RNase were analyzed according to the hypothesis that the slow refolding molecules contain one or two of the four proline of the molecule in an incorrect state of isomerization. Only the fraction of polypeptide chains which have the proline in the native configuration can fold rapidly. According to their model, Brandts and co-workers have calculated the fraction, α_2, of denatured protein which rapidly folds to native conformation, according to the equation:

$$\alpha_2 = [(1 + K_{tc})^{n-j}(1 + K_{tc})^{j}]^{-1} \tag{7.22}$$

K_{tc} being the cis–trans equilibrium constant, n the total number of proline and j the number of *cis*-proline in the native conformation. Brandts and co-workers concluded that cis–trans isomerization of proline could explain the slow process, since it occurs in the same time range; furthermore, it is consistent with the fact that slow and fast refolding species are spectroscopically equivalent and that the equilibrium constant for the $U_1 \rightleftarrows U_2$ reaction is independent of temperature.

To test their hypothesis, Brandts and co-workers (1977) studied the kinetics for unfolding and refolding of parvalbumin, a protein devoid of proline residues. Although these reactions occurred much faster, the results were not conclusive; complex biphasic kinetics were found, the slow phase being 100–500 times faster than the slow phase observed for proline-containing proteins. Comparative studies of three carp parvalbumins displaying great similarities in sequence and structure (two of these have no proline residues and one has proline residues) are more significant (Lin and Brandts, 1978). Kinetic measurements have shown that the protein, which has proline residues, exhibits an additional slow phase of refolding. The rate of this slow phase (10 sec at 25°C) and its activation energy were found consistent with proline isomerization (Fig. 7.9). Lin and Brandts (1978) concluded that

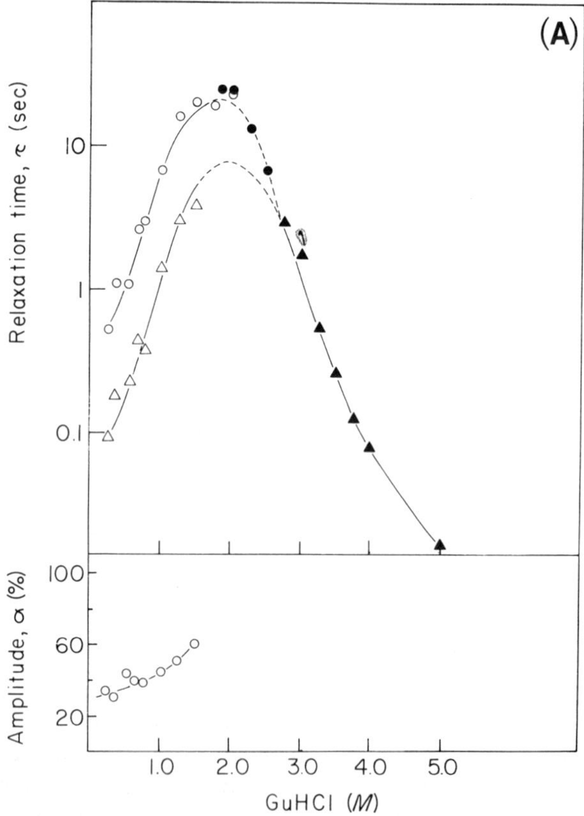

Fig. 7.9 (pp. 368–369). Comparative studies of unfolding–refolding of three carp parvalbumins (from Lin and Brandts, 1978) at pH 5.8, 25°C. Values of the kinetic parameters, relaxation time, and amplitude are given for each protein. Unfolding experiments are represented by filled symbols and refolding experiments by open symbols; (A), (B), and (C) correspond to different parvalbumins, only that studied in (B) contained proline residues (courtesy of Brandts).

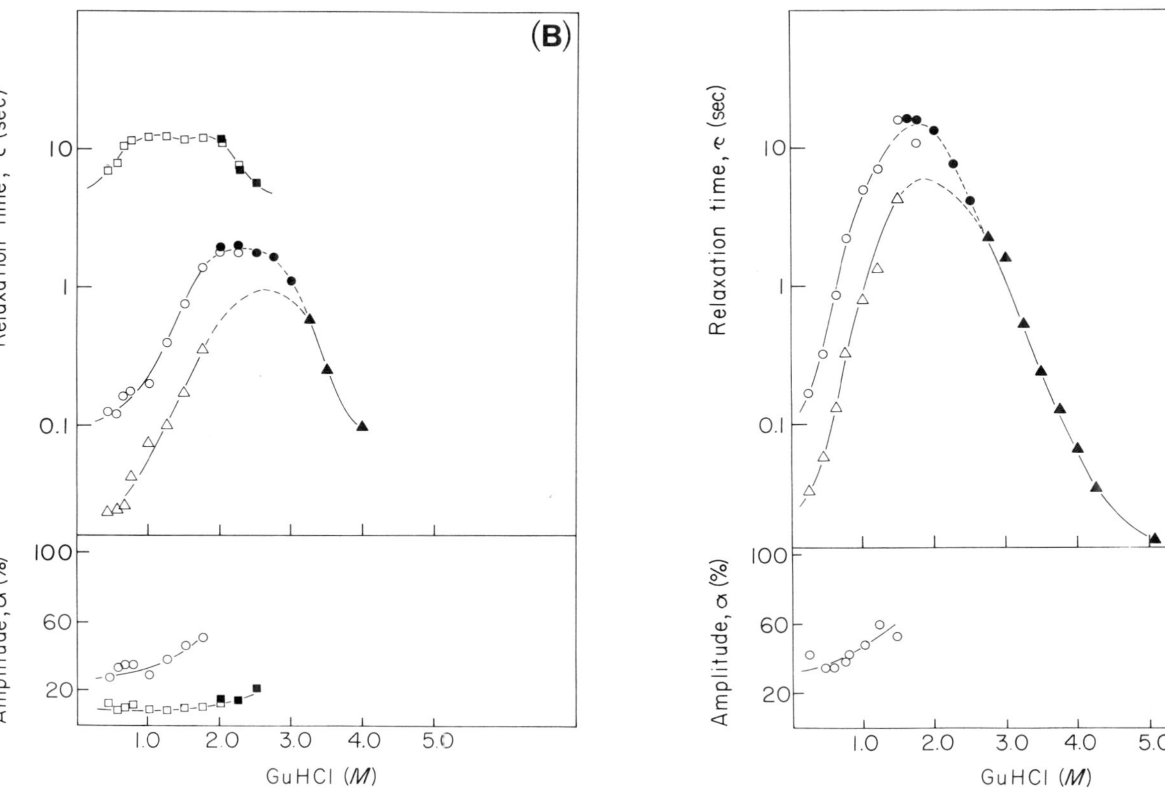

Fig. 7.9. *(continued)*

their results brought strong evidence for proline isomerism as responsible for the slow phase observed in refolding of proline containing proteins. Lin and Brandts (1979) reported the existence of a protease cleaving *X*-Pro bonds which seems exclusively specific to the trans form; Lin and Brandts will probably use this enzyme as a tool to check the proline isomerization hypothesis.

The proline isomerization hypothesis was also discussed and tested by Baldwin's group. In their first report on the subject, they mainly describe arguments which invalidated this hypothesis. The pH dependence of the slow phase, which follows the titration curve of a carboxyl group, was difficult to explain by proline isomerization unless the proline has a neighboring aspartyl or glutamyl. However, even with the pair (Glu 111–Pro 114), this hypothesis seemed unlikely, since the neighboring groups were described to have little influence on proline isomerism. A prediction of the proline isomerization model requires that the rate of unfolding of $U_1(U_1 \rightleftarrows N)$ is just the rate of proline isomerization. This prediction was invalid (Nall *et al.*, 1978). Several arguments, indicating the model is insufficient to explain the slow phase, were given: (1) the activation energy of $U_1 \rightleftarrows U_2$ reaction in refolding conditions is less than 5 kcal/mole; (2) rate and activation energy depend on GuHCl concentration below 2 *M*, whereas rate and activation energy for proline isomerization in models are not affected.

Another explanation was proposed in the same report to account for the slow phase: the existence of overlapping disulfide bonded loops in certain species in the unfolded state and the requirement of a loop threading reaction in the unfolding process.

On the basis of more recent data (Schmid and Baldwin, 1978a,b), the proline-isomerization model was reconsidered as probably accounting for the slow process. Schmid and Baldwin (1978a) used the properties of cis–trans isomerization of proline to be catalyzed by strong acids; protonation of the nitrogen at extreme acid pH allows the free rotation of the bond. To search for acid catalysis, the rate of formation of U_1 has been measured at different concentrations of $HClO_4$ (Fig. 7.10). The existence of two well separated reactions in the acid catalyzed range has suggested that at least two proline residues produce the slow refolding species. The slower reaction was ascribed to Pro 42, the proximity of Lys 41 making this bond more difficult to protonate. This new set of arguments is certainly more convincing. The interconversion between slow refolding and fast refolding forms is unaffected by GuHCl concentration because it exhibits the same characteristics as observed with peptide models (Garel and Siffert, 1979). It is a result of proline isomerization (Schmid and Baldwin, 1979a,b). In the S-protein, which contains all four proline residues of RNase, two classes of refolding molecules are also found (Labhardt and Baldwin, 1979a,b).

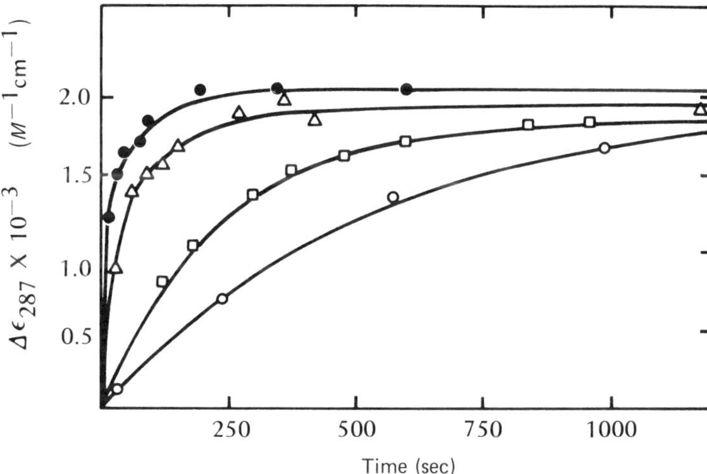

Fig. 7.10. Catalysis by $HClO_4$ of the formation of the slow refolding species U_1 (from Schmid and Baldwin, 1978a). At zero time a solution of native RNase A was diluted into concentrated $HClO_4$ at 0°C to give a final concentration of 3.3 *M* $HClO_3$ (○), 6.4 *M* $HClO_4$ (□), 9 *M* $HClO_4$ (△), and 10.6 *M* $HClO_4$ (●). Two kinetic phases are observed at 9 *M* and 10.6 *M* $HClO_4$. The amount of U_1 formed was measured by refolding assay (courtesy of R. L. Baldwin).

On the same basis, Stellwagen (1979) considered proline isomerization to be responsible for the slow phase reactivation of 11 denatured proteins. Alternative pathways were found for refolding of RNase (Hagerman *et al.*, 1979), new intermediates were detected. More complex schemes have been introduced.

Attempts to identify intermediates in the slow refolding reaction using hydrogen-isotope exchange (see Chapter 8) has shown hydrogen-bonded species in an early stage of refolding (Schmid and Baldwin, 1979a,b). This intermediate is stabilized by the S-peptide moiety in RNase *A*. Cook and co-workers (1979) working at low temperature, found a quasi-native intermediate in which proline was in the incorrect configuration. Recent data support this conclusion (Schmid, 1981; Schmid and Blashek, 1981). They concluded that not all proline need to be in the correct configuration before folding. Certain nonessential proline can be fitted either in the cis or in the trans configuration allowing the folding process before proline isomerization. Similar conclusion were reached by McPhie (1980) in the study on the refolding of urea denatured pepsinogen; if proline isomerization is involved, it must occur after the molecule has already undergone an extensive refolding.

However, the situation is not yet clear and the proline isomerization hypothesis is not generally accepted as responsible for the slow step. Contradictory results were reported. Other experimental data are not consistent with this model. It is particularly clear in the work on refolding of ferricytochrome *c* from different species (Babul *et al.*, 1978). Tuna heart and horse heart cytochrome *c* differ by one proline residue; horse protein has four *trans*-prolines while tuna has three. According to the model proposed by Brandts and co-workers (1975) (see Eq. 7.22), one can predict an α_2 of 0.51 and 0.41 respectively for tuna and horse proteins. Kinetic data obtained by stopped-flow measurements for the acid denaturation of both proteins indicated at least three phases, a very fast, a fast, and a slow one. If the very fast phase is omitted, an α_2 value greater than 0.80 is obtained for both proteins. If the fast phase is included, identical α_2 values equal to 0.97 are found for both proteins. Babul and co-workers concluded that the proline-isomerization model cannot account for the slow phase in the folding of cytochrome *c*. The slow kinetic phase was attributed to the coordination of His 18 and Met 80 to Fe(III) (Henkens and Turner, 1979). However, the possibility that cis-trans isomerization is responsible for the slow step was not discarded by Myer and co-workers (1981). Furthermore, they emphasized that, even in the work published by Brandts *et al.* (1975) on the folding of cytochrome *c*, the model can account quantitatively only for 7% of the total change in absorbance.

Not only for cytochrome *c*, but also for RNase, alternative explanations to proline isomerization have been proposed, for example, the cluster model of Kanehisa and Tsong (1978, 1979a,b). The cluster model is a two-state model and protein folding and unfolding are described by time evolution of the distribution function of microscopic states. In this model, the slow process is the conversion of the native and denatured population, the fast process corresponds to the redistribution within each population. Simulation of dynamics of protein folding according to the cluster model shows similar kinetics as those observed. However, this explanation was ruled out by Schmid and Baldwin (1979a,b) in the case of RNase since the rate of interconversion between slow and fast refolding species is insensitive to denaturant concentration.

It is impossible to reach any definitive conclusion concerning the cis–trans isomerization model for the moment. More data are needed on different proteins to clarify the situation. Furthermore, one cannot be convinced that it is the unique direction to search for an explanation for the slow kinetic process.

Plausibly, proline isomerization plays a role in protein folding (Freedman, 1979), and certain prolines probably have to be in the right configuration before the protein achieves its native structure. Some other proline are permissive. Although serious evidence has been provided that proline isomerization accounts for the slow kinetic phase in the case of RNase, nothing permits one to generalize this interpretation to other proteins. Even, when thermodynamic parameters are compatible with proline isomerization, it is difficult to extrapolate from model peptides to proteins. In this last case, many energetic compensations can give similar values. Other events may be responsible of a slow phase in protein folding (e.g., conformational readjustments, substructure assembly, and in hemoproteins, metal coordination).

In conclusion the following requires emphasis:

(*1*) *Kinetic studies have clearly demonstrated the existence of intermediates in the pathway of protein folding, even when equilibrium data seemed consistent with a two-state approximation.*

(*2*) *In several cases, kinetic schemes have been proposed; the phenomenological sequence of events has been determined.*

(*3*) *However, the nature of the different steps is generally not characterized by kinetic study and still remains a matter for discussion.*

8

Detection and Characterization of Intermediates

In the preceding chapters, the occurrence of intermediates in the refolding pathway was shown either by equilibrium studies with a deviation from a two-state behavior, or by kinetic studies with the existence of several steps in the refolding as well as in the unfolding process. In this chapter the methods used to detect and to characterize these intermediates are described. Of the different methods and approaches used, some allowed one to collect evidence of the presence of structured parts in the molecule; others, such as NMR and chemical methods, yielded point information from individual amino acid side chains.

8.1. DETECTION OF STRUCTURED SEGMENTS DURING UNFOLDING OR REFOLDING PROCESSES

8.1.1. Detection of Ordered Structures by CD or ORD Measurements

The use of CD and ORD and limitations of these methods to determine the content of ordered structures in proteins were discussed in Chapter 5. Nevertheless, these methods can be useful for following the variations of spectra during the unfolding or the folding process and can indicate what type of structure is affected even if the quantitative interpretation is often unsound.

The kinetics of refolding of reduced RNase as followed by variations of CD and rotatory dispersion led to the conclusion that secondary structure appears before the return of enzymatic activity (Anfinsen *et al.*, 1961; Schaffer *et al.*, 1975; Takahashi *et al.*, 1977). Following refolding of hen egg white lysozyme by the reappearance of the trough at 233 nm of the Cotton effect as measured by ORD and the negative bands centered at 213 nm and 222 nm as measured by CD, Yutani and co-workers (1968) concluded that about 70% of the original helical content of the native lysozyme is reformed immediately after dilution of the denaturant. Although the method is not reliable for determination of the exact percentage of helical content restored in the early stages, nevertheless, it clearly indicates that a great portion of the secondary structure is recovered before the complete refolding of the molecule which occurs gradually until the reappearance of the enzymatic activity. Wallevick (1973) described differential stabilities of segments of helical conformation as followed by optical rotation.

Similar deductions were made by Nitta and Sugai (1972), Kuwajima and co-workers (1976), Kita and co-workers (1976), Nitta and co-workers (1977a,b), and Kuwajima (1977) for the refolding of lactalbumin. This protein was shown to undergo a three-state denaturation, the intermediate being a helical state (see Fig. 6.18). This species has a content of 36% α helix (more than the native state) and 4% β structure. This was deduced from ultraviolet circular dichroism. The same restriction as for the previous example must be made on the quantitative deduction of the contents of each type of secondary structure. Qualitatively, it is interesting to note for this protein that there is a rapid formation of segments of secondary structures, mainly helices, in the early stage of protein folding. This behavior has been confirmed by other methods.

8.1.2. Detection of Ordered Structures by Raman Spectroscopy

The recent increase of interest in the use of Raman spectroscopy to investigate the structural properties of proteins arises largely from the technological improvement of the method and the availability of laser-equipped apparatus. Several reviews and reports appeared on this topic (see, e.g., Koenig, 1972; Koenig and Frushour, 1972; Tobin, 1972; Yu and Liu, 1972; Yu *et al.*, 1973; Chen *et al.*, 1973, 1974; Peticolas, 1975; Chen and Lord, 1976; Van Wart and Scheraga, 1978; Lippert *et al.*, 1975; Bandekar and Krimm, 1979; Porubcan *et al.*, 1978; Williams *et al.*, 1980; Williams and Dunker, 1981). The Raman spectrum of some proteins, but mostly of polypeptides, has been analyzed in several laboratories. In this chapter, the method is neither described nor are all the studies on model polypeptides reported. Only the data which are useful for the interpretation of the Raman spectra of proteins in determining the ordered structures are introduced.

The vibrational modes of the backbone are very sensitive to the conformation. The vibrational frequencies corresponds to the internal amide group and the stretching and bending modes of the N—C—C′ unit. These frequencies are very sensitive to the conformational state of the protein. (Table 8.1 gives the vibrational modes of the planar *trans*-amide group, —CONH.) This does not apply rigorously to the proline amide group. The frequencies used in the analysis of ordered structures are the amide I (1645–1675 cm^{-1}) and the amide III frequencies (1235–1295 cm^{-1}). Williams and Dunker (1981) determined secondary structure in proteins from the amide I band of the laser Raman spectrum with good accuracy. For 17 proteins samples, the difference between X-ray and Raman estimates of structure was less than 6%. Williams and co-workers (1980) used the amide III′ Raman spectrum intensity distribution of proteins in D_2O to determine secondary structure; the method was applied to four proteins, lysozyme, RNase, concanavalin A, and fd phage.

TABLE 8.1

Standard Vibrational Mode Frequencies[a] of the Planar *trans*-Amide Group (—CONH)

3300	NH stretching	Amide A and B
3280	NH stretching	
3100	NH stretching	
1650[b]	CO stretching	Amide I
1550	CN stretching + NH bending	Amide II
1250–1300[b]	CN stretching + NH in-plane bending	Amide III
750	NH + skeletal mode	Amide V
650	NH + C=O out-of-plane bending	Amide IV
500–600[b]	NH out-of-plane bending + C=O out-of-plane bending	Amide VI
200–300	C—N torsion	Amide VII

[a] From Koenig (1972).
[b] Strong band in the Raman spectrum.

The backbone vibrational frequencies are determined by: (1) the value of the dihedral angles ϕ and ψ; (2) the amide hydrogen bonding which differs according to the local structure (α helical, β parallel, β antiparallel); and (3) the interactions between amide units (e.g., dipole coupling) which also depend on the local structures.

A careful analysis of skeletal vibrational frequencies, correlated with specific topological states, was presented by Van Wart and Scheraga (1978). They considered four hydrogen bonded states in α helices: (1) $\alpha(-,-)$, (2) $\alpha(-,+)$, (3) $\alpha(+,-)$, and (4) $\alpha(+,+)$. The first symbol between parentheses indicates whether the CO group is hydrogen bonded to the NH of another amide group (+), or solvated (−), and the second symbol indicates the corresponding situation for the NH group. The same symbols are used for parallel (ε_p) and antiparallel (ε_{AP}). Table 8.2 gives the frequencies associated with several of the possible states. The vibrational frequencies were obtained from model compounds, homo- or copolypeptides. As for other methods, the application to protein conformation of the data obtained using polypeptides is not entirely satisfactory. Furthermore, experimental data are not available for all these possible topological states. The frequencies of amide I and amide III bands can be used to detect the presence of $\alpha_{RH}(+,+)$ and $\varepsilon_{AP}(+,+)$ in proteins. As emphasized by Van Wart and Scheraga (1978), the position of the frequency maxima of the bands in the unordered state is not useful for detecting the conformational states of proteins. In fact, (1) it represents different conformational states and probably different values for the vibrational frequencies; and (2) the resultant bands are broad and maxima are not well defined.

TABLE 8.2

Raman Vibrational Mode Frequencies (cm^{-1}) Characteristic of Different Ordered Structures[a]

Topographical State	Amide I		Amide III		
	In H_2O	In D_2O	In H_2O	In D_2O	$\nu(C'NC^{\alpha})$
$\alpha_{RH}(-,-)$	—		—		—
$\alpha_{RH}(-,+)$	—		—		—
$\alpha_{RH}(+,-)$	—		—		—
$\alpha_{RH}(+,+)$	1650–1660 (s)	1200	1260–1290 (w)	950–1000	890–940 (s)
$\alpha_{LH}(-,-)$	—		—		—
$\alpha_{LH}(-,+)$	—		—		—
$\alpha_{LH}(+,-)$	—		—		—
$\alpha_{LH}(+,+)$	~1663 (s)		—		~900 (s)
$\varepsilon(-,-)$	—		—		—
$\varepsilon_{AP}(-,+)$	—		—		—
$\varepsilon_{AP}(+,+)$	1665–1675 (s)		1235–1240 (s) ~1270 (w)		890–940 (w)
$\varepsilon_P(-,+)$	—		—		—
$\varepsilon_P(+,-)$	—		—		—
$\varepsilon_P(+,+)$	—		—		—

[a] Adapted from Van Wart and Scheraga (1978). Abbreviations: (s), strong; (w), weak.

Amide I and III bands of the Raman spectra can depend on contributions from side chain modes, such as amide modes from glutamine and asparagine, tyrosine ring modes, and proline modes.

Recently, a vibrational analysis was reported that assigns the characteristic amide bands of β turns (Bandekar and Krimm, 1979). An interesting conclusion is that amide I bands observed in proteins near 1690 cm^{-1} can be correlated with β turns as well as β structures. The amide II frequencies correlated with β turn type I are expected near 1550–1555 cm^{-1} and 1567 cm^{-1}, those for type II near 1545,1555, and 1560 cm^{-1}. The turns are characterized by amide II modes with frequencies higher than those of β structures and α helices.

Side chain vibrational modes are sensitive to the environment and therefore to the conformational variations. Attributions of several bands have been made; the S—S and C—S stretching of cystine in the 500–750 cm^{-1} range are the most extensively studied. The C—S stretching of Met, the aromatic ring modes of Phe, Tyr, and Trp, the COO^- symmetrical stretching, and the CH_2 bending modes also have been assigned. These vibrational modes can be used to detect variations of conformation in proteins.

Raman spectroscopy is not yet commonly used to follow conformational variations of proteins. Only a few reports are available. Brunner and Sussner (1972) and Brunner and co-workers (1974) studied the thermal denaturation of bovine pancreatic trypsin inhibitor (BPTI) with Cys 14–Cys 38 bond reduced and with this reduced molecule then carboxamidomethylated. The native molecule was first analyzed. The amide I band at 1664 cm^{-1} which is characteristic of β structure indicates that this structure is dominant in the molecule. The amide III band displays three well resolved peaks at 1243, 1265, and 1288 cm^{-1} which are characteristic of antiparallel β sheets, random coil, and α helical structures. From these data Brunner and Sussner (1972) and Brunner and co-workers (1974) evaluated the content of secondary structures to be ca. 45% β structure, 30% random coil, and 25% α helix; this is in satisfactory agreement with the X-ray data.

With increasing temperature no differences can be detected below 70°C. Above this temperature, the intensity of the band corresponding to the β structure decreases, whereas the α helix peak remains unchanged. Since it is known that BPTI is very stable, thermal denaturation experiments were performed on the protein after reduction of the Cys 14–Cys 38 bridge. On increasing the temperature, the peak of the β structure broadens and shifts to the random coil one, whereas the component representating the helical structure is barely detectable at 67°C. The thermal denaturation of carboxamidomethylated inhibitor is similar. In this study, the determination of secondary structures by Raman spectroscopy is in good agreement with the X-ray crystallography data. Contrary to Van Wart and Scheraga (1978),

Brunner and Sussner (1972) and Brunner and co-workers (1974) used attributions of random coil structures. Their results from thermal denaturation of BPTI indicate that the β structure disappears before α helical ones.

Laser Raman studies of lysozyme unfolding were reported by Lord and Mendelsohn (1972) and Porubcan and co-workers (1978). Denaturation was induced by various denaturing reagents: LiBr, GuHCl, sodium dodecylsulfate, and urea. The data recorded from Raman spectra were correlated to those obtained by ORD, enzyme activity and viscosity measurements. For this protein, Porubcan and co-workers (1978) were able to resolve three features in the amide region and they assigned the bands to the different ordered and coil structures. They used the relative integrated intensity at 1260 cm^{-1} to that at 1238 cm^{-1} as a parameter for evaluating denaturation. The results obtained by this method do not correlate with the results given by the b_0 value in ORD measurements. These discrepancies may arise from inaccuracy in evaluating the amount of ordered structures by ORD. However, Prorubcan and co-workers (1978) emphasized that relative intensities of amide III bands cannot be directly correlated with the relative content in α helix versus random coil and β structure, since transitions from coil or β sheet to helical structures are accompanied by a significant hypochromicity (Painter and Koenig, 1976). They may indicate whether or not there is more α helix than other structures in a protein.

Thermal unfolding of RNase was also studied by Raman spectrometry (Chen *et al.*, 1973, 1974; Chen and Lord, 1976) between 32° and 72°C. The variation of amide I and amide III band frequencies and intensities and those of the tyrosyl side chain (the doublet at 830 cm^{-1} and 854 cm^{-1}), and the S—S and C—S vibrations were followed as a function of temperature. Tentative assignments of the different frequencies and intensities for native and denatured RNase were given at pH 5.0 and in 0.1 *M* NaCl. They are presented in Table 8.3. The data indicate that thermal denaturation of RNase proceeds gradually by a step-wise unfolding process. Helical and pleated sheet conformations remain at a temperature of 70°C.

The persistence of ordered structures until the last events in denaturation and probably the persistence of α helical conformation later than β structures are suggested by these different studies. Perhaps during the reverse process, the refolding, α helices are formed before β sheets and all segments of ordered structures are formed in early stages of protein folding.

The scarcity of Raman studies detecting intermediary events during the unfolding–refolding of proteins prevents, at least for the moment, any generalization regarding the accuracy of the method. By Raman spectroscopy, one can expect to detect segments of ordered structures during the refolding of a protein. However, the correlation between the frequencies and the conformational states of the backbone was established from studies

TABLE 8.3

Raman Frequencies (cm^{-1}) and Intensities[a] of Native and Thermally Unfolded RNase A

Frequency (cm^{-1})			Frequency (cm^{-1})		
32°C	70°C	Tentative assignment	32°C	70°C	Tentative assignment
412 (0)	412 (0)		1003 (10)	1003 (10)	Phe
440 (0)	440 (0)		1015 (0 s)		
496 (0 s)			1030 (3)	1030 (4)	Phe
516 (3)	507 (3)	ν(S—S)		1047 (3)	
555 (0)	555 (0)		1062 (4)		$\nu(C_\alpha$—N)
594 (0)	598 (0)		1082 (4)	1080 (3)	
605 (0)			1106 (4)	1104 (4)	
622 (1)	622 (1)	Phe	1125 (3)	1125 (3)	
644 (3)	642 (4)	Tyr	1154 (1)	1154 (1)	
657 (4)	662 (2)		1180 (2)	1180 (2)	Tyr and Phe
675 (2 s)	675 (2 s)		1191 (0 s)		
	687 (2)		1210 (3)	1210 (4)	Tyr and Phe
	695 (2)		1239 (11)	1242 (11)	
	703 (2)	ν(C—S)	1263 (10)	1260 (11)	Amide III
714 (0 s)			1284 (0 s)	1290 (0 s)	
724 (2)	720 (2)		1315 (8)	1320 (8)	
	735 (1 s)		1324 (1 s)		$\gamma(CH_2)$(?)
755 (1)	750 (1)		1337 (3)	1337 (3 s)	
808 (1)	806 (1)		1399 (4)	1399 (3)	
834 (5)	829 (3)	Tyr	1412 (5)	1412 (4)	$\nu(CO_2^-)$
854 (4)	854 (5)		1420 (5)	1423 (4)	
892 (3)	886 (3)		1447 (10)	1447 (10)	$\delta(CH_2)$
902 (3)	902 (3)		1585 (0 s)	1585 (0 s)	
918 (0 s)		$\nu(C_\alpha$—C)	1603 (1 s)	1603 (1 s)	
937 (4)	937 (2)		1619 (3)	1619 (3)	Tyr and Phe
960 (0)			1668 (24)	1671 (21)	Amide I and H_2O
982 (9)	982 (12)	SO_4^{2-}			

[a] From Chen and Lord (1976). Conditions: aqueous solvent pH 5, 0.1 *M* NaCl. Abbreviations: (s) stretching; δ deformation; γ twisting. Peak intensities are given by the figures in parentheses.

on model compounds; therefore, the same limitations as for ORD apply for Raman spectroscopy. Infrared spectroscopy can give the same type of information as Raman spectroscopy. This method is usually coupled with the exchange hydrogen–deuterium, therefore it is presented in Section 8.1.3 Fourier transform infrared spectroscopy (Belasco and Knowles, 1980) can be very useful for detecting the reappearance of ordered structures during the refolding of a protein.

8.1.3. Detection of Structured Segments by Hydrogen–Deuterium or Hydrogen–Tritium Exchange

Measurements of the rate of hydrogen exchange are comparable to measurements of amino acid side chain reactivity in proteins. However, in hydrogen exchange, the hydrogen isotope has dimensions and properties very similar to the atoms it displaces in the native proteins. Since hydrogen atoms are present in great number, each hydrogen is more difficult to identify than the individual side chains such as tyrosine, and cysteine. This method is presented here since it allows the detection of helix-bonded structured segments in proteins.

The method was introduced by Linderstrøm-Lang (1955) and extensively detailed by Hvidt and Nielsen (1966), Englander and co-workers (1972), Englander and Englander (1972), and Woodward and Hilton (1979), and it was initiated by the original paper from Lenormant and Blout (1953). It is based on the following observations:

(1) Oligopeptides and randomly coiled polypeptides exchange labile H fast, whereas native proteins exchange very slowly.

(2) The rate of hydrogens slowly exchanging in the native form increases on denaturation.

The interpretation of protein hydrogen exchange presented by Linderstrøm-Lang (1955) is based on the assumption that, in a macromolecular conformation all peptide bond hydrogens fall into two categories: one in which the exchange rate is rapid and comparable to that of the model peptides, the other in which hydrogen exchanges extremely slowly. It is accepted the slowly exchanging atoms are those of the hydrogen-bonded peptide bonds, therefore, they are correlated with the ordered structure. In fact, most of the slowly exchangeable hydrogens in proteins are those of amide groups.

Two different models were proposed to account for amide proton exchange rates. The one, usually referred to as "the breathing of the protein" model (Linderstrøm-Lang, 1955; Linderstrøm-Lang and Schellman, 1959; Hvidt and Nielsen, 1966; Englander *et al.*, 1972) assumes time fluctuations of the structure. In the closed form C(H), interior protons are not accessible to the solvent and therefore cannot exchange; in the open structure O(H) in equilibrium with the previous one, the exchange becomes possible. This model involves a two-state description of the protein. It has been suggested that breathing processes in proteins are associated with physicochemical properties such as structure stability and spontaneous folding. Their biological role has been also discussed; it was proposed that they regulate interactions with other macromolecules, or that they act as a gate regulating flux of small ligands to and from the corresponding binding sites, especially in allosteric proteins (Englander *et al.*, 1972). Breathing segments may be very small. However, the entire protein may be involved in the breathing process; in this case, the fraction of time in which the hydrogens are accessible depends on the free energy of stabilization. For example, if the free energy of denaturation is 13 kcal/mole, the protein must spend about 10^{-10} ($13 = -RT\,In\,10^{-10}$) of the time in the unfolded state (Englander *et al.*, 1972). In agreement with this evaluation, Englander and Stanley (1969) found that in myoglobin the slowest hydrogens are retarded by a factor of 10^9 at pH 9.0 and 37°C. This is consistent with Tanford's evaluation of 13.5 kcal/mole for denaturation of myoglobin under the same conditions. However, most hydrogens appear to be slowed by factors ranging between 10^2 and 10^5. Therefore, it may be attributable to the breathing of smaller segments in the proteins. The ability of proteins to exchange hydrogen atoms with the solvent depends neither on the size of the protein nor on the number of S—S bridges (Hvidt and Pedersen, 1974).

An alternative model assumes diffusionlike penetration of the solvent at the interior of the protein; thus the exchange would occur (Englander *et al.*, 1972). A dynamic multistate model was proposed by Wüthrich and Wagner (1978).

On the basis of the two-state breathing model, analytical relationships were proposed. Conformations H_i and I_k were assumed in which respectively hydrogens are not rapidly exchangeable and are rapidly exchangeable; all the I_k conformations exchange with the same rate constant and each transformation is a first-order reaction. A general equation gives the degree α_m of the exchangeable hydrogen atoms:

$$\alpha_m = 1 - \sum a_n \exp(-\lambda_n t)$$

which corresponds to simultaneous first-order reactions with

$$\sum a_n = 1$$

A small number of exponentials are generally sufficient to describe the process.

If N and I are the average conformations, a simplified scheme may be proposed:

$$N \underset{k_2}{\overset{k_1}{\rightleftarrows}} I \xrightarrow{k_3} \text{exchange}$$

The first-order rate constant is

$$k_m = k_1 k_3/(k_1 + k_2 + k_3)$$

When $k_2 \gg k_3$, meaning that N and I are in rapid equilibrium, and if $k_1 \ll k_2$, the N conformations are mainly populated and

$$k_m = (k_1/k_2)k_3 = K_D k_3$$

K_D being the equilibrium constant between I and N forms. If $k_1 \gg k_2$, then $k_m = k_3$. The reaction mechanism is referred to as EX_2, a bimolecular exchange mechanism. The rate constant k_3 is strongly dependent on pH and temperature; it is proportional to catalyst concentration (H^+ or OH^-).

When $k_3 \gg k_2$ (with $k_1 \ll k_2$), the reaction mechanism is referred to as EX_1, a unimolecular exchange mechanism: $k_m \simeq k_1$. A decision in favor of one of these reaction mechanisms cannot be made from a single kinetic exchange. An EX_2 process could be distinguished from an EX_1 mechanism by the pH dependence of the exchange rate.

Several techniques were described to study hydrogen–deuterium exchange in solution. The Linderstrøm-Lang freeze drying method involved isolation of either isotopically enriched solvent water or protein. Methods substituting tritium for deuterium, thereby increasing the sensitivity, were developed. In this case, scintillation techniques were used for tritium analysis. Easier separation methods such as gel filtration or dialysis replace the freeze drying one.

Using infrared spectroscopy, it is possible to make direct measurements of the exchanging phase. The ^1H–^2H exchange can be followed by the decay of the amide II band at 1550 cm^{-1}. The use of infrared spectroscopic techniques, when adapted to a stopped-flow apparatus, as developed by Nielsen (1960), allows one to detect hydrogen–deuterium exchange in a very short time. Nuclear magnetic resonance techniques also can be used to study hydrogen–deuterium exchange. The corresponding results are discussed in Section 8.2.

Amide hydrogen atoms of a polypeptide chain in a random conformation do not exchange at the same rate, but 10^2-fold variations of rates have been observed (Molday *et al.*, 1972). For exchange of hydrogen atoms in native proteins, simple biphasic curves were obtained for several proteins (Nakanishi *et al.*, 1972, 1973; Takesada *et al.*, 1973). The fast and slow exchanging atoms corresponded to the fraction of exposed and shielded groups (Nakanishi and Tsuboi, 1974).

In erabutoxin B, the exchange kinetics indicate a number of slow protons consistent with the number of potential hydrogen bonds revealed by X-ray crystallography. In the rest of the molecules, distinct solvent accessibilities reflect different degrees of flexibility of the polypeptide chain (Thierry *et al.*, 1980).

However, many other studies indicate great variations in the exchange rates in folded proteins, the various parts of the protein having probabilities of exposure to the solvent ranging from 1 to 10^{-7} and even 10^{-9}. This suggests the occurrence of many limited conformational transitions in a folded protein. In BPTI, for example, Hvidt and Pedersen (1974) found probabilities of 10^{-3} for 50% of the potentially exchangeable hydrogen atoms and down to 10^{-7} for some others. In hemoglobin also, great variations in the rates of hydrogen exchange have been observed, even in the smaller number of hydrogens sensitive to the allosteric transition (Englander and Mauel 1972; Englander and Rolfe, 1973; Englander, 1975–1978; Malin and Englander, 1980; Liem *et al.*, 1980). All hydrogens in a set exchange at the same rate. Large variations in the rate of exchange were observed when the protein had undergone its allosteric transition. All hydrogens in a set were found to respond to allosteric transition as cohesive units. (Malin and Englander 1980; Liem *et al.*, 1980). These results indicate individual local unfolding reactions in hemoglobin in agreement with the breathing model. The difference hydrogen exchange methods developed by Englander and co-workers (Malin and Englander, 1980; Liem *et al.*, 1980) appear to be efficient in the detection of individual local changes in structure and in the evaluation of the free energy associated with each of them.

The hydrogen–tritium exchange was used to study variation of S peptide alone and bound to RNase (Shreier and Baldwin, 1976, 1977). Recently, this technique was utilized by Schmid and Baldwin (1978b, 1979a,b) to characterize intermediates in the refolding of RNase. A mechanism was previously proposed to explain different results (see Chapter 7):

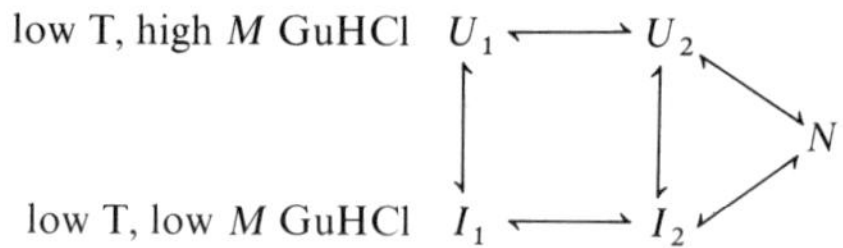

$U_1 \rightleftarrows U_2$ equilibrium was proposed to be the proline isomerization reaction.

To characterize the structure of I_1, the exchangeable amide protons of U_1 were labeled by ^3H and folding was allowed to occur under conditions where exchange reaction is fast compared to the refolding of U_1 into N. Since the unfolded protein contains both U_1 (80%) and U_2 (20%) species, an experimental procedure was developed to label U_1 without labeling U_2. The ^3H labeled U_1 was exposed to refolding. The ^3H exchanged before the refolding, but since I_1 formed rapidly, some ^3H labels were trapped by this formation and incorporated into the final native RNase. These trapped ^3H were then measured. The data gave direct evidence for the existence of the intermediate I_1 and indicated that it has an extensive H-bonded structure; therefore, it contains parts of ordered structure, probably α helix.

Although other reports exist that describe the use of hydrogen exchange to follow the unfolding or refolding of proteins, very few analyses were performed with the goal of detecting structured intermediates during the refolding process. The work of Schmid and Baldwin (1978b, 1979a,b) appears to be the only real attempt in this direction. Besides these studies, deuterium exchange was also used to study distorisons of the α helical structure in several proteins, by means of infrared spectroscopy in the region of the amide A band (Brazhnikov and Chirgadze, 1978).

Ideally this method might allow one to follow the exchange rate of each individual amide proton. This is sometimes possible by NMR, at least for several resolved amide proton resonances (see Section 8.2). Using chemical techniques of analysis, Rosa and Richards (1979) improved the method of assigning the exchange kinetics to known regions of the protein. In their procedure, after the out-exchange, the protein is immediately digested, the peptides are separated by high pressure liquid chromatography (HPLC), and the specific activity of each peptide is measured. The analytical process requires 20–30 min, and this corresponds to several half-lives of exchange since the peptides are exposed to solvent during the analysis. However, accurate analysis is possible with sufficient isotope in the starting material. With great care, it is possible to obtain reproducible results. Using this method, Rosa and Richards (1981) have studied the hydrogen exchange from identified regions of S-ribonuclease and have observed a decrease in the exchange from several sites on binding of the S peptide.

The results are encouraging for use of this method to detect segments of ordered structures in the early stages of protein folding. It may be more accurate than ORD or Raman spectroscopy for this approach. However, the procedure is very delicate and requires great care for the analysis. Data on a sufficient number of proteins is required to know if ordered structures are really formed in the early stages of protein folding and if α helical conformations appear in a faster time than β sheet during the folding process.

8.2. DETECTION OF INTERMEDIATES BY NUCLEAR MAGNETIC RESONANCE STUDIES

High resolution proton magnetic resonance is an important method that provides information about individual amino acid side chains and the variations of their environment during protein folding (Bradbury and King, 1971; Jardetzky *et al.*, 1971). This method has been used to detect intermediates during the refolding of several proteins including staphylococcal nuclease, RNase, BPTI, and lysozyme. Before considering the pathways of unfolding (or refolding) one must define the final state of unfolding. A limitation of the method is the concentration of protein normally needed for NMR measurements. It is generally 10- to 50-fold higher than that used for optical studies and very much higher than that used for enzymatic assays. It should be confirmed that the results do not depend on protein concentration.

For staphylococcal nuclease as well as for RNase, the C-2 proton resonance of the imidazole of histidine was used as a conformational probe to follow the transition. For staphyloccocal nuclease, Epstein and co-workers (1971b) and Anfinsen and co-workers (1972) measured the values of area, line-width shift of each of the four imidazole C-2 protons of His 8, 46, 121, and 124, and in the pH acid range of transition. Resonance H-3 and H-2 have been assigned respectively to His 124 and His 46, this last histidine being near the active site. The variations of H-1, H-2, and H-3 resonances and of their corresponding area on transition follow a single two-state equilibrium (see Chapter 5, Fig. 5.5). This transition is indistinguishable from that obtained using the variations of Trp 140 fluorescence. By contrast H-4 resonance follows a biphasic transition clearly indicating the existence of intermediates. For this protein, the change in hydrodynamic properties, tryptophan emission, and areas of H-1, H-2, and H-3 corresponds to the major conformational rearrangement of the refolding process. The H-4 resonance probably reflects some local rearrangement that occurs at slightly more alkaline pH.

Evidence for intermediates was also obtained using this method to study the unfolding of RNase. Thermal acid unfolding as well as unfolding induced by urea or GuHCl was studied by Benz and Roberts (1973, 1975a,b) and Westmoreland and Matthews (1973). The C-2 proton magnetic resonance of His 12, 48, 105, and 119 was used as a conformational probe. The assignments were made by Markley (1975a,b) and Markley and Finkenstadt (1975). The spectrum of folded and unfolded RNase and also the spectrum of the protein in the transition zone was obtained (Blum *et al.*, 1978). The changes in area of the corresponding resonance peaks were followed during transition. It was confirmed that the results are independent of protein concentration in the range from 0.2 to 2 m*M* RNase. The reversibility of the transition was observed for nearly all the conditions of unfolding. For thermal unfolding of RNase at 56°C, pH 1.5, the NMR spectrum is the one expected for a random coil. It is very similar to the spectrum obtained in 7.7 *M* urea or 4.1 *M* GuHCl.

At pH 5.5, thermal unfolding of RNase does not lead to a random coil, the remaining structures include two histidine residues and probably a phenylalanine. At this pH, a major transition is observed which affects His 48 and perhaps a methionine residue. At pH 1.5 and 2.9 where thermal unfolding may reach a random coil, evidence for the existence of an intermediate in the unfolding process was provided. In this intermediate, His 12 and His 119 are accessible to the solvent environment while His 48 and His 105 are not yet. When unfolding is induced by denaturants such as urea or GuHCl, similar results were obtained; His 12, then His 119 unfold before His 48 and His 108, these last two apparently unfolding together. Two intermediate states are therefore distinguishable for urea as well as for GuHCl denaturation, one in which only His 12 is in a solventlike environment, a second in which His 12 and His 119 are exposed to the solvent but His 48 and 105 are not (Benz and Roberts, 1975a). In their work, Blum and co-workers (1978) observed and characterized an intermediate by NMR in the early stage of the refolding of heat unfolded RNase at pH 2.0 and 10°C. This intermediate has one residue identified as His 12 in a partly folded environment and the other three His residues in unfolded environment. It apparently contains the helix located near the N-terminal end of the native protein. The properties of this intermediate are similar to those reported for the I_1 species.

Konishi and Scheraga (1980b) used NMR to analyze the intermediates during regeneration of RNase A from reduced protein and found that intermediates formed at an early stage of renaturation are rather disordered.

Reversible transition of lysozyme was also studied by NMR (McDonald *et al.*, 1971); temperature, pH, and GuHCl were used for denaturation. If reversibility of the transition was shown in all cases, no transition intermediates were detected.

Wooten and Cohen (1979) studying lysozyme by deuterium nuclear magnetic resonance of [ε^2H]imidazole found that His 15 is restricted in motion and that protein is predominantly dimerized at pH 7.5.

Transition of BPTI was widely studied by NMR (360 MHz) (Karplus *et al.*, 1973; Sykes *et al.*, 1975; Snyder *et al.*, 1975; 1976; Wüthrich *et al.*, 1974, 1975, 1978; Wüthrich and Baumann, 1976; Wüthrich and Wagner, 1975; Wagner and Wüthrich, 1978a,b, 1979a,b, Wagner *et al.*, 1975, 1976, 1979a,b, 1978a,b; Masson and Wüthrich, 1973). Certainly BPTI is the most extensively studied protein, not only by NMR, but also by diverse experimental approaches as those reported by Creighton (1974a,b,c; 1975a,b,c,d; 1977a,b,c; 1978a,b,c; 1979a, 1980a,b,c, Creighton *et al.*, 1980), and even by theoretical simulation of the structure (Levitt and Warshel, 1975; Levitt, 1976; Karplus and McCammon, 1979, 1980).

Individual resonance assignments were determined by various researchers. Labile protons have been observed in the low-field spectral region, between 6 and 11 ppm (Karplus *et al.*, 1973; Masson and Wüthrich, 1973); the resonance of the aromatic protons is also located in this region (Sykes *et al.*, 1975; Wüthrich and Baumann 1976). For several protons the exchange is very slow, with a lifetime of up to 1 year at room temperature and neutral pH; but complete exchange can be achieved by heating the protein at 80°C for several minutes without denaturation. At pH 4.6, BPTI contains 114 labile protons: 53 backbone amide protons; 8 amide protons of the side chains of asparagine and glutamine; 3 α amino protons of Arg 1; 12 ε amino protons of lysine; 30 guanidium protons of Arginine; and 8 hydroxyl protons of serine, theonine, and tyrosine. The amide protons of Tyr 21, Phe 22, and

TABLE 8.4

Individual Resonance Assignment of Protons in BPTI from the Proton Nuclear Magnetic Resonance Spectrum[a]

Residue	δ(NH)	$\delta(C^{\alpha}H)$	$\delta(C^{\beta}H)$	δ(Other Protons)	$^3J_{HN\alpha}$	$^3J_{\alpha\beta}$
Phe 4				$C^{\delta}H$ 6.99		
				$C^{\varepsilon}H$ 7.37		
				$C^{\zeta}H$ 7.06		
Leu 6				$C^{\gamma}H$ 1.73		
				$C^{\delta}H$ {0.95, 0.86}		
Tyr 10				$C^{\delta}H$ 7.32		
				$C^{\varepsilon}H$ 7.07		
Thr 11			4.04	$C^{\gamma}H$ 1.39		~12.0
Lys 15				$C^{\varepsilon}H$ 3.04		
Ala 16		4.3	1.19			7.0
Arg 17		4.29				
Ile 18	8.12 (16)	4.25	1.87	$C^{\gamma 1}H$ 1.07	9.0	
				$C^{\gamma 2}H$ 0.97		
				$C^{\delta}H$ 0.68		
Ile 19		4.30	1.96	$C^{\gamma 1}H$ 1.46		~12.0
				$C^{\gamma 2}H$ 0.73		
				$C^{\delta}H$ 0.68		
Arg 20	8.39 (11)	4.70	1.62		9.5	
Tyr 21	9.18 (6)	5.70	2.70	$C^{\delta}H$ 6.71	9.5	{6.5
				$C^{\varepsilon}H$ 6.78		6.5}
Phe 22	9.79 (3)	5.29	2.87	$C^{\delta 1}H$ 7.27	9.0	~3.0
			2.81	$C^{\delta 2}H$ 7.12		~3.0
				$C^{\varepsilon 1}H$ 7.07		
				$C^{\varepsilon 2}H$ 6.98		
				$C^{\zeta}H$ 7.31		
Tyr 23	10.55 (1)	4.31	{2.72	$C^{\delta}H$ 7.18	7.0	
			3.45}	$C^{\varepsilon}H$ 6.34		
Asn 24	7.78 (22)	4.41	2.69		9.0	
Ala 25		3.74	1.57			7.0
Lys 26				$C^{\varepsilon}H$ 3.04		
Ala 27		4.3	1.19			7.0
Leu 29				$C^{\gamma}H$ 1.47		
				$C^{\delta}H$ {0.86, 0.76}		
Cys 30	8.40 (g)	5.60	3.69		9.0	13.0
			2.67			2.5
Gln	8.77 (7)	4.87			10.5	~2.0
Thr 32	8.05 (18)	5.28	4.01	$C^{\gamma}H$ 0.59	9.0	~2.0

TABLE 8.4 *(continued)*

Residue	δ(NH)	$\delta(C^{\alpha}H)$	$\delta(C^{\beta}H)$	δ(Other Protons)	$^3J_{HN\alpha}$	$^3J_{\alpha\beta}$
Phe 33	9.37 (4)	4.89		$C^{\delta}H$ 7.14	7.0	
				$C^{\varepsilon}H$ 7.21		
				$C^{\zeta}H$ 7.33		
Val 34		3.92	1.96	$C^{\delta}H$ {0.81 0.71		12.0
Tyr 35	9.39 (5)	4.89	2.69	$C^{\delta 1}H$ 7.79	10.5	
			2.52	$C^{\delta 2}H$ 6.80		
				$C^{\delta 1}H$ 6.84		
				$C^{\delta 2}H$ 6.73		
Gly 36	8.61 (9)					
Ala 40		4.09	1.21			7.0
Lys 41				$C^{\varepsilon}H$ 2.96		
Asn 43				$N^{\delta}H$ {7.98 (19) 7.78 (23)		
Asn 44		4.88	2.78			
Phe 45	9.94 (2)	5.12	2.79	$C^{\delta}H$ {7.2 7.5	11	{13.0 2.5
				$C^{\varepsilon}H$ {7.6 8.2		
				$C^{\zeta}H$ 7.67		
Lys 46				$C^{\varepsilon}H$ 3.04		
Ala 48		3.14	1.04			
Glu 49	8.5 (f)					
Met 52	8.58 (10)			$C^{\varepsilon}H$ 2.16		
Thr 54			3.95	$C^{\gamma}H$ 1.61		~12.0
Cys 55	8.25 (15)	4.31			8.5	
Ala 58		3.99	1.31			

[a] From Wüthrich and Wagner (1979); see text.

Phe 33 were assigned by Marinetti and co-workers (1976), by using chemical modifications of BPTI and lanthanide shift reagents. The other amide protons and many α and β protons were determined by Dubs and co-workers (1979). The methyl groups and additional protons of aliphatic side chains were determined by de Marco and co-workers (1977) and by Wüthrich and co-workers (1978). The ε-methylene protons of Lys 15, Lys 41, and Lys 46 were determined by Brown and co-workers (1976), and the aromatic ring protons by Karplus and co-workers (1973), Snyder and co-workers (1975, 1976), Wagner and co-workers (1976, 1979a,b), and Perkins and Wüthrich (1978). The different resonance assignments are given in Table 8.4, examplifying the potentiality of NMR as a method of local identification. Table 8.4 gives different NMR parameters, the chemical shift of the amide protons

(δ-NH) and those of the coupled α protons (δ-C$^{\alpha}$H) which were determined by spin decoupling, and the spin–spin coupling constants $^3J^{\alpha}_{\text{HN}}$. The δ-C$^{\beta}$H chemical shifts and the $^3J_{\alpha\beta}$ coupling constants are also given when determined. The chemical shifts of the amide protons of Met 52 and Cys 55 are nearly identical to those of these two residues in a random coil conformation and the large coupling constant found for Cys 55 is in agreement with the fact that this residue deviates from a regular α helix. Another striking observation is that different chemical shifts prevail for the different hydrogen bonds of the antiparallel β sheet (Phe 33, Tyr 35, Glu 31, Arg 20, Ile 18, Asn 24); the amide proton chemical shifts reflect the distorsion of the β sheet (Wüthrich and Wagner, 1979).

From the study of the pH dependence of the amide proton chemical shifts, Wüthrich and Wagner (1979) were able to conclude that the titration shifts result from long-range interactions with the ionizable groups of the protein, and therefore they are mainly dependent on the protein conformation.

The amide proton rate of exchange was determined over a p^{2}H range from 0.1 to 10.9 and for temperatures ranging from 10° to 60°C (Richarz *et al.*, 1979). Different rates of exchange were observed for individual amide protons, but for all an EX_2 mechanism was obeyed with an acid–base catalyzed exchange. Figure 8.1 shows the dependence of k_m on p^{2}H for 8 amide protons. Slopes deviate from unity which is contrary to that found with peptide models, and broader minima are observed. Richarz and coworkers concluded that the two-state model is oversimplified even when the

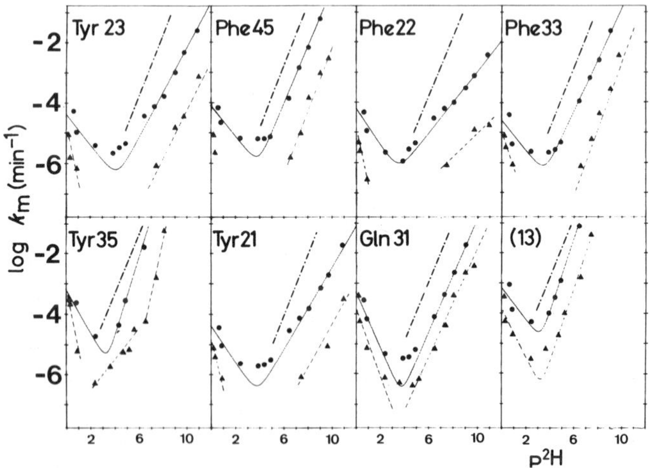

Fig. 8.1. The p^{2}H profiles of the exchange rates k_m for 8 amide protons of BPTI (from Richarz *et al.*, 1979, courtesy of Wüthrich); (—) data obtained at 45°C; (---) data obtained at 22°C; slope 1 is indicated on each curve by (-··-).

exchange of individual protons is considered and they proposed a dynamic multistate model for the interpretation of the results obtained with globular proteins (Wüthrich and Wagner, 1978; Wagner and Wüthrich, 1979a,b).

The thermal transition was studied; the amide proton exchange with solvent ^2H_2O and the intramolecular mobility of the aromatic rings were measured. The spectrum from [^{1}H]NMR was recorded at different temperatures. The amide proton exchange was found in good correlation with the thermal stability of the protein as observed in BPTI and in chemically modified forms of this protein as well as in homologous proteins (Wagner *et al.*, 1978a,b; Wagner and Wüthrich, 1979a,b). The thermal stability of a protein appears to be correlated with the breathing modes as manifested in the amide proton exchange rates.

However, the rotational motions of aromatic rings (tyrosine and tryptophan) are not correlated with amide proton exchange (Wüthrich and Wagner, 1975; Wagner *et al.*, 1976; Hetzel *et al.*, 1976). Rotational motions of aromatic groups are good probes of the microenvironment since their covalent structure is symmetrical with respect to a twofold axis through the C_α—C_β bond. This is shown in Fig. 8.2. The BPTI contains 4 tyrosine and 4 phenylalanine residues; resonance assignments are given in Table 8.4. The resonance of the aromatic rings was followed from 4° to 81°C. At 25°C, two tyrosine rings and one phenylalanine rotate rapidly on the NMR time scale; for the other rings the transition from slow to rapid rotation depended on the temperature. More precisely the rings of Tyr 10 and Tyr 21 rotate at a rapid rate even at 4°C, Tyr 23 is more hindered flipping at a rate of 5 per sec at 4°C, but more rapidly when the temperature increases. Tyr 35 is still more rigid and flips at a rate of less than 1 per sec at 4°C. The four Phe vary widely in their rates of rotation; 1 phenylalanine rotates rapidly on the NMR time scale, another is immobilized over the temperature range up to 80°C (Wüthrich and Wagner, 1975, 1978; Wagner *et al.*, 1976; Wagner and Wüthrich, 1978a,b).

From the various data, it was concluded that regions of hydrophobic stability are of great importance in protein structure. They probably are formed at an early stage of polypeptide folding and can play the role of nucleation sites. Aromatic rings buried in these regions remain immobilized even at high temperature. These experimental results are in good agreement with predictions reported by Matheson and Scheraga (1978).

Additional investigations were started on this protein, using high-field ^{13}C NMR at 90.5 MHz; individual resonance lines of methyl groups, carboxyl carbons, and of ionizable amino acid side chains of Asp 3, Asp 50, Glu 7, and Glu 49 were identified, and the ionization constants were determined (Richarz and Wüthrich, 1977, 1978). Until now these resonances were not used to follow conformational variations, but only to obtain more details

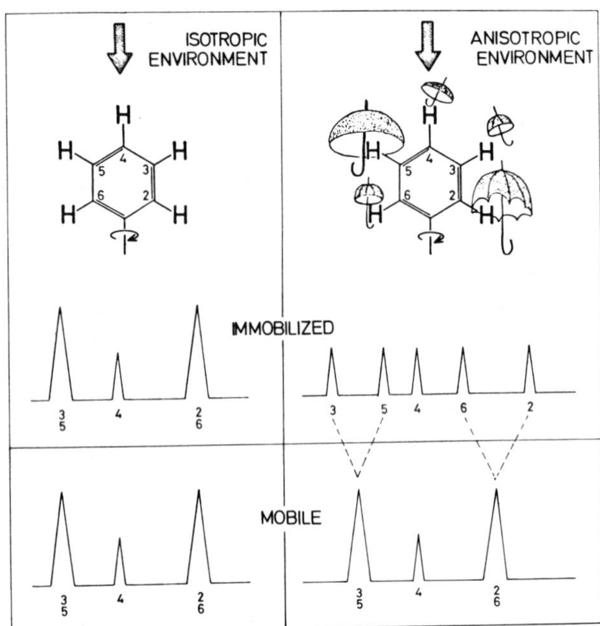

Fig. 8.2. Effects of the environment on the NMR spectrum of a phenylalanine ring (according to Wagner and Wüthrich, 1978b, courtesy of Wüthrich). Left: in an isotropic environment, protons 2 and 6, 3 and 5 are equivalent since there is a twofold symmetry. This equivalence exists whether the ring is immobilized or rotating. The NMR spectrum indicates equivalent chemical shifts in both cases. Right: the aromatic protons are in different environments because of the interactions with neighbor atoms and therefore are shielded differently against the magnetic field (as represented by the umbrellas). Consequently, the ring symmetry is marked by the assymetric environment and different chemical shifts are observed for the different protons. If the phenylalanine ring rotates rapidly, each of the 2, 6 and 3, 5 protons spend the same time in the different environments and their effects are averaged.

on the conformational description of BPTI in solution. It does provide an additional method for detecting local variations in proteins.

The refolding of reduced and denatured BPTI was studied by trapping intermediates with one or two disulfide bonds (see Section 8.3). These intermediates were studied by ^{1}H NMR (270 MHz) or aromatic and peptide NH protons (States *et al.*, 1980). Interpretation of the data assumed that upon refolding and reoxidation, a metastable isomer is formed that has a different conformation, especially in the region of β strand structure which passes through the molecule. However, reinvestigation has indicated that the major product of the refolding reaction was an intermediate with two correctly paired disulfide bridges (5–55 and 14–38); but lacking the 30–51 disulfide bond. (M. Karplus, personnal communication).

These different studies indicate that, in most cases, NMR studies of the unfolding–folding process have revealed the occurrence of intermediates. However, this method which provides local information has important limitations; it requires relatively high concentrations of protein, that are not commonly compatible with the unfolding–folding studies. This only applies to favorable cases where the aggregation process does not occur at the required protein concentration.

Rousslang and co-workers (1978, 1979) and Ross and co-workers (1980) used optically detected magnetic resonance (ODMR) of tryptophan triplet states to detect conformational variations in proteins and polypeptides. This method used in combination with phosphorescence appears to be an excellent probe for detection of conformational variations which affect the environment of tryptophan.

Besides the detection of intermediates in protein folding, one of the most striking results of NMR studies was the experimental evidence for the internal motion in globular proteins. Transitions with relaxation times of 10^{-5}–1 sec have been observed. In most proteins, internal aromatic rings flip by 180° rotation with rates greater than 10^4 times per sec. (Campbell *et al.*, 1972; Campbell *et al.*, 1975a,b; Snyder *et al.*, 1975). Transitions with times ranging from 10^{-11} to 10^{-8} sec were detected by [^{13}C]NMR (Allerhand *et al.*, 1971; Egan *et al.*, 1977; Wüthrich and Baumann, 1976; Richarz *et al.*, 1980).

8.3. TRAPPING OF INTERMEDIATES DURING REFOLDING OF UNFOLDED AND REDUCED PROTEINS BY BLOCKING THE SH GROUPS

During protein folding, disulfide bonds are formed or broken by thiol–disulfide interchange reaction (see Chapter 5, Section 5.4.3). This reaction can be quenched by blocking the thiol groups of the solution with appropriate reagent. Thus, it is possible to quench disulfide formation and to trap the species with disulfide bonds present at the time of quenching. This method was developed and detailed by Creighton (1974a,b,c, 1975a,b,c,d, 1977a,b,c, 1978a,b,c, 1979a,b); it was applied to the determination of the folding pathway of BPTI and RNase. It is one of the chemical methods that applies to the study of protein folding; this one is based on the redox properties of thiol–disulfide groups. The method implies first the quenching of disulfide formation, then the separation and characterization of trapped intermediates. The following description of the method is largely based on the review by Creighton (1978c).

8.3.1. Quenching of Disulfide Formation during Protein Folding

The chemical properties of thiol and disulfide groups have been described elsewhere (Chapter 5, Section 5.4.3). It has been reviewed by Overman and O'Connor (1976), Lapanje and Rupley (1973), Jocelyn (1972), Saxena and Wetlaufer (1970), Snyder and Carlsen (1977), Allinger *et al.* (1976), Van Wart and Scheraga (1977), and Creighton (1978c). The principle of the method for trapping intermediates during the refolding process is shown in Fig. 8.3. In the absence of appropriate nucleophile or electron donor, the disulfide bonds formed at the quenching time are stable enough to be present and can be used as conformational probes in the trapped intermediates. The quenching reaction must trap faithfully the species which are present at the time of quenching. Creighton emphasized that addition of the trapping reagent must react with the minimum alteration of the folding process. The first attempts to trap disulfide intermediates were described by Ristow

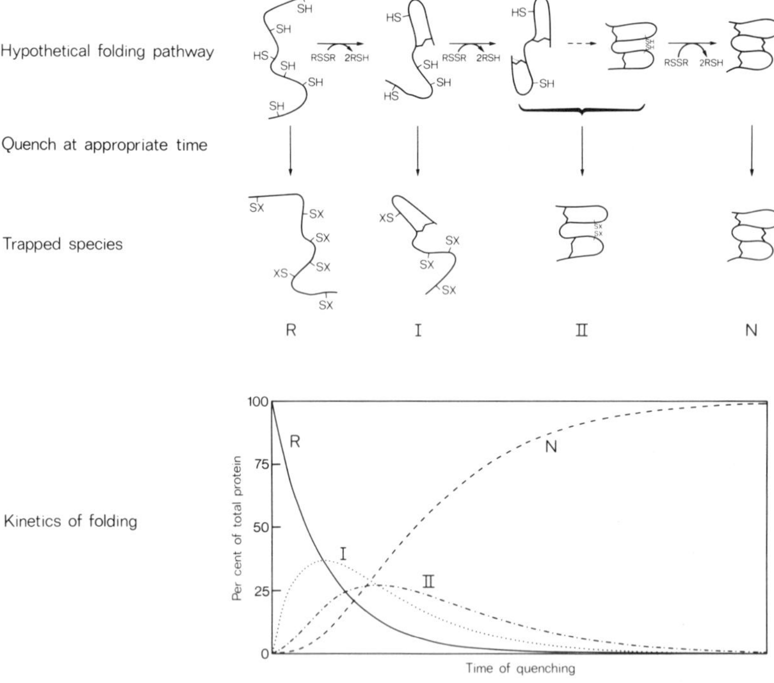

Fig. 8.3. Schematic representation of the method using disulfide bonds to trap intermediates in protein folding (courtesy of Creighton, 1978c). A simplified folding pathway is represented. In the quenching reaction each thiol is blocked with a moiety *X*. The kinetics of the appearance of each species is represented at the bottom.

and Wetlaufer (1973) and applied to the refolding of lysozyme by Anderson and Wetlaufer (1976) and by Acharya and Taniuchi (1976). The best documented studies were presented by Creighton as previously cited on the folding of BPTI and to a lesser extent RNase.

Reduced BPTI was reoxidized in the presence of oxidized DTT (DTT$<^{S}_{S}|$).

Reaction of free thiol groups with iodoacetate and iodoacetamide was used by Creighton (1979a,b,c) to quench their exchange. The reaction is completely irreversible and at pH 8.7, the second-order rate constant is 2–5 sec^{-1} M^{-1}. A high concentration of reagent (0.1 *M*) was used; it was thus possible to minimize the time of reaction with the highly reactive SH groups and therefore to minimize the reaction with other groups. The reagent can be removed by gel filtration. The validity of the method was confirmed by observing that the same intermediates are obtained by two (or more than two) different reagents, such as iodoacetate or iodoacetamide, or by lowering the pH.

8.3.2. Separation of Trapped Intermediates

Considering the folding pathway shown in Fig. 8.3, one can expect, at each time, to obtain a mixture of different species. It is thus necessary to separate and purify the intermediates for characterization. They can be separated by their conformational properties (shape or volume), or by variations in charge introduced by the chemical reagent, iodoacetate. Thus, Creighton (1974b,c) obtained the separation of the different species by gel electrophoresis. Figure 8.4 shows the relative electrophoretic mobilities of reduced and native BPTI and also those of the reduced protein in which the 6 SH groups have been blocked by iodoacetamide or iodoacetate. Figure 8.5 gives the densitometer traces of electrophoretic separation of the BPTI trapped by iodoacetate (Fig. 8.5A) and iodoacetamide (Fig. 8.5B) during its refolding. The same profiles were obtained with both chemical reagents.

A chromatographic separation of the species was also obtained. Two peaks, I and III, and a small peak, II, corresponded to species with a single disulfide bond. Peaks V, VI, and VII contained molecules with two disulfide bonds. A minor peak, VII, included at least three species with 3 disulfide bonds. The peak *N* contained molecules have the three native disulfide bonds. The stability of each trapped species was shown by a new chromatography of the corresponding peak.

When refolding was performed in urea with all other conditions kept identical, a different spectrum of species was obtained. That demonstrated the validity of the method, the observed species depended on the species

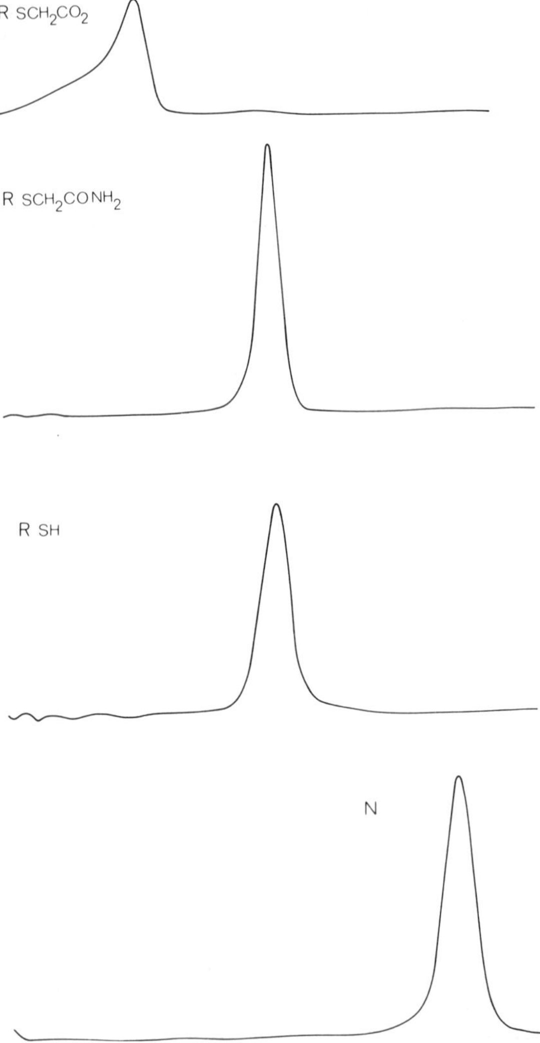

Fig. 8.4. Relative electrophoretic mobilities of the native, reduced BPTI and of reduced protein with the six SH blocked by iodoacetamide and iodoacetate (courtesy of Creighton, 1974b).

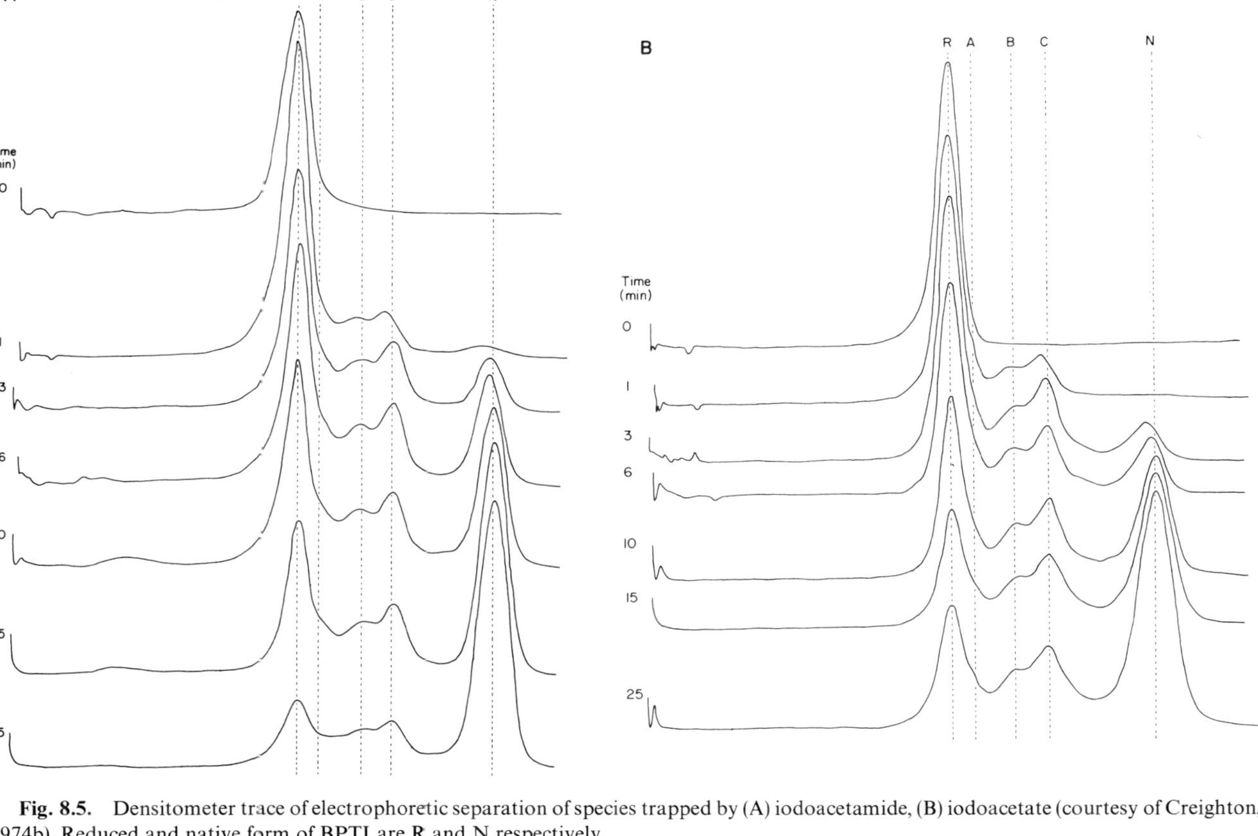

Fig. 8.5. Densitometer trace of electrophoretic separation of species trapped by (A) iodoacetamide, (B) iodoacetate (courtesy of Creighton, 1974b). Reduced and native form of BPTI are R and N respectively.

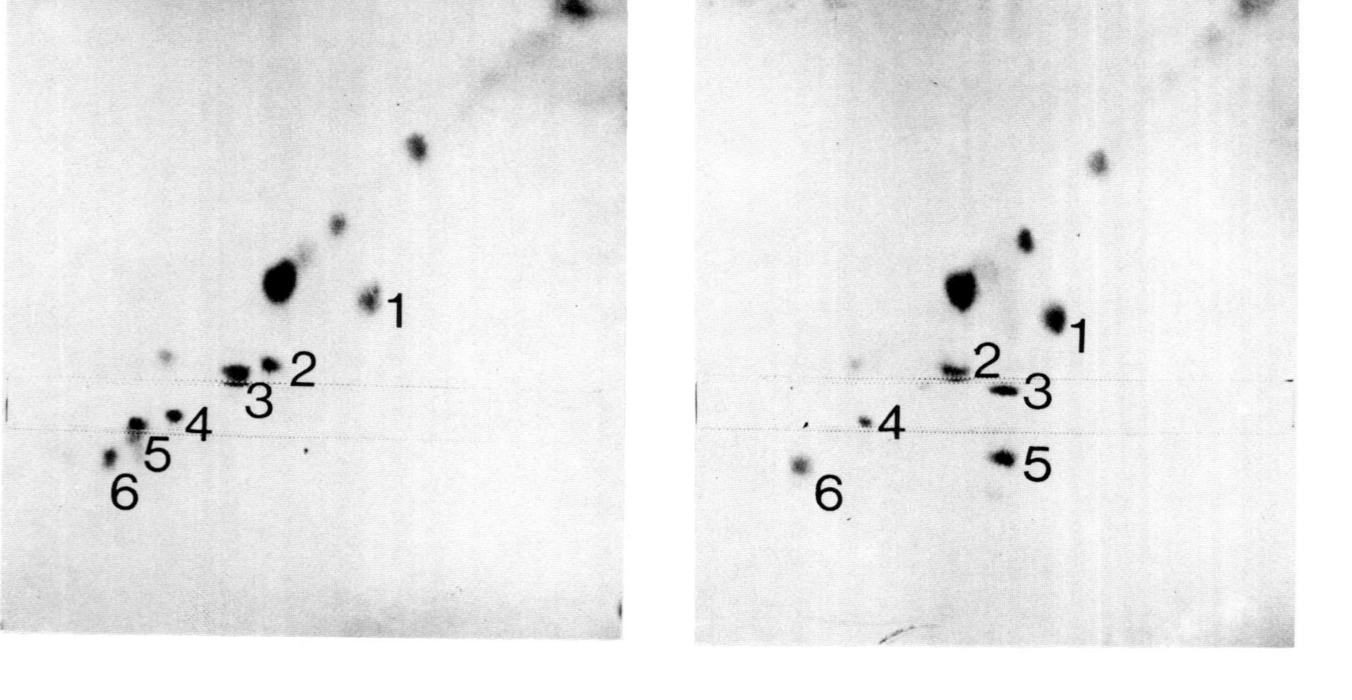

Fig. 8.6. (A) Two-dimensional diagonal maps of carboxymethylated reduced BPTI; (B) single disulfide peak isolated by ion-exchange chromatography. The numbered peptides define the CMCys diagonal, while the normal diagonal of peptides unaltered by performic acid is above it. Peptide I corresponds to residues 36–39; peptide 2, 1–15; peptide 3, 47–53; peptide 4, 5–15; peptide 6, 54–58. They result from digestion by trypsin followed by chymotrypsin (courtesy of Creighton, 1978c).

present at the time of trapping and did not result from the procedures employed for trapping and separation.

8.3.3. Characterization of Trapped Intermediates

Further analysis was necessary to determine the number of disulfide bonds present in the trapped and separated species. For this determination, Creighton (1974c) used the diagonal electrophoresis procedure introduced by Brown and Hartley (1966). The separate species were digested by proteolytic enzymes, and the peptides separated by paper electrophoresis under conditions which kept the disulfide bonds intact. Then the paper was treated with performic acid vapor to cleave the disulfide linkages and oxidize the Cys residues to cysteic acid. Then, the electrophoresis was repeated under the same conditions in a direction perpendicular to the first one. All peptides which were not modified by performic acid migrated identically in both directions and, therefore, were found on a diagonal line through the origin on the two-dimensional diagonal maps; peptides that were modified by performic acid migrated differently and were off the diagonal (Fig. 8.6).

With fully reduced and carboxymethylated BPTI, a second diagonal line of peptides appeared following electrophoresis at pH 3.5, the second diagonal line of peptide migrating as more acidic in the second direction. This second diagonal line of peptides did not appear when iodoacetamide was used. By this method, Creighton (1974c, 1975a,b,c) was able to identify the different disulfides existing in each peak. In peak I, the disulfide bond was Cys 30–Cys 51. The minor single disulfide peak II seemed to contain Cys 36–Cys 55 and Cys 5–Cys 51. Peak III contained Cys 5–Cys 30. Peak V contained Cys 30–Cys 51, Cys 5–Cys 55. Peak VI had similar amounts of Cys 30–Cys 51, Cys 5–Cys 14 and Cys 30–Cys 51, Cys 5–Cys 38. Peak VIII contained Cys 30–Cys 51 and Cys 14–Cys 38 and peak *N* contained the three native disulfide bridges Cys 30–Cys 51, Cys 5–Cys 55, and Cys 14–Cys 38. Peak VIII contained three disulfide bonds with incorrect pairing of cysteine residues; they have not been clearly identified.

8.3.4. Determination of Kinetic Pathway of Protein Folding and Unfolding Using Disulfide Bonds as Conformational Probe

From this identification and the kinetics of formation of the one, two, and three disulfide bonds with different disulfide reagents, a pathway of folding and unfolding of BPTI was proposed by Creighton (1975a, 1977b, 1978c) (Fig. 8.7). Single disulfide intermediate species Cys 30–Cys 51 are formed preferentially. They represent 60–70%; Cys 5–Cys 30, 20–30%, and as minor species Cys 30–Cys 55 and Cys 5–Cys 51 contribute 10%. In 6 *M*

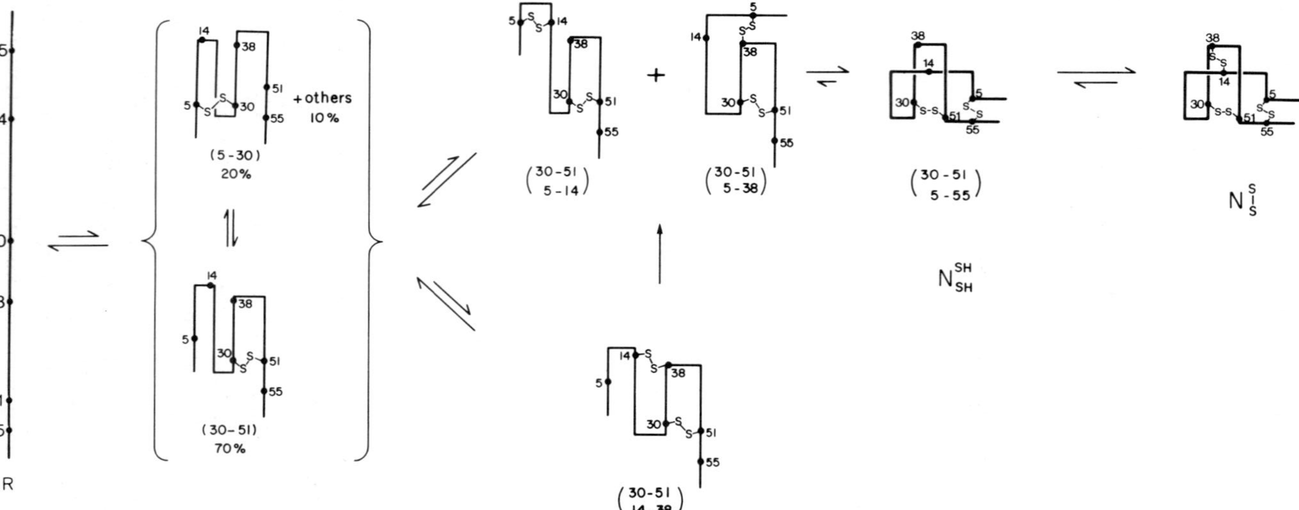

Fig. 8.7. Schematic diagram of the pathway of folding and unfolding of BPTI (courtesy of Creighton, 1977a).

GuHCl, a random species is formed. When removing the denaturant, an intramolecular rearrangement of the random disulfide bonds to the previous distributions rapidly occurs. Therefore, of the 15 possible single disulfide bonded species, one is predominant (Cys 30–Cys 51) which contains a native disulfide bridge and in a lower extent Cys 5–Cys 30 bridge. A rapid reequilibration of most of the single disulfide bonded intermediates prevents the determination of individual microscopic steps which contribute to the formation of the first disulfide bond. The Cys 5–Cys 55 species formed under denaturing conditions was not converted to normal species.

The predominant Cys 30–Cys 51 intermediate generate a second disulfide bonded species; the intermediate Cys 30–Cys 51, Cys 14–Cys 38 with two native disulfide bonds seems to be kinetically favored; however, two species, Cys 30–Cys 51, Cys 5–Cys 14 and Cys 30–Cys 51, Cys 5–Cys 38 are present in similar proportion. A fourth two-disulfide bonded intermediate was found to accumulate (Cys 30–51, Cys 5–Cys 55); it is apparently the most stable of the two disulfide species, but it was not formed directly from the single disulfide species, it results from the intramolecular conversion of species Cys 30–Cys 51, Cys 5–Cys 14 and Cys 30–Cys 51, Cys 5–Cys 38, each rearranging directly to Cys 30–Cys 51, Cys 5–Cys 55 at similar rate. The Cys 30–Cys 51, Cys 14–Cys 38 does not seem to rearrange directly to Cys 30–Cys 51, Cys 5–Cys 55, but the kinetics are consistent with its transformation to Cys 30–Cys 51, Cys 5–Cys 14 or Cys 30–Cys 51, Cys 5–Cys 38. The species Cys 30–Cys 51, Cys 5–Cys 55 or $\mathrm{N}\!\left\langle\begin{matrix}\mathrm{SH}\\ \mathrm{SH}\end{matrix}\right.$ have the properties of the nativelike conformation of BPTI obtained by reduction of the Cys 14–Cys 38 disulfide bond. This third disulfide bridge is formed very quickly from this species with a halftime of 12 μsec. The trapped intermediates have been analyzed by ultraviolet difference spectroscopy and suggest the same order for the decreasing expose to the solvent of the aromatic groups (Kosen *et al.*, 1980). The reverse reaction, reduction of disulfide bonds of native BPTI by $\mathrm{DTT}\!\left\langle\begin{matrix}\mathrm{SH}\\ \mathrm{SH}\end{matrix}\right.$ was also studied. The results are consistent with a reversibility of the pathway for folding and unfolding. Thus a complete study of the equilibrium and of the energetics of the unfolding–folding of BPTI was reported by Creighton (1977a,b,c). The apparent energy diagram of the transition is given in Fig. 8.8 (from Creighton, 1977b).

Creighton (1977d, 1979b) applied this method to the study of the refolding of reduced RNase. Using $\mathrm{DTT}\!\left\langle\begin{matrix}\mathrm{S}\\ |\\ \mathrm{S}\end{matrix}\right.$ as the disulfide reagent, he found that single disulfide intermediates are formed at the same rate as those in BPTI

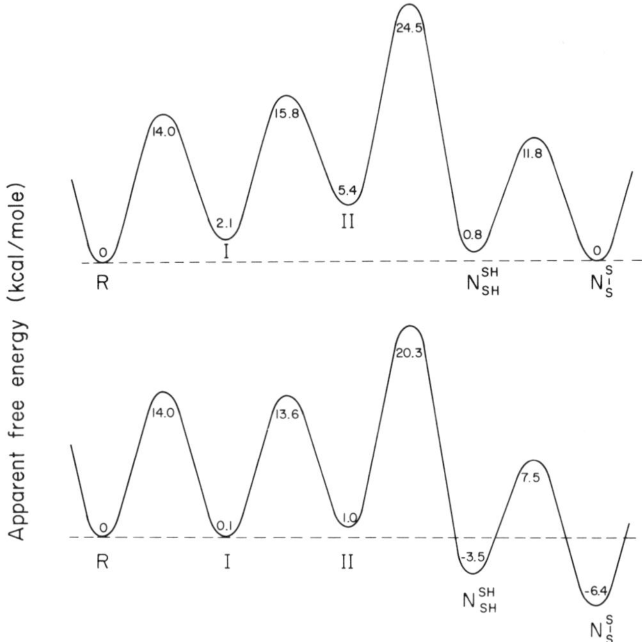

Fig. 8.8. Apparent energetic diagrams of the intramolecular transitions in the folding–unfolding of BPTI. Species I and II refer to the single and two-disulfide bond intermediates (see Fig. 8.7). The relative free energies of the reduced and native states depend on the stabilities of the disulfide bonds, which vary by changing the ratio of DTT$<^{S}_{S}|$ to DTT$<^{SH}_{SH}$. In the upper diagram this ratio is 26:1, in the lower diagram it is 10^3:1, each disulfide bridge is thus 2.1 kcal/mole more stable (courtesy of Creighton, 1977b).

and that two-disulfide bond species form at a slightly lower rate. However, the rate of formation of the third disulfide bond is extremely slow, and there is an accumulation of molecules with one and two disulfide bridges in equilibrium with the fully reduced protein. After 30 min, there are no molecules with more than two disulfide bridges. When the disulfide bonds are formed via thiol–disulfide exchange with oxidized–reduced gluthathione, intermediate species with one, two, three, and four disulfide bonds accumulate. Contrary to the case of BPTI, there is no formation of an initial correct disulfide bridge. The mixture of one-disulfide bond intermediates contains all 28 possible species indicating a random pairing. A large number of disulfide cross-linked species have been found for the two-disulfide intermediates, but not all possible isomers. For the three-disulfide bond intermediates, only a weak fraction of all possible isomers (420) have been found. The first and the second disulfide bridges are formed readily with half-times of a few milliseconds for the intramolecular steps while the third and fourth bridges are formed with half-time of 1 and 20 sec respectively for the fastest steps. All the trapped intermediates identified as unfolded by a wide variety of criteria (e.g., enzymatic activity, UV absorption, fluorescence and immunochemical probes). A three disulfide intermediate with a folded conformation was identified in the refolding of ribonuclease; the disulfide between Cys 40 and Cys 95 was missing (Creighton, 1980). During refolding, the stable conformation appears when the three or four disulfide bonds capable of stabilizing nativelike conformation are formed (Galat *et al.*, 1981).

Similar conclusions were reached by Konishi and Scheraga (1980a,b). In their experiments, regeneration of reduced RNase was achieved by regeneration mixtures of 0.5 m*M* reduced and oxidized glutathione at pH 7.7–7.85 and 22°–24°C. The reshuffling of incorrectly paired disulfide bridges was stopped by lowering the pH (but not by a chemical modification), and samples containing intermediates and regenerated native species were analyzed by measuring enzymatic activity, optical density, and optical rotation at various temperatures and by nuclear magnetic resonance. All methods indicated that dominant conformations of the intermediates were disordered.

From these experiments, the refolding of BPTI appears to follow a more restricted pathway than that of RNase. However, Creighton (1979b) concluded that neither BPTI nor RNase fold by a simple sequence of stepwise formation of the native disulfide bonds. Rearrangements must occur during the refolding; the final step, where all the correct stabilizing interactions take place, is the slowest step in refolding. Even in the case of RNase, Creighton (1979b) concluded that the protein does not acquire parts of its native structure during the intermediate steps, but in the last one. However, different conclusions were reached by other researchers, particularly Burgess and Scheraga (1975a) and Chavez and Scheraga (1977), who proposed that refolding of RNase occurs by a simple sequential formation of elements of final conformation. Their conclusions resulted from data obtained by various methods such as accessibility of peptide bonds to proteolytic enzymes, nonspecific surface-labeling reagents, and immunochemical methods. Results obtained by Raman spectroscopy (Chen and Lord, 1976) and those obtained by ^{13}C-NMR (Howarth, 1979) also supported a sequential pathway for the refolding of RNase. Nevertheless, conformational analysis indicated a rather disordered structure of the intermediates (Konishi and Scheraga, 1980a,b). Several researchers (see Section 8.1), on the basis of optical signals reported that the secondary structure reappears before the return of enzymatic activity (Anfinsen *et al.*, 1961; Schaffer *et al.*, 1975; Takahashi *et al.*, 1977).

One of the main conclusions, drawn by Creighton (1977d, 1979b) states that the simultaneous acquisition of specific stabilizing interactions is required for the folding of the polypeptide chain. He believed that, in this respect, a common mechanism for the kinetic process of protein unfolding and folding accounts for BPTI and RNase. This problem is discussed in Part III. At this stage, one should note the discrepancies in the conclusions reached by different investigators from different criteria for RNase. It should be emphasized that BPTI is a particularly small protein with 58 amino acid residues, and it might behave as a domain or more generally as a microstructure. Furthermore, this protein is one with a great number of S—S bridges per amino acid. In RNase the situation may be different.

Different attempts to determine the folding pathway of reduced lysozyme by blocking thiol groups have been reported. Anderson and Wetlaufer (1975, 1976) stopped regeneration by lowering the pH and then blocked thiol groups by alkylation with *N*-ethylmaleimide (NEM). From their results, they proposed a direct pathway of disulfide bond formation. However, their experimental procedure was criticized by Creighton (1978c) who suspected that the trapped disulfide bonds were not those initially present, since the treatment used addition of 4 *M* urea before adding low concentration of NEM. Acharya and Taniuchi (1976, 1977, 1978, 1980) trapped the intermediate species by alkylation of free thiol groups with [1-^{14}C]iodoacetate. They did not observe any evidence for obligatory intermediates under their experimental conditions (i.e., air reoxidation for 36 hr in the presence of β-mercaptoethanol). In lysozyme, as in RNase, disulfide bonds are formed before the return of enzymatic activity; a very weak proportion of molecules (0.4% in RNase) are active before re-formation of disulfide bridges (Garel, 1978). For chymotrypsinogen, DTT$<\begin{matrix}S\\|\\S\end{matrix}$ cannot generate native protein (Orsini *et al.* 1975); it seems that in this case as for RNase, there is no preferential pathway of folding.

The best documented report of folding–unfolding of BPTI demonstrates the potentiality of using disulfide bonds as conformational probes. Nevertheless, this method requires many controls and the use of different blocking reagents to compare the results. The rapidity of quenching is a problem of great importance. This approach is experimentally very accurate and very elegant. However, conclusions reached by Creighton (1975a, 1977b, 1978c) were criticized (Scheraga, 1980). Indeed, the analysis of these average intermediate species provide no information about the pathway of refolding, but only on the way of disulfide formation. They may not be directly on the folding pathway, but result from formation of incorrectly paired species at dead ends, which have to go backward before they can refold to the native form. Thus, the trapped intermediates may be on bypasses and not on the direct pathway of folding. As is the case of other methods, the validity will be demonstrated when the same conclusions are reached by at least one completely different technique. In the case of RNase, some discrepancies must be noted and are discussed later. However, in all cases the occurrence of intermediates of protein folding has been demonstrated.

8.4. DETECTION OF INTERMEDIATES IN PROTEIN FOLDING BY DIFFERENTIAL LABELING

The previous method has several limitations. First, it detects only the intermediates in which disulfide bonds have been formed. Second, it applies only to proteins with disulfide linkages. Other chemical methods may be used as local probes of different parts of the polypeptide chain. There are other methods that use chemical labeling of amino acid side chains to follow the conformational variations in proteins. However, in the study of protein folding, an important prerequisite is that the label does not interfere in any way with the folding process. That prevents, except in favorable cases, the use of reporter groups such as fluorescent or paramagnetic probes. This point discussed in Section 8.6. Another important requirement is the rapidity of labeling with regards to the folding process. These conditions are fulfilled in the method of differential labeling for determining the variations in reactivity of individual groups in proteins developed by Ghélis (1980) which is adapted from competitive labeling introduced by Hartley (1970) and Kaplan *et al.* (1971).

8.4.1. Method of Differential Labeling

The reactivity of amino acid side chains in proteins is very sensitive to the nature of their environment. Thus, this can be used as an intrinsic probe of the conformational state of a protein. In fact, the apparent reactivity of individual amino acid side chains in protein reflects different effects:

(1) The nucleophilicity of the amino acid side chain is directly related to its pK of ionization.

(2) The shielding effect occurs because certain amino acid side chains are not accessible to reagents or to solvent in the native structure. They become accessible on denaturation of the protein. Some of them might be more or less shielded from the solvent depending on their location in the protein and also on the breathing effect as for ^{1}H–^{2}H exchange. Evidence for this breathing effect is demonstrated by the existence of a residual chemical

accessibility of apparently fully shielded groups. An example is the possibility of iodinating the buried tyrosine groups of RNase (Cha and Scheraga, 1963).

(3) The effect of the microenvironment (polarity, local interactions) can modify the reactivity of a group even in a denatured polypeptide chain compared to that of a model compound.

The principle of the method is the following: in the presence of a trace of radioactive label, the different reactive groups in a protein will compete for the label according to their reactivity which depends on different factors (i.e., nucleophilicity, accessibility, and microenvironment). Both accessibility and microenvironment of a great number of groups vary according to the folded or unfolded state of the protein. Thus, the relative amount of label incorporated into the different groups at different times allows one to evaluate the kinetics of refolding in the different parts of the polypeptide chain.

The reactive function of an individual amino acid A reacts with an electrophile reagent RX with a rate constant k_A such that

$$\mathrm{A + RX \xrightarrow{k_A} RA + X}$$

the following other reactions may take place in water:

$$\mathrm{H_2O + RX \longrightarrow R{-}OH + HX}$$

$$\mathrm{OH^- + RX \longrightarrow R{-}OH + X^-}$$

The reaction, made in the presence of a trace of labeled reagent, is completed by unlabeled reagent. The intrinsic reactivity of each individual amino acid side chain is evaluated in the unfolded chain and thus considered as the reference. The fraction of accessible groups, α, can be estimated by the following ratio:

$$\alpha = \frac{\text{specific radioactivity of RA}}{\text{specific radioactivity of RA}_\mathrm{U}} \tag{8.1}$$

RA and $\mathrm{RA_U}$ refer to the labeled groups A in the refolding and totally unfolded states respectively; α, the fraction of accessible groups varies from 1 in the unfolded state to a minimum value which can reach zero if the group is totally shielded and unreactive.

Experimentally, the protein is treated with a very limiting quantity of a suitable radioactive electrophile reagent. The protein is then treated under denaturing conditions (8 *M* urea) by an excess of unlabeled reagent to obtain a complete reaction and therefore a chemically homogeneous material although it is heterogeneously labeled. Times of labeling were varied between 100 and 800 msec and the reaction was stopped by isotopic dilution on a multimixing apparatus. The fraction of any particular group which is thus labeled is very small with regard to the total amount of unsubstituted groups, so that the reaction of an individual side chain cannot be affected by substitution of neighboring groups. It is even possible to perform extremely low proportions of labeling and to obtain accurately measurable values by using reagent with a high specific radioactivity.

This method can be used for mapping native proteins. Its rapidity also allows one to follow kinetics of protein folding and unfolding. It was used by Ghélis (1971) and Ghélis and co-workers (1975) to detect local conformational differences resulting from activation of chymotrypsinogen into chymotrypsin and from the alkaline conformational transition of the enzyme.

8.4.2. Identification of Labeled Residues

After reaction with the labeled reagents, the protein is treated with proteolytic enzymes; thermolysin, chymotrypsin, and trypsin have been used for proteolysis of the labeled macromolecule. The resulting peptides are separated by high-pressure liquid chromatography (HPLC), and the amount of radioactivity of each eluted peak is measured. Then the labeled peptides are analyzed after acid hydrolysis and the amount of radioactive label incorporated into individual amino acid side chains is measured.

8.4.3. Study of Protein Topography by Differential Labeling

The method of competitive labeling was used by Kaplan and co-workers (1971) to determine the p*K* of several ionizing groups in elastase. It was also utilized by Bosshard and co-workers (1978) to characterize the amino groups implicated in a tRNA synthetase–tRNA complex formation.

In elastase, several amino acid side chains were used as conformational probes by Ghélis (1980). Labeling by [^{14}C]acetic anhydride, [^{14}C]dimethyl sulfate, [^{14}C]iodoacetate, and [^{14}C]ethoxyformic anhydride was performed, allowing the measurement of reactivity of different amino acid side chains (N-terminal Val; ε-NH_2 of lysine groups, methionine, tyrosine, histidine) located in various positions in the polypeptide chain of elastase. Thus Ghélis (1980) was able to build a conformational map of elastase on the basis of the differential reactivity of these groups. Figure 8.9 shows the data obtained by this approach. The accessibility of side chains to chemical reagents follows a regular curve in which two buried areas are well delineated; strikingly, they correspond to the structural domains of the protein, as observed in the three-dimensional structure (Sawyer *et al.*, 1978).

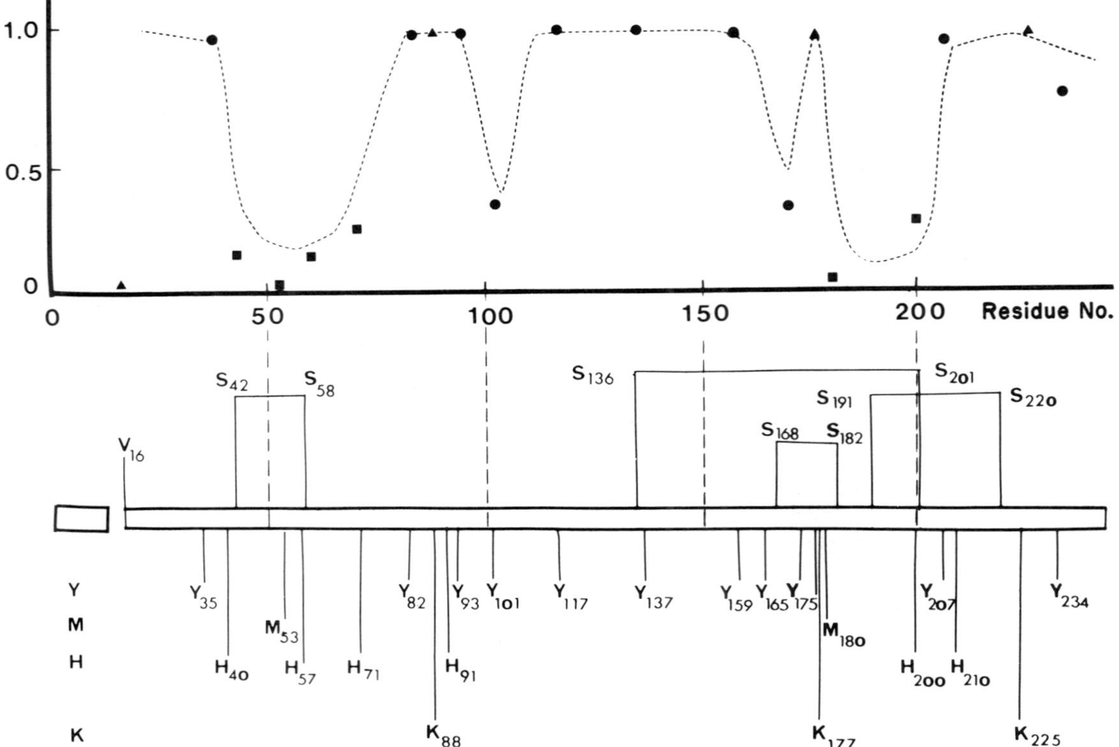

Fig. 8.9. Chemical reactivity of individual amino acid side chains in native elastase. The reagents used were [^{14}C]acetic anhydride (▲); [^{14}C]dimethyl sulfate (●); and [^{14}C]iodoacetate (■). The reference value for a totally accessible group was the chemical reactivity determined in the denatured and reduced form of the protein. The lower part of the figure indicates all the modified side chains along the sequence (from Ghelis, 1980).

8.4.4. Detection of Intermediates during Protein Folding or Unfolding by Differential Labeling

To follow the kinetics of protein folding or unfolding it is necessary to choose, as conformational probes, several groups accessible to chemical reagents in the denatured form and buried in the native form. The method was used to follow the refolding of reduced elastase denatured by 6 *M* GuHCl or by 8 *M* urea. The reactivity of Met 180, Val 16, and Tyr 234 toward [^{14}C]iodoacetamide, [^{14}C]acetic anhydride, and [^{14}C]dimethyl sulfate

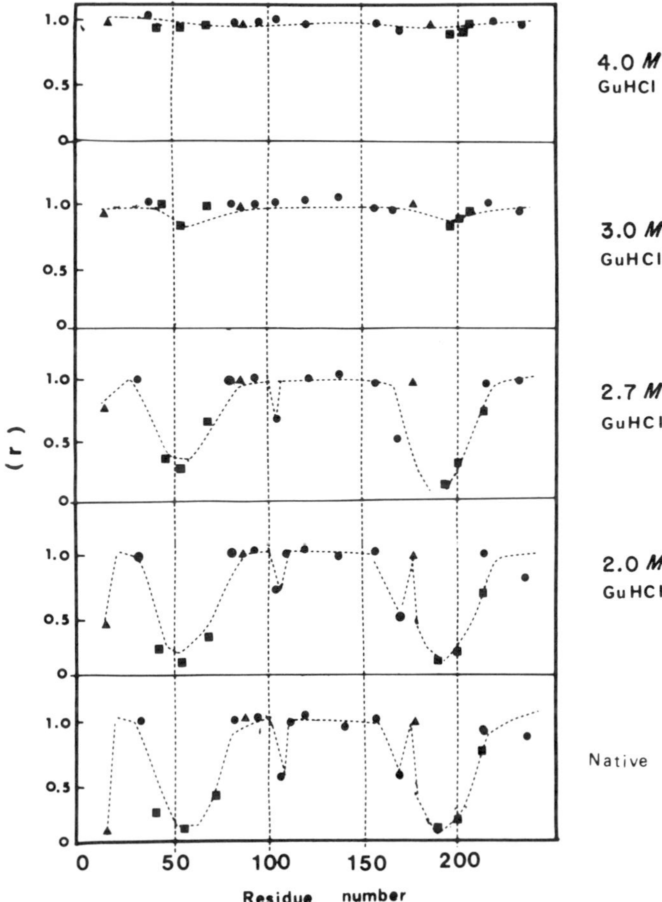

Fig. 8.10. Titration of the chemical reactivity of elastase amino acid side chains during this refolding in varying denaturant concentrations. The contration of denatured elastase was 4 μM, in 10 mM acetate buffer pH 5.5. Labeling time was 400 msec. The symbols are those of Fig. 8.9 (from Ghélis, 1980).

were respectively estimated at different times of refolding. The observed rates of decrease in the chemical reactivity were compared with the refolding rates observed by different other methods such as optical signals, return of enzymatic and antigenic activity. The data obtained showed comparable rates of refolding when observed by the decrease of chemical reactivity of these different side chains and by optical signals (circular dichroïsm). The return of both enzymatic and antigenic activities takes place at a slower rate. These results are shown in Fig. 5.20. These data indicate folded but inactive forms, in which the side chain probes are buried and have the circular dichroïc signal of the native protein, appear before the native and functional form.

This procedure has been also used to analyze the conformational states of elastase in intermediate concentrations of denaturant. The data are presented in Fig. 8.10. They indicate that in distinct regions of the elastase molecule, structural elements are stabilized even in 3 *M* GuHCl, a denaturant concentration above the transition zone. These structural elements might correspond to nucleation centers; one in each domain was observed.

The measurement of chemical reactivity of amino acid side chains is a potent tool in the study of protein folding; it allows one to follow the folding process in different parts of a polypeptide chain. The sensitivity of the method allows one to work with amount of protein ranging from 1 to 10 nmoles for complete analysis. It does not require the protein to be pure if the corresponding peptides can be purified. Therefore, it could be used under heterogeneous environments such as in ribosome or in cells.

8.5. DETECTION OF INTERMEDIATES BY ACCESSIBILITY OF PEPTIDE BONDS TO PROTEOLYTIC ENZYMES

The accessibility of the different peptide bonds to proteolytic enzymes was used in the study of the different stages of protein folding and unfolding. It is not significantly different from the previously discussed method in the sense that it is a measurement of accessibility. The chemical reagent is replaced by specific proteases and definite peptide bonds are cleaved when they are accessible and correspond to the specificity of the enzyme. An analysis of the peptides and of the N-terminal groups, that appear by proteolysis is required in order to determine the position of the cleavage. This method, inspired by the discussion of Linderstrøm-Lang (1952), was relatively little employed. Mainly thermal unfolding of RNase was investigated by this method. Chymotrypsic hydrolysis (Rupley and Scheraga, 1963) as well as tryptic hydrolysis (Ooi *et al.*, 1963; Ooi and Scheraga, 1964) and also carboxypeptidase (Klee, 1967; Burgess *et al.*, 1975) were used. From the obtained data and also from results obtained by other methods, Burgess and Scheraga (1975a) were able to reconstitute an hypothetical pathway for the thermal unfolding of unreduced RNase. In these experiments, either chymotrypsin or trypsin were incubated with RNase at pH 6.5 and temperatures ranging from 25°C to 60°C. At this final temperature, successive additions of chymotrypsin or trypsin were needed because of the inactivation of the protease (Ooi and Scheraga, 1964; Ooi *et al.*, 1963; Rupley and Scheraga, 1963). The use of immobilized carboxypeptidase (Burgess *et al.*, 1975) avoided the denaturation of the proteolytic probe. Peptides obtained by proteolysis were isolated by chromatography and the positions of the splits were determined by chemical analysis.

From all the data, the following tentative pathway was proposed by Burgess and Scheraga (1975a) for the thermal unfolding of RNase. In a first stage (i.e., between 15°C and 35°C), Tyr 92, which is buried in the native protein, normalizes according to the spectral probes. However, the backbone conformation in this region is not altered and the Lys 91–Tyr 92 remains inaccessible to trypsin even at 60°C (Ooi *et al.*, 1963). Between 35°C and 45°C, in a second stage, the region from Thr 17 to Tyr 25 unfolds and Tyr 25–Cys 26 becomes accessible to chymotrypsin (Rupley and Scheraga, 1963). The peptide bond between Ala 20–Ser 21 is split by subtilisin even under nondenaturing conditions (Richards and Vithayathil, 1959). The helical region between residues 4 and 12 does not unfold at this stage; Lys 7–Phe 8 bond is not split by trypsin; Phe 8–Glu 9 is not cleaved by chymotrypsin; His 12 and Ser 16 are stabilized by hydrogen bond to Val 47 and His 48 respectively and by hydrophobic interactions with Met 13 and Val 47. In the following stage (stage III), at temperatures between 40°C and 50°C, two regions of the backbone unfold, those from Asn 27 to Asn 34 and from Ser 75 to Ser 80. Trypsin hydrolyzes Lys 31–Ser 32 bond and Arg 33–Asn 34 bond (Ooi *et al.*, 1963), these last residues being at the C-terminal end of the helix segment (25–33). The N-terminus of this helix unwinds in stage II (Tyr 25). The whole helical segment probably unfolds at stage III. At this stage also (45°C, pH 6.5) the Met 79–Ser 80 and Tyr 76–Ser 77 (43°C, pH 6.8) peptide bonds are hydrolyzed by chymotrypsin. Under the same conditions, Tyr 73–Gly 74 bond remains unaccessible to the enzyme. In stage IV, the C-terminal loop from Lys 104 to Val 124 unfolds between 50°C and 60°C. This loop is stabilized in the native structure by interaction with the α helix 51–60 which is linked to the C-terminal end of the chain by a disulfide bridge (Cys 58–Cys 110). The unwinding of the helix is proposed from the optical rotation study. The unwinding of the C-terminal end is demonstrated by the accessibility of Val 124 to immobilized carboxypeptidase (Burgess *et al.*, 1975) and the susceptibility of peptide bond 120–121 to pepsin even at 36°C (Anfinsen, 1956). Val-124 is stabilized by its interaction with the α-carbonyl

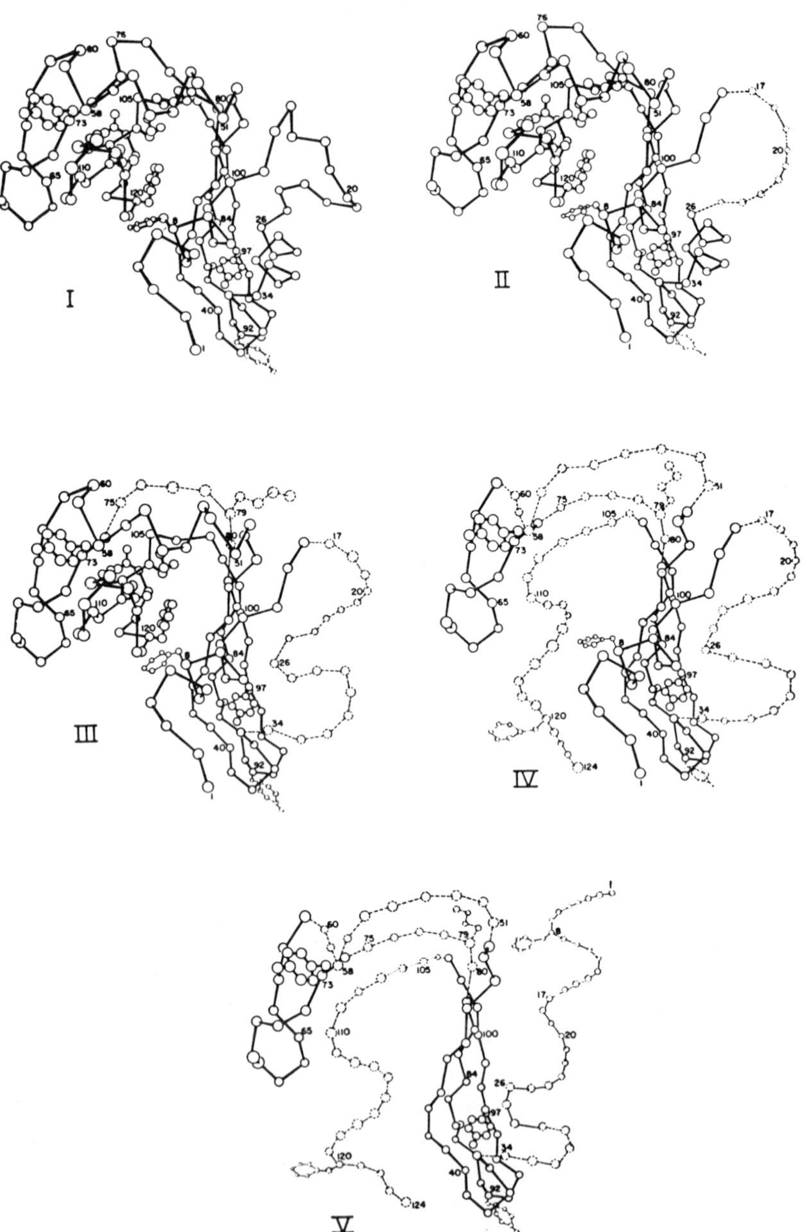

Fig. 8.11. Pathway of thermal unfolding of unreduced RNase (according to Burgess and Scheraga, 1975a) (see text).

group of Lys 104 in the native structure. Therefore, all this part of the chain probably unfolds at this stage, whereas the N-terminal α helix remains folded. In stage V (55°–65°C), the proteolytic liberation of the N-terminal residue (Lys 1) by aminopeptidase (Klee, 1967) at pH 8 and 58°C indicates the unwinding of the N-terminal region. Three loops seem to retain their structure even at high temperature, 62–74, 81–87, and 96–102. Indeed neither of the bonds close to Lys 66 or Trp 73 are split by trypsin or chymotrypsin even at 60°C. Similarly Lys 91–Tyr 92 is not attacked by trypsin. In fact 81–87 and 96–102 are forming an antiparallel β pleated sheet. These regions unfold in a last stage. Figure 8.11 shows the proposed pathway of the thermal unfolding of RNase, the pathway of folding is assumed to pass through the same intermediates. From these data a sequential pathway of folding is proposed.

The use of proteases as conformational probes has allowed the detection of the native–denatured transition in lysozyme (Imoto *et al.*, 1974, 1976). Lysozyme was digested by pronase at 40°C, pH 8.0. Chromatographic analysis of the digest showed the existence of unmodified proteins and the production of only small peptides, no intermediate-sized products were found. For a chemically modified form of lysozyme which has an oxindolyl ester of Glu 35 (Glu 35–ester lysozyme) no digestion was observed. The proteolytic probe was useful for detecting the transition on the unmodified enzyme, also observed by hydrogen exchange, but difficult to detect by physicochemical methods under mild conditions. In this case, the proteolytic probe was used to detect an equilibrium difficultly observable by classical methods, but it was not used in the research of intermediates during an unfolding process. In this case, the proteolytic probe was an alternative method to hydrogen exchange.

8.6. DETECTION OF INTERMEDIATES BY USE OF NONSPECIFIC SURFACE-LABELING REAGENTS

Matheson and co-workers (1977a) in Scheraga's laboratory developed a method similar in its basic principle to the differential labeling. This method uses nonspecific labeling of all exposed residues in a protein, the reagent being generated by flash photolysis, therefore in a very short time. From reagent *N*-(4 azido-s-nitrophenyl)2-amino ethylsulfonate, flash photolysis generates within 500 μsec, the aryl nitrene *N*-(4-nitreno-2-nitrophenyl)2-aminoethyl sulfonate according to Staros and Richards (1974):

$$N_3\text{—}C_6H_3(NO_2)\text{—}NH\text{—}CH_2\text{—}CH_2\text{—}SO_3^- \xrightarrow{h\nu} N\text{—}C_6H_3(NO_2)\text{—}NH\text{—}CH_2\text{—}CH_2\text{—}SO_3^- + N_2$$

The reactive nitrene is covalently inserted into exposed carbon–hydrogen bonds of the protein in about 2 msec. This method was applied to RNase A. From the amino acid analysis of labeled protein, only the exposed residues were shown to react with the nitrene; those buried at the interior of the molecule (according to the X-ray data) do not react.

This method was thus applied to the study of the thermal unfolding of the protein at pH 5.0. The number of modified (or unmodified) residues was estimated during the unfolding by amino acid analysis. At very high temperature, the precision of the data was evaluated to be about $\pm 25\%$; the particular contribution of each type of amino acid was determined, except for His and for Cys for which the method does not seem reliable. The temperature profile of each types of amino acid and its dependence on ionic strength was analyzed (Matheson and Scheraga 1979a,b). However, the labeled amino acids were not identified in the polypeptide chain, the attempts of the authors to obtain and isolate fragments of labeled RNase were unsuccessful.

It is difficult to understand the real meaning of the data averaged over all residues of a particular type in terms of protein unfolding; it is unlikely for them to be able to reveal unfolding of particular region of the protein since none of them contains particular clusters of one type of amino acid. Even nonpolar groups are not all located inside the proteins. The data were interpreted with reference to research described previously in Section 8.5. They were considered consistent with the previous interpretations by Burgess and Scheraga (1975a) with some slight modifications proposed for the folding pathways (1) in stage I (15°–35°C) Tyr 92 unfolds; (2) in stage II (30°–45°C) segment 13–25 unfolds and the degree of exposure to solvent of residues 1–12 changes; (3) stage III (40°–50°C) is marked by unfolding of segments 27–34 and 75–80; (4) in stage IV (50°–60°C), there are conformational changes in segments 51–60 and 104–124; (5) in stage V, segment 1–12 unfolds; and (6) in stage VI (60°–70°C) segments 25–50, 62–74, and 81–102 unfold. This pathway is in good agreement with those of Chavez and Scheraga (1977) who used an immunochemical approach (see Chapter 9).

Matheson and Scheraga (1979a,b) considered that the structure of the outer shell, near the surface of RNase, begins to unfold and then, when the temperature reaches 60°C, the bulky apolar residues become accessible. In the final unfolded protein, some residual and local compact structures persist. These clusters of nonpolar residues are considered, in the folding process, to be possible nucleation sites. Bayley and Knowles (1978) discussed the method employed by Matheson and Scheraga (1979a) and emphasized possible artifacts: during photochemical labeling, not only the nitrene insertion, but also UV damage to the mechanism occurs, because of the formation of a large number of free radicals which can attack the protein. The method of analysis by difference with unlabeled RNase used by Matheson and Scheraga (1979a) does not allow one to distinguish between residues modified by the nitrene and residues modified by free-radical reactions. Even in the estimation of differences in the exposure of residues to solvent after unfolding, this represents a source of possible artifacts. The low degree of accuracy obtained at high temperatures also makes the method questionable.

The three methods described in Sections 8.4, 8.5 and 8.6 are based on the estimation of accessibility of side chains or peptide bonds to reagents or proteolytic enzymes. The hydrogen exchange is also a method which allows one to determine the accessibility of hydrogen atoms. Besides this last method, only differential labeling and proteolytic probes were really quantitated with a satisfactory accuracy.

8.7. USE OF EXTRINSIC PROBES ATTACHED TO SIDE CHAINS FOR DETECTION OF INTERMEDIATES IN PROTEIN FOLDING

In the study of protein folding, extrinsic probes such as paramagnetic or fluorescent probes, covalently attached to definite amino acid side chains, have been used in several cases. Spectroscopic or paramagnetic probes must be attached in the denatured protein to amino acid side chains which are buried in the native structure. However, the validity of the method requires that the modified residue not be essential for protein structure and the attachement of the probe must not perturbate the refolding process. This last condition is often difficult to fulfil, since the thermodynamic properties of the molecule will be at least slightly modified by the presence of the probe. In the case of human carbonic anhydrases (Carlsson *et al.*, 1975), the attachment of the spin-labeled reagent (*N*-1-oxyl-2,2,5,5-tetramethyl-3-pyrrolidinyl) iodoacetamide and of a fluorescent probe (7-chloro-4-nitro benzofurazan) on the SH group of Cys 204 in the C enzyme and of Cys 212 in the B enzyme, does not significantly impair the refolding process of the molecules, but it introduces a small conformational change and a reduction of the esterasic activity of the C enzyme. Even in this case, which seems favorable, when the probe is attached, the modified polypeptide chain which refolds does not have exactly the same properties as the initial chain. However, in this case, intermediates were not detected.

Matheson and co-workers (1977b) used two different spin labels to modify Lys 1 and His 105 in RNase. They used the paramagnetic probe to study the thermal unfolding of the protein over a temperature range from 3°C to 65°C at pH 5.5, conditions identical to those of the previous studies of thermal unfolding of RNase. The electron paramagnetic resonance (EPR) spectra were recorded at different temperatures. The data obtained were interpreted by a sequential unfolding of the protein. In the region around His 105 a local increase in the motional freedom of the spin label at 40°C was noted; at 55°C there was a complete unfolding of this region. The N-terminal α helix tail is not affected until 55°–65°C. The investigators concluded that it was a sequential mechanism consistent with the data obtained with other methods. In this research, the probes were attached to the native protein and used to follow the unfolding process. It was not confirmed by another method whether the unfolding process is exactly the same when the two spin-label molecules are attached to the protein. Whatever is the potentiality of EPR spectroscopy for detecting intermediates, the method requires many controls. Each modification, even very small, introduced in the protein can modify its properties and the unfolding–folding process.

A small modification, such as nitration of tyrosine in RNase (Garel and Baldwin, 1975b) was successful. Neither the enzymatic activity, nor the kinetics of refolding were modified (see Chapter 7, Section 7.2.2).

8.8. DETECTION OF INTERMEDIATES IN PROTEIN FOLDING BY OTHER METHODS

By gel electrophoresis, Hawley and Mc Leod (1976) were able to separate three different species associated with the thermal transition of chymotrypsinogen A. Besides the native and unfolded protein, a species X appeared in amount varying according to temperature, from 5% to 20% of the total protein. Although the data were still qualitative, no conclusion as to the nature of the X species, whether it is an intermediate in protein unfolding, was possible.

Creighton (1979a, 1980) and Creighton and Pain (1980) used electrophoresis on polyacrylamide gels with a gradient of urea concentration, the gradient being perpendicular to the direction of the electric field. This procedure gave a two-dimensional pattern of the action of urea on the shape of the protein and allowed to detect conformational microheterogeneity of the protein. It was used with various proteins. With BPTI, RNase, staphylococcal nuclease, lysozyme, chymotrypsinogen, and cytochrome c, the results are consistent with the presence of only the native and the fully denatured states. With serum albumin, heterogeneity was observed. Figure 8.12 shows the profiles obtained for serum albumin and RNase, which represents very favorable cases.

An exhaustive review of all approaches allowing detection of intermediates in protein folding is not given here. Detection of intermediates during protein

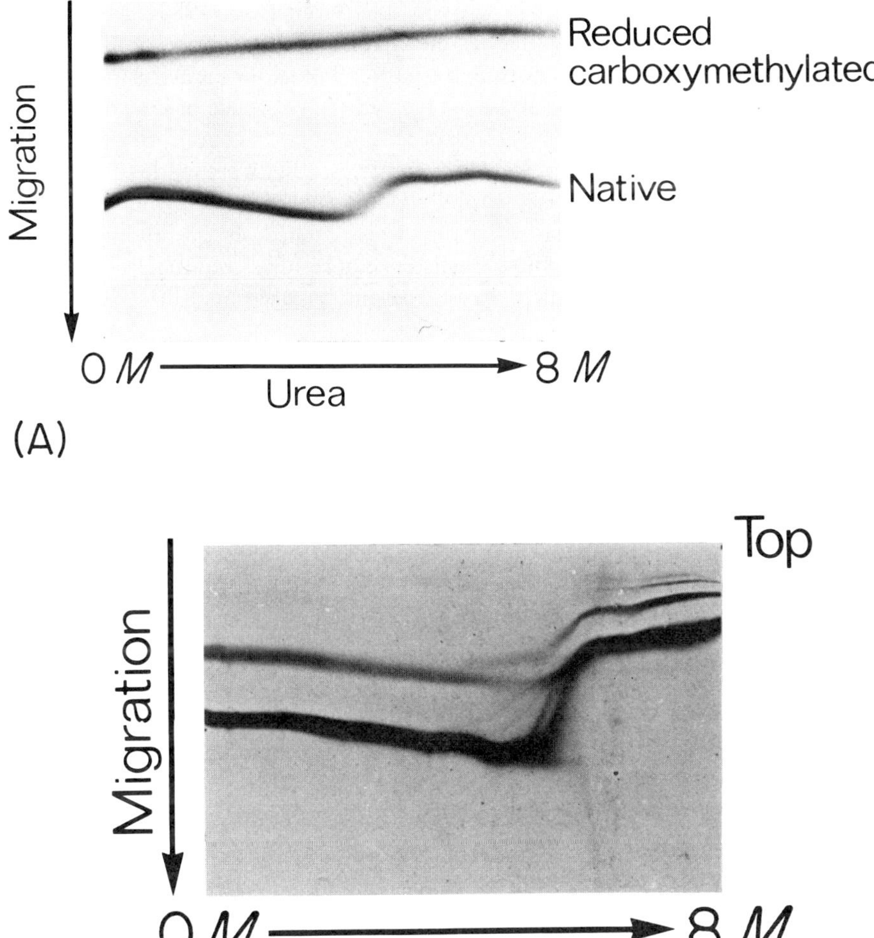

Fig. 8.12. Urea gradient electrophoresis of (A) native and reduced carboxymethylated bovine RNase A (0.05 *M* Tris-acetate buffer, pH 4.0) (B) bovine serum albumin (Tris-borate EDTA buffer, pH 8.6). The linear gradient of 0–8 *M* urea was superimposed on the inverse linear gradient of 15–11% acrylamide (courtesy of Creighton, 1979a).

folding has been reviewed by Baldwin and Creighton (1980). The most significant studies were presented in this chapter with exception of immunochemical studies which are reported in Chapter 9. Various methods were used in very different approaches, all have revealed the existence of intermediates in either the unfolding or the refolding process, even when equilibrium studies indicate only two states. Most well documented studies were made with small proteins such as BPTI, RNase, staphylococcal nuclease, lysozyme, and also serine proteases. Convergent conclusions were reached on several points: the occurrence of intermediates and the existence of residual and local structures in thermal denaturation which may represent possible nucleation sites for the folding process.

Great divergences also exist: sequential pathways of folding, parts of the folded structure being formed at each stage were proposed particularly for RNase (Burgess and Scheraga 1975a). However, for the same protein, a different interpretation was reached by Creighton using disulfide bonds as a conformational probe. He concluded that for RNase, as for BPTI folded conformation is not reached until all the stabilizing interactions take place, emphasizing the importance of long-range interactions and the cooperativity of the folding process. This point which poses the question of the minimal stable structural unit is discussed in Chapter 10 and in Part III. It represents an important aspect of the understanding of the mechanism involved in protein folding.

9

Immunochemical Approaches to Protein Folding

9.1. ANTIBODY PROBE AS A TOOL TO STUDY PROTEIN STRUCTURE

9.1.1. Antigenic Properties of Proteins

The use of antibodies as a probe of protein conformation was originated by Sela and co-workers (1967). The characteristics of protein antigenicity was reviewed by Gally (1973), Crumpton (1974), Reichlin (1975), and by Kabat (1978). When injected into a vertebrate (most frequently a goat or a rabbit) a foreign protein induces the biosynthesis of antibodies directed against different parts of the molecule localized at the surface, in contact with solvent. Proteins are immunogenic and antigenic. Immunogenicity refers to the induction of antibody biosynthesis, whereas antigenicity refers to the interaction with antibodies. The relationship between protein conformation and antigenicity was known from the beginning of immunochemistry. It has been shown that antigenicity of proteins decreases as a result of denaturation or even after reduction of disulfide bridges. Antigenic reactivity is very sensitive to alterations, even very small, in protein conformation. Many examples of the sensitivity of antigenic reaction to protein conformation have been given (see Crumpton, 1974). A striking example is the discrimination between oxyhemoglobin and deoxyhemoglobin by methemoglobin antibodies (Reichlin *et al.*, 1964).

The antigenicity of a protein depends on its structure and size. It is also related to the degree of phylogenetic difference between the origin of the immunogen and the animal species used for immunization.

Many immunological studies have revealed that protein antigens are multivalent and that several antigenic determinants are distributed over the surface of the molecule (Atassi, 1975). The definition of antigenic determinant remains in many cases an operational one and refers to the number of antibody molecules an antigen molecule can bind (Sachs *et al.*, 1972b). Sela and co-workers (1967) distinguished between sequential and conformational antigenic determinants of a protein molecule. The former are defined as determinants "due to an aminoacid sequence in a random coil" conformation, whereas the later "result from the steric conformation of the antigenic macromolecule." Antibodies specific to conformational determinants generally react poorly with unfolded proteins; conversely, antibodies specific to sequential determinants react poorly with native proteins. Thus, it is possible to use both types of antibodies to study protein folding. However, random-coil polypeptide chains are generally much less antigenic and also immunogenic than native proteins. Antidenatured-protein antibodies are certainly more adequate for determining protein homologies. The number and characteristics of antigenic sites were determined with antibodies directed against native structure of proteins.

9.1.2. Characterization of Antigenic Sites

For a globular protein, there is a relation between the molecular weight and the number of antigenic determinants. There is a limited number of antigenic determinants accessible to specific antibodies. This number is defined by the maximum number of antibody molecules which can bind simultaneously and cover the antigen surface. Benjamin and Teale (1978) assumed that the number of determinants is directly proportional to molecular weight. From an evaluation on different proteins, an empirical relationship which is slightly different, was found by Louvard and co-workers (1975, 1976). A satisfactory correlation was obtained by plotting the logarithm of molecular weight versus the logarithm of the number of antigenic determinants. Figure 9.1 shows the linearity obtained for several proteins. The portion of antigen surface covered by the binding of an antibody or F_{ab} fragment was estimated to be ca. 2000 Å^2 from X-ray analysis. The number of antigenic determinants found from different proteins is consistent with this data.

From more detailed studies using several proteins, antigenic determinants appear to generally consist of structural elements involving 6–7 amino acid side chains (Atassi, 1975; Atassi and Stavitsky, 1977). Antigenic determinants

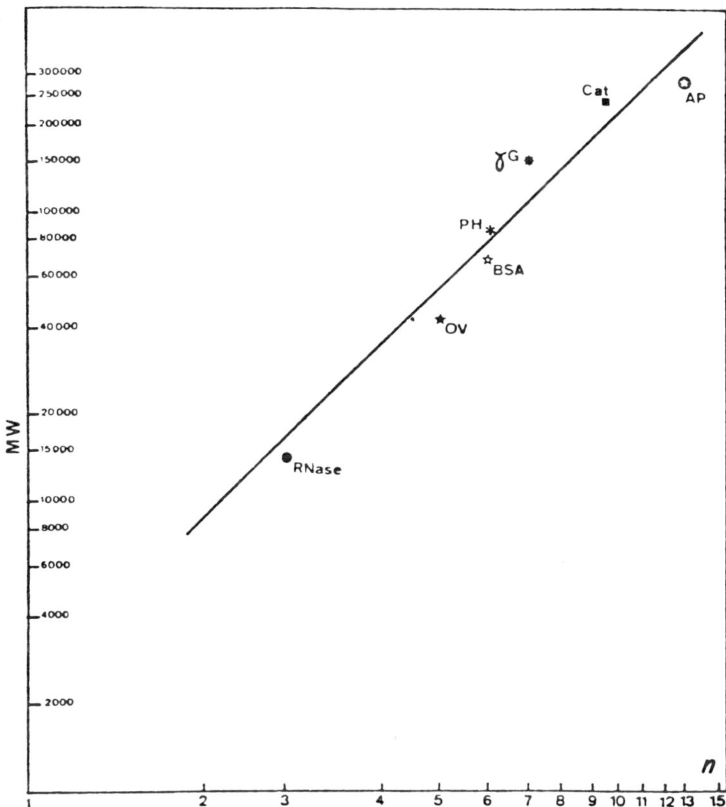

Fig. 9.1. Linear relationship between molecular weight logarithm of globular proteins and log n (number of antigenic determinants) (from Louvard *et al.*, 1976). Abbreviations are as follows: RNase; pancreatic ribonuclease; OV, ovalbumin; BSA, bovine serum albumin; γG, human IgG immunoglobulin; PH, *E. coli* alkalin phosphatase; Cat, liver catalase; AP, porcine intestinal aminopeptidase. The following equations apply: $s = k\ M^{2/3}$, $s = k_2 n$, $\log M = \frac{2}{3} \log nc$ where s is the surface; M, the molecular weight; n, the number of antibody molecules bound to antigen.

of proteins were examined in great detail only for few proteins. Accurate delineation of antigenic sites was made through chemical modifications of amino acid side chains by the study of different fragments, including synthetic peptides. Precise and entire antigenic structure of hen egg white lysozyme was reported by Atassi and Lee (1978a,b). All the residues forming each of the three antigenic sites were determined. They are represented in Fig. 9.2. Refinement in the topology of antigenic sites was obtained by Atassi and co-workers (1976a,b) using the method referred to as surface simulation

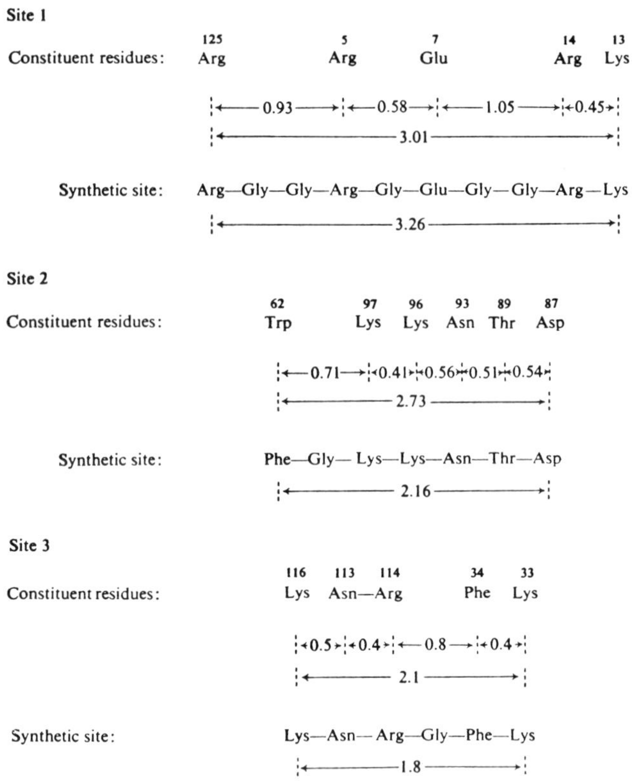

Fig. 9.2. The antigenic sites of hen egg white lysozyme and the corresponding synthetic peptides used in the method of surface simulation synthesis (from Atassi and Lee, 1978b).

synthesis. It consists of synthesizing a single peptide which contains contiguous surface residues of an antigenic site, these residues being placed at the same distance from each other as in the protein. Figure 9.2 indicates the corresponding peptides for each antigenic site.

9.1.3. Study of Protein Folding by Antibody Probe

Since antibody probes are extremely sensitive to variations, even subtle ones, in protein conformation, they are potent but rather delicate tools for the study of protein folding or unfolding. For the first time, Anfinsen's group developed a quantitative immunochemical approach to study the refolding of staphylococcal nuclease (Sachs *et al.*, 1972a,b,c, 1974; Eastlake *et al.*, 1974; Furie *et al.*, 1974, 1975). From that, different methods using specific antibodies for detection and characterization of intermediates in protein folding were developed. In some methods, antibodies which are purified to monospecificity (i.e., antibodies directed against distinct regions of the protein) were used to study the refolding in local parts of the molecule. For the others, refolding was studied by the interactions of the protein with the unfractionated specific antibody population.

The interest in immunochemical approach arises from its extreme sensitivity which allows one to detect very small quantities of antigens. For this reason, it is detailed in a full chapter. Here are reported different aspects of the approach, whether or not success in finding intermediates in the folding pathway has been achieved.

9.2. IMMUNOCHEMICAL APPROACHES USING MONOSPECIFIC ANTIBODIES

The first quantitative immunochemical approach of protein folding reported by Anfinsen's group, used antibody population fractionated to monospecificity (i.e., directed against a distinct antigenic determinant of staphylococcal nuclease). Then, such a method was employed to follow the refolding of different proteins: serum albumin (Teale and Benjamin, 1976a,b, 1977; Benjamin and Teale, 1978; Chavez and Benjamin, 1978), elastase (Ghélis *et al.*, 1978, 1982), and RNase (Chavez and Scheraga, 1977, 1979, 1980a,b).

In this kind of work, monospecificity was operationally defined by the experimental demonstration that only one molecule of the antibody population can bind to one molecule of the antigen. Under these conditions, the complex formed is soluble and a radioimmunoassay or a double precipitation must be performed to quantitate the determination.

9.2.1. Preparation of Antibodies Specific for a Distinct Antigenic Determinant of a Protein

This method was developed by Sachs and co-workers (1972a) who prepared antinuclease antibodies by repeated immunization of a goat with staphylococcal nuclease. Selective fractionation of antinuclease antibodies was achieved on columns of Sepharose 4B on which either entire nuclease or a definite fragment had been linked (through lysine ε-amino groups). Figure 9.3 shows the fractionation of antinuclease antibodies on immunoabsorbent columns. First, immunoabsorption on a Sepharose–nuclease column was followed by a second immunoabsorption on a Sepharose–peptide (99–149) fragment column. Further purification on Sepharose–peptide (127–149) fragment column was necessary to obtain a monospecific antibody population. After the second column fractionation, more than one antibody

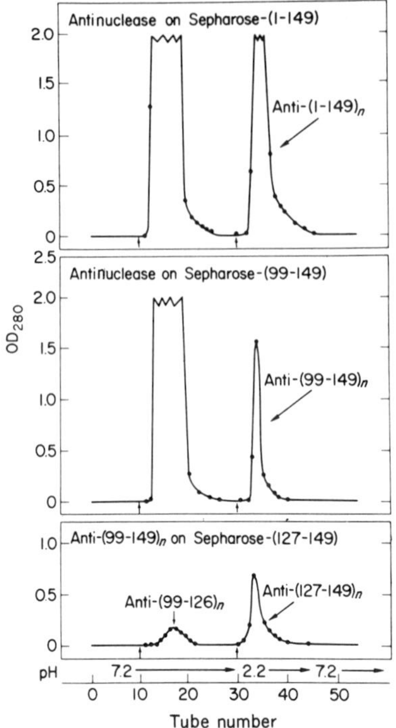

Fig. 9.3. Fractionation of antinuclease antibodies to monospecifity (from Sachs *et al.*, 1972b, courtesy of A. Schechter).

molecule could bind to the antigen. All purified antibody populations consisted of immunoelectrophoretically pure IgG. In this study, antibodies were eluted with 0.2 *M* sodium citrate at pH 2.2 after washing with 0.15 *M* NaCl. Different elution procedures have been reported. Because of the great affinity of antibody for antigen, it is often necessary to elute the column with 2 *M* or even 4 *M* GuHCl. Using the fractionated antibody population (after the third fractionation), analytical velocity centrifugation indicated the binding of at most one molecule of antibody per molecule of nuclease. In experiments conducted by Sachs and co-workers (1974), 11.5% of the antibody present in anti–$(1–149)_n$ peptide fragment were removed by fractionation on Sepharose $(99–149)_n$ peptide fragment and 74% of this population showed specificity for region (127–149), 26% for region (99–126). Thus, if a given fractionated antibody population is not monospecific, further immunoabsorption on columns bearing peptides of decreasing length might be performed until monospecificity is reached. All these antibody populations inhibited the enzymatic activity of nuclease.

Teale and Benjamin (1976a,b, 1977), Benjamin and Teale (1978), and Chavez and Benjamin (1978) were also successful in fractionating antibody

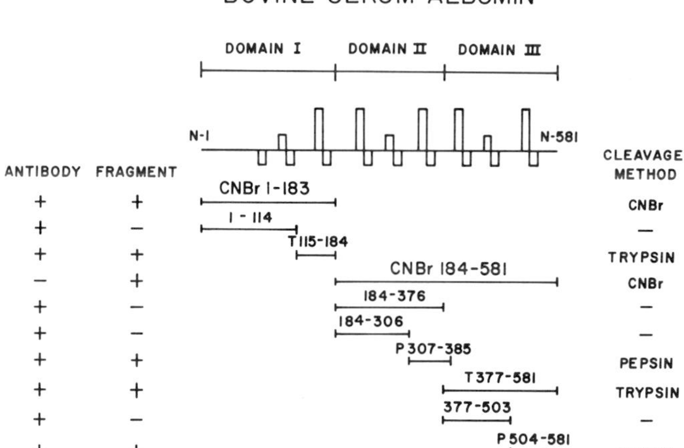

Fig. 9.4. Cleavage of serum albumin into antigenically active fragments (from Teale and Benjamin, 1978, courtesy of Benjamin).

populations specific for different parts of the serum albumin molecule. The polypeptide chain has 581 amino acids and peptide fragments were obtained either by proteolytic digestion by trypsin, or by pepsin, or cleavage by cyanogen bromide (Fig. 9.4). The molecule is organized into three structural domains (see Chapter 10). Fragments 1–183, 184–376, and 377–381 correspond respectively to domain I, II, and III. Other peptides represent subdomain regions of the molecule. Using these different fragments, for fractionation of antiserum to native serumalbumin, Teale and Benjamin (1977) and Benjamin and Teale (1978) were successful in isolating antibodies directed toward each of the three domains and antibodies directed toward several subdomains of the molecule (Fig. 9.5). Using the same procedure, Chavez and Scheraga (1977) prepared antibodies specific for different regions of RNase. Antibodies were fractionated on Sepharose coupled with fragments 1–13, 31–79, and 80–124.

Antielastase antibodies obtained by hyperimmunization of rabbits were purified on a Sepharose–elastase column after previous fractionation on DEAE–cellulose allowing separation of the inhibiting and nonprecipitating antibody population. Antibody elution was achieved using 4 *M* GuHCl, after washing with phosphate buffer saline (PBS) (Ghélis *et al.*, 1982).

Purified antibody populations directed toward distinct antigenic determinants were also obtained for various other proteins [e.g., sperm whale myoglobin (Atassi, 1975) and lysozyme (Atassi and Lee, 1978a,b)] (see Section 9.1.2).

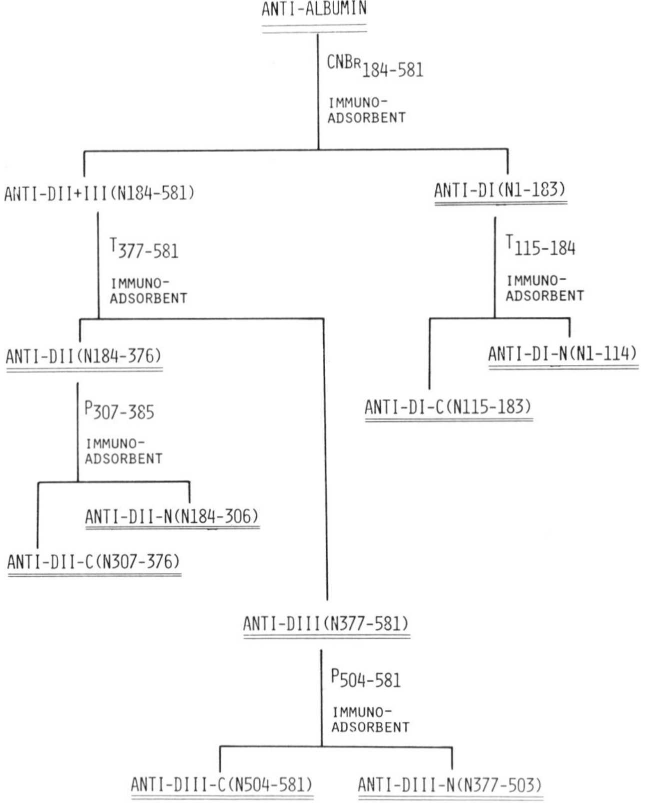

Fig. 9.5. Representation of the sequence of chromatography on different immunoabsorbants used to fractionate antialbumin into subpopulations directed against each of the three domains and subdomains of the molecule (from Benjamin and Teale, 1978, courtesy of Benjamin).

Antibodies directed against hemoglobin S poorly discriminated between hemoglobin A and S. When fractionated by gel filtration on a Sepharose column to which a synthetic peptide corresponding to N-terminal segment (1–13) of hemoglobin S had been covalently attached, these monospecific antibodies demonstrated a very restricted specificity for hemoglobin S (Young *et al.*, 1975, 1976). Also in this case, antibodies were obtained by elution with 4 *M* GuHCl. This antibody population was proposed for detection of sickle cell hemoglobin.

Antibodies specific to the conformation of the chain COOH-terminus of hemoglobin A_0 were isolated by purification on Sepharose–(129–141) peptide. These antibodies could distinguish between deoxyhemoglobin and carboxyhemoglobin; they bind preferentially to carboxyhemoglobin (Dean and Schechter, 1979).

9.2.2. Study of Antigen–Antibody Interaction

Since with antibodies purified to monospecificity, the antigen–antibody complex is expected to be soluble, study of antigen-antibody interaction requires an assay to quantitate the extent of binding. Different methods are available for a quantitative estimation of soluble antigen–antibody complex:

(1) Titration of residual enzymatic activity with inhibiting antibodies directed toward an enzyme does not require separation of the antigen–antibody complex. Nevertheless, there are some limitations because that antibody population is directed only against a region limited to the active site. This method works only when the enzyme has an affinity higher by several orders of magnitude for the antibodies than for its substrate.

(2) In radioimmunoassay (RIA) the antigen molecule is labeled by a radioactive compound covalently attached to amino acid side chains. An important requirement is that labeling not affect immunochemical properties of the antigen. In some cases, where it is only important to have an order of magnitude of the affinity, it is possible to label the antibody as routinely done in all the biological tests. Iodination with ^{125}I of tyrosine groups in protein antigen is commonly used, but external side chains can be labeled by appropriate ^{14}C or tritiated compound. Since monospecific antigen–antibody is soluble, a precipitation method must be used for separation of the complex. Precipitation by polyethylene glycol or ammonium sulfate is often used and determination of radioactivity of the precipitate provides a value of bound antigen (B) and therefore a B/F ratio (where F is the free antigen). Quantitative data can be obtained with a satisfactory accuracy when experiments are carefully performed and when many controls are done.

The inhibition (or displacement) method is frequently employed. Antibody solution sufficient to bind 50% of the labeled protein is added. Varying amounts of unlabeled protein are also added in buffer saline solution. Unlabeled protein inhibits binding of labeled protein. Radioactivity of the precipitate in each sample is determined. The inhibition is estimated to be

$$\%\ \text{inhibition} = [(A - B)/A] \times 100$$

A being the percent precipitation in the binding control (in the absence of unlabeled protein), and B, the percent of precipitation in the presence of unlabeled protein.

Solid-phase procedures are also applied for separation of free antigen (or free antibody). It is also possible to precipitate antibodies by double immunoprecipitation technique in the presence of an excess of anti-IgG. However, this method is not fully reliable since a dissolution of the complex can take place in the presence of an excess of antibodies.

The discovery of the great affinity of staphylococcal protein A for IgG provides a method for precipitation of all antibodies, since protein A binds specifically to the F_c region of IgG without inhibiting antigen–antibody binding (Kronvall *et al.*, 1970; Langone, 1978; Langone and Levine, 1979; Langone *et al.*, 1979). The use of protein A in radiommunoassay has been described for solid-phase assays (Langone, 1978; Langone *et al.*, 1979; Lessard *et al.*, 1979).

(3) Immunofluorescence takes advantage of intrinsic fluorescence of the antigen or fluorescence of a compound covalently attached to the protein (such as a fluoresceine isothiocyanate, or a dansyl group) and can be used as well as radioactive labeling to evaluate free antigen after precipitation of the complex. Fluorescence quenching as a result of antibody binding provides a method that allows evaluation of bound antigen when quenching is sufficiently effective. The method has an advantage: it does not require separation of bound from free antigen.

(4) Agglutination techniques are very sensitive and can be used even when the antigen–antibody complex is soluble. In passive hemagglutination, antigens are bound on erythrocytes (usually sheep erythrocytes) treated with tannic acid and thus used as a matrix (Kabat and Mayer, 1961). Reaction of antibody with bound antigens leads to agglutination of the red cells. Inhibition of passive agglutination by addition of soluble antigen provides a sensitive assay for antigens.

(5) Complement fixation assay allows better precision for quantitative assay. Usually sheep red cells are coated with rabbit antibodies to sheep cells. The addition of complement in the presence of Mg^{2+} and Ca^{2+} causes the cells to lyse. The extent of lysis is evaluated by measuring the concentration of supernatant hemoglobin after sedimentation of intact cells and stroma. Complement fixation on antibodies only occurs when the antigen–antibody complex is formed. Complement fixation assays are performed in two stages. First, antigen and antibodies are incubated in the presence of a definite amount of complement for 30 min at 37°C or overnight at 4°C. Then sensitized red cells are added to measure remaining unfixed active complement.

All the techniques used for quantitative evaluation of free and bound antigen or antibodies require many controls, especially radioimmunoassay because of the great sensitivity of the method which allows detection of few picomoles of complex.

In the study of staphylococcal nuclease, binding of antibodies specific to the region (99–126) inactivates the enzyme. Therefore, enzymatic activity was used to evaluate free nuclease concentration (Sachs *et al.*, 1974). In this case not only was the affinity constant determined from equilibrium measurements, but also kinetics of antibody binding to nuclease were followed:

$$\text{Ab} + \text{nuclease} \underset{k_{\text{off}}}{\overset{k_{\text{on}}}{\rightleftharpoons}} \text{Ab–nuclease}$$

The kinetics of nuclease activation by antipeptide $(99–126)_n$ antibodies was found to be a first-order reaction in nuclease concentration:

$$d(\text{nuclease})/dt = -k_{\text{on}}(\text{Ab})(\text{nuclease})$$

in the presence of varying concentrations of antipeptide $(99–126)_n$ antibodies.

For equilibrium studies, nuclease and antibodies were incubated at 25°C for 15 min; this time was found sufficient for attainment of equilibrium. Aliquots were then assayed for residual enzymatic activity. The data were analyzed, on the one hand by plotting the activity as a function of nuclease concentration, on the other hand by the Scatchard plot. The first method indicates the formation of an inactive Ab–nuclease complex involving one molecule of antigen and one molecule of antibody (Fig. 9.6). This unexpected result was not in agreement with the valence of 2 observed for IgG. Scatchard's plot also reveals a discrepancy. In this case the curve indicates

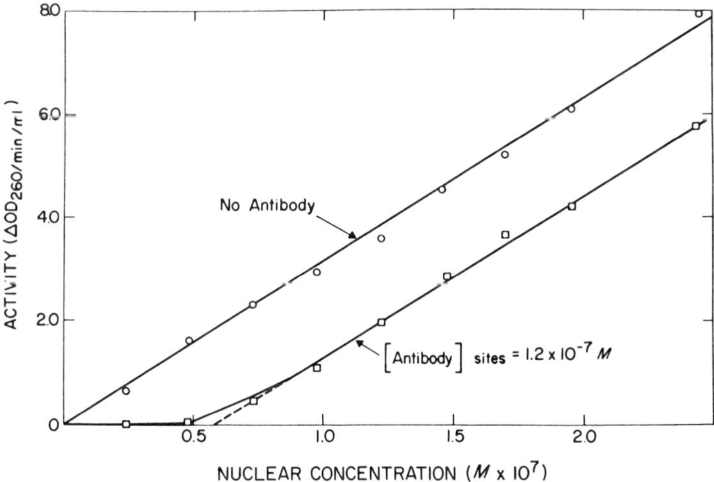

Fig. 9.6. Equilibrium study of the antipeptide $(99–126)_n$ antibody–nuclease complex (from Sachs *et al.*, 1972a). Free nuclease was evaluated by the enzymatic activity. (Courtesy of A. Schechter.)

two categories of sites, some which extrapolate to $r = 1$ with an affinity constant of $8.3 \times 10^8\ M^{-1}$, the other portion of the curve intercepting $r = 2$ allows one to determine an affinity constant $K = 1.7 \times 10^7\ M^{-1}$ (Fig. 9.7). From these experiments, it was not possible to conclude if this difference in affinity constants reflects an antibody heterogeneity, or an anticooperativity in binding of nuclease on the two sites of antibodies. The use of F_{ab} fragments supported the first explanation.

Equilibrium constant and kinetic constants for the antibody–antigen interactions ($K_{ass} = 8.3 \times 10^8\ M^{-1}$ for the higher affinity sites; $k_{on} = 4.1 \times 10^5\ M^{-1}\ sec^{-1}$, and $k_{off} = 4.9 \times 10^{-4}\ sec^{-1}$) have shown the high affinity of these monospecific antibodies and the rapidity of the complex formation. In this case, since inhibiting antibodies were used, enzymatic assay allowed the quantification of complex formation. With antibodies prepared against an unfolded fragment $(99–149)_R$ of staphylococcal nuclease, a similar Scatchard plot was obtained, suggesting that the antibodies contained a high affinity population with an average $K_{ass} = 1.2 \times 10^7\ M^{-1}$ and a lower affinity population with an average $K_{ass} = 10^5\ M^{-1}$. There is a difference of two orders of magnitude in the affinity constants when antibodies are prepared with unfolded nuclease fragment (Sachs *et al.*, 1974).

In the study of elastase, a population of inhibiting antibodies purified on Sepharose–elastase were used. Also, in this case, enzymatic activity allowed

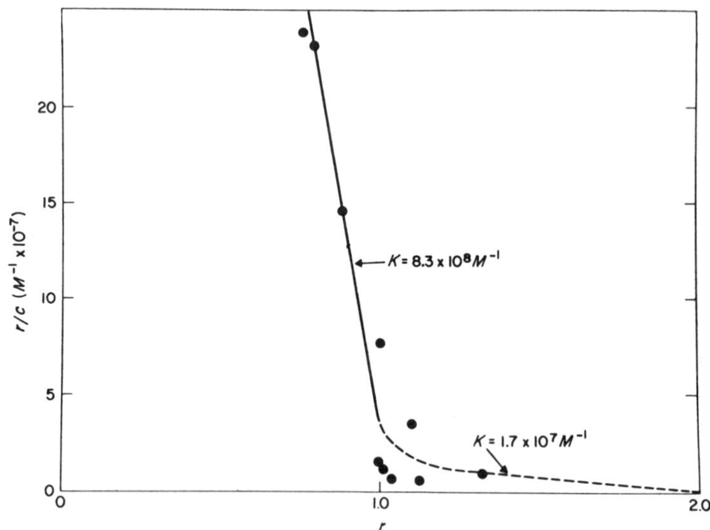

Fig. 9.7. Scatchard plot from the equilibrium study of the interaction antipeptide $(99–126)_n$ antibodies and staphylococcal nuclease; r is the concentration of bound nuclease per total antibody in equilibrium mixtures (from Sachs *et al.*, 1972a, courtesy of A. Schechter).

an estimation of free antigen (Ghélis *et al.*, 1978). A radioimmunoassay using elastase acetylated by [^{14}C]acetic anhydride on its three lysine groups (Lys 87, 177, 224) was developed. The labeled protein was confirmed to be antigenically identical to the unlabeled native one, indicating that these lysine groups are not involved in antigenic sites. With purified inhibiting antibodies, both enzymatic assay and radioimmunoassay were used and gave the same results. An association constant of $2 \times 10^8\ M^{-1}$ was found for antielastase–elastase complex. Specific carbamylation of the NH_2-group of fragment (99–149) of nuclease by [^{14}C]cyanate was performed by Furie and co-workers (1974) for a radioimmunoassay in the study of the unfolded form of this fragment. For the immunochemical approach of the refolding of bovine serum albumin, Teale and Benjamin (1976a,b) used [^{125}I]albumin and evaluated the displacement of [^{125}I]albumin from antigen–antibody complex by the refolding protein. For complex formation, the amount of antibodies added was just sufficient to bind 50% of the labeled protein in absence of inhibitor. Aliquots of renatured protein were added to the complex which was precipitated by ammonium sulfate and washed. Radioactivity was counted in the precipitate. The same type of radioimmunoassay and procedure was used to study refolding of RNase (Chavez and Scheraga, 1977). However, in these cases, the analysis of antibody–antigen interaction was not performed. The radioimmunoassay was only used to estimate the degree of refolding of the denatured protein.

Kinetics of association and dissociation of albumin and antialbumin indicate a rapid formation of complex. At 22°C, more than 80% of the antibody population specific to definite regions of serum albumin were bound immediately and 100% in less than 10 min. Dissociation of the complex is very low (Benjamin and Teale, 1978).

Although the goal was not to study refolding of a protein, but to discriminate small conformational variations, monospecific antibody specific of the N-terminal fragment of β^s-globin were studied. Quantitative analysis of the antibody–antigen complex was reported using [^{125}I]-β^s-globin (1–55) with anti-β^s(1–55) antibodies. An affinity constant of $5 \times 10^7\ M^{-1}$ was found (Curd *et al.*, 1976). These different quantitative studies indicate the high affinity of monospecific antibodies for antigen and the rapidity of the binding process (Sachs *et al.*, 1974).

9.2.3. Use of F_{ab} Fragments

To escape the divalent character of immunoglobulin, F_{ab} fragments which possess only one binding site can be used. Two different methods were described for purification of monovalent F_{ab} fragments from IgG. One is a digestion by papain (Porter, 1959) at an enzyme-to-IgG final ratio of 1:25

(wt:wt), for 27 hr at 37°C, pH 7.0. The other one used trypsin as proteolytic enzyme at pH 8.1, 37°C with the same final enzyme-to-protein ratio. This digest produces mainly $F_{ab'2}$ fragments which can be purified on an immunoabsorbant column and then treated with a reducing agent to give F'_{ab} fragments. The following treatment was described by Eastlake and co-workers (1974). The purified $F_{ab'2}$ divalent fragments were treated with 5 m*M* dithiothreitol at pH 8.1 (0.1 *M* Tris-HCl buffer), 25°C for 40 min to achieve complete reduction. Then iodoacetamide was added (final concentration 10 m*M*). After 15 min, chromatography on Sephadex G150 was performed. A major peak was obtained, consisting mainly of F'_{ab} fragments.

The F_{ab} univalent fragments of antibodies specific to the region 99–126 of staphylococcal nuclease were used to study the antibody–antigen interaction at equilibrium and to determine whether the previous results describing two different average association constants for the interaction of nuclease with specific antibodies were attributable to an anticooperativity or to the existence of different categories of antibodies (Eastlake *et al.*, 1974). The same type of Scatchard plot as in Fig. 9.7 was observed with F_{ab} fragments discarding the hypothesis of anticooperativity. It indicates antibody heterogeneity, with the existence of two antibody populations within antipeptide $(99\text{–}126)_n$ region of nuclease, one with $K_{ass} = 10^9\ M^{-1}$, the other with $K_{ass} = 10^7\ M^{-1}$.

The use of monovalent F_{ab} fragments can be very useful since equimolecular antibody–antigen complexes are expected with monospecific antibodies. The F_{ab} fragment was also used in the determination of the number of antigenic determinants of soluble and of membrane-bound intestinal aminopeptidase (Louvard *et al.*, 1976).

9.2.4. Study of Conformational Equilibrium

An immunochemical approach was used to study conformational equilibria of isolated polypeptide fragments as well as entire staphylococcal nuclease (Sachs *et al.*, 1974; Furie *et al.*, 1975). An antibody population directed against the unfolded form of fragments $(99\text{–}126)_R$ was isolated to study the equilibrium between unfolded nuclease and the spontaneously refolded form. Figure 9.8 is a schematic representation of the presumed two-state equilibrium between native and unfolded forms. It may be described by the following expression:

$$P_U \xrightleftharpoons{K_{conf}} P_F$$

with $K_{conf} = P_F/P_U$. Antibodies obtained by immunization with native protein $(Ab)_F$ are specific to P_F and antibodies produced by immunization with

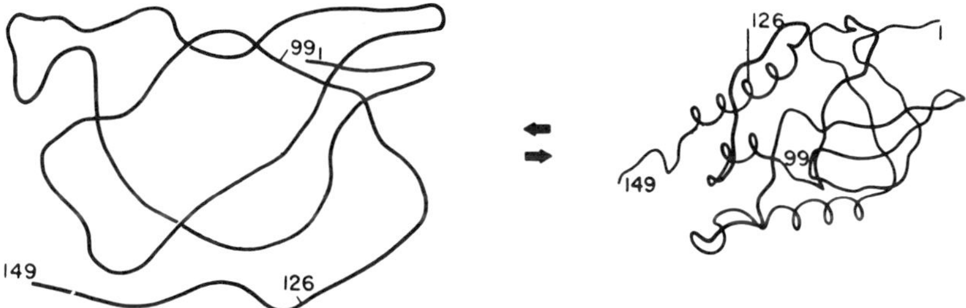

Fig. 9.8. Schematic representation of the conformational equilibrium between folded and unfolded state of staphylococcal nuclease (from Furie *et al.*, 1975, courtesy of A. Schechter).

unfolded polypeptide fragments (Ab_U) are expected to have affinity for P_U. Thus, with both types of antibodies, the following equilibria take place:

$$P_U + Ab_U \rightleftharpoons P_U \cdot Ab_U$$

$$P_F + Ab_F \rightleftharpoons P_F \cdot Ab_F$$

cross interactions with heterologuous antigen were assumed to be negligible.

The equilibrium was studied by competition of binding of antipeptide $(99\text{–}126)_R$ to fragment (99–149) labeled with [^{14}C]KNCO and to unfolded protein P_U, according to the following equilibria:

$$\text{antipeptide } (99\text{–}126)_R + P_U \overset{K_2}{\rightleftharpoons} P_U\text{–antipeptide } (99\text{–}126)$$

$$\text{antipeptide } (99\text{–}126)_R + \text{peptide } (99\text{–}149) \overset{K_1}{\rightleftharpoons} (99\text{–}149)\text{–antipeptide } (99\text{–}126)_R$$

Furie and co-workers (1975) postulated the end effect in antigen–antibody complex negligible, so that antipeptide $(99\text{–}126)_R$ binds with equal affinity to (99–149) and to unfolded ribonuclease and K_1 is taken equal to K_2. Thus by a combination of these equilibria and the conformational equilibrium, K_{conf} is evaluated as

$$K_{conf} = \text{antipeptide } (99\text{–}126)_R(P_F) / [\text{antipeptide } (99\text{–}126)_R(P_U)\text{–peptide } (99\text{–}149)]$$

A typical competitive inhibition curve of the formation of the soluble complex antipeptide $(99\text{–}126)_R$–(99–149) by nuclease is indicated in Fig. 9.9. An average value of 2,900 ± 1,400 was thus determined for K_{conf} at 25°C. The effect of temperature on K_{conf} of nuclease was also evaluated by an immunochemical approach, allowing the determination of an enthalpy variation ΔH of -10.3 kcal/mole. The presence of ligands, Ca^{2+} and pdTp increases K_{conf} up to 30,000–50,000 showing a strong stabilization by specific ligands.

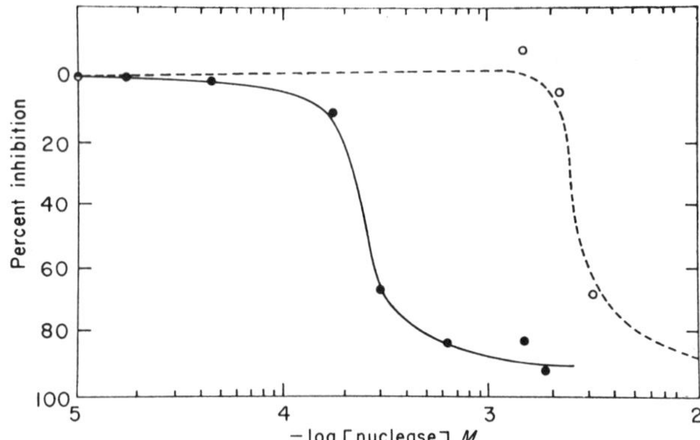

Fig. 9.9. Inhibition by nuclease of the formation of the soluble complex of antipeptide $(99-126)_R$ and fragment (99–149): (●), experiments performed without ligands; (○) experiments performed in the presence of 20 mM $CaCl_2$ and 10 mM pdTp at 25°C. Double immunoprecipitation by rabbit antigoat immunoglobulin was used to separate antibody-bound and unbound labeled fragments (from Furie *et al.*, 1975, courtesy of A. Schechter).

The accuracy of the method is limited by several simplifying assumptions, the equality of K_1 and K_2, the absence of binding with heterologous antigen, this last assumption being more probable. Besides these limitations, the immunochemical approach allowed a direct study of conformational equilibria without using denaturant and extrapolating to zero (see Chapter 6). The results indicate that staphylococcal nuclease is a rather flexible molecule in dynamic equilibrium with the unfolded form. By a similar approach, polypeptide fragments of nuclease [(99–149), (50–149)] were found in equilibrium with a weak but significantly folded fraction. An average K_{conf} of 2×10^{-4} was determined. The last experiments were interpreted as an indication of the spontaneous refolding of fragments to a nativelike structure, but the folded state is not stable enough. It requires that all long chain interaction be present in the whole protein for stability. In the presence of a synthetic fragment (6–43), the value of K_{conf} significantly increases.

The same approach was used by Chavez and Scheraga (1980b) and Scheraga (1980), to determine the stability of various protein derivatives and fragments of bovine pancreatic ribonuclease. Fractionated antibodies specific for antigenic determinants existing in segments 1–13, 31–79, and 80–124 were used to determine the values of K_{conf}. A value of K_{conf} of ca. 6×10^{-2} was obtained for RNase after reduction of disulfide bonds. As reported in Chapter 5, Garel (1976) found an activity of 0.04% for renatured but not reoxidized RNase, also indicating the presence of native structure before reformation of disulfide linkages. The population of molecules of reduced RNase with their native antigenic determinants seems greater than the population of molecules with a functional catalytic site. The K_{conf} value decreases to ca. 10^{-3} after S-carboxymethylation of the resulting sulfhydril groups. When residues 1–20 from the N-terminus are removed, K_{conf} also decreases to 6×10^{-2}. The loss of the C-terminal tetrapeptide resulting from proteolysis by pepsin lowers K_{conf} to 3×10^{-2} whereas it decreases to 6×10^{-3} after removal of the C-terminal hexapeptide. The K_{conf} was also evaluated for cyanogen bromide peptides 1–13, 31–79, and 80–124 to evaluate the extent to which smaller fragments are able to adopt a nativelike conformation (see Chapter 10). An approximate value of K_{conf} of 10^{-4} was determined for these three peptides. Furthermore, the temperature dependence of K_{conf} was evaluated for three different regions of Des(121–124) RNase, yielding different transition curves for the three antigenic sites. This last result shows the occurrence of detectable intermediates in the folding of RNase. Hurrell and co-workers (1971) also found that an isolated peptide of myoglobin containing residues 132–153, which is largely helical in the native protein, has a value of K_{conf} equal to 10^{-2} at 36°C. That indicates a relatively high stability of this small helical segment.

These results reveal that the entire protein (as well as fragments) is able to fluctuate between at least two conformational states, even if one of them is poorly populated. The determination of K_{conf} was of great importance in determining conformational fluctuations of either native or unfolded proteins.

These studies emphasize the potentiality of the immunochemical approach in the study of conformational equilibria, in which one state is very little populated precluding the use of physicochemical methods. By immunochemistry as well as by hydrogen exchange, it is possible to reach the equilibrium constant in the absence of denaturant (see Chapter 6).

9.2.5. Kinetic Study of Protein Folding Using Fractionated Antibodies

The obtention of specifically purified antibody to different regions of the surface of a protein makes immunochemical methods particularly adequate to follow the return of native structure in different parts of a protein. Such kinetic approaches to the folding pathway were reported for serum albumin (Teale and Benjamin, 1976a,b, 1977; Benjamin and Teale, 1978), RNase (Chavez and Scheraga, 1977, 1980a,b), and elastase (Ghélis *et al.*, 1978).

Refolding of reduced serum albumin either in the presence of a redox regeneration mixture (0.3 mM reduced glutathione, 0.03 mM oxidized glutathione) or in the presence of a disulfide interchange enzyme, was investigated using different antibody populations (as summarized in Fig. 9.4 and 9.5) (i.e., antibodies directed against domains I, II, and III and also

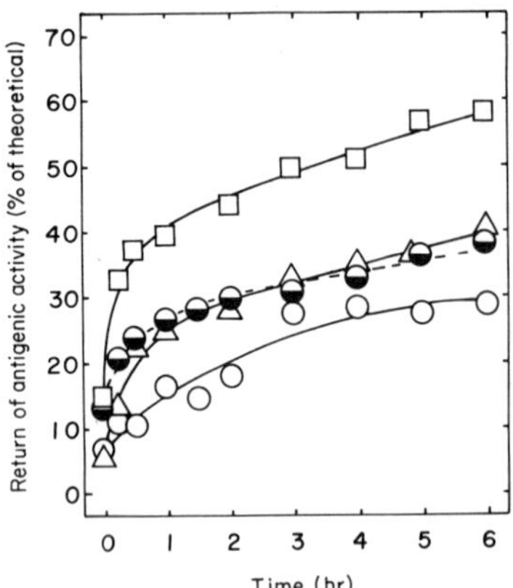

Fig. 9.10. Kinetics of regain of native antigenic structure in the three domains in reduced serum albumin, within the whole molecule as followed with restricted antibody populations. Reoxidation was performed in the presence of disulfide interchange enzyme: (◒), kinetics followed with whole antialbumin; (△), kinetics followed with antidomain I; (○), kinetics followed with antidomain II; (□), kinetics followed with antidomain III (according to Teale and Benjamin, 1977, courtesy of Benjamin).

subdomains). Most of these antibody populations were not monospecific, but only specific to a restricted region of the protein. With these purified antibodies, an apparent pathway of folding was proposed in which the third domain (C-terminal) seems to refold faster than the first domain (N-terminal), the second domain refolding the least rapidly (Fig. 9.10). In serum albumin, homologies between domains suggested that the protein arose by gene triplication of a single primordial gene coding for a small protein of MW 22,000. The folding pathway of serum albumin deduced from immunochemical studies is consistent with the proposed evolutionary pathway of the molecule. Furthermore, a sequential pathway within each of the three domains was also reported. The C-terminal one-third refolds faster than the N-terminal two-third of each respective domain. Figure 9.11 shows this sequence of events for domain I. In these experiments, the refolding molecule was intact serum albumin. In other types of experiments, refolding of isolated fragments corresponding to each domain was investigated.

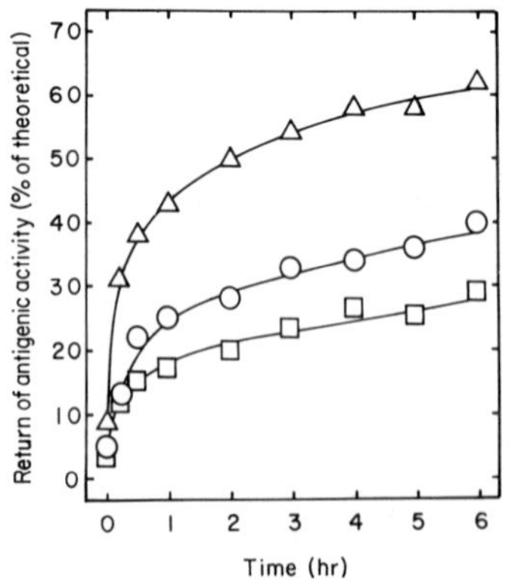

Fig. 9.11. Kinetics of regain of antigenic structure within the first domain in whole serum albumin with restricted antibody populations. Conditions were the same as for Fig. 9.10: (△), kinetics followed with anti-$T_{115-184}$; (□), kinetics followed with anti-N_{1-114}; (○), with anti-N-fragment (antidomain I) (from Teale and Benjamin, 1977, courtesy of Benjamin).

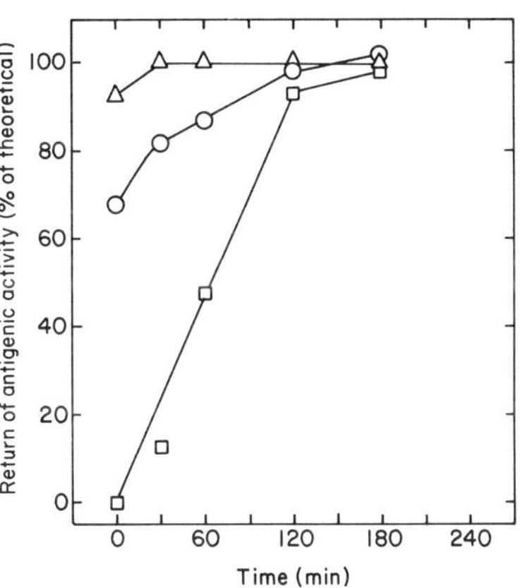

Fig. 9.12. Kinetics of return of antigenic activity in different regions of the molecule during air reoxidation of reduced ribonuclease (from Chavez and Scheraga, 1977). Kinetics followed by (○), antipeptide $(1-13)_n$, (□), antipeptide $(31-79)_n$, and (△), antipeptide $(80-124)_n$.

Using the same procedure for RNase, Chavez and Scheraga (1977, 1979, 1980a) reported a sequential order of refolding by air oxidation of the reduced protein. Kinetic studies indicated that antigenic determinants reappeared in the following order, 80–124, 1–13, and 31–79, as indicated in Fig. 9.12. Consistent with other data, these results indicate the importance of the C-terminal region in early folding of RNase. Refolding of RNase, S-protein, and Des(121–124) RNase by gluthathione oxidation leads to similar conclusions (Chavez and Scheraga, 1980a). These data are consistent with those reported by Burgess and Scheraga (1975a), modified by Matheson and Scheraga (1979a,b). Furthermore, they are in good agreement with a theoretical pathway deduced from the contact map (Némethy and Scheraga, 1979).

The great sensitivity of the antibody probe allows one to discriminate what happens in different regions of a protein. However, the use of the method for kinetic studies was criticized. Kinetic studies of the refolding process requires a slower rate for refolding than for antibody binding. Although antibody binding is a rapid process as shown by Sachs and coworkers (1974) (see Section 9.2.2), displacement of labeled antigen by a refolded one in the inhibition method requires an incubation time. In a typical experiment, an aliquot of the protein solution in regeneration mixture is removed at increasing time during the refolding process, and then assayed for its ability to inhibit binding of labeled native protein or fragment to specific antibodies. Samples are generally incubated for a certain time. In experimental procedure, a 30-min incubation period is usually chosen before precipitation of the antibody–antigen complex. A 30-min incubation time was used for serum albumin as well as for RNase studies. A serious limitation of the method for kinetic studies is certainly this 30-min incubation period in which a varying extent of refolding might occur. This procedure was particularly criticized by Creighton (1978c), specially in the case of RNase for which different amounts of refolded species were occurring during incubation as indicated by the value of the zero time. The researchers were certainly aware of the limitation in time of the method which restricts kinetic measurements to relatively slow processes, and on the possibility of refolding during the incubation time. For serum albumin, Teale and Benjamin (1976a,b) incubated their samples at pH 6.0 in the presence of normal rabbit serum to minimize refolding during the assay. For RNase, normal rabbit serum was also added, pH being maintained at 7.2. For this last protein, whether refolding processes have occurred before or during the 30-min incubation time in the presence of specific antibodies, differences are observed in the refolding of the different regions. Furthermore, it was verified that results are independent of the time required for the assay between 5 and 30 min (Chavez and Scheraga, 1980a). Thus, differences can be attributed to sequential regain of antigenic determinants of RNase, or to different induction of the native structure by the different antibody populations during the incubation time.

Induced fit caused by antibody binding was observed for different proteins. One of the most striking examples is probably the induction of functional structure of β-galactosidase after binding of an inactive mutant with antibody specific to the wild-type active enzyme (Celada and Strom, 1972). Chavez and Benjamin (1978) also found that the presence of a mixture of antibody in the refolding mixture greatly enhances the refolding rate of reduced serum albumin. The effect is very specific, since this enhancement is restricted to that domain against which the added antibody was directed. Figure 9.13 shows this result. Furthermore, it was observed that for each domain the enhancement produced by the presence of specific antibody is delayed in the N-terminal portion of the domain compared to that observed in the C-terminal region. It was concluded that a sequential pathway of refolding within each domain was directed from the C-terminal to the N-terminal part. The investigators suggested the existence of a nucleation center near the C-terminal in each domain. These nuclei are stabilized when bound to specific antibody and subsequently return to native structure is fastened. They proposed this kind of immunochemical approach to search the location of nucleation centers in proteins.

Similarly, Dean and Schechter (1979) reported that the presence of specific antibodies in stoiechiometric amount increases the affinity of hemoglobin A_0 for oxygen and they concluded that antibody and oxygen binding are linked functions.

With elastase, the regain of antigenic activity, detected by using purified monospecific inhibiting antibody, followed the same kinetics as the return of enzymatic activity. Both signals are delayed compared with the return of

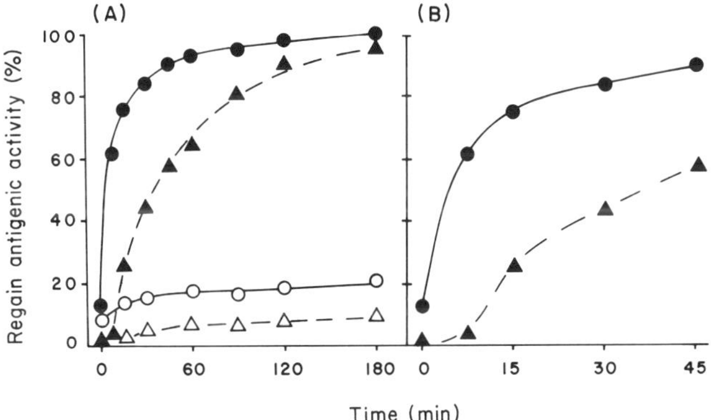

Fig. 9.13. Enhancement of the rate of reappearance of antigenic properties within domain I of RNase in the presence of specific antibodies in the regeneration mixture (from Chavez and Benjamin, 1978): (A) (▲), antipeptide (1–114) plus reduced albumin in the refolding mixture; (●), antipeptide (115–183) plus reduced albumin in the refolding mixture; (△), no antibody in the refolding mixture, but antipeptide (1–114) used for the assay; (○), no antibody in the regeneration mixture, but peptide (115–183) used for the assay; (B) the same as for (A) in early time.

physicochemical properties (dichroic signals for example) and decrease of reactivity of several side chains buried in the native structure. The coincidence of the two signals (enzymatic and antigenic properties) indicates, in this case, the validity of antibody probes even for kinetic studies (Ghélis *et al.*, 1978).

In the different examples of kinetic studies analyzed in this section, the antibody probe had allowed the detection of intermediates in the folding pathway and a sequential refolding of the different parts of RNase as well as serum albumin was reported. For both proteins, the importance of the C-terminal region of either whole molecule, or even within each domain in serum albumin, was emphasized by the experimental results. Another interesting point was the use of antibody probe in the search of nucleation centers.

9.3. IMMUNOCHEMICAL APPROACHES USING UNFRACTIONATED ANTIBODIES AGAINST THE ENTIRE PROTEIN

Different studies on protein folding using unfractionated antibodies or even whole antiserum were reported, among them kinetic studies or characterization of intermediates trapped by other methods.

9.3.1. Study of Antigen–Antibody Interactions

Proteins generally have several antigenic determinants and elicit a mixture of antibodies directed against these different determinants, which can have various affinities for their specific antibodies, leading to a complex multiequilibria system. A theory of the interactions between multideterminant antigens with multispecific antisera was developed by Berzofsky and co-workers (1976). They used probability analysis to predict the fraction of free antigen and of antigen bound to antibody. An antiserum consists of a mixture of antibodies, each population of antibody being of one of the n determinants. The fraction b_i of the ith determinant bound by antibody is specified by $[Ab_i]$, the concentration of specific antibodies, and K_i the affinity constant where n is the number of distinct antigenic determinants, and T is the antigen concentration.

The probability F for the whole antigen to be free is given by the following expression:

$$F = \prod_{i=1}^{n} (1 - b_i)$$

and the probability B that at least one of the n determinants on a molecule has bound an antibody is

$$1 - F = B = 1 - \prod_{i=1}^{n} (1 - b_i)$$

That implies the independency of antigenic sites. The ratio

$$R = B/F = \prod_{i=1}^{n} [1/(1 - b_i)] - 1$$

The bound to free ratio, r_i, for each individual determinant is $b_i/(1 - b_i)$ therefore

$$R = \prod_{i=1}^{n} (r_i + 1) - 1$$

The slope of the curve of bound to free as a function of antigen concentration is given by

$$\frac{\partial R}{\partial T} = \frac{\partial}{\partial T} \prod_{i=1}^{n} (r_i + 1)$$

When all the sites have similar affinities for antibodies, then all the b_i are approximately equal to b, therefore

$$\frac{\partial R}{\partial T} = n\left(\frac{1}{1 - b}\right)^{n-1} \frac{\partial}{\partial T}\left(\frac{1}{1 - b}\right)$$

$$\frac{\partial r}{\partial T} = \frac{\partial}{\partial T}\left(\frac{1}{1 - b}\right)$$

and the ratio

$$\frac{\partial R}{\partial T}\bigg/\frac{\partial r}{\partial T} = n\left(\frac{1}{1-b}\right)^{n-1}$$

for $0 < b < 1$, this ratio increases rapidly with n. For example, the ratio is 4 for two sites, and 32 for four sites.

When the binding functions are known for each determinant, it is possible to predict the experimental binding curve for the whole antiserum. Such an analysis was applied to the amino terminus of the chain of hemoglobin S, β^s(1–55) which contains known nonoverlapping antigenic sites β^s(1–13) and β^s(40–53). However, it is not always possible to dissect out all the individual determinants of an antigen molecule. However, this treatment emphasizes the fact that the shape of the binding curves can be affected by the various interactions.

9.3.2. Kinetic Studies of Protein Folding

In their first study of the kinetics of refolding of reduced bovine serum albumin, Teale and Benjamin (1976a) used unfractionated antialbumin. By this approach, Teale and Benjamin determined the best conditions for refolding, with respect to pH conditions, the presence of redox mixture, or disulfide interchange enzyme. Refolding of serum albumin is rather slow with an initial rate of 5.3×10^{-3} min^{-1} at pH 8.0 and room temperature in the presence of a mixture of 0.3 mM GSH and 30 mM GSSG. This rate is comparable to that of refolding of RNase (6.6×10^{-3} min^{-1}) according to Hantgan and co-workers (1974); 50% regeneration is obtained in approximately 2 hr.

The kinetics of refolding of completely reduced human serum albumin have been followed by rocket immunoelectrophoresis (according to Laurell, 1966), using antihuman serum albumin immunoglobin from rabbit (Wichman *et al.*, 1977). It was confirmed that the completely reduced protein does not react with antibodies, but a rapid regain of immunoprecipitability was obtained after reoxidation. About one-third of the antigenicity is recovered within 10 min. A new consumption rocket electroimmunoassay was employed in this research. This method measures residual antibodies which have not reacted with refolding protein. In the experimental procedure, the reoxidized samples are applied to the plate and allowed to diffuse for 15 min and run for 2 hr. The partially folded protein precipitates and consumes antibodies directed against the determinants that have reappeared. Then, it is allowed to continued for 16 hr after a fixed amount of native albumin was applied in all sample wells. The other antialbumin which have not precipitated with refolded protein can thus react with native albumin. The height of the rocket permits the evaluation of the amount of antibodies consumed by the partially refolded protein.

Immunochemical studies using whole antibody populations elicited against denatured forms of proteins have been used to detect structural homologies (Zakin *et al.*, 1978; Sundaram *et al.*, 1980; Chaffotte *et al.*, 1980).

9.3.3. Studies of Intermediates in Protein Folding

Immunochemical method was employed to examine the conformational properties of intermediates trapped during folding or unfolding of BPTI (Creighton *et al.*, 1978). In this approach, the conformational properties of

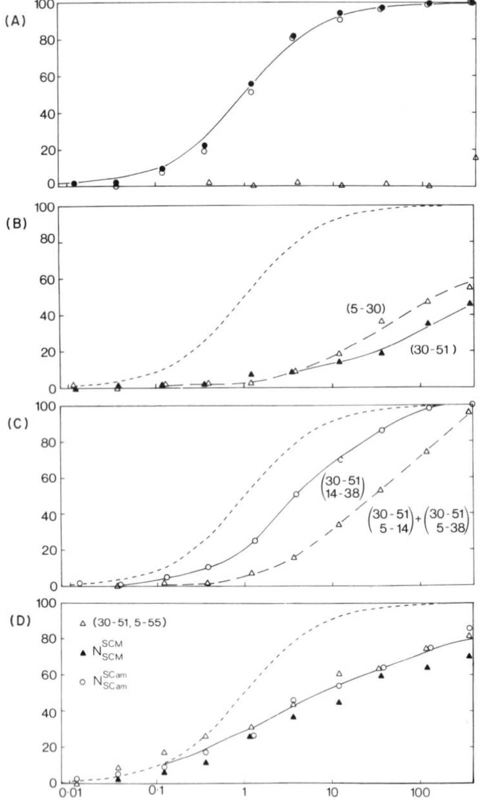

Fig. 9.14. Competition between [^{125}I]-BPTI and reduced intermediates and refolded forms of BPTI for anti-N-antibodies: (A) (○), native; (●), refolded; (△), reduced-carboxymethylated BPTI; (B) (△) single-disulfide bond intermediate (5–30) and (▲), (30–51); (C) (○) two-disulfide bond intermediate (30–51, 14–38) and (△), the mixture of (30–51, 5–14) and (30–51, 5–38); and (D) (△) nativelike two-disulfide bond species (30–51, 5–55) and (▲) N_{SCM}^{SCM} and (○) N_{SCam}^{SCam} (from Creighton *et al.*, 1978, courtesy of Creighton).

the entire protein were estimated by their interaction with unfractionated antibodies directed against either the entire native protein (anti-N-antibodies) or the reduced unfolded form (anti-R-antibodies). Radioimmunoassay using ^{125}I-labeled BPTI was used. Competition between [^{125}I]-BPTI and reduced, intermediate, and refolded unlabeled BPTI for antibodies was analyzed. The experiments are shown in Fig. 9.14. Refolded BPTI was indistinguishable from native BPTI. The reduced form did not bind significantly to anti-N-serum (Fig. 9-14A). The single disulfide intermediates (Cys 5–Cys 30) and (Cys 30–Cys 51) bind to a lower extent with anti-N-antibodies than native protein (Fig. 9.14B). The two-disulfide bond intermediates (Cys 5–Cys 31, Cys 14–Cys 38) with two native disulfide bonds gave a normal inhibition curve with a lower affinity for antibodies than the native protein. The mixture (Cys 30–Cys 51, Cys 5–Cys 14) and (Cys 30–Cys 51, Cys 5–Cys 38) was shifted to higher concentrations of competitive protein (Fig. 9.14C). The nativelike two-disulfide bond species (Cys 30–Cys 51, Cys 5–Cys 55) gave the same inhibition curve as chemically modified protein [carboxymethylated (S^{CM}_{CM}) or carboxamidomethylated (S^{Cam}_{Cam})] after reduction of the Cys 14–Cys 38 disulfide bridge (Fig. 9.14D). Although this species behaves the same as the native form, surprisingly it seems less antigenic toward anti-N-antibodies than the other two-disulfide bond species.

Analogous competition experiments using anti-R-antibodies gave clearer results, although quantitative analysis was complicated by the low solubility of reduced BPTI and the weak affinities of the anti-R-antibodies. Nevertheless competition curves showed that the affinity for anti-R-antibodies decreases at each step of the folding pathway. One-disulfide bond species (Cys 5–Cys 30) and (Cys 30–Cys 51) have significant affinity for antibodies; these two species are indistinguishable (Fig. 9.15B,C). The two-disulfide bond intermediate (Cys 30–Cys 51, Cys 14–Cys 38) has a little less affinity for antibodies than the previous species. But affinity decreases in the mixture of (Cys 30–Cys 51, Cys 5–Cys 14) and (Cys 30–Cys 51, Cys 5–Cys 38) (Fig. 9.15E). The native and carboxymethylated form have no significant affinity for antibodies.

Both the native and reduced forms of BPTI gave precipitation with their respective specific antibodies indicating that at least three antigenic determinants exist in this protein. The correlation between the kinetic role of intermediates in the folding pathway was not very good in the case of anti-N-antibodies, but much better with anti-R-antibodies. Creighton and co-workers (1978) concluded that unfolded forms are still present until all stabilizing interactions are established in the last step of the process.

Ghélis and Zilber (1982) studied reversibility of unfolding–folding of unreduced elastase denatured by GuHCl, using unfractionated but purified antielastase antibodies by quantitative radial immunodiffusion (according to

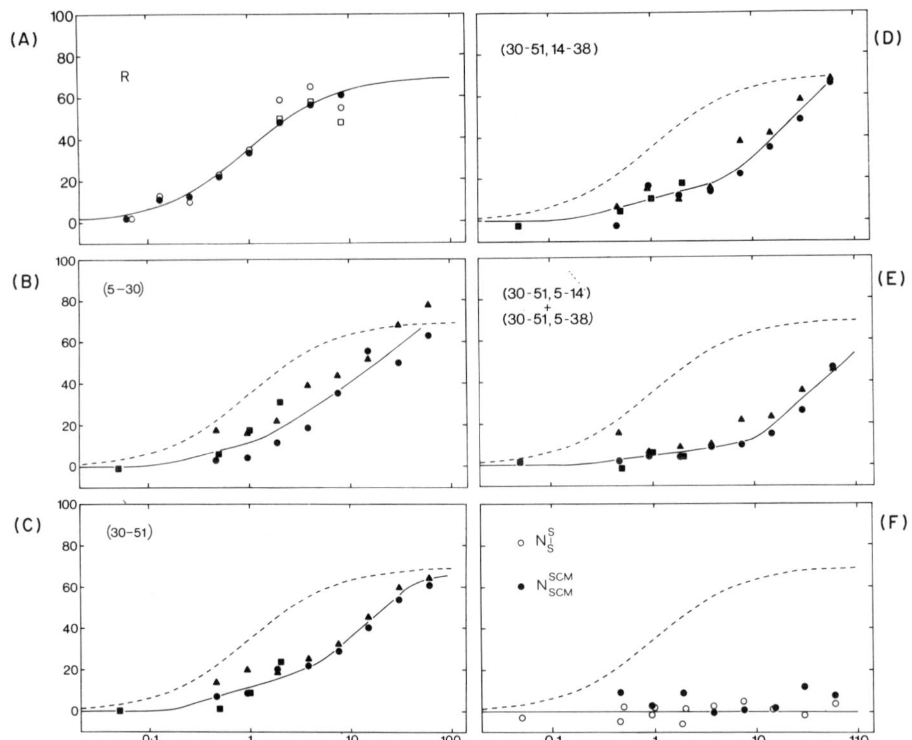

Fig. 9.15. Competition between ^{14}C-reduced carboxymethylated BPTI and reduced intermediate, and native forms of BPTI for anti-R-antibodies: (A) unlabeled reduced protein; (B) one-disulfide bond (Cys 5–Cys 30) intermediate; (C) one-disulfide bond (Cys 30–Cys 51) intermediate; (D) two-disulfide bond (Cys 30–Cys 51, Cys 14–Cys 38) intermediate; (E) two-disulfide bond mixture (Cys 30–Cys 51, Cys 5–Cys 14) plus (Cys 30–Cys 51, Cys 5–Cys 38); (F) N^{SCM}_{SCM} (●), and N^{SCam}_{SCam} (○) (from Creighton *et al.*, 1978, courtesy of Creighton).

Mancini *et al.*, 1965). The reversibility of the process, when protein was incubated in intermediate concentrations of GuHCl, was analyzed by this method and compared with the return of enzymatic activity (see Chapter 5). Both methods indicate the same decrease in reversibility when elastase was incubated at critical concentration of denaturant. By contrast, for hen egg white and phage T_4 lysozyme, a perfect reversibility was observed (Desmadril *et al.*, 1982b).

The various immunochemical approaches of protein folding indicate the potentiality of the method to (1) study dynamic conformational equilibrium, (2) measure the probability of fragments to reach the conformation they have in the native protein, and (3) detect intermediates in protein folding or to analyze trapped intermediates. The advantages and disadvantages of the method have been emphasized. Its great sensitivity and specificity make it a potent tool for studying protein folding of the nascent polypeptide chain in the cellular environment.

In the future, a new potentiality of immunochemical approach certainly will be provided by monoclonal antibodies. This method which remains laborious and delicate needs further improvement for use in the study of protein folding.

Different methods have been used to detect and to characterize intermediates in protein folding.

(1) Several of them, such as Raman or IR spectroscopy or isotope exchange techniques, appear to be promising methods for the detection of the early formation of ordered structures. However, this is just the beginning of detailed analysis in this direction and it is too early to generalize concerning the chronology of their appearance during protein folding.

(2) Various powerful techniques allow local investigation of the folding process in different parts of a polypeptide chain: (a) high resolution NMR; however, the high concentrations of proteins which are required represent a serious limitation of the method; (b) chemical methods, such as differential competitive labeling, photolabeling, blocking of SH groups before reformation of S—S linkages. The first of these methods may principally be a potent tool in the study of protein folding even in heterogeneous environment.

(3) Immunochemical methods are the most sensitive for the detection of the reappearance of structural motifs.

Data obtained from these methods agree for the existence of intermediates and also have given evidence for motion in folded proteins. Elements of folded structures have been characterized and their sequence of re-formation determined. But interpretations are greatly divergent. Several authors deduce that folding is a simple sequential formation of final conformations, whereas others conclude that rearrangements during the refolding are due to formation of wrong-folded species, a concerted process occurring when right interactions take place. Presumably, both types of situation do exist for the same protein and either one aspect or the other becomes predominant according to the experimental conditions and the trapping method. The following scheme can probably account for the unfolding–folding process and reconcile divergent points of view.

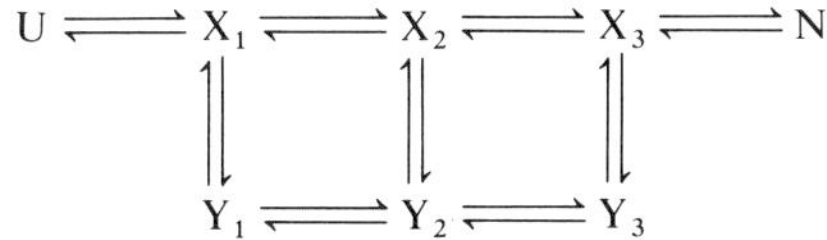

whatever the number of stable and correctly partially folded intermediates X_i. *It is assumed that at each step before the native structure is formed, side reactions leading to incorrectly partially refolded species compete with the simple sequential pathway. Depending upon the conditions, rate constants may be such that side reactions can or cannot significantly occur. Furthermore, these dead-end species* Y_i *may be trapped by a method used in the procedure to detect* X_i *species, and a by-pass such as*

$$U \longrightarrow X_1 \longrightarrow Y_1 \longrightarrow Y_2 \longrightarrow Y_3 \longrightarrow N$$

may be artificially deduced as the folding pathway.

Various arguments converge to accept that the early refolded species X_1 *are nuclei, resulting from hydrophobic interactions.*

10

Folding and Assembly of Building Blocks

In the preceding chapters, evidence for intermediates in protein folding was analyzed and discussed. However, divergent interpretations have arisen from the different experimental data. Some researchers considered that the native structure can occur only when all stabilizing interactions are present simultaneously. Others proposed that folding proceeds by stages in which various parts of the polypeptide chain are folded in a sequential order. In fact, the clearest results indicating the cooperative formation of native structure in the last step of folding were obtained with BPTI which is a very small protein with 58 amino acids. Such a folding mechanism also seems plausible for another small protein, staphylococcal nuclease with 149 amino acids. However, even for ribonuclease which has 124 amino acids, divergent observations and conclusions were reported. For larger proteins such as serum albumin with its three structural domains, return to a nativelike structure in a stepwise process seems to encounter more general agreement.

What remains is the examination of the set of experimental data actually available to answer the questions: (1) what are the smallest stable folding units? (2) Do they persist in the native structure of a protein? The answers to these questions were discussed in Part I on the basis of crystallographic data and of theoretical approaches. Now, consideration is given to the experimental support for the different theories concerning a possible hierarchy in the formation of protein structure, in which small folding units might interact each with the others to generate the native structure.

10.1. FOLDING OF POLYPEPTIDE FRAGMENTS

10.1.1. Folding of Small Fragments

It was demonstrated that for polypeptides shorter than 15–20 amino acids, even ordered structures such as α helices cannot be stabilized in solution (Schellman, 1958). Nevertheless, attempts to determine whether small fragments can exist as any significant folded nativelike structure were undertaken with short synthetic oligopeptides that reproduce sequences of known structures in proteins. As an example, conformation of β bends was investigated with a tetrapeptide having the sequence of region 35–38 in chymotrypsin known to form a chain reversal:

(Asp–Lys–Thr–Gly) (1)

and with the following other variants:

(Gly–Thr–Asp–Lys) (2)

(Asp–Lys–Gly–Thr) (3)

(Lys–Thr–Gly–Asp) (4)

C- and N-terminal ends were blocked respectively with methyl ester and acetyl groups to avoid end effects. Circular dichroism signals and also NMR (Nuclear Overhauser Effect, NOE) spectra were recorded at 5° and 26°C. At 5°C dichroic signals of peptides 1, 3, and 4 differed from that of peptide 2 (Fig. 10.1). This difference disappears quite entirely at 26°C (Howard *et al.*, 1975). Both CD and NOE data indicate that a small but significant fraction of these peptides are in a chain reversal conformation. These observations were found in agreement with the relative probabilities of bend occurrence in these four peptides as determined by Lewis and co-workers (1971), which were 20/1/40 and 40, respectively, for the four peptides.

Not all the studies using homo- or heteropolypeptide models to determine the type of structure they adopt in particular environment (pH, temperature, solvent, and ionic strength, see Singer, 1962) are discussed here. The only example reported here indicates the tendency of even small oligopeptides to adopt their preferred structure dictated by short-range interactions; however, the very small extent of structure they adopt in solution clearly indicates the requirement of long-range interactions for stabilization. Even short oligopeptide hormones, or enzyme inhibitors, have in solution fluctuating conformations, one of them being selected and stabilized by their interactions with specific receptors.

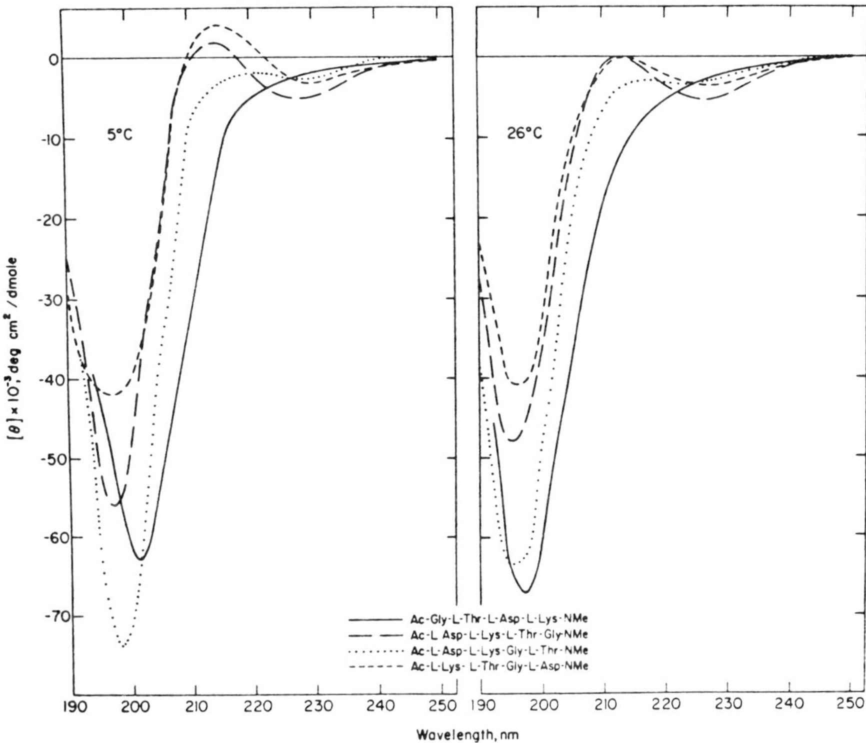

Fig. 10.1. Circular dichroism spectra of the four oligopeptides (1), (2), (3), and (4) (see text) (from Howard *et al.*, 1975).

10.1.2. Folding of Larger Fragments

In this subsection, examination is made of the fragments which are portions of well characterized proteins with exception of those which correspond to structural domains; these latter are considered in Section 10.2. Several investigations have been reported concerning the folding of fragments from different proteins obtained either after proteolytic degradation or by semisynthetic methods. The most complete studies were conducted with ribonuclease, myoglobin, staphylococcal nuclease, and hen egg white lysozyme.

Fragments of various proteins were found to contain less (sometimes much fewer) nativelike structures than intact native protein in aqueous solvent. This finding has led to the hypothesis that the entire polypeptide chain is required for folding into native conformation.

Fragments of myoglobin obtained by cleavage with cyanogen bromide give a N-terminal fragment which contains the first 55 amino acids; 44 of these amino acids are known to possess a helical conformation in the intact native protein. This peptide was allowed to renature after denaturation in 6 *M* GuHCl. It was found that 60% of the α helix was regained (Crumpton, 1968). Antibody probes indicate that the peptide is able to react with anti-myoglobin antibodies. However this N-terminal peptide has a tendency to aggregate in water. Peptide 1–55, and also C-terminal peptide containing 22 amino acids and another peptide with 76 amino acid residues were investigated by Epand and Scheraga (1968). The researchers concluded from ORD and CD measurements that a markedly lower helix content was assumed by the isolated fragments in water as compared to the native protein in aqueous solutions. In other solvents, such as 95% aqueous methanol, a higher helix content was observed. A relatively high amount of helix was reported in two central peptides of sperm whale myoglobin, it was still much lower than in native intact protein (Atassi and Singhal, 1970; Singhal and Atassi, 1970). The helical structures investigated become less stable in water when long-range interactions, mainly hydrophobic interactions, are missing.

The N-terminal peptide consisting of the first 20 (S peptide) and another of the first 13 amino acid residues of ribonuclease were found to fold into helical

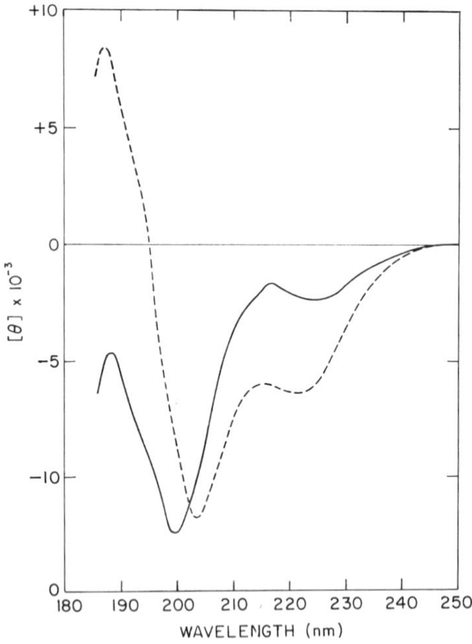

Fig. 10.2. Circular dichroism spectra of peptide 1–13 from ribonuclease in 0.033 *M* Na_2SO_4 at 26°C (—) and 1°C (---). Peptide concentration was 0.254 mg/ml at 1°C and 0.254 mg/ml at 26°C from 186 to 215 nm and 0.507 mg/ml for 215 to 250 nm; path length 0.02 cm (from Brown and Klee, 1971).

conformation in aqueous solution (Klee, 1968; Brown and Klee, 1969, 1971; Silverman *et al.*, 1972). They contain all the amino acid residues which comprise the helix segment of the N-terminal part of the molecule as observed by X-ray crystallography (Kartha *et al.*, 1967; Wyckoff *et al.*, 1967). In fact, peptide 1–13 is the smallest segment that can be induced to fold; the folding is avoided after proteolysis of the Phe 8–Glu 9 peptide bond by chymotrypsin. The structure of the peptide is sensitive to different factors such as solvent and ionic strength. Helix formation markedly increases when ionic strength reaches 0.1 or greater. Figure 10.2 shows the dichroic spectra of peptide 1–13 in 0.033 *M* Na_2SO_4 at two temperatures 1°C and 26°C, the change in the form of the CD spectrum on raising the temperature up to 26°C indicates a decrease in helix formation. Figure 10.3 indicates the variation in structure induced by the presence of GuHCl. At 26°C, a large change in ellipticity is observed. At high concentrations of GuHCl the CD spectrum is characteristic of a random coil.

Although this peptide tends, under definite conditions, to assume the helical structure it has in crystalline ribonuclease, it does not have great stability and it is extremely sensitive to the experimental conditions. The investigators (cited previously) all emphasized not only the importance of

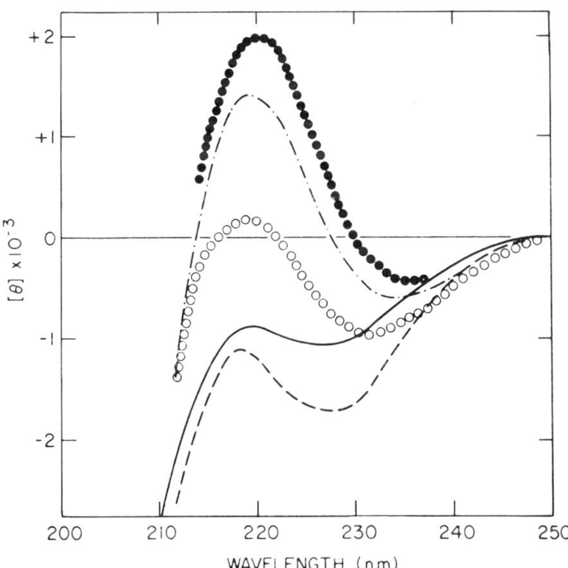

Fig. 10.3. Variation of circular dichroism spectra of peptide 1–13 from ribonuclease, in presence of guanidine hydrochloride at different concentration at 26°C: (—) peptide in water; (---) peptide in 1 *M* GuHCl; (○) peptide in 3 *M* GuHCl; (- · - · -) peptide in 5 *M* GuHCl; (●) peptide in 8 *M* GuHCl (from Brown and Klee, 1971).

short-range interactions to the initiation of folding, but also the requirement of long-range interactions to stabilize short helices in solution.

In spite of the dichroic spectra, Taniuchi and Anfinsen (1969) did not accept, from the available data, that peptide S had a significant helical character in solution. It is certainly difficult to draw conclusion on the basis of circular dichroism alone. When small fragments were able to undergo a significant folding, these fragments had helical structures in the intact protein.

When a part of the C-terminal end is removed by proteolysis in ribonuclease as well as in staphylococcal nuclease, the polypeptide chain is prevented from returning to the native structure. For staphylococcal nuclease, the removal by proteolysis of residues 127–149 (P 1–126) suppresses the ability to refold, whereas even after removal of five first residues at the N-terminal, the enzyme remains fully active (Taniuchi and Anfinsen, 1968, 1969; Bohnert and Taniuchi, 1972). Different fragments of nuclease were obtained (Fig. 10.4). Nuclease was cleaved by trypsin in the presence of a competitive inhibitor (3′5′ thymidine diphosphate, pdTp). The bond between Lys 5 and Lys 6 was cleaved first and yielded nuclease (6–149) which had the same enzymatic activity, physical, and folding properties as the complete protein. Then fragments nuclease T (6–48) and nuclease T (49–149), or nuclease T (50–149) were obtained. None of these fragments alone in solution showed ordered structure. Fragment 1–126 was obtained by action of trypsin on the trifluoroacetylated nuclease in which lysine groups were protected. Other fragments were obtained by CNBr cleavage (nuclease 99–149). Among these different fragments, only 6–149, exhibited nativelike structure in solution (Taniuchi and Anfinsen, 1968, 1969, 1971; Taniuchi *et al.*, 1969).

For ribonuclease, removal by pepsin proteolysis of only a tetrapeptide sequence at the C-terminus (Asp–Ala–Ser–Val) lead to an inactive enzyme and unstable structure (Anfinsen 1956; Taniuchi, 1970). When this shortened enzyme, the so-called des(121–124) ribonuclease or pepsin inactivated ribonuclease (PIR), is reduced, it cannot reoxidize to yield the native pairing of disulfide bonds (Taniuchi, 1970). Removal of six C-terminal residues to form RNase 1–118 also yields a structureless and inactive enzyme (Lin, 1970; Andria and Taniuchi, 1978). The importance of the C-terminal end to the folding of these two proteins was also emphasized by the data reported in Chapter 9. The information contained in the C-terminal sequence of the two nucleases appears to be crucial for their refolding. It was proposed that the polypeptide chain of nuclease and ribonuclease cannot achieve the native structure, during biosynthesis, until the termination of the polypeptide chain. For RNase even the N-terminal end seems to be important for folding. After removal by proteolytic action of subtilisin of the first 20 amino acids (Richards, 1958), the RNase S protein is unstable and cannot refold correctly when disulfide bridges are reduced.

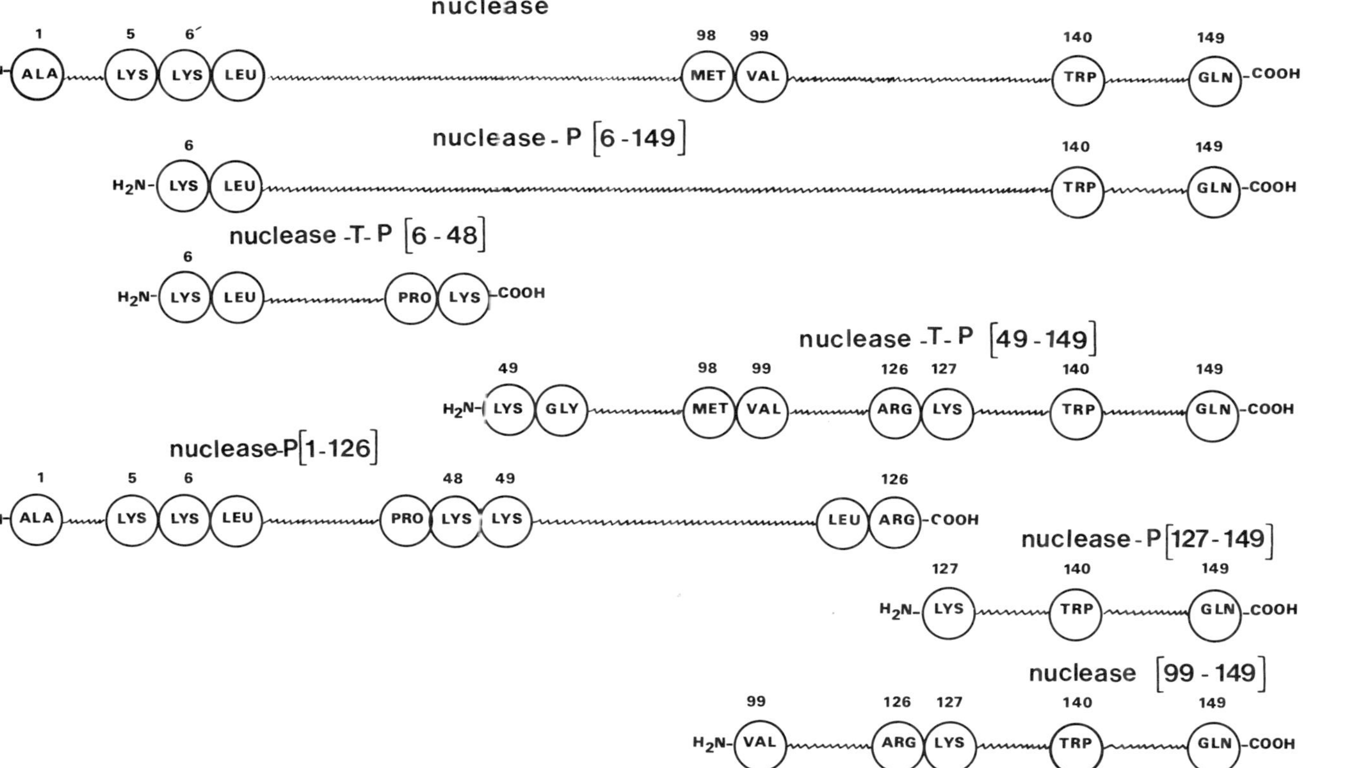

Fig. 10.4. The different fragments obtained from staphylococcal nuclease (adapted from Taniuchi and Anfinsen, 1969).

In contrast, for hen egg white lysozyme removal by carboxypeptidase of the C-terminal dipeptide Arg–Leu to produce fragment 1–127 does not affect significantly the refolding process. Both rate and yield of regeneration evaluated in the same conditions by, respectively, the rate and the extent after 120 min of regain of enzymatic activity were found very close to those of intact lysozyme (Johnson *et al.*, 1976). From these data, the C-terminal end of lysozyme does not seem to contain essential information for the formation of native structure. Smaller fragments of this protein were obtained by cleavage with CNBr of reduced lysozyme (1–12, 13–105, and 106–129). The reduced peptide 13–105 was regenerated in the presence of reduced and oxidized glutathione and the return of native properties was investigated with regard to cysteine content, immunochemical reaction with antisera to native lysozyme, CD, and the ability to bind specifically on an affinity column of carboxymethyl–chitine. Most criteria provided support to indicate that reduced fragment 13–105 regains a nativelike structure (E. R. Johnson *et al.*, 1976, 1978). This region contains the totality of the helical content and β-structure of lysozyme. However after reoxidation, a substantial difference in dichroic signal was observed indicating that there was a more disordered structure than in native lysozyme or even than in initial 13–105 peptide. These results were interpreted as supporting the idea of independent refolding of continuous regions in proteins as predicted by Wetlaufer (1973; see also, Wetlaufer, 1981). It is difficult to consider the 13–105 region in lysozyme as one isolated domain, because it represents a larger part of the molecule. However, based on dichroic signal data, this region does not refold to a structure completely identical to the native one. Therefore, some interactions are missing that ensure stabilization of a nativelike structure.

As revealed by CD studies, smaller fragments such as (57–107) and (57–85) + (91–107) in lysozyme are less ordered structures than native protein (Matthyssens *et al.*, 1972, 1974). In contrast, the N–C fragment (i.e., the region (1–27) + (122–129) linked through disulfide bridge Cys 6–Cys 127 seems to retain a high content of ordered structure (Matthyssens and Kanarek, 1974). Matthyssens and Kanarek (1974) concluded that stabilization of native structure in lysozyme takes place between the N- and C-terminal parts of the molecule. This interpretation does not agree with a sequential folding from N-terminal to C-terminal during biosynthesis as postulated by Phillips (1967). It also diverges from conclusions on refolding of fragment 13–105.

The number of well-documented examples is still insufficient to infer whether the entire polypeptide chain is required for folding and stabilization of native proteins. Many arguments support this assumption for small proteins, the situation being different for proteins built up of several structural domains.

10.2. FOLDING OF FRAGMENTS CORRESPONDING TO STRUCTURAL DOMAINS

The existence of domains in proteins has strongly suggested that the corresponding parts of the polypeptide chain are able to fold independently from separate nucleation centers (Wetlaufer, 1973). Despite the scarcity of experimental data, several observations are consistent with this assumption.

Elastase, which consists of two continuous domains was acetylated at its lysine groups. Then limited proteolysis by trypsin in the presence of its substrate, elastine, yielded two fragments, elastase T (16–125) and elastase T (126–245) (the numbering is that of chymotrypsinogen). These fragments correspond to the N- and C-terminal domains respectively, the former with one disulfide bridge, the latter with three disulfide bridges. Figure 10.5 shows the cleavage. Independent refolding and stabilization of fragment 126–245 (which was previously denatured by GuHCl or urea and reduced) was demonstrated by various methods. The refolded fragment exhibits a dichroic spectrum of structured protein. One equivalent of refolded fragments displaces one molecule of native elastase from its complex with purified specific antibodies (Fig. 10.6). The refolded fragment regains its ability to bind solubilized elastine with an affinity identical to that of the refolded complete protein ($K_D = 2.5\ \mu M$) (Ghelis *et al.*, 1978).

Spontaneous refolding of domains was also described for bovine serum albumin. As reported in Chapter 9, the three structural domains of bovine serum albumin were isolated (see Fig. 9.4). Corresponding reduced fragments were tested for return of antigenic activity during reoxidation. Independent refolding of separated domains was demonstrated by the return of their antigenic activity. An interdomain influence was indicated by variation in the rate of refolding of a particular domain when isolated or in the complete molecule (Fig. 10.7) (Teale and Benjamin, 1976a,b, 1977). Azuma and coworkers (1972) reported independent refolding domains in a Bence–Jones protein.

Interdomain influence on the refolding rate was also reported for aspartokinase II–homoserine dehydrogenase II (AK II–HDH II). A fragment having only HDH II activity was obtained by limited proteolysis by subtilisin (MW 2 × 37,000) (Dautry-Varsat and Cohen, 1977; Dautry-Varsat and Garel, 1978). This fragment after denaturation by GuHCl is able to refold and to regain dehydrogenase activity just as well as the complete bifunctional enzyme. The rate of refolding, as measured by the return of enzymatic activity, is 20 times faster when isolated (Dautry-Varsat and Garel, 1978). In this case, however, the HDH II fragment corresponds to a compact globule endowed with one of the two enzymatic activities, but its rather large size suggests that it contains more than one domain. With AK1–HDH1,

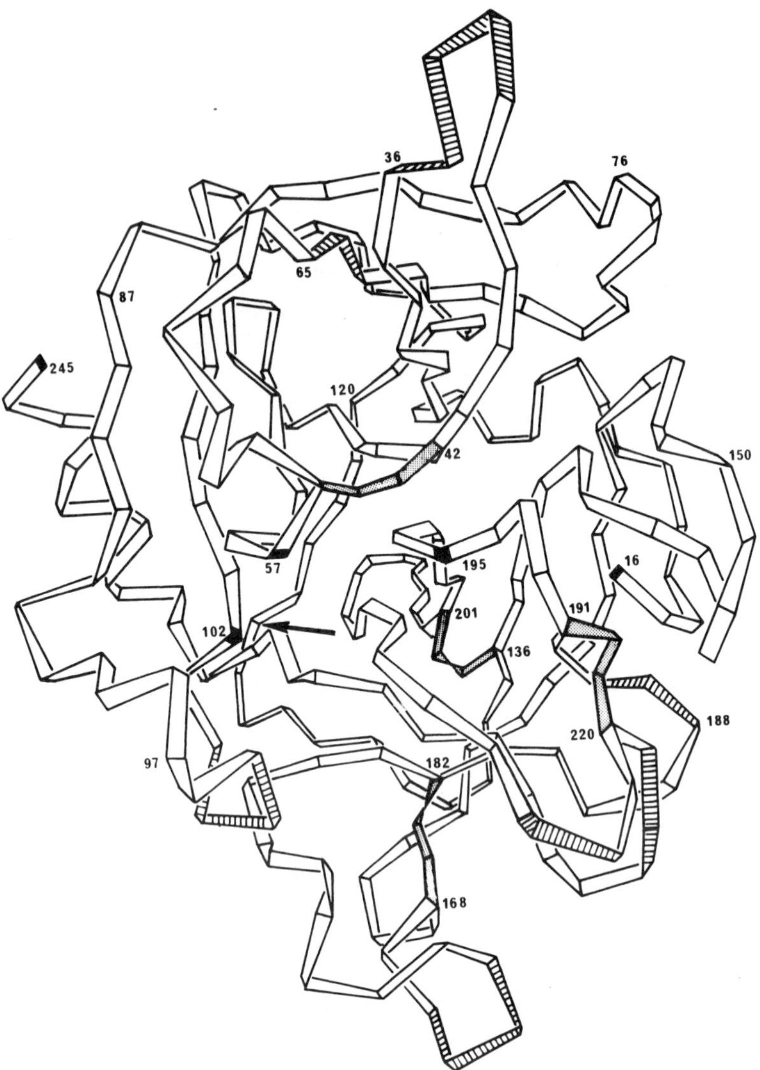

Fig. 10.5. Three-dimensional structure of elastase (according to Sawyer *et al.*, 1978) indicating the two domains. An arrow indicates the cleavage by trypsin (courtesy of H. C. Watson).

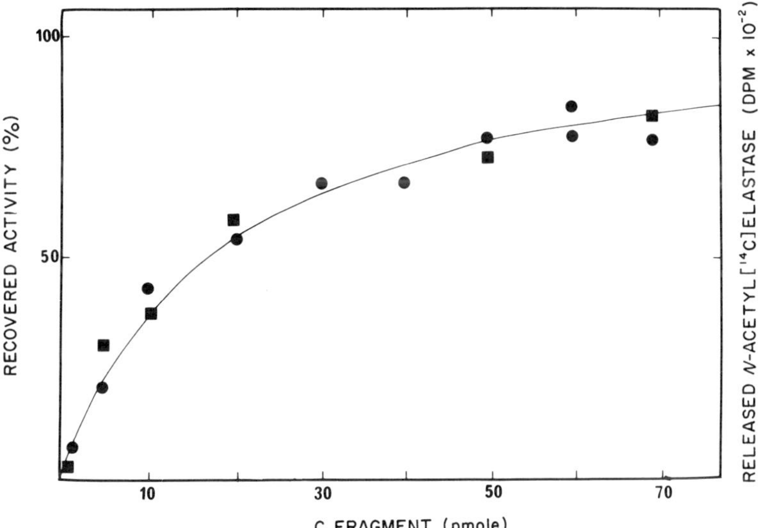

Fig. 10.6. Competitive displacement of native *N*-acetyl-[^{14}C]elastase from the specific elastase–antielastase complex by increasing amount of restructured elastase T (126–245) fragment. Concentration of elastase in antigen–antibody complex: (●) amount of *N*-acetyl [^{14}C] fragment displaced from antigen–antibody complex; (■) enzymatic activity of the released enzyme (from Ghélis *et al.*, 1978).

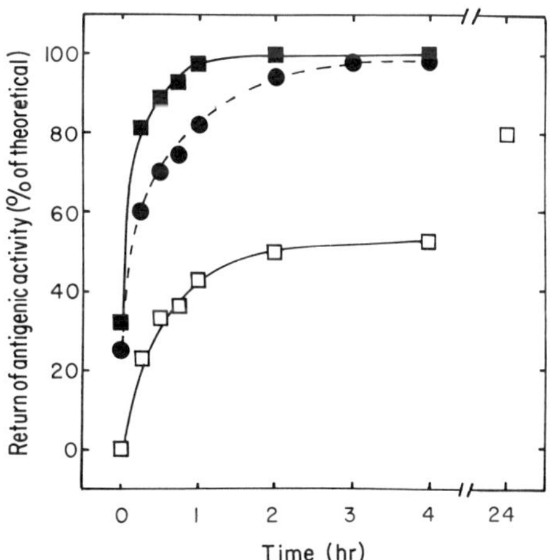

Fig. 10.7. Interdomain influence on the rate of refolding of domain III in serum albumin: (■) domain III; (●) domain (II + III); (□) entire albumin [i.e., domains (I + II + III)]. Refolding of domain III was followed by return of antigenic activity with purified antibodies specific to domain III (from Teale and Benjamin, 1977b).

it was also demonstrated that the two monofunctional regions in the intact enzyme (Garel and Dautry-Varsat, 1980), and when separated by limited proteolysis (Dautry-Varsat and Garel, 1981), are able to refold independently to produce functional species. However the structure is still unknown. Also the β_2 subunit of tryptophanase, whose three-dimensional structure is also unknown, was cleaved by trypsin into two large polypeptide fragments of MW 29,000 and 12,000. The denatured isolated fragments refold spontaneously and independently as indicated by various physical and chemical data (circular dichroism, fluorescence, cysteine reactivities) (Högberg-Raibaud and Goldberg, 1977a; Goldberg and Högberg-Raibaud, 1979; Zetina and Goldberg, 1980a,b). The α subunit of tryptophan synthase was also separated into two fragments. These fragments regain ordered structure after removal of urea as evinced by CD data (Higgins *et al.*, 1979). In these cases, there is no evidence that the fragments obtained correspond to domains; the size of each of them is larger than is generally found for a single domain and possibly the fragments obtained are not the smallest folding unit of these molecules.

The data presented in this section demonstrate independent refolding and stabilization of large fragments corresponding to structural domains or larger globules. The only two examples where fragments correspond exactly to domains are elastase and serum albumin although three-dimensional structure is still unknown for the latter protein. Similar observations of two different proteins indicate that isolated fragments refold faster when isolated than the corresponding region in the intact protein suggesting the influence on refolding of interaction between independent regions within the molecule.

Conversely, evidence was obtained that suggested the independent unfolding of domains in multidomain proteins. Two quite independent transitions were found by calorimetry for the unfolding of papain which is composed of two distinct domains (Tiktopulo and Privalov, 1978). Also using calorimetry, Pabo and co-workers (1979) observed two transitions in λ repressor, one near 50°C, the other near 70°C. The protein was cleaved by papain producing two fragments, the N-terminal (1–92) and the C-terminal (132–236). These isolated fragments unfold respectively near 50°C and 70°C. By calorimetric measurements, Ploplis *et al.* (1981) have obtained indications that in prothrombin domains unfold separately. Similar conclusions were proposed by Irace *et al.* (1981) on apomyoglobins by analysis of the tryptophanyl fluorescence. In this case, and also for other examples, the authors certainly have obtained evidence for separate unfolding of distinct regions, but nothing allows assimilation of these regions into structural domains. An increasing number of examples tend to demonstrate the independent folding or unfolding of structural domains.

Thus fragments which correspond to structural domains or larger *segments* behave as independent building units. All interactions are present within each of them to direct the refolding process and to stabilize the folding unit. This is in contrast with smaller fragments which have in the most favorable cases a significant probability for refolding, but are not stable enough to get complete nativelike structure.

10.3. NUCLEATION AND FOLDING

As suggested, folding of a polypeptide chain may be initiated in preferential sites that act as nucleation centers. At least two different models were proposed for nucleation process. The sequential model of nucleation propagation was proposed by Tsong and co-workers (1972a) to account for kinetic data of protein folding. In this model, the first step is a random search nucleation assumed to be a rate limiting process. It is followed by chain propagation in which propagation steps occur very quickly. Thus rapid folding is achieved by a cooperative process. Chain propagation steps were evaluated to be as fast as 10^{-6}–10^{-8} sec. In the model proposed by Matheson and Scheraga (1978, 1979a,b), nucleation is dominated by hydrophobic interactions. Occurring in the most hydrophobic regions of a molecule, nucleation is not a random search. In the dynamic model of diffusion–collision proposed by Karplus and Weaver (1976), several short parts of the polypeptide chain rapidly search their conformation, diffuse together, collide, and coalesce into a structural entity with the native conformation. In this model nucleation is expected to be fast, since the random search occurs in very small segments. The time required for coalescence may be longer. Nuclei are expected to persist as structures in the final folded protein according to the nucleation propagation model, whereas they may or may not persist in the diffusion–collision model.

Whether one accepts one hypothesis or the other, the concept of nucleation is clear. However, experimental support for the role of nucleation and the location of nucleation sites in proteins is only indirect evidence. Nucleation sites were predicted by Matheson and Scheraga (1978) for several proteins on the basis of hydrophobic interactions. In earlier interpretation of kinetic data, the slow step in refolding was attributed to the nucleation process (Tsong *et al.*, 1972a) according to nucleation model. Schmid and Baldwin (1978a,b) and Labhardt and Baldwin (1979a,b) have considered this step to be caused by proline isomerization. Scheraga (1980) proposed that proline cis–trans isomerism could possibly be included in the nucleation process. However, Schmid and Baldwin (1978a,b) and Schmid (1980) showed that RNase can fold whether or not cis–trans isomerization of proline was achieved. The cis–trans isomerism of proline represents a very particular aspect and has to be considered separately.

Experimental approaches by immunochemical methods using fractionated monospecific antibodies to determine location of nucleation sites were employed by Chavez and Benjamin (1978) for serum albumin, and by Chavez and Scheraga (1980a) for ribonuclease. For this latter protein, it was found that the antigenic determinant in segment 87–104 folds at an early stage and that its folding precedes folding of antigenic determinant in peptide segments 40–61, 63–75, and 1–10 (see Chapter 9). Matheson and Scheraga (1978) have proposed that the primary nucleation site for RNase is contained in residues 106–118. It was thus suggested that this nucleated region induced early folding of antigenic determinant in segment 87–104.

After thermal denaturation of BPTI, remaining microstructures consisting of hydrophobic pockets, as reported by Wagner and Wüthrich (1978a,b) (see Chapter 8), may be also considered as nucleation centers. Whether or not nucleation centers persist during the whole folding process may be questioned. For ribonuclease, antigenic determinants of native protein contained in segment 87–104 reappear in the early stage of the folding process. Therefore, native structure in this segment is formed from the beginning of the process and persists in the folded molecule. Nevertheless, more information is required to resolve the matter satisfactorily.

When nucleated, the folded segment very rapidly grows in size to achieve a globular form, within either a domain or the whole protein. However, it has not been demonstrated whether only one or more than one nucleation sites exist within a structural domain. Complementary information obtained from both experiment and theory certainly suggest that folding is initiated by nucleation in a limited portion of the polypeptide chain. This portion probably consists of hydrophobic residues, although this has not been proven. The nucleation process is of importance to the selection of the folding pathway(s). Accurate detection and location of nucleation sites remains to be performed for different proteins. There is insufficient data for understanding the nucleation process, but an experimental approach is not easy since nucleated and not yet folded species probably have a very short lifetime.

10.4. *IN VITRO* COMPLEMENTATION OF FRAGMENTS

Although slightly different from complementation of fragments, the ancient experiment of Fruchter and Crestfield (1965) on dimers of ribonuclease is the first example of *in vitro* complementation and must be cited to introduce this section. By complementation of ribonuclease inactivated by carboxymethylation of either His 16 or His 119, but not both, formation of an active dimer was obtained in which 1 mole with free His 12 associates with 1 mole with free His 119.

With ribonuclease, complementation of peptide S (1–20) with RNase S (21–124) yields an active protein. The addition of peptide S during refolding of reduced RNase S allows the correct disulfide pairing (Kato and Anfinsen, 1969). It is likely that this small peptide S has a role in directing the correct refolding of RNase by providing interactions which stabilize the structure. The helix character of peptide S is assumed when it is combined non-covalently with protein S by stabilization of its folded form as a result of long-range interactions (Kato and Anfinsen, 1969). Complementation was also achieved with synthetic fragment 1–20 and yielded an active semi-synthetic complex after binding with native RNase S (Chaiken *et al.*, 1973; Pandin *et al.*, 1976). However, when the last five residues of the carboxyl end are removed, only a very weak complementation with peptide S can be obtained and an inactive complex is formed (Potts *et al.*, 1964); this indicates the importance of the C-terminal region even for complementation.

RNase 1–118 complements with synthetic C-terminal fragments of various lengths (from residues 111, 113, 114, 115, or 116 to residue 124) to yield an enzymatic activity of 60–98% (Lin *et al.*, 1970; Gutte *et al.*, 1972). X-ray crystallographic study of such a complemented molecule is under study (Sasaki *et al.*, 1979). Complementation of three fragments, peptide S (1–20), des (121–124) RNase S (21–120), and synthetic peptide (111–124) yielded approximately 30% of enzyme activity (Lin *et al.*, 1970; Gutte *et al.*, 1972).

In a similar manner, complementation was achieved with different fragments of staphylococcal nuclease. Andria and co-workers (1971) formed a complemented enzyme with the three fragments (6–48, 49–126, and 99–149) obtained as shown in Fig. 10.4, with a stoichiometry of 1:1:1, yielding a weak but significant enzymatic activity of 0.1% of native enzyme. With staphylococcal nuclease, complementation of nuclease T (6–48) and nuclease T (49–149), or nuclease T (50–149) led to a protein which displayed about 8% of native activity in the presence of Ca^{2+}. The overall conformational properties were very close to those of native protein, 80% of helix content was regained as estimated by CD and ORD (Taniuchi and Anfinsen, 1968, 1969, 1971). Another type of enzymatically active structure was formed by complementation of nuclease (1–126) and nuclease T (49–149) with a yield of approximately 50%. When the two complementing and overlapping fragments are added in equal amounts, two alternative enzymatically active structures can form (Fig. 10.8). These two types of complemented structures are formed simultaneously in approximately equal amounts. In type I, sequence 1–48 of fragment 1–126 binds to nuclease T (49–149). In type II, sequence 1–126 binds portion 111–149 of fragment 49–149. The redundant portions protrude flexibly, being accessible to trypsin even in the presence of specific ligands. These two types of structures were identified by analysis

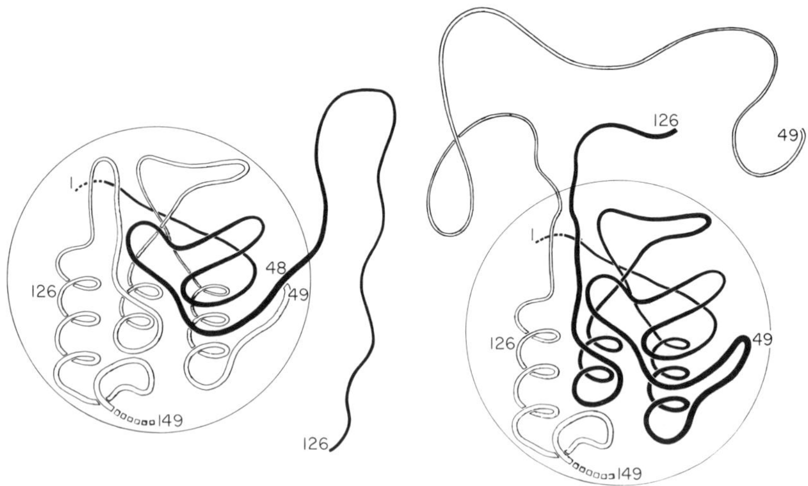

Fig. 10.8. The two alternative complementing structures formed by complementation of two overlapping fragments of staphylococcal nuclease, nuclease 1–126 and nuclease 49–149 (from Taniuchi and Anfinsen, 1971, courtesy of Taniuchi).

of trypsin digests. Complementation was found quantitatively with an association constant of 10^6–10^7 M^{-1}. A slightly greater thermodynamic stability was reported for type I structure (Taniuchi and Bohnert, 1973, 1975).

A kinetic analysis of complementation has been done (Light *et al.*, 1974). Apparent first-order kinetics with respect to each fragment concentration and variation of time were observed with rate of 0.03–0.05 sec^{-1}. From kinetic data, a mechanism of complementation was suggested. A model was developed in which prefolding of each fragment gave productive collision. This collision, thought to be a rate limiting step, was discarded because of the results. However, Light and co-workers did not make any decision between either prefolding of one of the two fragments followed by binding to the other, or binding of the two disordered fragments followed by folding of the unfolded intermediate. Taniuchi and co-workers, (1977) in a quantitative analysis of equilibration of the complementing structures, found that 2 min after mixing of fragments a ratio of type I to type II of 0.3 is obtained. This ratio is independent of temperature and presence of ligands. Equilibrium is reached through unfolding and refolding: at equilibrium the ratio of type I to type II is 1.1 at 6°C and 2.4 at 23°C. Since the initial ratio is independent of temperature, it seems related to the entropic barrier of folding. In contrast, the equilibrium is temperature dependent and probably reflects the characteristics of the energy barrier of unfolding. Taniuchi and co-workers, taking into account these data and the deductions from kinetic analysis, assumed a hypothetical disordered intermediate complex of fragments 1–126 and T (50–149), and supposed that a dynamic event taking place in this intermediate governs the ratio of type I to type II.

Similar experiments were reported with cytochrome *c* (Hantgan and Taniuchi, 1977, 1978; Parr *et al.*, 1978; Parr and Taniuchi, 1979, 1980a,b, 1981). The heme containing fragment 1–65 (or 1–38) of horse heart cytochrome *c* combined with apoprotein 1–104. Redundant portions of the polypeptide chain were submitted to limited proteolysis by trypsin and complementing fragments were identified. Two alternative structures were also described (1–25)H + (23–104), and (1–38)H + (56–104) with a ratio of 2:1. A dissociation constant of about 3×10^{-8} *M* was determined. These structures possess 60% and 46% respectively of the biological activity of cytochrome *c*. The importance of sequence 26–38 in stabilizing ordered structure was emphasized by the weaker affinity of the heme fragment (1–25)H for apofragment 36–104 which gives a nonproductive complex.

In a more detailed study from the same group (Juillerat *et al.*, 1980), a three fragment complex was obtained from an equimolecular mixture of heme fragment (1–25)H and apofragments (1–65) and (39–104) with a yield higher than 90%. Redundant portions were removed by proteolysis. Two resulting three-fragment complexes were characterized. The one endowed with activity (i.e., reducible by ascorbate and lactate dehydrogenase when in the oxidized form) was identified as (1–25)H:(28–38):(56–104). It has a conformation resembling native cytochrome *c*. The overall apparent dissociation constants were found to be 10^{-12} M^2 and 10^{-15} M^2 for the ferric and the ferrous form respectively. It can be reconstituted within 1 sec. Stopped-flow kinetic studies suggest a pathway of assembly to be

$$(1\text{–}25)\text{H} \xrightarrow{(56\text{–}104)} (1\text{–}25)\text{H}:(56\text{–}104) \xrightarrow{(28\text{–}38)} (1\text{–}25)\text{H}:(28\text{–}38):(56\text{–}104)$$

In a later phase Met 80 is liganded to the heme iron.

In a detailed kinetic study of recombination of S peptide with S protein during the refolding of RNase S, Labhardt and Baldwin (1979a,b) found that the combination of the two fragments occurs before folding of S protein. Furthermore, RNase S is formed more rapidly than S protein alone. However, S protein gives a stable folding intermediate at low temperatures and at pH 1.7 but not at pH 6.8.

These examples indicate that a variety of combinations can give a functional structure. Other experiments describing complementation of fragments to yield nativelike structure have been reported. In most cases overall conformation of protein as examined by a physicochemical or an immunochemical probe was fully regained. Return of biological activity, while significant, was often very weak. It is the case for staphyloccocal nuclease and for ovine prolactine (Birk and Li, 1978) where 2% of activity was obtained by complementation of the two fragments PRL (1–53) and PRL (54–199). Full growth promoting activity was regained by complementation of N-terminal fragment of human somatotropin with synthetic analogs of the 47–52 amino acid sequence of the C-terminal fragment (Li *et al.*, 1978).

The possibility of obtaining the values of association constants during the complementation of fragments, and also of studying the kinetics of the process is of great importance in the understanding of the conformational state of the individual fragments and of the structural adjustments required for their stabilization in the native structure. This aspect was particularly developed by Taniuchi and Bohnert (1975).

In the experiments using *in vitro* complementation of fragments, several features must be emphasized: (1) even after proteolytic cleavage of a limited number of peptide bonds, noncovalent interactions are sufficient for retention of structure; (2) by reassociation of separated fragments, the interactions required for folding do exist and allow return to nativelike conformation; (3) it seems unlikely that associations result from interactions of folded fragments (A complemented intermediate with a still disordered structure has been postulated. Thus independent refolding of fragments derived from small proteins is unlikely. Refolding might arise within the unstructured complemented intermediate.); (4) overlapping fragments can reassociate in different ways to give alternative complemented structures.

10.5. *IN VITRO* COMPLEMENTATION OF FRAGMENTS CORRESPONDING TO STRUCTURAL DOMAINS

There is a significant number of proteins which can be separated in several folded fragments following limited proteolysis. In most cases, sufficient non-covalent interactions maintain native structure and often dissociating reagents are required for separation. However, experimental data which refer to complementation of fragments corresponding to globular domains are limited to a small number of examples (Yon, 1978; Ghélis and Yon, 1982). The reassociation of domains was achieved with elastase (Ghélis *et al.*, 1978), thioredoxin (Holmgren, 1972; Slaby and Holmgren, 1975, 1979), tryptophan synthase (Högberg-Raibaud and Goldberg, 1977b; Goldberg and Högberg-Raibaud, 1979; Zetina and Goldberg, 1980a,b), and β-galactosidase (Goldberg, 1972) although in these two last cases complemented fragments did not really correspond to structural domains, but are larger. The two fragments of elastase, elastase T (16–125) and elastase T (126–245), obtained by limited proteolysis with insolubilized trypsin, when mixed in equal amounts under conditions where disulfide bonds are not reduced, reassociate to yield a complemented protein which displays physico-chemical characteristics of native elastase (absorbance, dichroic signal, elution profile from Sephadex G 100). Furthermore a weak but significant enzymatic activity was measured on specific substrate acetyltrialanine methyl ester that corresponds to about 2% activity compared to native enzyme ($k_{cat} = 1.4\ sec^{-1}$, whereas native enzyme has a $k_{cat} = 63\ sec^{-1}$). Figure 10.9 shows separation, purification, and reassociation of fragments. The nicked protein before separation of the domains keeps about 60% activity.

Holmgren (1972) and Slaby and Holmgren (1975, 1979) described the same kind of experiments with thioredoxin. This small protein can be split by trypsin digestion of the citraconylated protein, in two fragments corresponding to its two domains, thioredoxin T (1–73) and thioredoxin T (74–108). Cleavage by CNBr at Met 37 gives two other peptide fragments. An equimolecular mixture of thioredoxin T (1–73) and thioredoxin

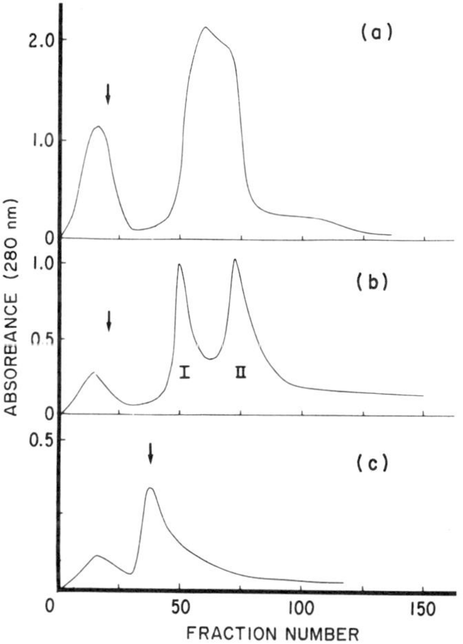

Fig. 10.9. Separation, purification, and complementation of fragments elastase T (16–125) and elastase T (126–245). Elution profile of proteins from a Sephadex G 100 column: (a) *N*-acetyl [^{14}C]elastase after limited proteolysis in a column equilibrated in 6 *M* GuHCl (100 m*M* acetate buffer, pH 5.0); (b) Purification of protein fragments (obtained as illustrated in a) by repeated chromatographic runs in the same conditions; (c) Reassociation of fragments obtained in b. After incubation at pH 8, the proteins were fractionated in the column without GuHCl, at pH 8.0 (borate buffer, 50 m*M*). The arrows indicate the position corresponding to the elution of the intact protein (from Ghélis *et al.*, 1978).

T (74–108) formed a reconstituted protein which had full immunoprecipitation activity with antithioredoxin antibodies, a significant but very weak enzymatic activity with NADPH and thioredoxin reductase of only 0.4%. Higher activity was regained by another complementing system of thioredoxin C (1–37) and thioredoxin C (38–108) (20% of activity was reported). Enzymatic activity of 4% was obtained after complementation of two overlapping fragments, thioredoxin T (1–73) and thioredoxin C (38–108). The three complemented systems had immunochemical properties similar to native thioredoxin.

In immunoglobulins, it was shown that heavy and light chains are able to fold independently of each other (Jaton *et al.*, 1968). With the F_v fragment derived from myeloma protein 315 which retains the binding site, it was shown that complete dissociation and unfolding of V_L and V_H fragments can be followed under suitable conditions by independent refolding of each of the variable domains which reassociate to form an active F_v fragment (Hochman *et al.*, 1973, 1976).

Human γ-thrombin, which consists of three noncovalently linked domains derived from a single chain (α-thrombin), by proteolytic cleavage is also able to refold after urea denaturation yielding an active enzyme (Bauer *et al.*, 1980; Chang *et al.*, 1980).

Complementation of large fragments was also achieved with tryptophan synthase (Högberg-Raibaud and Goldberg, 1977a,b; Goldberg and Högberg-Raibaud, 1979; Zetina and Goldberg, 1980a,b). The renatured fragments, as described in the preceding section can reassociate with a stoichiometry of one to one fragment. The reconstituted protein is very similar to native protein. Similar dichroic spectrum and fluorescence very close (80–90%) to that of initial nicked protein were observed. The sedimentation coefficient was not significantly different from that of the nicked protein. The dissociation constant for substrate L-serine ($K_D = 13 \pm 5$ mM) was found very similar to that reported for the nicked protein ($K_D = 20 \pm 5$ mM).

With soybean Bowman–Birk proteinase inhibitor, Odani and Ikenaka (1973) succeeded in separating two small fragments, one with 38 amino acids residues having a trypsin inhibitor activity, the other with 29 residues having a chymotrypsin inhibitor activity. However, reconstitution was not achieved.

Mevel-Ninio and co-workers (1977) and Gervais and co-workers (1980) succeeded in splitting cytochrome b_2 into different fragments, one with the heme group, the other with the flavin group, and to obtain reconstitution of the enzyme by complementation of fragments.

Although no kinetic studies have been performed for complementing domains, the mechanism of complementation for fragments of the size of structural domains probably differs from that described for smaller fragments. For small fragments, an unfolded intermediary complemented protein was postulated as follows:

$$\begin{array}{c}\text{unfolded fragment 1}\\+\\\text{unfolded fragment 2}\end{array}\longrightarrow\begin{array}{c}\text{unfolded complemented}\\\text{protein}\end{array}\longrightarrow\begin{array}{c}\text{folded complemented}\\\text{protein}\end{array}$$

With complementing domains, a different folding pathway might be postulated although the sequence of events is still not known:

$$\begin{array}{rcl}\text{unfolded fragment 1} & \longrightarrow & \text{folded domain 1}\\\text{unfolded fragment 2} & \longrightarrow & \text{folded domain 2}\\\text{folded domain 1} + \text{folded domain 2} & \longrightarrow & \text{complemented folded protein}\end{array}$$

Although not proven, this mechanism is plausible.

Ptitsyn (1978) assumed that intradomain interactions are stronger than interdomain interactions. The independency of refolding and stabilization of separated domains generally observed is consistent with this statement. However the role of the interdomain interactions in the folding process and in the energetics of stabilization of the native structure probably differs for each protein. Here, emphasis has been placed on the role of interdomain interactions in providing a strong coupling between domains giving rise to full activity of elastase (Ghélis *et al.*, 1978), and related proteins (Yon, 1978; Ghélis and Yon, 1979, 1982). The stabilizing effect of a domain resulting from interaction with another domain in β_2 subunit of tryptophan synthetase has been demonstrated by the investigations by Zetina and Goldberg (1980b), who concluded that there is a strong energetic coupling between the two domains. In T_4 lysozyme, the influence of one domain on the other is so strong that the entire molecule is rather resistant to proteolytic cleavage which is not able to separate the two domains. However, when the splitting starts, it leads to a complete break down of the molecule. Both equilibrium and kinetic data agree with a sequential pathway of folding; the C-terminal domain may refold independently in an early step, and the other domain requires the presence of the folded C-terminal domain for folding and stabilization (Desmadril and Yon, 1981, 1982; Desmadril *et al.*, 1982a).

Further studies are needed on a larger number of domain proteins to determine the respective energies of intradomain versus interdomain interactions.

10.6. ROLE OF DOMAINS AND OF SMALLER SUBSTRUCTURES IN PROTEIN FOLDING

The importance of domains as smaller units by which a protein can reach its three-dimensional and functional structure is experimentally supported by the previously cited examples. The functional role of domains in proteins and their evolutionary meaning are discussed in Part I. Their structural independency has received experimental support. However, it is not proven whether only one nucleation center per domain exists. No experimental data really evinces the detection of nucleation centers. Residual structures (probably in the hydrophobic part of the molecule) observed in thermal denaturation of BPTI, arguments developed from the acceleration of folding induced by the binding of specific antibodies to a limited region of serum albumin, and also some other experimental data were interpreted as arguments for the existence of nucleation centers. However, nobody has clearly detected nor ever isolated and analyzed nucleation centers in early refolding of proteins. The instability of such nucleus probably prevents any possibility of direct analysis. Subdomain structures resulting from interactions between segments of ordered structures have not been isolated. The smallest independent and relatively stable substructures isolated and analyzed with respect to folding and assembly properties are structural domains. Their existence as structural building blocks, very similar in many respects to subunits in oligomeric proteins, probably has evolutionary reasons. In early times, they could have plausibly existed as independent proteins and, with a selective advantage in several cases, gene fusion produced monomeric proteins from independent units. Covalent association of two functions represents, on the one hand, a functional advantage, on the other hand, a structural advantage since there is an increase in stability when compared with the same substructures noncovalently associated. For these two advantages, multidomain proteins might have emerged from selective pressure. During evolution, from the necessity to have differentiated functions, various proteins might have emerged from common ancestors by different combinations and associations of a functionally similar building block with another differentiated and variable building unit. According to that mechanism, Rossmann and co-workers (1975) proposed an explanation for the evolution of NAD^+ dehydrogenases.

The organization of large proteins in continuous structural domains certainly represents an advantage for the folding process with each domain being able to find in itself the correct interactions to direct the folding. Thus the process can be accelerated by a simultaneous folding of several smaller units.

Another important aspect of the existence of domains, is their interactions during folding and assembly. The importance of these interactions was emphasized by different experimental arguments. On the one hand, refolding of a domain is made slower by the presence of another. On the other hand, when reassociated after limited proteolytic cleavage and dissociation, only a weak enzymatic activity was regained in the cited examples. This activity is weak, even though all overall conformational properties, even in some cases binding of specific ligands, were identical to native proteins. Certainly enzymatic activity is the most sensitive property of a protein, since variation as small as a fraction of an Ångstrom in the position of catalytic groups might be dramatic. It was suggested (Ghélis and Yon, 1979), that the last and decisive step in protein folding consists of the correct coupling of domains in a multidomain protein as well as the correct coupling of subunits in oligomeric proteins. This coupling induces structural refinements which may be of very small amplitude, but which are required for the expression of full activity of an enzyme. Probably this conformational coupling cannot be optimal in the nicked protein, which behaves as desensitized oligomeric proteins.

It was also suggested (Ghélis and Yon, 1979) that differences in conformational coupling between structural domains in zymogens and in active enzymes, at least for serine proteases, such as chymotrypsin, elastase, and trypsin, can explain quantitatively the difference in activity, since zymogens have a very weak activity (Gertler *et al.*, 1974). The correct coupling between domains might ensure an important amplification of enzymatic activity as does the correct coupling through protomer interaction in oligomeric enzymes.

It would be of interest to obtain more experimental data on the folding and assembly of other multi-domain proteins to determine accurately the role of domains in the folding process. In the next step in the study of protein folding, one must study experimentally the mechanism of formation and the role of smaller substructures.

11

Folding–Unfolding of Oligomeric Proteins

11.1. SPECIFIC PROBLEMS CONCERNING FOLDING OF OLIGOMERIC PROTEINS

In preceding chapters analysis was made of several problems concerning the folding of monomeric as well as oligomeric proteins and experimental attempts to answer the main questions (as stated in Chapter 1). In this chapter consideration is given to problems specific to oligomeric proteins that arise from noncovalent assembly of subunits (identical or not). During refolding of oligomeric proteins, early steps are presumably very similar to those involved in refolding of single polypeptide chains. Independent refolding of each subunit probably must occur before assembly. However, the refolded structure of an isolated protomer might be different from the structure it has in the oligomer. Thus, two questions are posed at the starting point of any study concerning the refolding of an oligomeric protein: (1) Does the isolated subunit have full enzymatic activity? and (2) What are the conformational refinements induced by subunit assembly? Clear answers to these questions are of importance to the understanding of the folding pathway and therefore the mechanism of folding and assembly. These questions are closely related to the functional role of oligomeric structure. One may ask why have oligomers been selected for during evolution? It is now well understood from the example of allosteric proteins. In these proteins, the quaternary contraints through protomer interactions allow regulatory properties such as amplification, or contrarely, attenuation of the functional answer to a signal represented by variation of metabolite concentration. For allosteric proteins, one can expect structural modifications induced by protomer assembly, and that the subunit will be devoid of any activity. Other oligomeric proteins have no allosteric properties and selection of oligomeric structure is not clearly understood. However, maximum conformational stability of structure may be reached by subunit assembly. Plausibly, in this category of proteins, activity can be expected in the monomer.

Similarities between domains and subunits have been emphasised several times, and similarities in the folding pathway are expected. For oligomeric proteins, principally allosteric ones, conformational coupling between protomers probably occurs in the last step of folding and this correct coupling is crucial for functional properties.

11.2. KINETICS OF UNFOLDING–FOLDING OF OLIGOMERIC PROTEINS

Oligomeric proteins as well as monomeric ones are able to refold spontaneously under suitable regeneration conditions (see Chapter 5, Section 5.6), but the yield following reconstitution is often low because of the formation of wrong aggregates. The process of denaturation and reconstitution was described by the simplest sequence of events:

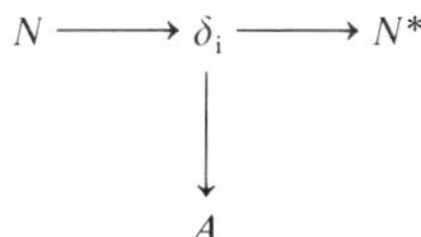

in which side reaction $\delta_i \rightarrow A$ refers to the formation of aggregates from unfolded state δ_i, N being the initial state and N^* the refolded protein (Jaenicke, 1979). These aggregates consist of incorrect intrachain interactions between either unfolded or partially refolded molecules. These aggregates can be converted back to unfolded form in the presence of 6 M GuHCl, EDTA, and dithiothreitol (Rudolph *et al.*, 1979). A kinetic competition between renaturation and aggregation was accepted from the results, since kinetics are indentical for each cycle of regeneration.

Under conditions of the equilibrium for unfolding–refolding and dissociation–association processes, side reactions become preponderant. For this reason, conditions for kinetic studies are generally chosen far from this equilibrium so that the reactions are practically irreversible (i.e., the rate constants for reverse reactions are negligible). In the critical range of the monomer–oligomer transition, a kind of hysteresis, i.e., a noncoïncidence of, on the one hand, deactivation–denaturation, on the other hand,

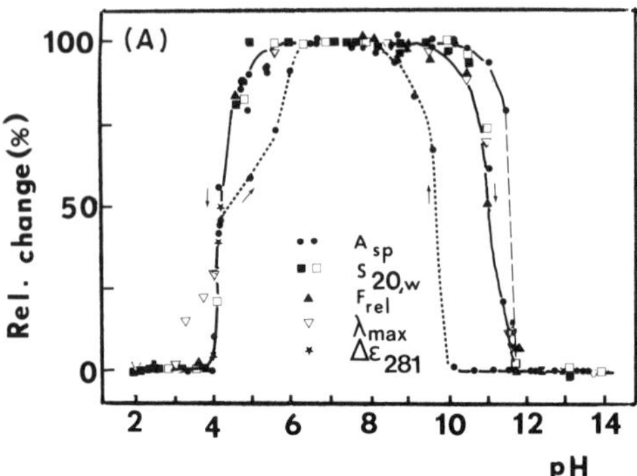

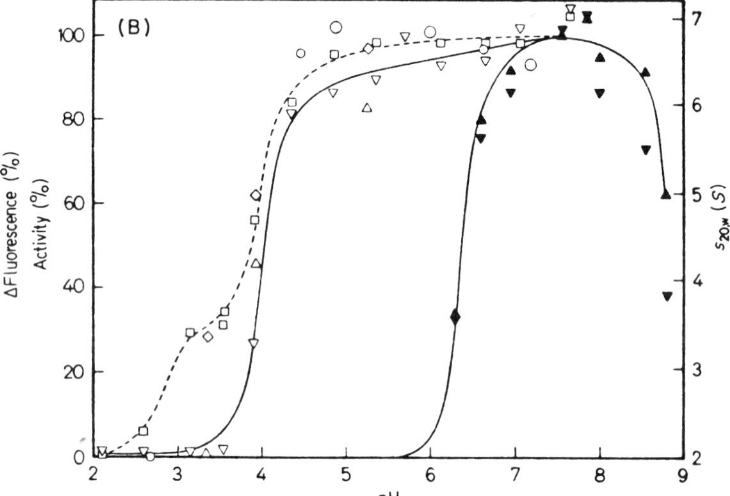

Fig. 11.1. The pH-dependent dissociation, deactivitation, and unfolding of two oligomeric enzymes. (A) Rabbit muscle aldolase (according to Jaenicke, 1979): (—) deactivation, denaturation, and dissociation of the tetramer, (· · ·) reconstitution from denatured monomers. Various parameters were measured: sedimentation coefficient s_{20w} (■, □) (c = 0.3 mg/ml), enzymatic activity (c = 0.02 mg/ml) A_{sp} (●), relative fluorescence at 320 nm F_{rel} (c = 0.02 mg/ml) (▲), maximum fluorescence λ_{max} (c = 0.02 mg/ml) (▽), circular dichroïsm at 281 nm $\Delta\varepsilon$ (c = 0.3 mg/ml), (∗) conditions 0.1 *M* phosphate + 1 m*M* EDTA + 0.1 m*M* DTT at 20°C. (B) Yeast GPDH (according to Rudolph *et al.*, 1977b) at 20°C in various conditions: phosphate 0.05 *M* or 0.1 *M*, 0.05 m*M*–1 m*M* EDTA, 0.05 or 0.1 m*M* DTT, (△) deactivation after 4 hr incubation (c = 6 μg/ml), (▼) reactivation after 20 hr, (▲) reactivation after 44 hr; denaturation was performed in citrate buffer pH 2.3 in the presence of 6 *M* GuHCl, □ ◇ denaturation measured by changes in protein fluorescence after 4 hr incubation (c = 6 μg/ml).

reactivation–renaturation is observed. Figure 11.1 indicates pH-dependent unfolding and folding of two enzymes, rabbit muscle aldolase (Fig. 11.1A) and yeast GPDH (Fig. 11.1B) (according to Jaenicke, 1979).

A prerequisite to the study of the kinetics of refolding and reassembly of oligomeric enzymes, is a good determination of the initial and final states. In several cases, the denatured form was certainly not entirely unfolded and residual structure was observed in the isolated monomer. In most studies, careful determination of the properties of the nativelike state obtained after regeneration was performed. Separation of aggregates by gel chromatography has made possible the study of renatured protein and the determination of physical characteristics. For most investigated proteins, renatured protein was found to be indistinguishable from the initial native one by many criteria (spectral, hydrodynamic, and enzymatic properties). This was observed for (1) aldolase (Teipel, 1972; Jaenicke, 1974; Engelhard *et al.*, 1976), (2) yeast GPDH (Rudolph *et al.*, 1977b), (3) for LDH (Zettlmeissl *et al.*, 1979a), (4) malate dehydrogenase (Jaenicke *et al.*, 1979), and (5) histidine decarboxylase which is a $\alpha_5\beta_5$ oligomeric protein (Yagamata and Snell, 1979).

In most studies of oligomeric enzymes, kinetics of renaturation is followed by the observation of several signals. Teipel and Koshland (1971a) reported a very rapid return of optical rotation and fluorescence signals for fumarase, enolase, aldolase, GPDH, and malate dehydrogenase (MDH). For all of them, physical properties of the native structure return in less than 1 min, whereas activity is regained very slowly. For several of these proteins, the gross structural change appears within 1 min; smaller conformational rearrangements occur later. Figure 11.2 shows the reactivation as a function of time of these enzymes and Fig. 11.3 shows the variations of fluorescence during renaturation of the same enzymes. In these experiments, kinetics of renaturation were not found to follow single first or even second order kinetics suggesting a higher order process. For these oligomeric enzymes, the return of enzymatic activity is generally very slow, but the rate of reactivation differs markedly from enzyme to enzyme. The half-time of reactivation varied from 4 min for aldolase to 75 min for fumarase and malate dehydrogenase (at pH 7.5 and 25°C).

A comparative study of the kinetics of refolding, subunit reassociation, and regain of enzymatic activity of rabbit muscle aldolase, a tetrameric enzyme was reported by Teipel (1972). In this study, the enzyme was denatured by 6 *M* GuHCl. The results indicate significant differences in the rates of variation of these different parameters. The return of native structure, as measured by optical rotation was completed within 30 sec. Kinetics of reassociation were followed by Rayleigh light scattering. Biphasic kinetics were recorded. A rapid reassociation was observed in phase I producing a mixture of monomers and dimers. In phase II, further association occurred

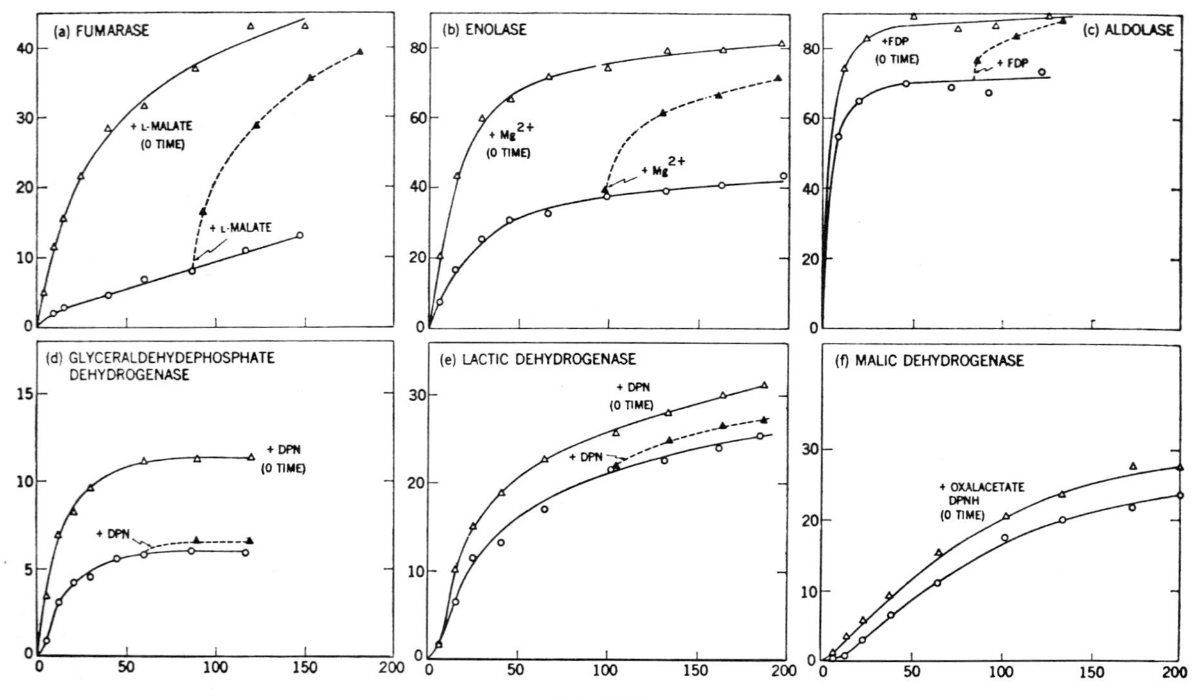

Fig. 11.2. Reactivation of several enzymes in the presence or absence of substrate or coenzyme (according to Teipel and Koshland, 1971a) after denaturation in 6 *M* GuHCl. Refolding was performed by a 101-fold dilution into regeneration mixture 0.05 *M* Tris acetate, 0.01 *M* DTT, 0.01 *M* EDTA pH 7.5, 25°C. Enzyme concentration was 0.028 mg/ml fumarase, 0.027 mg/ml enolase, 0.048 mg/ml aldolase, 0.01 mg/ml GPDH, 0.01 mg/ml LDH, and 0.02 mg/ml MDH.

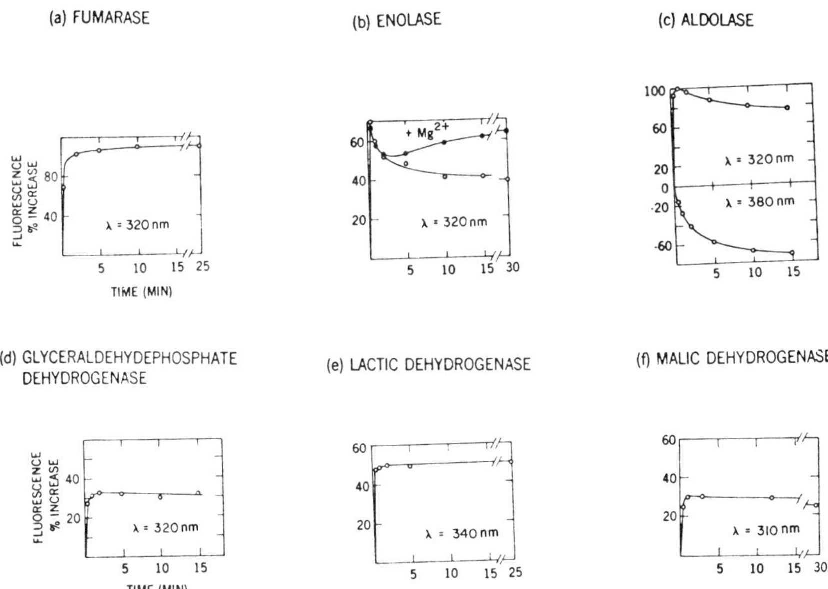

Fig. 11.3. Renaturation of the same enzymes as in Fig. 11.2 followed by variations of fluorescence emission (excitation at 280 nm) (according to Teipel and Koshland, 1971b).

with a slow rate to yield the nativelike tetramer. Kinetics of both reassociation and return of activity were described by two first-order limiting reactions. The apparent first-order dependence of association reaction was explained by assuming a rate-limiting isomerization followed by a rapid second or higher order process. The data did not allow one to distinguish whether these two first-order steps are sequential or parallel. Vimard and co-workers (1975) proposed a sequential pathway for the renaturation of denatured rabbit muscle aldolase to account for the renaturation kinetics.

More detailed studies using slow and fast kinetic techniques were performed to follow refolding of several enzymes after dissociation [e.g., rabbit muscle aldolase (Rudolph *et al.*, 1976), yeast GDPH (Rudolph *et al.*, 1977b), lactate dehydrogenase from pig skeletal muscle (isoenzyme M_4) (LDH-M_4) (Rudolph and Jaenicke, 1976; Rudolph *et al.*, 1977a, 1979; Zettlmeissl *et al.*, 1979a,b), and from pig heart (isoenzyme H_4) LDH-H_4 (Rudolph *et al.*, 1977c), malate dehydrogenase (*m*-MDH) (Jaenicke *et al.*, 1979), for rabbit muscle phosphofructokinase (Parr and Hammes, 1976)]. In most cases fluorescence emission was described as a multi-step process. For rabbit muscle aldolase after acid dissociation, a fast increase in fluorescence (first-order constant 15 sec^{-1}) is followed by a slow decrease. This last phase, which is concentration dependent, parallels the recovery of enzyme activity (Rudolph *et al.*,

1976). Rudolph and co-workers concluded that reassociation is a prerequisite for full catalytic activity and native fluorescence. Similar qualitative conclusions were reached by Rudolph and co-workers (1977c) for reactivation of aldolase after denaturation by GuHCl. Their results are not consistent with the activity reported for the monomer (Chan, 1970; Chan and Mawer, 1972; 1973a,b). This is discussed further in Section 11.4. A complex profile of fluorescence recovery was observed for *m*-MDH. For several dehydrogenases a multistep process is involved. For LDH-M_4 return of native fluorescence obeys a single second-order process.

Generally, the kinetics of return of native fluorescence are described by a biphasic process: a fast process which probably reflects the refolding of the isolated monomers and a slower process which, in most cases, parallels the reactivation process with similar reaction order at comparable enzyme concentration.

Jaenicke and co-workers (see review, Jaenicke, 1979) conducted careful kinetic analyses using different parameters, physical signals, and enzymatic activity recovery. Data were reported for GPDH, *m*-MDH, LDH-H_4, LDH-M_4, alcohol dehydrogenase (ADH), and triose phosphate isomerase (TIM). Groha and co-workers (1978) published a first report on β_2 dimer of tryptophanase. Kinetics of reactivation of most oligomeric enzymes displayed sigmoidal traces, and were described by an apparent unimolecular–bimolecular mechanism. Second-order reactivation kinetics were thus obtained at low enzyme concentration. When the enzyme concentration increased, it progressively shifted to approximately first-order kinetics. Figure 11.4 shows typical kinetic runs for reactivation of three different enzymes, LDH-H_4, MDH, and yeast GPDH.

The data obtained are generally consistent with unimolecular–bimolecular, multi-step-order reaction; unimolecular folding is followed by a bimolecular reaction, regardless of the number of protomers in the oligomeric enzyme.

Fig. 11.4. Kinetics of reactivation of three dehydrogenases: (A) glyceraldehyde-3-phosphate dehydrogenase (GPDH) (according to Rudolph *et al.*, 1977b). Deactivation by incubation for 5 min in 0.1 *M* citrate/HCl pH 2.3 + 5 m*M* EDTA + 10 m*M* DTT + 6 *M* GuHCl at 25°C. Reactivation by dilution at 15°C in 0.05 *M* triethanolamine/HCl pH 7.6 buffer + 5 m*M* EDTA + 10 m*M* DTT + 10 m*M* NAD^+ at varying enzyme concentrations. (B) Malate dehydrogenase (MDH) (according to Jaenicke *et al.*, 1979). Deactivation after 5 min in 1 *M* glycine/H_3PO_4 pH 2.3 1 m*M* EDTA, 1 m*M* DTE at 20°C. Renaturation in 0.2 *M* phosphate buffer pH 7.6 10 m*M* EDTA, 10 m*M* DTE at varying enzyme concentration. (C) Lactate dehydrogenase-H_4 (LDH-H_4) (according to Rudolph *et al.*, 1977a). Denaturation was produced either by 6 *M* GuHCl pH 2.3, or by 6 *M* GuHCl pH 7.6 or by 6 *M* urea pH 2.3. Reactivation in 0.2 *M* phosphate pH 7.6 + 1 m*M* EDTA + 0.1 m*M* DTT at 20°C at various enzyme concentrations.

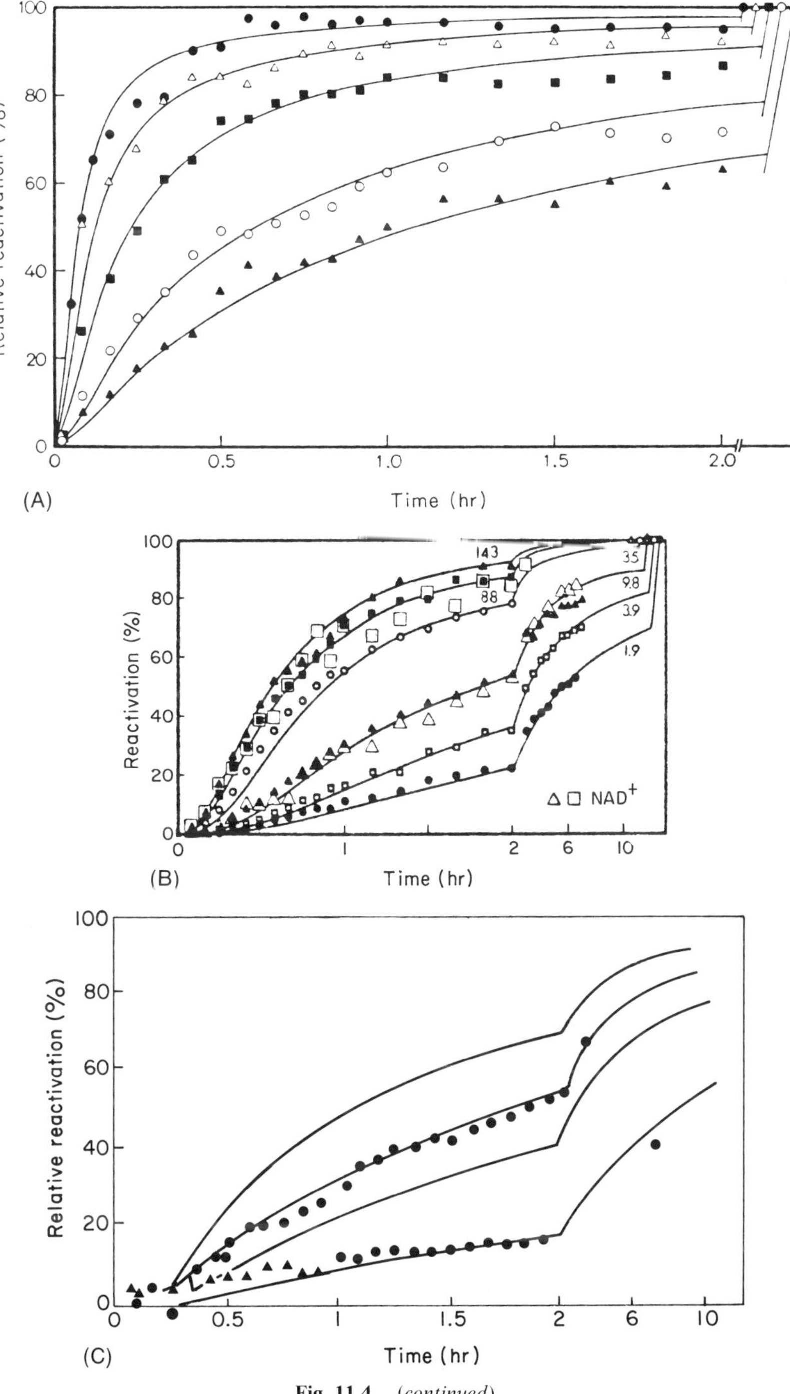

Fig. 11.4. *(continued)*

For a dimeric enzyme, the simplest kinetic pathway proposed for renaturation is

$$2\delta \xrightarrow{\text{uni}} 2D \xrightarrow{\text{bi}} D_2 \xrightarrow{\text{uni}} N^*$$

unimolecular refolding followed by a bimolecular step, then a final isomerization generating nativelike structure and functional properties.

For tetrameric proteins, a more complex sequence of unimolecular–bimolecular kinetics was proposed:

$$4\delta \xrightarrow{\text{uni}} 4D \xrightarrow{\text{bi}} 2D_2 \xrightarrow{\text{uni}} 2D'_2 \xrightarrow{\text{bi}} D'_4 \xrightarrow{\text{uni}} N^*$$

Different competitive side reactions may arise from δ and intermediate species to form incorrect aggregates as discussed previously.

A satisfactory fit was obtained by assuming the same phenomenological behavior for various oligomeric enzymes. Figure 11.5 shows the reaction order obtained for five different enzymes. The reactivation process for LDH-M_4, TIM, and Zn^+-ADH approaches a single two-order mechanism. For tryptophanase β_2 dimer reactivation obeyed first-order kinetics, indicating that an isomerization process is probably the rate-limiting step (Groha *et al.*, 1978).

For aldolase renaturation, the simplest scheme that describes the process includes a sequence of two rate-determining reactions of first and second order:

$$2D \xrightarrow{k_1} 2D^* \xrightarrow{k_2} N$$

with $k_1 = (7.4 \pm 1.4) \times 10^{-4}\ \text{sec}^{-1}$ and $k_2 = 1.4 \pm 0.4\ \text{m}M^{-1}\ \text{sec}^{-1}$. A residual activity of 50% ± 10% for the D^* form allowed a good fit of the data.

Vimard and co-workers (1975) studying renaturation of acid-denatured rabbit muscle aldolase, have characterized a folded monomer with a sedimentation coefficient $s_{20,w} = 3.1$ S. This monomer is more heat labile than the tetramer.

Denaturation of LDH-H_4 was induced by various methods including 6 *M* GuHCl, 6 *M* urea, and low pH. An irreversible unimolecular–bimolecular kinetic mechanism correctly describes refolding and reactivation (Fig. 11.5), with only a first-order rate constant $k_1 = (1.45 \pm 0.45) \times 10^{-3}$ sec^{-1} and a second-order reaction rate constant $k_2 = (5 \pm 1)\ \text{m}M^{-1}\ \text{sec}^{-1}$. These two constants are identical regardless of the denaturant employed. Irreversible steps in the kinetic mechanism are only operational; that means rate constants of the reversible process are very small under the experimental conditions. For the formation of an active tetrameric enzyme the following pathway was proposed:

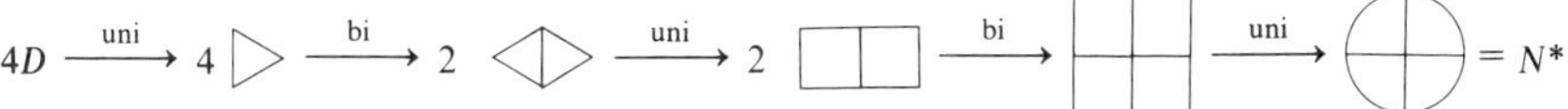

Formation of aggregates led to 60% regain of active tetramer under conditions of regeneration i.e., phosphate buffer, final pH 7.6 in the presence of 1 or 10 m*M* EDTA and of 0.1, 1, or 10 m*M* dithiothreitol. Inactive aggregates display spectral properties similar to that of denatured monomers.

With the M_4 isoenzyme, a single irreversible bimolecular process was sufficient to describe the kinetics of reactivation (Rudolph and Jaenicke, 1976). Competition between reactivation and formation of aggregates has been particularly well studied for this enzyme. The aggregates have been analyzed and their reactivation has been achieved after previous degradation by 6 *M* GuHCl (Zeittlmeissl *et al.*, 1979a,b; Rudolph *et al.*, 1979).

Refolding of yeast GPDH was also achieved by a unimolecular–bimolecular mechanism (Fig. 11.5A). Formation of irreversible aggregates competes with regeneration process. To obtain a high yield of reactivation, experimental conditions far from the transition, (tetramer ⇄ monomer) were chosen. Denaturation was performed by citrate pH 2.3 in the presence of 6 *M* GuHCl. Regeneration conditions chosen to ensure maximum recovery were pH 7.6, 0.1–10 m*M* dithiothreitol in the presence or in the absence of 10 m*M* NAD^+. High aggregates were separated by gel chromatography for analysis of the final regenerated enzyme. Figure 11.4A shows a typical kinetic run of reactivation. For this enzyme, regain of fluorescence parallels reactivation and is described by the same kinetics (see Fig. 11.5A).

Reconstitution of mitochondrial porcine malate dehydrogenase which is a dimeric enzyme was also described by a unimolecular–bimolecular mechanism with $k_1 = 6.5 \times 10^{-4}\ \text{sec}^{-1}$ and $k_2 = 3 \times 10^{-4}\ M^{-1}\ \text{sec}^{-1}$ regardless of the dissociation procedure applied (either pH 2.3, or 6 *M* urea, or 6 *M* GuHCl) (Rudolph *et al.*, 1979).

Complex kinetics of reactivation were observed for liver alcohol dehydrogenase in the presence of Zn^{2+}. No association was detectable in the absence of Zn^{2+}. However, excess of metal led to the formation of incorrectly refolded species (see Section 11.5) (Jaenicke, 1979).

A sequential unimolecular–bimolecular process was proposed to account for refolding and reactivation of tryptophan synthetase β_2 subunit previously denatured in 4.5 *M* GuHCl at pH 2.3. The return of enzymatic activity can be described by first-order kinetics over a large concentration range (3–0.04 μM) with a kinetic rate constant $k = 6 \pm 1 \times 10^{-4}\ \text{sec}^{-1}$. This was explained by a slow reshuffling process occurring after the first association

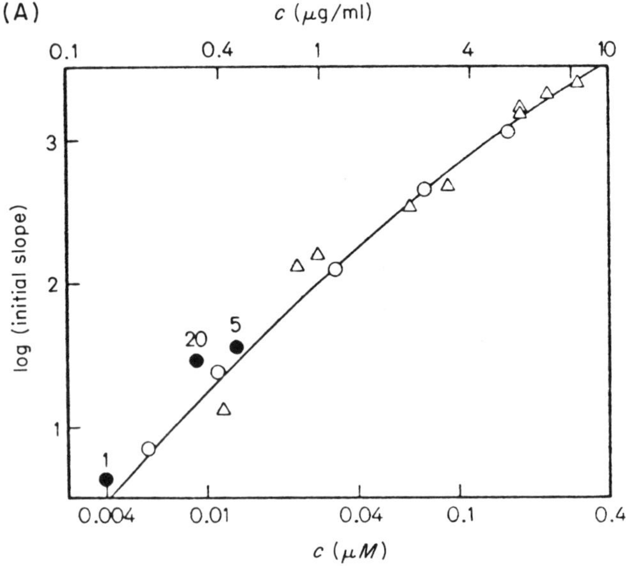

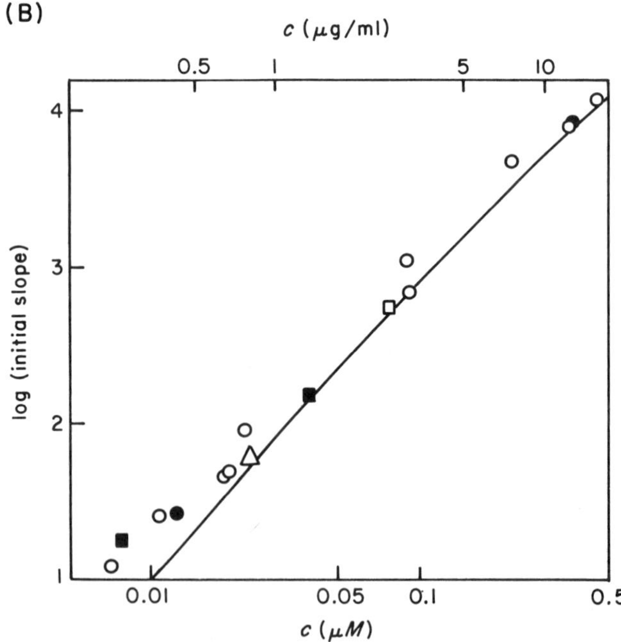

Fig. 11.5 (pp. 479–480). Reaction order for reactivation of several enzymes, determined from initial slopes: (A) GDPH (Rudolph *et al.*, 1977b); (B) LDH-H_4 (Rudolph *et al.*, 1977a); for both enzymes plots were calculated for irreversible unimolecular bimolecular model; (C) LDH-M_4 reaction order $n = 2.0 \pm 0.1$ (Rudolph *et al.*, 1976); (D) TIM reaction order $n = 2.0 \pm 0.2$, at high protein concentration the first-order reaction becomes significant (Jaenicke, 1979); (E) Zn^{2+} liver alcohol dehydrogenase, $n = 1.9 \pm 0.2$ (Jaenicke 1979).

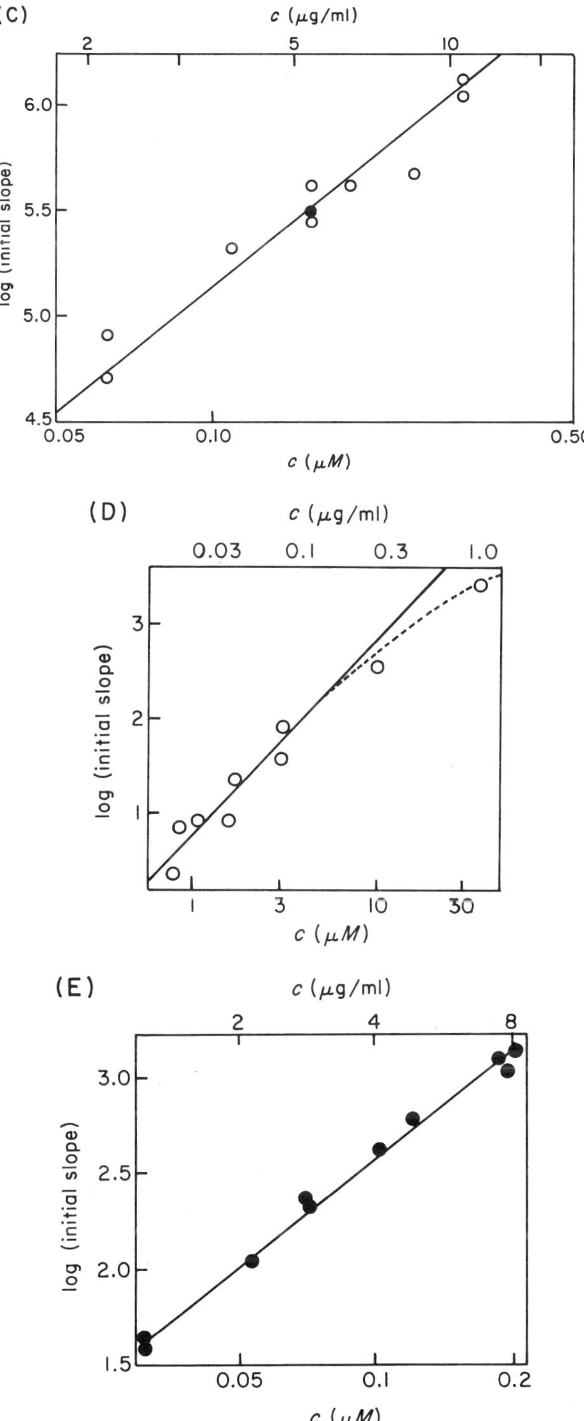

Fig. 11.5 (C), (D), (E) *(continued)*

of refolded monomers to form inactive precursor of native β_2 dimer (Groha *et al.*, 1978).

Denaturation of skeletal rabbit muscle phosphofructokinase was induced by GuHCl. As for other oligomeric enzymes, study was complicated by aggregation. Kinetics of denaturation and renaturation indicated the occurrence of several stages including refolding, reassembly, and regain of activity. The reactivation process was very slow and depended on the time of incubation in the denaturant. Detailed kinetic study of dissociation and reassociation of protomers was performed using stopped-flow fluorescence and activity measurements. For the denaturation process, three kinetic phases were found when changes in the protein fluorescence emission were measured and the following mechanism was proposed:

$$P_4 \underset{}{\overset{\tau_1}{\rightleftarrows}} 2P_2 \underset{}{\overset{\tau_2}{\rightleftarrows}} 4P \underset{}{\overset{\tau_3}{\rightleftarrows}} 4P' \rightleftarrows \text{random coil}$$

Dissociation of tetramer to dimer occurred with a relaxation time, τ_1, of a few milliseconds, the second relaxation time, τ_2, corresponding to transformation of dimers into monomers had a magnitude of a few seconds. The longest relaxation time, τ_3, was 4.8×10^{-2} sec at 23°C, and was associated with a conformational change of the monomer. Reassembly from 0.8 *M* GuHCl by 1 to 10 dilution of denaturant yielded about 70–94% of the original activity depending on the protein concentration. The reactivation process follows second-order kinetics (see Fig. 11.6 for the second-order profiles for return of enzymatic activity) (Parr and Hammes, 1975, 1976).

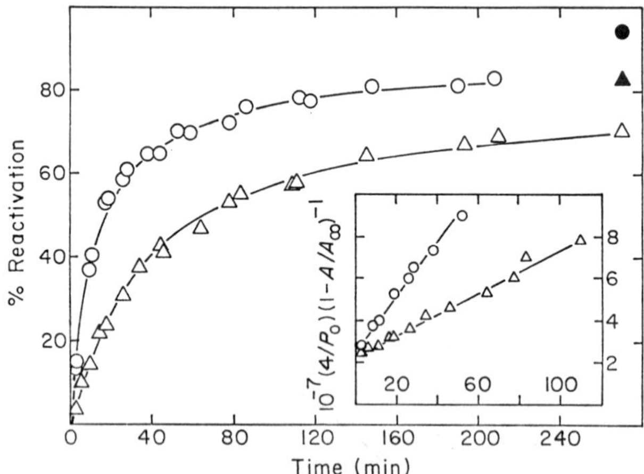

Fig. 11.6. Kinetics of reactivation of phosphofructokinase dissociated in 0.8 *M* GuHCl at 5°C (according to Parr and Hammes, 1976). Reactivation was started by a tenfold dilution of the denatured protein in phosphate buffer pH 8.0 (1 m*M* EDTA, 5 m*M* DTT) in the presence of 5 m*M* ATP (○) and in the absence of ATP (△). Insert: second-order kinetic plots.

TABLE 11.1

Kinetic Constants for Reactivation of Oligomeric Enzymes[a]

Enzyme	Source	M_r	n	Denaturation[b] Conditions	Denaturation[b] H	Renaturation–Reactivation[b] Conditions	$10^3 \times k_1$	$10^{-4} \times k_2$	A_{sp}
Alcohol dehydrogenase	Horse liver	80 000	2	A	0	J (25°C)	2.0[c]	0.16[c]	0
Aldolase	Rabbit muscle	160 000	4	B	50	K (0°C)	1.6	0.18	50
				A, C, D	0	L (4°C)	1.2		
				B	50	L (4°C)	1.3		
				E	70	L (4°C)	1.1		
D-Glyceraldehyde-3-phosphate dehydrogenase	Yeast	144 000	4	D	0	M (15°C)	11	4.5	0
				F (0°C)	100	F (25°C)	9	7.0	0
Lactate dehydrogenase	Pig heart H_4	140 000	4	A, C, G	0	K (20°C)	1.5	0.5	0
				B	40	K (20°C)	1.5	0.5	0
	Pig muscle M_4	140 000	4	B		K (20°C)	—	2.3[d]	0
				H (2 kbar)[e]		H (1 bar)	15	0.7	0
Malate dehydrogenase	Pig heart (mitochondrial)	70 000	2	B, C, G		K (20°C)	0.7	3.0	0
Triosephosphate isomerase	Rabbit muscle	50 000	2	A		L (0°C)	20	30	0

[a] According to Jaenicke (1979). M_r, native molecular weight; n, number of subunits per oligomer; H, % residual helicity (native state = 100%; k_1 (sec^{-1}), k_2 (M^{-1} sec^{-1}), kinetic constants for (apparent) first- and second-order reactions; A_{sp} (%), monomer activity fitting the irreversible unimolecular–bimolecular model.

[b] Optimum conditions of denaturation–renaturation: A, 0.1 M phosphate or Tris + 6 M Gdn-HCl, pH 7.8; B, 1 M Gly/H_3PO_4 pH_2; C, 0.05 N HCl + 7 M Gdn-HCl, pH 2; D, 0.1 M citrate + 6 M Gdn-HCl, pH 2; E, 5 M $MgCl_2$, pH 7.6; F, 0.1 M Tris, pH 8.5 + 5 mM ATP + 0.1 M 2-mercaptoethanol; G, 7 M urea, pH 2–8; H, 0.2 M Tris, pH 7.6 + 0.1 mM EDTA + 10 mM dithioerythritol; J, 0.1 M phosphate pH 7.6 + 0.1 mM NAD^+ + 5 μM Zn^{2+}; K, 0.2 M phosphate pH 7.6; L, 0.03 M TEA pH 7.6; M, 0.05 M TEA pH 7.6 + 10 mM NAD^+.

[c] Equilibrium constant K of the formation of the Zn^{2+} complex: $K = 0.17\ \mu M^{-1}$.

[d] The regain of native fluorescence at 340 nm ($\lambda_{exc} = 275$ nm) obeys the same rate constant. No anomalous temperature dependence is observed: activation energy 58 kcal/mole.

[e] Dissociation and deactivation at high hydrostatic pressure under anaerobic conditions (Schade, Lüdemann, Rudolph, and Jaenicke, unpublished data).

For aspartate transcarbamylase, denaturation of the catalytic trimer by 6 *M* GuHCl was obtained under the following conditions: pH 7.2, temperature 25°C, and enzyme concentration ranging from 10–30 μg/ml. Native fluorescence and circular dichroism return faster than enzymatic activity (Ghélis and Hervé, 1978).

The results from studies of the kinetics of refolding of various oligomeric proteins are summarized in Table 11.1. Also listed are the kinetic constants for renaturation of the different enzymes studied by Jaenicke and co-workers as well as the activity of the monomeric unit (if any) fitting with the irreversible unimolecular–bimolecular model (Jaenicke, 1979 Jaenicke and Rudolph, 1980).

As pointed out by Jaenicke and co-workers (1979), the similarity of the second-order rate constant for reactivation of different dehydrogenases (dimeric *m*-MDH, a tetrameric LDH-M_4, LDH-H_4, and yeast GPDH) (see Table 11.1) indicates that, for tetrameric dehydrogenases, the formation of the Q-axis dimer is rate limiting. In a recent paper, Jaenicke *et al.* (1981) have shown directly that the limiting process in reactivation of lactate dehydrogenase is the dimerization of inactive dimers to produce an active tetramer.

Certain common features can be noted from kinetic data concerning refolding of oligomeric proteins: (1) regain of activity is a relatively slow process; (2) most kinetics are consistent with a unimolecular–bimolecular mechanism indicating intramolecular folding, reassembly of protomers, and structural readjustments generating activity; (3) formation of incorrect aggregates kinetically competes with the regeneration process as a side reaction; and (4) the similarity in refolding and assembly process for dimeric and tetrameric enzymes clearly emphasizes that tetramers are formed on assembly of dimeric entities, in agreement with a D_2 symmetry observed for most tetrameric proteins (see Part I).

11.3. TRAPPING OF INTERMEDIATES DURING REFOLDING OF OLIGOMERIC PROTEINS

An analysis of kinetic studies using fluorescence signals and regain of activity indicated that tetramers are formed in two second-order reactions; the first one concerns formation of dimers; the other one, the associations of dimers to give tetramers. More direct analysis was provided by cross-linking of intermediates at different times during reconstitution of oligomeric enzymes which allows identification of the species and further quantitative determination by sodium dodecyl sulfate (SDS) polyacrylamide gel electrophoresis. Parallel measurements of enzymatic activity were made to correlate oligomer formation and regain of functional properties. Such a study was reported by Hermann and co-workers (1979) for LDH-M_4 from pig skeletal muscle. In this case, quantitative determination of kinetics of association by light scattering measurements was prevented by an unfavorable signal-to-noise ratio. Cross-linking was realized with glutaraldehyde as the bifunctional reagent, and the various products separated by SDS gel electrophoresis were quantitatively evaluated even at low enzyme concentration. The enzyme was denatured at pH 2.3 and renatured at pH 7.6. Figure 11.7 shows gel scanning at 560 nm. The bands corresponding respectively to tetramer (*T*), dimer (*D*) and monomer (*M*) of LDH were correlated by calibration with proteins of known molecular weight. In Fig. 11.7A are presented cross-linked native tetramer and SDS-denatured monomer. In Fig. 11.7B results of cross-linking on reactivation of LDH by dilution at 20°C and pH 7.6 are shown. No higher molecular weight species are formed under the conditions used.

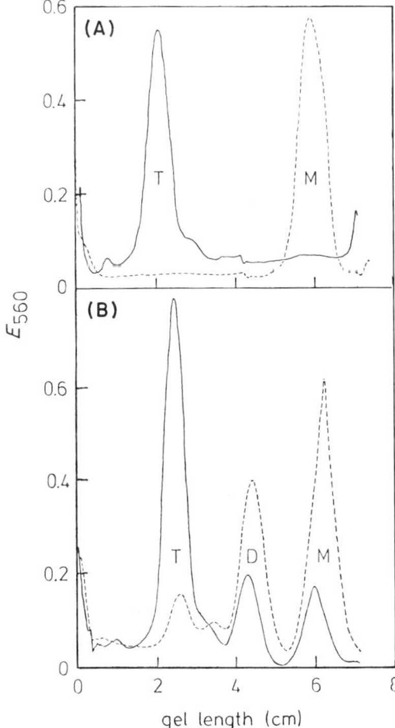

Fig. 11.7. Sodium dodecyl sulfate (SDS) polyacrylamide gel electrophoresis of cross-linked LDH-H_4 (according to Hermann *et al.*, 1979): (A) cross-linking of native tetramers (T) and SDS-denatured monomers (5 M); (B) cross-linking after 210 sec (---) and 1 hr (—) reactivation at 20°C by dilution at pH 7.6 after acid dissociation at pH 2.3.

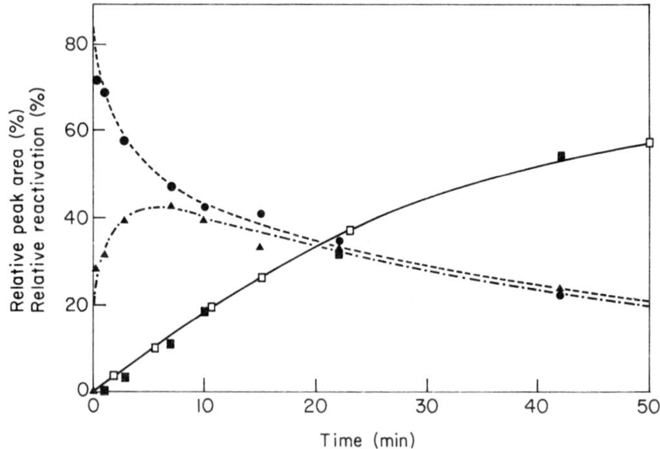

Fig. 11.8. Kinetics of reassociation as measured by cross-linking of LDH-H_4 (according to Hermann *et al.*, 1979); monomers (●); dimers (▲); tetramers (■); (□) regain of enzymatic activity (□).

Figure 11.8 shows the kinetics of disappearance of monomeric species, and of formation of dimers and tetramers. The identification of these three species only indicate that at least two reactions are limiting in the overall reconstitution process. Since a single second-order rate constant was sufficient to describe reconstitution of LDH-M_4 (Rudolph and Jaenicke, 1976; Jaenicke, 1979), only one of the reactions, either formation of dimer from monomer or formation of tetramers from dimers is rate determining.

11.4. ACTIVITY OF ISOLATED SUBUNITS

The difficulties encountered when obtaining isolated monomers without denaturing oligomeric proteins make it very difficult to prove whether an isolated monomer is really active even when dissociation is realized, since substrates and specific ligands generally favor reassociation. Under functional conditions, even at very weak protein concentration, an enzyme could have reassociated in the presence of substrate.

Nevertheless, the activity of an isolated monomer has been shown for aldolase (Chan and Mawer, 1972; Chan *et al.*, 1973a), and for yeast transaldolase (Chan *et al.*, 1973b), which is a dimeric protein, by the method of matrix-bound derivative (Chan, 1970). This method consists of binding native enzyme to a Sepharose-activated matrix, followed by exhaustive washing with 6 *M* GuHCl which allows the dissociation and removal of protomers which were not bound. Most of the linked molecules are covalently bound to the matrix through only one subunit. Then renaturation conditions restore 9.8% activity of the original protein for aldolase, and 63% for yeast transaldolase. Fully regained enzymatic activity is achieved in both cases by association with nascent subunits generated *in situ* (Fig. 11.9 shows these experiments schematically). However, Rudolph and co-workers (1977d)

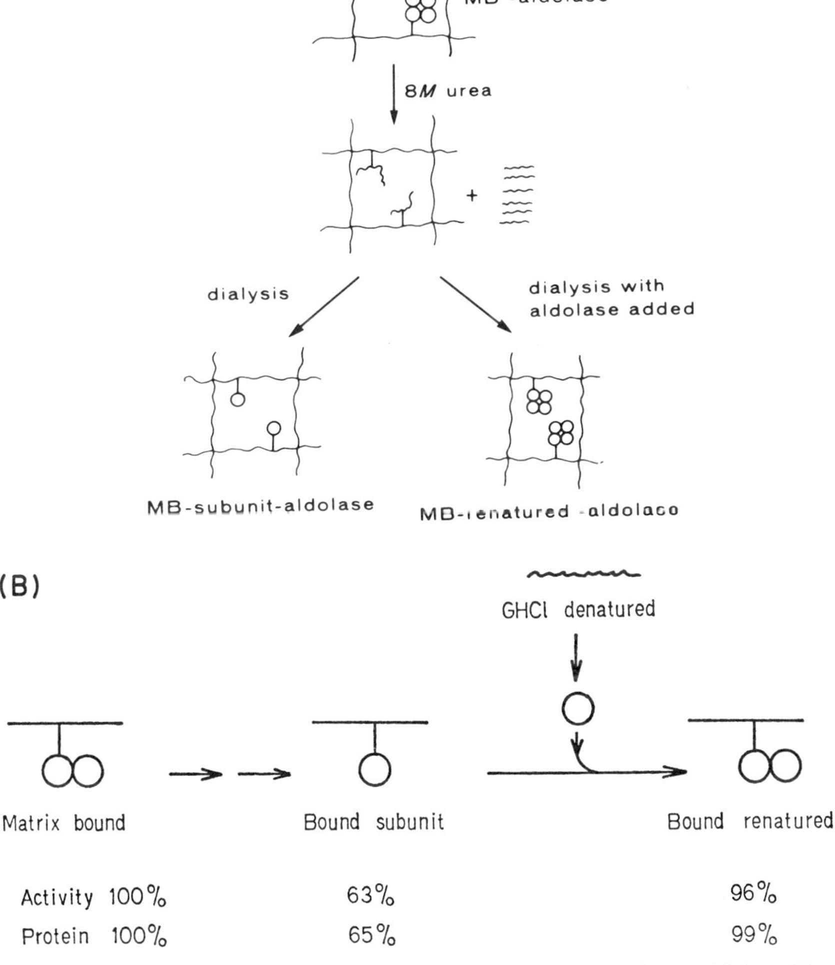

Fig. 11.9. Matrix-bound derivatives of aldolase (Chan, 1970) and transaldolase (Chan *et al.*, 1973a).

by assuming 50% of activity for the aldolase monomer in solution, obtained a good fit of their kinetic data and concluded that quaternary structure is necessary for full expression of enzymatic activity even when the monomer retains a great part of the activity. It was concluded for most oligomeric enzymes that the free subunit is devoid of activity. Such a conclusion was reached from kinetic data by Jaenicke and co-workers who used different enzymes, since in most cases the regain of enzymatic activity involved a second-order process.

Tetrameric structure is required for enzymatic activity of tryptophanase. For this enzyme, the dimer was inactive (Raibaud and Goldberg, 1976a,b). Apotryptophanase can be easily and reversibly dissociated into dimers at low temperature (Morino and Snell, 1967; Raibaud and Goldberg, 1973). The enzyme reassociates on binding of coenzyme pyridoxal-5′-phosphate (PLP). Raibaud and Goldberg (1976a,b) used tricine-KCl buffer pH 7.8 to dissociate tetramer into dimers and studied the equilibrium. The dissociation constant was determined:

$$K_{\mathrm{D}} = 16 \pm 4\ \mu M$$

Iodide was found to favor the dimeric form and citrate favored the tetramer. Kinetics of the appearance of enzyme activity induced by addition of PLP was found to be second order with respect to enzyme concentration and did not depend on PLP concentration. The second-order rate constant was $5 \times 10^4\ M^{-1}\,\mathrm{sec}^{-1}$. It is of interest to note that this value is of the same order of magnitude as those reported for reactivation of several dehydrogenases (see Table 11.1) (Jaenicke *et al.*, 1979). Therefore reassociation of dimers is the rate limiting step in the regain of enzymatic activity indicating that only the tetramer is functional.

With another pyridoxal phosphate enzyme, aspartate amino transferase from pig heart cytosol, it was shown by immobilization of the enzyme on a collagen film (Arrio-Dupont and Coulet, 1979), that quaternary structure is required for enzymatic activity. The immobilized monomers remaining after alkaline dissociation were devoid of any enzymatic activity. However, full activity of immobilized enzyme was regained after incubation with either native enzyme, or reduced inactive holoenzyme, or apoenzyme. Therefore, even binding of an inactive subunit allows the bound subunit to regain enzymatic activity indicating the importance of quaternary constraints in the structure of the active site of the enzyme. Thus, in most oligomeric enzymes, even in those which have no allosteric properties, quaternary structure is required for activity. When significant activity is retained in an isolated subunit in solution, as in the case of aldolase, it seems that quaternary structure enhances functional properties. Conformational coupling between protomers in oligomeric enzymes leads to amplification of catalytic properties.

11.5. CONFORMATIONAL CHANGES UPON SUBUNIT ASSEMBLY

In the different kinetic studies presented in Section 11.2, it was not possible to follow the reassembly process by a direct measurement of molecular weight as a function of time using light scattering (because of the limit of signal-to-noise ratio). From available data it is not possible to determine to what extend variations in fluorescence are caused by association. It is often very difficult to achieve dissociation of a protein under nondenaturing conditions, such as dilution, the high affinity of each subunit for the others might constrain one to work at very dilute concentration, so that in many cases changes in optical and other signals would be too weak for detection. In certain cases, it was possible to determine directly or indirectly the conformational changes attributable to subunit assembly.

Since quaternary constraints are driving regulatory properties of allosteric proteins, one can reasonably expect strong subunit interactions and consequently significant conformational changes in isolated monomers and in protomers in the oligomeric molecule and in homologous as well as in heterologous oligomers. More generally, when quaternary structure is required for the activity, one can expect conformational variations, at least near the active site, by protomer association, whether the protein is allosteric or not. Such conformational changes are not expected when the isolated monomer is found active, since, in this case they do not seem necessary.

The case of allosteric proteins and of their conformational properties is interesting with many respects. For proteins obeying the model of Monod *et al.* (1965), a pre-equilibrium between two active conformations is involved with an allosteric constant L_0 such that $L_0 = T_0/R_0$. This value indicates the relative conformational stability of each form compared to the other in the absence of ligand. Although it was determined only for a small number of proteins, several values are given in Table 11.2. In most cases one of the two forms is favored over the other. However, it is noteworthy that yeast phosphofructokinase has an allosteric constant of 3, indicating that under the same conditions and in the absence of ligand 25% of the protein is in the *R* form and 75% in the *T* form. An appreciable population of molecules is thus in a slightly less stable form. We can remark that allosteric proteins do exist under two different conformations, *R* and *T* which, in *K* systems have the same catalytic activity i.e., identical structure of the active

TABLE 11.2

Allosteric Constants for Several Proteins

Protein	L_0	Reference
Phosphofructokinase (*E. coli*)	4×10^6	Blangy *et al.* (1968)
Phosphofructokinase (yeast)	3	Laurent *et al.* (1979)
Aspartokinase I–Homoserine dehydrogenase I (*E. coli*)	0×016	Janin (1971)
Pyruvate kinase (yeast)	6×10^4	Johannes and Hess (1973)
Glyceraldehyde-3-phosphate dehydrogenase (yeast)	60	Kirschner *et al.* (1971)
Aspartate transcarbamylase	250	Howlett *et al.* (1977)
Phenylalanine hydroxylase (rat liver)	100	Dhondt *et al.* (1978)
Hemoglobin	9×10^3	Monod *et al.* 1965

site, but different structures of the binding site since each has a different affinity for the same ligand; one of these conformations is more stable than the other).* In this case, it is not possible to say that a polypeptide sequence determines a unique structure of the native protein under definite conditions, since the protein oscillates between two equally active but different conformations. In hemoglobin, the changes in substructure associated with the allosteric transitions are tightly localized rather than generally distributed. Each of the individual changes that occur on ligand binding represents structural destabilization (Malin and Englander, 1980).

Since it is generally difficult to obtain a stable monomer, the amount of data concerning the conformational changes induced by subunit assembly is limited. Aspartate transcarbamylase is an allosteric enzyme and an heterologous oligomer consisting of 6 catalytic (2×3) and 6 regulatory (3×2) subunits (Kantrowitz *et al.*, 1980a,b). It is possible to dissociate catalytic and regulatory subunits (Bothwell and Schachman, 1980). Several data indicate variation in conformation of catalytic subunits resulting from binding of the regulatory ones. Catalytic subunits are more stable to temperature when associated with regulatory subunits (Kerbiriou and Hervé, 1972). Differential spectra indicate small but significant variations of structure when catalytic and regulatory subunits are isolated or associated (Kirschner and Schach-

* However, it is difficult to accept the reality of a pure K system; that is, two conformations leading to different binding sites but identical catalytic properties, since catalysis is much more sensitive to variation of conformation than binding properties. For catalysis, a fraction of an Ångstrom may be determinant, whereas it may be indifferent for binding.

man, 1973). The same conclusions were reached by ellipticity measurements (Cohlberg *et al.*, 1972).

In *E. coli* tryptophan synthase, which is also an heterologous $\alpha_2\beta_2$ tetramer, conformational changes induced by the complementary subunit have received experimental support. The isolated β_2 dimer exhibits only 2% of enzymatic reactivity of the $\alpha_2\beta_2$ molecule (Janofsky and Crawford, 1972). Interactions between α_2 and β_2 subunits mutually induced conformational changes in both kind of protomers (Faeder and Hammes, 1971). Mechanism of reconstitution have been investigated by kinetic studies (Bartholmes *et al.*, 1980).

For tryptophanase, conformational variations in isolated protomers and protomers in the oligomeric enzyme have not been demonstrated directly. However, they can be deduced indirectly. A very strong interdependence of protomers in the tetrameric protein was evinced during binding of PLP. Binding of PLP causes an important conformational change in the protein accompanied by (1) an increase in sedimentation coefficient, (2) a decrease in intrinsic viscosity (Morino and Snell, 1967), and (3) a burying of four accessible cysteines (Högberg-Raibaud *et al.*, 1975). Using hybrid molecules between apoprotomers and irreversibly saturated holoprotomers (in which Schiff base with PLP was reduced by sodium borohydride), it was shown that structural changes are transmitted from holoprotomers to apoprotomers (Skrzynia *et al.*, 1974; Raibaud and Goldberg, 1976a,b). The presence of three reduced holoprotomers imposes a holo conformation to the last protomer. It binds PLP much slower and cystein reacts much slower than in the apoenzyme conformation. Furthermore the presence of two or three reduced holoprotomers in hybrids (H_2A_2) or (H_3A_1) stabilizes the neighboring protomers against thermal denaturation or cryodissociation. Certainly these data did not prove variations in conformation between monomers when isolated or in tetrameric tryptophanase; however, they clearly demonstrate a strong coupling between protomers in the oligomeric molecule. This coupling allows transmission of conformational change from a protomer to the others through quaternary constraints. It seems very likely that such strong interactions between subunits reflect conformational changes induced by protomer assembly.

Similar results were obtained in Yon's laboratory with aspartate amino transferase from pig heart cytosol, which is a dimer with one PLP bound per subunit. A strong anticooperativity for binding of pyridoxamine phosphate (PMP) was reported (Arrio-Dupont, 1972). In the haloenzyme–apoenzyme hybrid an important decrease of reactivity of Cys 190 (which is entirely accessible in apoenzyme and buried in holoenzyme) also indicated strong coupling between protomers in the enzyme molecule (Cournil, 1975; Cournil and Arrio-Dupont, 1975). Dissociation of the apoenzyme by dilution was studied and a constant of 0.8 μM was found for dissociation. Ligands such as PLP, PMP, and specific substrates favored dimerization ($K_D = 0.18\ \mu M$ for PLP–holoenzyme) (Cournil *et al.*, 1975). Nevertheless, it was possible to observe differences between monomeric and dimeric molecules using fluorescence quenching by iodide. In monomeric apoenzyme, a greater number of tryptophan groups are accessible to iodide molecules (Arrio-Dupont, 1978).

For aldolase, activity of the monomer was reported; therefore, only very small changes in conformation, if any, are expected by association of subunits. Native aldolase did not dissociate into monomers even at high dilution since very strong interactions between protomers stabilize the oligomeric structure. However, in 1.2 *M* $MgCl_2$, it dissociates into dimers and monomers (Hsu and Neet, 1973). Different spectroscopic parameters were studied in various $MgCl_2$ concentrations from 0.3 *M* to 3.2 *M* (Hsu and Neet, 1975). Very small variations occurred in these parameters in 0.3–1.2 *M* $MgCl_2$. A small fluorescence quenching was recorded and also small variations in absorption

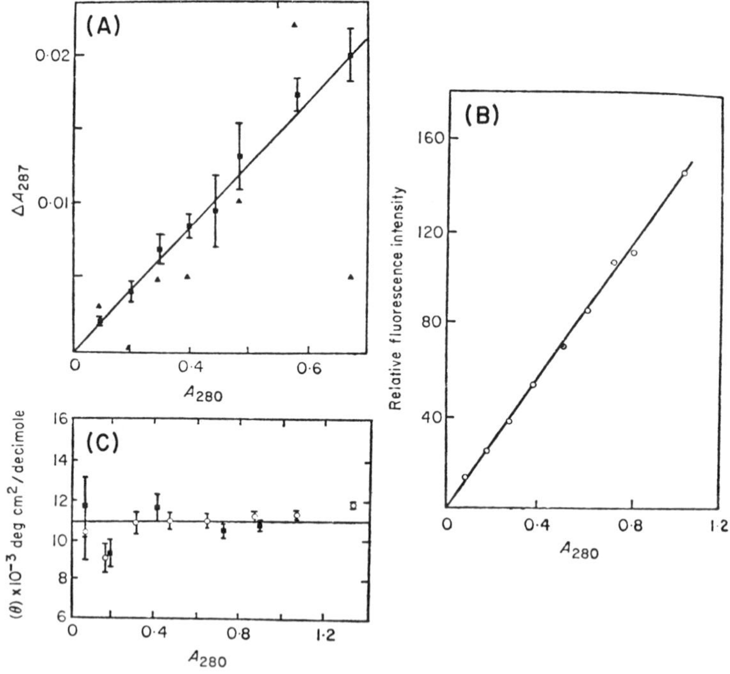

Fig. 11.10. Dependence of several spectroscopic parameters on protein concentration of aldolase in 1.2 *M* $MgCl_2$ which dissociates the enzyme (according to Hsu and Neet, 1975): (A) absorbance at 287 nm; (B) fluorescence intensity; (C) mean residue ellipticity.

whereas ellipticity at 220 nm did not vary. In 1.2 *M* $MgCl_2$, all spectroscopic parameters remained proportional to protein concentration (Fig. 11.10). Since the protein dissociates under these conditions, changes in optical properties do not depend on the dissociation process. Hsu and Neet concluded, from the absence of spectroscopic changes correlated with subunit dissociation, that each subunit (when isolated) remains in the same conformational state as in the tetramer (within the limits of detection). However, even a very small conformational change of each protomer, unstabilizes the tetramer and dissociation can occur.

Formation of oligomeric structures, when required for activity, certainly induces conformational changes in each protomer. These variations may be of a very small amplitude in the overall conformation, but decisive for enzymatic activity or regulation of this activity.

11.6. EFFECTS OF LIGANDS IN REFOLDING OF OLIGOMERIC PROTEINS

The role of specific ligands in the refolding process was mainly investigated with oligomeric proteins. Teipel and Koshland (1971a,b) reported that when specific substrates or coenzymes were added at zero time of refolding higher yields of reactivation were obtained. The extent of stimulation differed with each enzyme. Figure 11.2 shows the effect of specific ligands on reactivation of six different enzymes. The greatest effect is observed with the addition of substrate L-malate, either at zero time or during refolding of fumarase. The presence of Mg^{2+} increased considerably refolding of enolase.

Teipel and Koshland (1971a,b) and to a greater extent Jaenicke and coworkers investigated the role of coenzymes NAD^+ or NADH at saturating concentration on several dehydrogenases .In most cases, NAD^+ and NADH have no detectable effect on the kinetics of reactivation. It appears that specific cofactors only increase the yield of renaturation by stabilizing the native form of the enzyme. For LDH-M_4 reduced or oxidized coenzyme has no effect either on kinetics or on the yield of refolding. The kinetics of LDH-H_4 reactivation were not affected by the presence of coenzyme. However, a slight increase in the yield of renaturation was observed in the presence of NAD^+. By contrast NADH had no action after 3-hr incubation, but it reduces considerably the yield of reactivation after a longer period (5–6 days). Figure 11.11 shows these different effects. The deactivation effect induced by reduced coenzyme was observed for both isoenzymes.

Under certain conditions, for GDPH the rate constants of the unimolecular as well as of the bimolecular process are not influenced by the presence of NAD^+. The presence of NADH yields more complex kinetic profiles

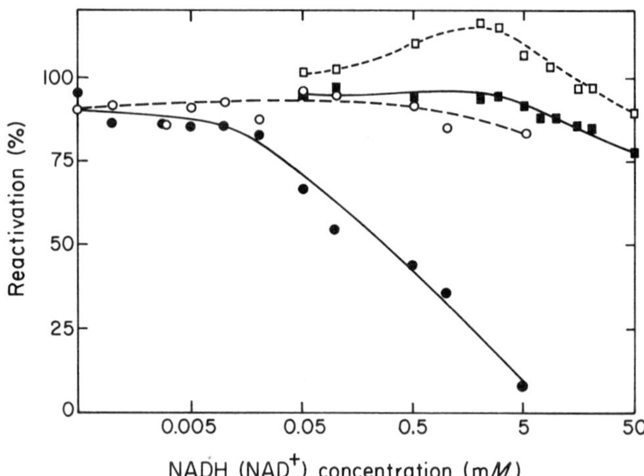

Fig. 11.11. Effects of NAD^+ and NADH on the yield of reactivation of LDH-H_4 (according to Rudolph *et al.*, 1977a): reactivation in the presence of NAD^+ after 3 hr (□) and after 5–6 days (■); reactivation in the presence of NADH after 3 hr (○) and after 5–6 days (●).

(Fig. 11.12); in this case, it is attributable to an instability of the enzyme in the presence of NADH (Rudolph *et al.*, 1977b; Krebs *et al.*, 1979). According to Teipel and Koshland (1971a,b) for GPDH the effect of specific ligand depends on the time of addition. When added at zero time of refolding, a twofold increase of reactivation by NAD^+ was reported. However, no effect was observed when addition of the coenzyme was delayed. This be-

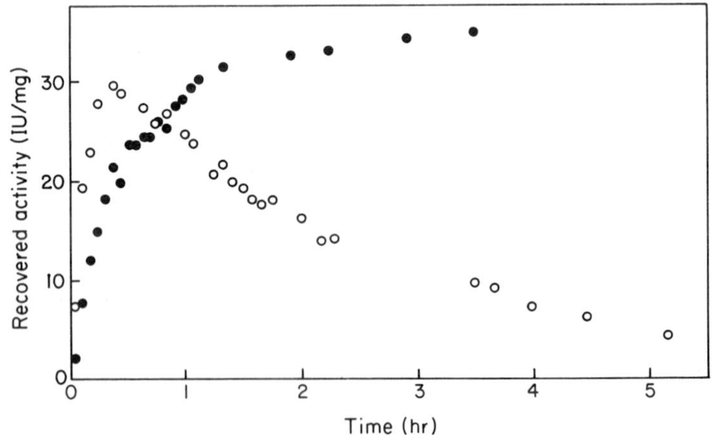

Fig. 11.12. Effects of NAD^+ and NADH on kinetics of reactivation of GDPH (according to Rudolph *et al.*, 1977b): 1 m*M* NAD^+ (●); 5 m*M* NADH (○).

havior depends on the enzyme concentration. The importance of time for adding ligands was particularly well illustrated during refolding of β-galactosidase, which is a tetramer. The enzyme dissociates and unfolds in 8 *M* urea. It can refold and regain full activity after removal of denaturant, only if Mg^{2+} ions are added after completion of refolding. When Mg^{2+} is present in the renaturation mixture, it prevents regain of activity by formation of illicit interactions (Ullmann and Monod, 1969).

Concentration of ligands as well as concentration of enzyme may modify the effect of ligands. The complex action of Zn^{2+} on alcohol dehydrogenase refolding depends strongly on concentration. Excess of Zn^{2+} drives the reaction to the formation of inactive species, thus lowering the reactivation yield. Addition of Zn^{2+} before refolding produces a reactivation which depends on enzyme concentration. When Zn^{2+} is added after completion of the first-order processes, these reactions are no longer limiting steps for reactivation. The action of Zn^{2+} was interpreted by a side reaction to form inactive species and by an effect on the last step of refolding which originates native enzyme from inactive folded and reassociated oligomer (Jaenicke, 1979).

ATP affects only the second process during the dissociation of phosphofructokinase (Parr and Hammes, 1976). It only delays dissociation of dimer into monomers at the concentration of 50 m*M*, but neither dissociation of tetramer into dimer nor the conformational change which occurs in the monomer are influenced by the presence of ATP.

A very curious activation of renaturation of pyruvate kinase (PK) from *Saccharomyces carlsbergensis* by L-valine was reported (Bornmann *et al.*, 1974). Pyruvate kinase is a tetrameric enzyme which consists of four identical subunits. It was found that each protomer noncovalently binds 1 molecule L-valine which is removed during denaturation by 6 *M* GuHCl. Renaturation can be also induced by the presence of Mg^{2+} or Mn^{2+} ions in the absence of L-valine. Kinetics of reactivation were described by a pseudofirst-order reaction, with a first-order constant of 0.019 min^{-1} with respect to the monomer species in the presence of L-valine. D-Valine has no effect on PK reactivation (Fig. 11.13). When both L-valine and Mg^{2+} or Mn^{2+} ions are present at optimal concentrations in a regeneration mixture, a pseudo-first-order rate constant of 0.05 min^{-1} is obtained indicating additive effects of cations and L-valine. The effectors must be present at zero time of refolding. From these data, Bornmann and co-workers (1974) concluded that induction of refolding by amino acids and ions operates through the first monomolecular refolding step and that they act as nucleation primers. They have no effect on reassociation of protomers, which is not a rate-limiting step for refolding of PK. The activation of PK folding by L-valine is certainly not well understood. A unique report was published on this topic and no more

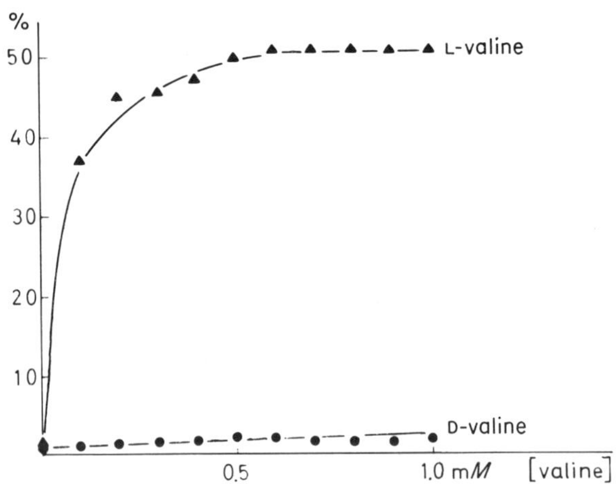

Fig. 11.13. The effect of L-valine and D-valine on the yield of renaturation of pyruvate kinase (according to Bornmann *et al.*, 1974, courtesy of Hess).

detailed study is available. Valine is not a direct effector of PK. The meaning of this action is not clear (Bornmann *et al.*, 1974).

From the scattered data concerning effectors in refolding, coenzyme or specific ligands, one can provisionally conclude that, when they affect refolding process, they act by stabilizing partially or completely refolded structures during the process directing folding toward the correct folding and thus minimizing formation of incorrect aggregates or incorrect species. Stabilization of native structure by specific ligands has been known for some time (see Yon, 1969) and many examples are available in literature. It is well known that holoenzymes are generally more stable than apoenzymes. During refolding, ligands may stabilize either small pieces of substructures (nucleus or structured segments), or whole refolded structure. More data are needed to understand the detailed mechanism of stabilization by ligands. The action of ligands as a primer of nucleation seems unlikely; a minimum structure is certainly required to produce a binding site that has a significant affinity for ligands.

Common features appear in refolding of oligomeric proteins. Subunits refold to a global structure before reassembly. However, association induces structural refinements which can be of great importance for the expression of the activity. In several proteins, however, reappearance of enzymatic activity parallels the reassembly process. This behavior generally occurs at a low concentration of protein. Conformational variations which occur during reassembly may be of very weak amplitude. Full activity requires quaternary structure even when significant activity does exist in the isolated subunit. Importance of conformational coupling through the interactions between subunits, for the expression of functional properties and for its amplification is recognized for oligomeric proteins. This coupling probably occurs in the last step of protein folding; the one that allows the conformational refinements induced by subunit assembly.

III

Concluding Comments

The main approaches to determining the mechanisms involved in protein folding have been presented. Many conclusions were deduced from three-dimensional structures of folded proteins. A large variety of techniques were used in the experimental approaches. Despite the tremendous progress in the resolution of protein structure, the great number of experiments, and the variety and the potentiality of the techniques employed, many questions remain unanswered. Until now, only a few proteins have been extensively investigated. A large field of investigation still requires exploration. This last part contains only a concluding chapter which comments on the main results discussed in both Parts I and II. While examining and summarizing what has really been established, Part III also discusses what remains to be proven.

Particular attention is brought to the dynamic aspect of protein structure and to the last events which achieve protein folding or modulate protein structure in response to an external stress.

Further, conformational dynamics will retain the attention of most researchers interested in protein structure and folding, who are trying to understand the mechanisms by which the biological activity of a protein is generated. Studies in this direction will increase very rapidly. These aspects remain mostly in the field of physical and chemical methods.

Another direction actually in progress consists of studies of protein folding in the cellular environment. The biological aspect is presented in

Chapter 1 to introduce the main problems concerning the formation of functional proteins. It has not been reviewed in this volume since very little information on the mechanism of protein folding has been obtained until now under *in vivo* conditions. It must be examined in this last part as a field of investigation for the near future. Research concerning protein folding in the cellular environment is discussed.

12

Mechanisms of Protein Folding

12.1. PATHWAY(S) OF FOLDING

The most important, and at least the first point in understanding the mechanisms by which a polypeptide chain reaches its functional structure, concerns the pathway of folding. Taking into account the most significant information reported in this book, we have to examine which elements of the problem are well established, and whether it is possible to decide among the different models proposed. The main models to describe the folding process: nucleation–propagation, diffusion–collision and a hierarchic folding by stages, have been presented in Chapter 1; These models display several similarities and differences. Now it is necessary to evaluate the validity of the arguments in favor of a particular model. It is necessary to discuss if the experimental data available actually provide sufficient information to decide whether a particular model is relevant or not. First, the existence of intermediates and their nature when detected and characterized are examined. Then, a survey of well-established results concerning each level of structure, from nuclei to the overall conformation is presented. The plausibility of the different models of folding is discussed in the light of these arguments.

12.1.1. Cooperativity of the Folding Process and Evidence for Intermediates in the Folding Pathway

The reversibility of the unfolding–folding process was established in earlier studies for many proteins, even for the disulfide-linked molecules. The equilibrium transition was investigated for several proteins by a great number of techniques. In many cases investigations have revealed a very cooperative process with the transition very closely following a two-state mechanism. Transitions of chymotrypsinogen, ribonuclease, and hen egg white lysozyme are classic examples of such behavior. However, several examples of equilibrium studies have revealed the occurrence of intermediates. Phosphorylase *b*, carbonic anhydrase, penicillinase, papain, and lactalbumin are not described by a two-state process. In all cases, kinetic studies have shown the existence of intermediates in the folding pathway. In several examples abortive species were also detected.

It is now well established that the folding pathway involves intermediary steps. The transient states are populated to some degree. According to their degree of population, two-state or more than two-state behavior may be observed. The direct evidence of intermediates was provided either by methods used to detect the appearance of local structures or by trapping these species. The different methods, which include NMR, differential labeling, accessibility of peptide bonds to proteases, immunochemistry using monospecific antibody populations, and those that allow detection of local changes in protein conformations, have yielded valuable information. All these methods have shown that intermediates are formed during the folding process. However, divergent conclusions concerning the folding pathway have been proposed for the same protein. For ribonuclease, a sequential pathway of folding was proposed by Burgess and Scheraga (1975a) on the basis of data obtained by a great variety of techniques. Creighton (1979b) for the same protein as well as for BPTI proposed a common mechanism, i.e., a very cooperative overall process in which folded conformation occurs only when all stabilizing interactions takes place; wrong species are formed which have to be destroyed during the search for the right interactions.

A preferential pathway of folding was shown for elastase. The same folded but inactive intermediates being formed and trapped in critical concentration of denaturant, in the denaturation as well as in the renaturation process (Ghélis and Zilber, 1982; Ghélis, 1980).

The occurrence of intermediates, whether or not they are unstable, and the existence of a preferential pathway, or at least a restricted number of pathways in the folding of a protein are concepts which seem rather well established. Until now only a very small number of proteins was investigated.

12.1.2. Nucleation Process

The experimental results obtained regarding the existence of preferential sites acting as nucleation centers and their persistance until the end of the folding process is examined first. In fact, the concept of nucleation was mainly deduced from theoretical studies. These initial structures are certainly not stable enough to be directly detected as intermediates. Their lifetime is certainly very short. However, detection may become possible if they persist until almost the end of the folding process.

Different hypotheses have been proposed for the initial structures acting as nucleation centers. These hypotheses include a compact nativelike structure occurring by a random folding (see Baldwin, 1980), a cluster of hydrophobic residues (Matheson and Scheraga, 1978; Wüthrich and Wagner, 1978), a cluster of α helices (Lim, 1978), or a cluster of β sheet hairpin helices (Ptitsyn *et al.*, 1979).

Several attempts were made to detect and identify nucleation centers, but until recently no direct evidence for their existence was obtained. The slow step observed in earlier fast reaction kinetic studies was at first attributed to a nucleation process (Tsong *et al.*, 1972a), and subsequently to cis–trans isomerization of proline residues. However, this last statement is still controversial.

Evidence obtained from immunochemical studies (Chavez and Scheraga, 1980a,b) and the use of photochemical labeling reagents (Matheson and Scheraga, 1979a,b) on the one hand, and predictions based on hydrophobic interactions (Matheson and Scheraga, 1978) on the other hand, have suggested that a primary nucleation site in ribonuclease is contained in residues 106–118 inducing early folding of antigenic determinant 87–104.

Also considered as possible nucleation centers were the microstructures which form hydrophobic pockets and remain after thermal unfolding of BPTI, as probed by NMR studies (Wüthrich and Wagner, 1978).

The decrease of accessibility of several side chains to chemical reagents in two regions of the elastase molecule suggested that even in 3 *M* GuHCl (which is above the transition zone) structural elements are stabilized. Since, there is one residual structure in each domain, these structural elements possibly correspond to nucleation centers (Ghélis, 1980).

These data which are described in detail and discussed in chapter 10 are the only significant experimental arguments in favor of the existence of nucleation sites to initiate the folding process. They indicate the particular role of hydrophobic interactions in nucleation. Until recently nucleation centers have been deduced rather than really observed, and their persistence has not been proved. Their detection is not expected to be easy because they are short-lived intermediates. Nevertheless, it should be noted that theoretical and thermodynamic arguments support the formation of nucleation centers by hydrophobic interactions.

12.1.3. Formation of Initial Structures

At a higher level, very little information was obtained on the initial formation of segments of secondary structures, the α helices or β structures. Once formed, do they persist until the native conformation of the whole protein is achieved? Or, as proposed by Ptitsyn and co-workers (1979), are the long hairpin helices formed initially? Then, does a reorganization of the overall structure including the transformation of hairpin helices into other ordered structures with different mechanisms according to the type of protein take place?

Experimental data on the formation of secondary structures during the folding of a protein are scattered in the literature. It would certainly be difficult to answer the preceding questions. Studies conducted either by ORD, CD, or Raman spectroscopy indicate a rapid formation of secondary structures, principally α helices which seem to be formed more rapidly than β sheets. In ribonuclease and lysozyme, helical structures appear before the enzymatic activity. In this case the data obtained did not clarify whether they are formed before the overall conformation of the molecule forms. In phage T4 lysozyme, about 60% of helical structure is formed before the appearance of intrinsic fluorescence properties of the native structure (Desmadril and Yon, 1981). Hydrogen exchange measurements supported the view of an extensive hydrogen-bonded structure, probably an α helix of an early intermediate in the refolding of RNase (Schmid and Baldwin, 1978a,b, 1979a,b). Certainly among secondary structures, at least α helices are formed rapidly in early steps of protein folding. Nevertheless, experimental data are still needed to reveal the mechanism by which segments of ordered structures are formed in globular proteins.

The studies on lactalbumin published by Kuwajima (1977) indicate a higher content of α helix, as measured by CD in early refolding species of the protein, than in the native one. The difficulties of interpretation of data obtained for proteins, in terms of helix content, have been discussed previously. It is very difficult to conclude from this example whether the model proposed by Ptitsyn *et al.* (1979) applies for the folding of lactalbumin.

12.1.4. Formation of Substructures

Nothing is known from experimental studies about the formation of supersecondary structures, i.e., about the mechanisms by which segments of secondary structure interact. Characteristics and tentative rules concerning supersecondary structure formation were deduced from the examination of three-dimensional structures.

The most significant results concerning the folding process of smaller elements within a protein are related to domains. Until recently, domains were considered to be the smallest independent and stable substructures isolated and studied with respect to the folding and assembly process. Experimental evidence has been presented for the independent refolding of protein domains as reported in Chapter 10, in the discussion of elastase, thioredoxin, serum albumin, tryptophanase, immunoglobulins, aspartokinase homoserine dehydrogenase, and thermolysin. In these cases, separation of domains by proteolytic cleavage has been successful. Furthermore, for most of these molecules, complementation experiments have yielded evidence of the formation of a protein with the conformational properties of the native one and with small but significant enzymatic activity.

Although isolated domains are able to fold independently, interdomain influence was indicated by a difference in the rate of refolding of a particular domain when isolated, or in the entire molecule. Depending on the extent of interactions between domains, this mutual influence may be more or less marked. In T4 lysozyme, for example, the interdomain influence prevents the separation of the two domains; furthermore, the folding and stabilization of the N-terminal domain requires the previous refolding of the C-terminal domain (Desmadril and Yon, 1981, 1982; Desmadril *et al.*, 1982).

In most cases, domains may be considered as building blocks acting independently during the folding process and interacting in one of the final steps. The existence of structural domains represent a kinetic advantage for the folding process which can occur in a short time even for large molecules. Indeed, it allows a simultaneous folding of several building blocks. It also represents a structural advantage: a great stability is expected since the domains are linked, not only by noncovalent interactions, but also by at least one covalent peptide bond. Domain assembly in a monomeric protein or subunit is a process very similar to subunit assembly in an oligomeric protein. There is no fundamental difference between folding of monomeric and oligomeric proteins; the same mechanisms of folding and assembly are involved. Conformational readjustments on "building blocks" assembly are of great importance for the expression or the amplification of biological activity. The functional consequences of a correct coupling between domains in a monomeric unit as well as between protomers in an oligomeric protein has been emphasized (Ghélis and Yon, 1979). This coupling allows transmission of signals from one building block to another through the interdomain (or intersubunit) constraints that induce structural refinements and therefore permits optimization of functional properties. Thus, the organization of molecules into domains represents a functional advantage; it is probably more efficient and allows subtle regulations. Furthermore, the existence of domains as "building blocks" can explain the emergence of novel proteins during evolution (Schulz, 1981). The existence of structural domains seems to be a rather general characteristic in the organization of biological molecules. Transfer RNAs also display such structural patterns (Quigley and Rich, 1976), and the molecule folds according to a sequential process (Boyle *et al.*, 1980). It seems, that in the building of molecules, nature proceeds by steps, forming a small number of subelements which assemble according to definite patterns.

12.1.5. Precursors as Stable Intermediate States in Protein Folding

Another interesting aspect in the study of protein folding is the comparison between precursors and functional proteins, zymogens and enzymes. It has been often proposed that spontaneous refolding occurs in precursors and not in proteins after the proteolytic cleavage has activated the precursor. We have already discussed this aspect in Chapter 5. Certainly it depends upon the molecule. The case of proinsulin and insulin is not very representative, since a great part of the molecule is removed upon activation.

In contrast, in the case of chymotrypsinogen–chymotrypsin, proelastase–elastase, and trypsinogen–trypsin, activation of zymogens proceeds either by removal of a small peptide, or even by the simple cleavage of a peptide bond (e.g., δ-chymotrypsin). Conformational rearrangements occurring as a result of activation of chymotrypsinogen into δ-chymotrypsin were analyzed by chemical modification of Tyr 171 in *N*-acetyl-chymotrypsinogen (C. Ghélis and R. Caillet unpublished data). This molecule and the corresponding enzyme have very close related structures. Differences are on very localized properties. It seems that tyrosine 171 is involved in an intermediate occurring during the activation of the zymogen.

Feldhammer *et al.* (1977) emphasized that in trypsinogen, the charge relay system is quite similar to the equivalent system in trypsin, but does not adopt the optimal conformation during catalysis. The Ile–Val dipeptide induces conversion of the trypsinogen to the trypsin conformation, but only in the presence of covalently bound acyl ester groups like *p*-guadinino benzoate (Bode, 1979, Perkins and Wüthrich, 1980). Activation of zymogen, or Ile–Val dipeptide binding to the acyl enzyme results from subtle differences in flexibility in localized regions of the protein; these regions become more rigid in the active enzyme. Based on these results and those obtained by C. Ghélis and R. Caillet (unpublished data), it is possible that, at least for serine proteases, zymogen corresponds to a stable intermediate in protein folding, with a structure very close to that of the active enzyme. The failure of electrostatic interaction prevents the protein from reaching its final conformation, which in the enzyme is stabilized through the salt bridge. This interpretation is also supported by the observation that high ionic strength significantly increases reactivity of trypsinogen and chymotrypsinogen (Lonsdale-Eccles *et al.*, 1979). The completely folded and active structure of zymogens has a low probability and is very unstable, but it can be stabilized by increasing ionic strength in the absence of the salt bridge.

We have proposed that functional differences between chymotrypsinogen and chymotrypsin are related to differences in conformational coupling between domains in these two molecules. As already mentioned from the data obtained on the refolding of elastase (Ghélis *et al.*, 1978; Ghélis and Yon, 1979, and unpublished data), the last step in protein folding ensures an optimal coupling between domains, the energy of interactions being utilized for catalysis. In zymogen, because salt bridges cannot be formed, the last conformational refinements that ensure an optimal coupling between domains are prevented. Consequently, the enzymatic activity of chymotrypsinogen is extremely weak, 10^6–10^7 smaller than the activity of chymotrypsin (Gertler *et al.*, 1974). A tentative evaluation to quantitate the energetic aspect of conformational coupling and enzyme catalysis plausibly supports the idea that the energy required for the amplification of the catalytic response can be provided by the salt bridge (the energy of the salt bridge being about 10 kcal/mole) (Ghélis and Yon, 1979).

These different results strongly suggest that at least chymotrypsinogen and probably zymogens in the same family are frozen in a step preceding the last one in the folding pathway that is accompanied by conformational refinements allowing the right coupling between domains. In this respect, the experimental study and computation of hinge-binding fluctuations in the zymogen and the corresponding enzyme might lead to very useful information about the coupling between domains.

12.1.6. Comments on the Folding Pathways

It is worthwhile to ask how the different pieces of the puzzle are relevant to a particular model? The nucleation–propagation mechanism (which requires a single-search nucleation process followed by a rapid propagation) is the only which can be discarded, even if the process occurs within each domain for large proteins. This model certainly accounts, with a good approximation, for helix formation in polypeptides. Nevertheless the complicated structure of globular proteins, made of segments of different ordered and nonorder structures, involves a conformational search at each step of propagation. Such a mechanism requires a time longer than the time really needed by a polypeptide chain to fold in a globular protein.

Beside this first mechanism, nothing for the moment authorizes a decision in favor of the diffusion–collision model or of a folding-by-stage model. Both mechanisms require nucleation as a starting point for the folding process. Nucleation centers may or may not be persisting structures. In the diffusion–collision model, the initial structures are expected to persist only if they are in the right conformation. A folding-by-stages, involves a well-defined sequential pathway in which, at each step of the process, structured elements have a relative stability; even in this case, however, conformational readjustments may occur upon interactions of smaller substructures and initial structures may be rearranged.

Although it has not been demonstrated, it seems plausible that both models are operating. At each stage of the process, substructures may be formed by a diffusion–collision mechanism. At each level of organization and in in different parts of the molecule, a random search can conceivably lead to the formation of either right or wrong structures; these last ones have to be corrected to achieve the right structure of the whole molecule.

Folding of a polypeptide chain into a biologically active protein proceeds by a search between a sequence of right processes and also of errors that have to be corrected and thus obtains a defined molecule only after series of approximations rather than through a strict determination. Folded proteins are also the product of "chance and necessity."

12.2. STABILITY OF THE NATIVE STRUCTURE

The importance of short, medium, and long-range interactions in protein folding was discussed in terms of many theoretical aspects. It was stated that short-range interactions determine protein folding and long-range interactions stabilize the folded structure. Deductions were made from comparisons of conformational predictions of local structures and their existence in the protein from X-ray data. Recent studies reveal medium and perhaps long-range interactions which are believed to play a role, even in a determination of local structures in proteins. The importance of long-range interactions was supported by different types of studies.

Experimental results have indicated the stabilizing role of long-range interactions. Fragments smaller than part of a polypeptide chain corresponding to domains are able to form local structures fluctuating with the unfolded state. The weak proportion of folded species, as evaluated by the value of K_{conf} obtained from immunochemical methods indicates the role of long-range interactions. For three cyanogen bromide fragments from ribonuclease, an approximate value of K_{conf} of 10^{-4} was obtained by Chavez and Scheraga (1980b). The same order of magnitude was found for polypeptide fragments of staphylococcal nuclease (Furie *et al.*, 1975). Complementation of unstructured fragments for this protein and for several others, such as cytochrome *c* and lysozyme (Taniuchi *et al.*, 1977, Parr and Tanuichi, 1980a,b, 1981; Juillerat *et al.*, 1980), yields folded proteins, clearly showing the importance of long range interactions in the stability of the native conformation.

The studies by Labhardt and Baldwin (1979a,b) on the refolding of RNase S (resulting from complementation of S-protein with S-peptide) have shown that RNase S folds more rapidly than S-protein. They have indicated that S-peptide combines with S-protein before its complete folding and therefore stabilizes an intermediate which directs the correct refolding of the complemented protein.

For larger proteins, stabilization occurs within each domain and also by interactions between domains. No quantitative data indicating the relative contribution of intra- and interdomain stabilizing interactions for any proteins are available. Certainly, this relative contribution depends upon the size, the interface interactions between domains, and the existence of disulfide bonds. Domains which contain disulfide linkages are certainly better stabilized. Such an evaluation would be interesting in order to predict the possibility of obtaining stable isolated domains from a multidomain protein.

The native structure results from a delicate balance between stabilizing and unstabilizing contributions. Many noncovalent and cooperative stabilizing interactions are partially compensated for by conformational entropy which tends to destabilize the structure. Under conditions of maxima stability, the total energy of stabilization is very weak. For the different proteins studied, ΔG varies between -5 and -10 kcal/mole. Thus, very weak variations in external conditions are able to drive the molecule out of its energy well and disorganize its structure. Further stabilization can be provided by prosthetic groups, or upon ligand binding.

12.3. CONFORMATIONAL FLUCTUATIONS OF NATIVE PROTEINS

Folded proteins are not rigid molecules. Their structure fluctuates spontaneously around equilibrium. The range of motions in proteins appears to be on different time scales. Time scales of dynamic events in proteins are indicated in Table 12.1. Conformational fluctuations have been evaluated

TABLE 12.1

Time Events in Proteins and Enzymes[a]

Determinants	Time (sec)
Protein surface	
Bound water relaxation	10^{-9}
Side chain rotational correlation	10^{-10}
Proton transfer reaction of ionizable side chains	10^{-7}–10^{-9}
Protein conformation	
Local motion	10^{-8}–10^{-9}
Isomerization process	10^{-2}–10^{-7}
Folding–unfolding transition	10^{+2}–1
Enzyme substrate complex in solution	
Encounter rate	diffusion controlled
Estimated lifetime of the transition state in covalent reactions	10^{-10}
Change in metal ion coordination sphere in metalloenzymes	10^{-6}–10^{-9}
Enzyme–substrate local conformational motion	10^{-9}
Covalent enzyme–substrate intermediate lifetime	10^{-2}–10^{-4}
Enzyme–substrate complex conformation isomerization	10^{-2}–10^{-4}
Enzyme–substrate complex unfolding transition	10^{+2}–1

[a] From Careri *et al.* (1979).

both by computation–simulation and by experimental studies. Several reviews have been published recently (Lakowicz and Weber, 1973; Weber 1975; Careri *et al.*, 1975, 1979, Gurd and Rothgeb, 1979; G. B. L., 1979; Karplus and McCammon, 1980). Recent studies have brought a new insight into conformational mobility of proteins and its biological importance. It might be interesting to know, if there are concerted motions in proteins: What parts in a protein are mobile? Does denaturation begin from those parts? What is the functional significance of the different internal motions? Further, experimental investigation is required to answer these questions.

12.3.1. Packing Density and Defects in Proteins

Packing defects and the existence of cavities inside a protein allow motions in certain parts of the molecule even when localized at the interior. Packing density of the atoms ensures transmission of motion through the protein molecules. The relative rigidity of helix segments, the packing of α helices in sperm whale myoglobin allow the vectorial transmission of motions (Schoenborn, 1971; Richmond and Richards, 1978).

The internal cavities may provide space that can become rearranged into channels allowing penetration of small molecules. Many arguments support the penetration of oxygen inside proteins. Quenching of tryptophan fluorescence was studied by Lakowicz and Weber (1973) in 14 proteins. Even tryptophans buried mainly in nonpolar regions were found accessible to oxygen, leading to a quenching of their fluorescence. In the study of the quenching of fluorescence by a larger molecule, acrylamide, some tryptophans in several proteins were found shielded from collisions.

12.3.2. Local Motions in Proteins

The existence of packing defects in proteins allows the mobility of internal side chains although motions of external side chains are more probable. Internal mobility of protein structure arise from vibrational and rotational motions. The theoretical arguments have been discussed in Chapter 4. Vibrational motions have a time range from 3.10^{-14} to 3.10^{-12} sec. Different spectroscopic methods, such as infrared absorption, Raman spectroscopy and quasi-elastic neutron scattering have been used in these studies. The studies have generally revealed high frequency vibrations that correspond to bond-length and bond-angle deformations. Changes in the environment of proteins are accompanied by shifts in the frequencies of vibrations. Low vibrations (ca 50 cm^{-1}) have been reported in several proteins; since these motions have a collective character they will be considered in the following discussion.

Rotational motions are associated with rotations around dihedral angles (ϕ_i, ψ_i) of the backbone or dihedral angles of side chains (χ). Rotational motions generally display time scales of $1–5 \times 10^{-8}$ sec. They can be probed by different methods such as NMR, fluorescence depolarization, fluorescence or phophorescence quenching by oxygen or iodide, or acrylamide. Rotational motions of aromatic rings of tyrosine and phenylalanine as probed by NMR studies, have been extensively studied in BPTI at different temperatures by Wüthrich and co-workers, as detailed in Chapter 8. As previously mentioned, several aromatic groups have been found immobilized even at high temperature and buried in regions of hydrophobic stability which are assumed to be possible nucleation sites. Rapid ring rotations have been reported in different native proteins (Campbell *et al.*, 1975a,b, 1976; Cave *et al.*, 1976; Williams, 1978; Hull and Sykes, 1975; Gurd and Rothgeb, 1979), lysozyme, cytochrome *c*, and parvalbumin (see Table 12.2). McCammon *et al.* (1977) concluded from their computation simulation that torsional motions which involve substantial displacement will have a collective diffusion-like character, whereas those with smaller steric interactions will have a more local character.

More recent results concerning protein dynamics were found from X-ray crystallography; these new refinements allow evaluation of small displacements. The displacement parameters $\langle X^2 \rangle$ result from several contributions.

TABLE 12.2

Motions of Aromatic Amino Acid Side Chains and Observed Rotations of Aromatic Rings in Proteins[a]

A.

Ring Side Chain	Motion Observed
Indole (tryptophan)	Usually none, occasionally rocking
Phenyl (phenylalanine)	Restricted rotation
Phenol (tyrosine)	Restricted rotation
Imidazole (histidine)	Restricted rotation

B.

	Rotation Observed	
Protein	Tyrosine	Phenylalanine
Lysozyme	All rotate	Probably all rotate
Cytochrome *c*	Approximately half rotate	Approximately half rotate
Parvalbumin		All rotate
Troponin C	Most rotate	Most rotate
Phospholipase A_2	Most rotate	Most rotate
Neurotoxin	Some rotate	—
Trypsin inhibitor	Some rotate	Some rotate

[a] From Gurd and Rothgeb (1979).

In the study of sperm whale ferrimyoglobin crystal at different temperatures, Frauenfelder *et al.* (1979) assumed $\langle X^2 \rangle_{cv} = \langle X^2 \rangle_c + \langle X^2 \rangle_v$, c and v referring to conformational and vibrational substates respectively. With $\langle X^2 \rangle_{cv}$ reaching values larger than typical vibrational ones, evidence is given for conformational stubstates. An X-ray diffraction study by Frauenfelder *et al.* (1979) of sperm whale myoglobin, one by Sternberg *et al.* (1979), and one by Artymiuk *et al.* (1979) on lysozyme, indicate smaller displacement inside the protein than outside. Charged and polar groups display larger displacements than nonpolar residues. Movements of helices also show rippling and breathing modes. The active center of lysozyme (i.e., the "lips of the active site cleft") is among the regions of greatest displacement. In myoglobin, a rather great mobility is observed in the region around the heme.

It is certainly interesting to note the great mobility of the active site, which is probably related to the biological function of the protein.

12.3.3. Concerted Motions in Proteins

There is strong evidence for concerted motions such as the hinge-bending mode and breathing motions involving more than one residue in proteins McCammon and Karplus (1977), Karplus and McCammon (1980), and McCammon *et al.*, (1976) have studied hinge-bending modes to describe the relative motion of two domains in proteins. They applied their method to lysozyme and immunoglobulins. In lysozyme motions of the two globules separated by a cleft which contains the active site might allow a coupling between the opening and closing of the cleft, the entrance and the exit of the substrate. Experimental support indicates the possibility of hinge bending in several proteins; it was assumed in hen egg white and phage T4 lysozyme (Timchenko *et al.*, 1978). Hinge-bending was described in phosphoglycerate kinase (Banks *et al.*, 1979), hexokinase (Shoham and Steitz, 1980), and perhaps also in dehydrogenase, the relative motion of the two domains being hindered in the enzyme–substrate complex. Probably this kind of motion is important for the enzyme action. Low frequencies are expected to be associated with hinge bending. They can be also associated with vibrations of structural elements such as helices. Breathing motions in proteins are probed by isotope exchange of internal hydrogen. They can occur in small elements, segments of α helices for example (see previous discussion), or within the entire protein molecule. Breathing of the entire molecule is probed by isotope exchange (see Chapter 8), conformational equilibrium as evaluated by ΔG_0, or immunochemical approaches (K_{conf}). Breathing motions are slow types of relaxing processes occurring on a larger time scale.

12.3.4. Functional Role of Conformational Fluctuations in Proteins

Internal mobility and flexibility of proteins are of great importance for the expression of biological function. Conformational adaptability on ligand binding through an induced-fit mechanism (Koshland and Neet, 1968; Koshland, 1970) allows better orientation of the substrate relative to the catalytic groups conditioning the efficiency of enzymatic catalysis. This aspect will be considered in the next section.

The significance of conformational fluctuations for catalysis has been particularly emphasized by Careri *et al.* (1975, 1979). They used an approach based on the Onsager phenomenological equations. They assumed that weakly bound water molecules play an important role in inducing fluctuations of the backbone by which the protein can exchange free energy with its environment. These fluctuations propagate to the active site and help catalysis mainly by entropic effects. Kinetic coupling between the solvent protein surface interactions and the chemical changes occurring at the active site of the enzyme is expressed as a function of two statistical macrovariables that are independently fluctuating: the bound solvent concentration and the reaction coordinate.

Allosteric communication consisting of a vectorial transmission of information from one protomer to another is an example of concerted motion in proteins occurring through the close packing of the atoms. The role of the process in modulating the activity by amplification or attenuation of the catalytic response is fundamental for regulation of cell metabolism.

An extension of the concept of allostery has been proposed i.e., an interdomain transmission in multidomain protein. It has been accepted that in complemented elastase, the interactions between domains are not strong enough (compared with the native enzyme) to ensure allosteric transmission and amplification. The complemented enzyme behaves as a desensitized enzyme and therefore has only a weak activity. This extension is consistent with all other similarities between domains in multidomain proteins and subunits in oligomeric enzymes (Ghélis and Yon, 1979).

The hinge-bending motion of domains has been reported for several kinases. Thus, it is tempting as a working hypothesis to assume a correlation between this motion and the characteristic "enzyme memory" described for wheat germ hexokinase acting as a "mnemonic enzyme" (Ricard *et al.* 1974).

It is likely that conformational fluctuations also have an important role in protein folding. Fluctuations are certainly less restricted in the unfolded form than in the native one. Their role is particularly emphasized in the diffusion–collision model.

12.4. PROTEINS ACHIEVE THEIR FOLDING UPON SPECIFIC BINDING PROCESS OR DURING CATALYSIS

As a consequence of its flexibility, the structure of native proteins undergoes conformational changes of variable amplitude, which may often be very weak as a result of ligand binding or during catalysis. In a sense, one can say that native protein is not entirely folded and stabilized in the absence of substrate and this folding is achieved along the catalytic pathway, with various changes occurring at each step of the process. Although arbitrary, one may distinguish non covalent and covalent events.

12.4.1. Conformational Changes upon Ligand Binding

As previously mentioned, the induced-fit theory formulated by Koshland involves an instructive role of the substrate in inducing an optimal conformation of the enzyme. Conformational changes upon ligand binding have received much experimental support (see review in Citri, 1973). Another formulation consists in assuming that the enzyme fluctuates between various conformational states; most of these states are very sparsely populated in the absence of ligand and are stabilized on ligand binding. According to this model, which is the generalized model of Adair–Weber, proposed by G. Weber (1965), and written as follows:

$$\begin{array}{ccccccccccc}
P_1 & \rightleftharpoons & P_2 & \rightleftharpoons & \cdots P_{j-1} & \rightleftharpoons & P_j & \rightleftharpoons & P_{j+1} & \rightleftharpoons & \cdots P_m \\
\upharpoonleft\downharpoonright & & \upharpoonleft\downharpoonright & & \upharpoonleft\downharpoonright & & \upharpoonleft\downharpoonright & & \upharpoonleft\downharpoonright & & \upharpoonleft\downharpoonright \\
P_1X & \rightleftharpoons & P_2X & \rightleftharpoons & \cdots P_{j-1}X & \rightleftharpoons & P_jX & \rightleftharpoons & P_{j+1}X & \rightleftharpoons & \cdots P_mX \\
\vdots & & \vdots & & \vdots & & \vdots & & \vdots & & \vdots \\
P_1X^{i-1} & \rightleftharpoons & P_2X^{i-1} & \rightleftharpoons & \cdots P_{j-1}X^{i-1} & \rightleftharpoons & P_jX^{i-1} & \rightleftharpoons & P_{j+1}X^{i-1} & \rightleftharpoons & \cdots P_mX^{i-1} \\
\upharpoonleft\downharpoonright & & \upharpoonleft\downharpoonright & & \upharpoonleft\downharpoonright & & \upharpoonleft\downharpoonright & & \upharpoonleft\downharpoonright & & \upharpoonleft\downharpoonright \\
P_1X^i & \rightleftharpoons & P_2X^i & \rightleftharpoons & \cdots P_{j-1}X^i & \rightleftharpoons & P_jX^i & \rightleftharpoons & P_{j+1}X^i & \rightleftharpoons & \cdots P_mX^i \\
\vdots & & \vdots & & \vdots & & \vdots & & \vdots & & \vdots \\
P_1X^N & \rightleftharpoons & P_2X^N & \rightleftharpoons & \cdots P_{j-1}X^N & \rightleftharpoons & P_jX^N & \rightleftharpoons & P_{j+1}X^N & \rightleftharpoons & \cdots P_mX^N
\end{array}$$

the ligand has only a guiding effect. Specific substrates stabilize that conformation which is optimal for binding and catalysis. This fluctuating view of enzyme structure is more realistic and has the advantage of reconciling the allosteric concerted model (from Monod *et al.*, 1965) with the induced-fit model of Koshland *et al.* (1966), and further, it applies for monomeric as well as for oligomeric proteins. Stabilization of protein structure as a result of ligand binding has received a great amount of experimental support.

Conformational adaptability, which is a selective mechanism allowing specific recognition of a molecule can be easily explained by such a scheme in which a functional state is selected and frozen amongst several fluctuating ones, by a thermodynamic stabilization of a protein-ligand complex. This dynamic view of protein specificity certainly agrees with the significant observation of a great mobility of the active site in an enzyme molecule.

12.4.2. Conformational Changes During the Catalytic Process

A conformational step may determine a catalytic pathway. A very significant aspect of the functional role of conformational adaptability of an enzyme was observed for *E. coli* β-galactosidase. One of the kinetic steps of the catalytic pathway, which was shown to be the limiting process in the hydrolysis of specific substrates with the wild type enzyme, was a conformational step but not a catalytic one. (Viratelle and Yon, 1973; Yon, 1976; Deschavanne *et al.*, 1978a,b). This was proved by the absence of correlation of the catalytic constant k_2, with the chemical reactivity of the substrates and by the study of the isotopic effect. It was assumed that this conformational effect is probably related to the great efficiency of catalysis, with a turn over of 1,200 sec^{-1}, which represents a high reactivity for such an hydrolytic enzyme. The k_2 becomes a purely chemical step with unspecific substrates. This conformational step is also prevented in a point mutant enzyme which has a weaker catalytic efficiency. In β-galactosidase which is a particularly large protein, these conformational changes are of very weak amplitude. Nevertheless they determine the efficiency of the catalytic process.

The remarkable stability of acyl enzymes also indicates that protein conformation is stabilized in several steps in the catalytic process (Bender and Kezdy, 1964; Zerner and Bender 1964; Bernhard and Lau, 1972; Bernhard and Rossi, 1968). Acyl proteases are locked in a conformation which prevents acid and alkaline transitions, (Hess *et al.*, 1970, Ghélis, 1971). Stabilization of the acyl enzyme (DIP or tosyl elastase) is indicated by a shift of the transition curve, transition being induced by GuHCl, toward higher concentrations of denaturant (c_m varies from 1.6 for elastase to 2.3 for DIP elastase at pH 5.4 and from 2.6 to 3.25 for these two forms respectively at pH 8.0) (Ghélis and Zilber, 1982).

Noncovalent, and to a larger extent, covalent binding of a specific substrate prevents conformational mobility of the active site, thus freezing and stabilizing one conformational state (or a very limited population of conformational states).

Rossi and Bernhard (1971) attempted to show the close relation between catalysis and the conformational variations of an enzyme molecule. Their study indicated that the rate of nucleophile-catalyzed hydrolysis of an acyl chymotrypsin (the *o*-β-(3-indole)-acryloyl]-Ser 195-chymotrypsin), is the same as the rate of denaturation of the acyl enzyme when evaluated from the optical properties of the acyl group in the enzyme. Thus, Rossi and Bernhard (1971) concluded that a covalently modified acyl enzyme–nucleophile complex is an obligatory intermediate in both catalysis and denaturation of this acyl enzyme.

The active site of an enzyme is certainly, from a conformational point of view, a strategic locus where disorganization of the molecule can occur, but also where specific binding of a ligand may introduce a stabilization of the whole molecule.

12.5. CONCLUDING REMARKS ON THE DEVELOPMENT OF STUDIES ON PROTEIN FOLDING

A conclusion is often provisional depending on the amount of progress in a field. A great number of studies have been reported and the performance of the methods employed has increased remarkably. Nevertheless, this is still the beginning of the study of protein folding. In the near future it is evident that an important development concerning *in vitro* studies can be expected; but it is also important to pursue the problem at the cellular level. Taking into account the rapid progress and the introduction of biochemical methods in cell biology, such studies will progress rapidly.

Certainly, *in vitro* studies still need further development. Indeed, very few proteins have been investigated with sufficient insight and a large number remains to be explored. Much information has been obtained on protein structure, but the way this structure is achieved has not been dealt with. There is still a lot of outstanding information to be gathered on a great number of different proteins to allow generalization. It is not obvious that the mechanisms involved in the folding process are exactly the same for all proteins. Perhaps it will be important to such an approach to look at the differences and similarities in the refolding and even unfolding of all α, all β, and α/β proteins, these three classes being significantly different in their structural organization.

Attempts to detect initial structures either using isotope exchange followed by a chemical or NMR analysis in order to identify the formation of local structures, or methods such as differential labeling or photochemical labeling are accessible approaches. The existence of intermediates in the folding pathway is rather well established; however, the nature of these intermediates is generally not yet identified.

Another important aspect of the recent research is the study of conformational fluctuations using all possible methodologies, computation, very high resolution crystallography at various temperatures, NMR, chemical and immunochemical methods, and isotope exchange. The study of motions in the entire molecule or in substructures, and determination of parts of the molecules displaying the greatest fluctuations may be very valuable information. As previously described, one of the most significant results obtained from these studies concern the great mobility observed in the active site. This indicates a particular instability in this region of the protein, instability that disappears upon binding of specific substrates. Such a lack of stability was observed in two proteins lysozyme and chymotrypsin; the question arises whether it is a general phenomenon. Only repetition of this type of experiment with different proteins will allow a generalization.

As already proposed, a careful analysis of the hinge-bending mode of domains in proteolytic enzymes and corresponding zymogens and other proteins will be very helpful to the understanding of conformational coupling which probably occurs in the last step of protein folding.

In vitro experiments of denaturation–renaturation are very useful and for the moment quite the only approach accessible with a physicochemical precision. However one must keep in mind the biological aspect of the problem as described in the introduction. It was not a concern of a particular chapter since there is not enough data for the moment. There is increasing interest in the topic which allows one to predict progress in the near future.

The special effect of the microenvironment on the nascent polypeptide chain might be helpful in selecting a preferential pathway. However nothing is known on this aspect. Very rare data were obtained on the folding of nascent proteins and they remain mostly qualitative. Very recently, Bergman and Kuehl (1979a,b,c) reported an interesting investigation on cotranslational modification of nascent immunoglobulins, light and heavy chains. The sequence of events, glycosylation, and the formation of disulfide bonds were described. Figure 12.1 indicates this sequence. It was shown that nascent heavy chains are glycosylated very soon after the asparaginyl acceptor site passes through the membrane; nonglycosylated completed heavy chain cannot be glycosylated probably because of the folding. Formation of the correct intrachain disulfide bond within the first light chain domain also occurs rapidly as soon as these corresponding SH residues passes in the cisterna. Formation of disulfide bonds in the second or constant region of the light chain only occurs after completion of the chain, but one of the cysteine residues is close to the C-terminal end of the light chain. The sequence of covalent events is rather well described in this case, but little is known about the folding of the nascent chain. From their results, Bergman and Kuehl (1979a,b) concluded that domains fold independently as shown by the quantitative formation of disulfide bond (Cys-35—Cys-100) and reaction with antibodies, a nucleation process likely having brought these residues in close proximity. They proposed a sequential initiation of folding from amino to carboxy terminal end, but complete folding would require the entire molecule. Even if some conclusions remain hypothetical, this approach is very significant and sheds new light on the folding problem at the cellular level.

Other aspects must be investigated, in the near future, such as folding and insertion of proteins in the membrane. The particular environment of lipid bilayer, the role of hydrophobic interactions with the apolar parts of lipids certainly determine a particular conformation. The Folch-pi apoprotein from brain myelin offers an interesting model for such a study, since it is soluble in both aqueous and organic solvents, allowing one to study the conformational changes from water to organic solvents (Lavialle *et al.*, 1979).

However, many membrane proteins span the membrane and have a part in the lipid environment and other parts surrounded by the aqueous solvent. For such membrane proteins, the problem is not to find an adequate hydrophobic solvent to study how they can fold, but a new degree of complexity results because of the presence of various local environments. One can seek

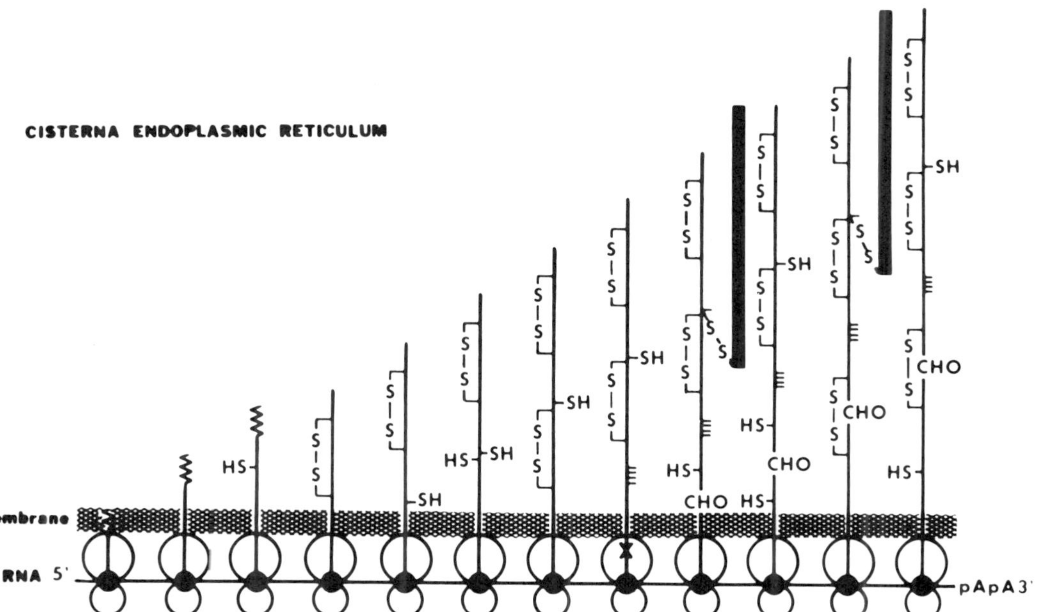

Fig. 12.1. Model of nascent immunoglobulin heavy chain polyribosomal complex in mouse myeloma cells (MPC 11) indicating several cotranslational modification events. (1) Cleavage of the amino terminal signal peptide (〰〰); (2) formation of intrachain disulfide bonds; (3) formation of interchain disulfide bonds with light chain (indicated by the thick line); (4) transfer of core oligosaccharide (CHC) to asparaginyl acceptor (X). About half of the nascent heavy chains form an interchain disulfide bond with a complete light chain before heavy chain completion and release from the polysomal complex (from Bergmann and Kuehl, 1979c, courtesy of Kuehl).

information on how the insertion of well-defined parts of the polypeptide chain into the membrane takes place and what dictates the other parts to be outside. If a special sequence is required to determine the location of part of the polypeptide chain, then what is the degree of specificity of the interaction with the membrane components. The study of the refolding of an integral membrane protein, bacteriorhodopsin (Huang *et al*., 1981) is very significant. Huang and co-workers obtained complete denaturation of the delipidated protein in trifluoroacetic acid; subsequently, they succeeded in renaturing the protein in the presence of phospholipids. They also succeeded in the reconstitution of the protein from two chymotryptic fragments (1–71 and 72–248).

Nothing is known for the moment about the conformational variations of proteins during translocation and migration through different parts of a cell. These modulations of the local environment probably constitute a particular mode of regulation.

In the different chapters, the most significant results obtained from each particular aspect of the studies of protein folding have been emphasized. In many cases difficulties have been encountered partly because of (1) a lack of sufficiently sound data, (2) the existence of scattered data which were not always obtained with the same method and under comparable experimental conditions, and (3) the fact that when sound results were obtained, it was only from a very limited number of proteins and therefore did not allow generalization. Premature conclusions have been avoided as much as possible. Nevertheless, it has seemed very urgent to gather the most significant results obtained to date. Through the reported data, the possibility is offered to readjust perspectives in light of new facts.

Despite these difficulties, three aspects have been emphasized. First, nature seems to have built protein molecules by assembling smaller building blocks at different levels and using a limited number of structural patterns. The interactions between these structural units, the conformational readjustments upon their correct coupling seem to be essential for the expression and optimization of biological activity. Second, the native structure is not a rigid one and it fluctuates between several conformation states. The most specific substrate of an enzyme probably stabilizes the most reactive conformation, perhaps the one in which conformational coupling between building blocks is optimized. Study of the different motions in proteins in relation to their activity is certainly a very promising field of investigation which will progress rapidly in the near future.

The third point concerns protein folding and conformational dynamics in the cellular environment. In fact this field of research is not very well delineated at the moment due to the complexity of the systems and the heterogeneity of the environment. It remains a fundamental approach to the understanding of the genesis of enzymatic activity under biological conditions.

Many problems remain to be solved at the molecular level, and still more at the cellular level where the folding and conformational studies are just beginning. These studies certainly require a great variety of approaches and methodologies; new methodologies probably have to be introduced and developed.

It is possible that this book has asked more questions than it has answered. One of the goals in writing such a book is to stimulate the interest of young scientists for this type of study, which includes fundamental steps to understanding teleonomic performances of living organisms.

References

Acampora, G., and Hermans, J., Jr. (1967). *J. Am. Chem. Soc.* **89**, 1543–1552.

Acharya, A. S., and Taniuchi, H. (1976). *J. Biol. Chem.* **251**, 6934–6946.

Acharya, A. S., and Taniuchi, H. (1977). *Proc. Natl. Acad. Sci. U.S.A.* **74**, 2362–2366.

Acharya, A. S., and Taniuchi, H. (1978). *Biochemistry* **17**, 3064–3070.

Acharya, A. S., and Taniuchi, H. (1980). *J. Biol. Chem.* **255**, 1905–1911.

Adams, M. J., Brundell, T. L., Dodson, E. J., Dodson, G. G., Vijayan, M., Baker, E. N., Harding, M. M., Hodgkin, D. C., Rimmer, E., and Sheat, S. (1969). *Nature* (*London*) **224**, 491–495.

Adler, A., Greenfield, N. J., and Fasman, G. D. (1973). *In* "Enzyme Structure," Part D (C. H. W. Hirs and S. N. Timasheff, eds.), Methods in Enzymology, Vol. 27, pp. 675–735. Academic Press, New York.

Adman, E. T., Saken, L. C., and Jensen, L. H. (1973). *J. Biol. Chem.* **248**, 3987–3996.

Ahmed, A. K., Schaffer, S. W., and Wetlaufer, D. B. (1975). *J. Biol. Chem.* **250**, 8477–8482.

Alden, R. A., Birktoft, J. J., Kraut, J., Robertus, J. D., and Wright, C. J. (1971). *Biochem. Biophys. Res. Commun.* **45**, 337–344.

Allerhand, A., Doddrell, D., Glushko, V., Cochran, D. W., Wenkert, E., Lawson, P. J., and Gurd, R. R. (1971). *J. Am. Chem. Soc.* **93**, 544–546.

Allinger, M. J., Hickey, M. J., and Kao, J. (1976). *J. Am. Chem. Soc.* **98**, 2741–2745.

Almassy, R. J., and Dickerson, R. E. (1978). *Proc. Natl. Acad. Sci. U.S.A.* **75**, 2674–2678.

Ambles, R. P., and Scott, G. K. (1978). *Proc. Natl. Acad. Sci. U.S.A.* **75**, 3732–3736.

Amzel, L. M., and Poljak, R. J. (1979). *Annu. Rev. Biochem.* **48**, 961–997.

Amzel, M., Poljak, R. J., Saul, F., Varga, J. M., and Richards, R. F. (1974). *Proc. Natl. Acad. Sci. U.S.A.* **71**, 1427–1430.

Ananthanarayanan, U. S., and Bigelow, C. C. (1969a). *Biochemistry* **8**, 3717–3723.

Ananthanarayanan, U. S., and Bigelow, C. C. (1969b). *Biochemistry* **8**, 3723–3728.

Anderson, C. M., McDonald, R. C., and Steitz, T. A. (1978). *J. Mol. Biol.* **123**, 1–13.

Anderson, C. M., Zucker, F. H., and Steitz, T. A. (1979). *Science* **204**, 375–380.

Anderson, W. L., and Wetlaufer, D. B. (1975). *Anal. Biochem.* **67**, 493–502.
Anderson, W. L., and Wetlaufer, D. B. (1976). *J. Biol. Chem.* **251**, 3147–3153.
Andrews, L. J., and Forster, L. S. (1972). *Biochemistry* **11**, 1875–1879.
Andria, G., and Taniuchi, H. (1978). *J. Biol. Chem.* **253**, 2262–2270.
Andria, G., Taniuchi, H., and Cone, J. L. (1971). *J. Biol. Chem.* **246**, 7421–7428.
Anfinsen, C. B. (1956). *J. Biol. Chem.* **221**, 405–412.
Anfinsen, C. B. (1962). *Brookhaven Symp. Biol.* **15**, 184–198.
Anfinsen, C. B. (1966). *Harvey Lect.* **61**, 95–116.
Anfinsen, C. B. (1972). *Biochem. J.* **128**, 737–749.
Anfinsen, C. B. (1973). *Science* **181**, 223–230.
Anfinsen, C. B., and Haber, E. (1961). *J. Biol. Chem.* **236**, 1361–1363.
Anfinsen, C. B., and Scheraga, H. A. (1975). *Adv. Protein Chem.* **29**, 205–300.
Anfinsen, C. B., Haber, E., Sela, M., and White, F. H. (1961). *Proc. Natl. Acad. Sci. U.S.A.* **47**, 1309–1314.
Anfinsen, C. B., Sela, M., and Cooke, J. P. (1962). *J. Biol. Chem.* **237**, 1825–1831.
Anfinsen, C. B., Schechter, A. N., and Taniuchi, H. (1972). *Cold Spring Harbor Symp. Quant. Biol.* **36**, 249–255.
Anson, M. L. (1945). *Adv. Protein Chem.* **2**, 361–384.
Anson, M. L., and Mirsky, A. E. (1934a). *J. Gen. Physiol.* **17**, 393–398.
Anson, M. L., and Mirsky, A. E. (1934b). *J. Gen. Physiol.* **17**, 399–408.
Argos, P., Schwarz, J., and Schwarz, J. (1976). *Biochim. Biophys. Acta* **439**, 261–273.
Argos, P., Rossmann, M. G., and Johnson, J. E. (1977). *Biochem. Biophys. Res. Commun.* **75**, 83–86.
Argos, P., Rossmann, M. G., Grau, U. M., Zuber, H., Franck, G., and Tratshin, J. D. (1979). *Biochemistry* **18**, 5698–5703.
Arnone, A., Bier, C. J., Cotton, F. A., Day, V. W., Hazen, E. E., Richardson, D. C., Richardson, J. S., and Yonath, A. (1971). *J. Biol. Chem.* **246**, 2302–2316.
Arrio-Dupont, M. (1972). *Eur. J. Biochem.* **30**, 307–317.
Arrio-Dupont, M. (1978). *Eur. J. Biochem.* **91**, 369–378.
Arrio-Dupont, M., and Coulet, P. (1979). *Biochem. Biophys. Res. Commun.* **89**, 345–352.
Artymiuk, P. J., Blake, C. C. F., Grace, D. E. P., Oatley, J. J., Philipps, D. C., and Sternberg, M. J. E. (1979). *Nature (London)* **280**, 563–567.
Aschaffenburg, R., Green, D. W., and Simmons, R. M. (1965). *J. Mol. Biol.* **13**, 194–201.
Astbury, W. T. (1940). *J. Soc. Leather Trades Chem.* **24**, 69.
Astbury, W. T., and Bell, F. O. (1940). *Nature (London)* **145**, 421–422.
Astbury, W. T., and Bell, F. O. (1941). *Nature (London)* **147**, 696–699.
Atassi, M. Z. (1975). *Immunochemistry* **12**, 423–438.
Atassi, M. Z. and Stavitsky, A. B., (1977). "Immunochemistry of Proteins." Plenum, New York.
Atassi, M. Z., and Lee, C. L. (1978a). *Biochem. J.* **171**, 419–427.
Atassi, M. Z., and Lee, C. L. (1978b). *Biochem. J.* **171**, 429–434.
Atassi, M. Z., and Singhal, R. P. (1970). *J. Biol. Chem.* **245**, 5122–5128.
Atassi, M. Z., Koketsu, J., and Habeeb, A. F. S. A. (1976a). *Biochim. Biophys. Acta* **420**, 358–375.
Atassi, M. Z., Lee, C. L., and Pai, R. C. (1976b). *Biochim. Biophys. Acta* **427**, 745–751.
Aune, K. C., and Tanford, C. (1969a). *Biochemistry* **8**, 4579–4585.
Aune, K. C., and Tanford, C. (1969b). *Biochemistry* **8**, 4586–4590.
Aune, K. C., Salahuddin, A., Zarlengo, M. H., and Tanford, C. (1967). *J. Biol. Chem.* **242**, 4486–4489.
Austen, B. M. (1979). *FEBS Lett.* **103**, 308–313.
Azuma, T., Hagamuchi, K., and Migita, S. (1972). *J. Biochem. (Tokyo)* **72**, 1457–1467.

Babul, J., and Stellwagen, E. (1972). *Biochemistry* **11**, 1195–1200.

Babul, J., Nakagawa, A., and Stellwagen, E. (1978). *J. Mol. Biol.* **126**, 117–121.

Baldwin, J., and Chothia, C. (1979). *J. Mol. Biol.* **129**, 175–220.

Baldwin, R. L. (1975). *Annu. Rev. Biochem.* **891**, 453–475.

Baldwin, R. L., (1980). *In* "Protein Folding" (R. Jaenicke, ed.), pp. 369–386. Elsevier/North-Holland, Amsterdam.

Baldwin, R. L., and Creighton, T. E. (1980). *In* "Protein Folding" (R. Jaenicke, ed.), pp. 217–260. Elsevier/North-Holland, Amsterdam.

Balestrieri, C., Colonna, G., Giovane, A., Irace, G., and Servillo, L. (1976). *FEBS Lett.* **66**, 60–64.

Bandekar, J., and Krimm, S. (1979). *Proc. Natl. Acad. Sci. U.S.A.* **76**, 774–777.

Banks, R. D., Blake, C. C. F., Evans, P. R., Haser, R., Rice, D. W., Hardy, G. W., Merrett, M., and Phillips, A. W. (1979). *Nature (London)* **279**, 773–777.

Banner, D. W., Bloomer, A. C., Petsko, G. A., and Phillips, D. C. (1972). *Cold Spring Harbor Symp. Quant. Biol.* **36**, 151–155.

Banner, D. W., Bloomer, A. C., Petsko, G. A., Phillips, D. C., Pegson, C. I., and Wilson, I. A. (1975). *Nature (London)* **255**, 609–614.

Barden, R. E., Darke, P. L., Deems, R. A., and Dennis, E. A. (1980). *Biochemistry* **19**, 1621–1625.

Barker, J. A., and Watts, R. O. (1969). *Chem. Phys. Lett.* **3**, 144.

Bartholmes, P., Balk, H., and Kirschner, K. (1980). *Biochemistry* **19**, 4527–4533.

Bassford, P., and Beckwith, J. (1979). *Nature (London)* **277**, 538–541.

Bauer, R. S., Chang T. L., and Berliner, L. J. (1980). *J. Biol. Chem.* **255**, 5900–5903.

Bayley, H., and Knowles, J. R. (1978). *Biochemistry* **17**, 2414–2419.

Béchet, J. J., and d'Albis, A. (1969). *Biochim. Biophys. Acta* **178**, 561–576.

Beck, K., Gill, S. J., and Doromy, M. (1965). *J. Am. Chem. Soc.* **87**, 901–904.

Beissner, R. S., Quiocho, F. A., and Rudolph, F. B. (1979). *J. Mol. Biol.* **134**, 847–850.

Belasco, J. G., and Knowles, J. R. (1980). *Biochemistry* **19**, 472–477.

Bello, J. (1969). *Biochemistry* **8**, 4542–4550.

Bender, M. L., and Kezdy, F. J. (1964). *J. Am. Chem. Soc.* **86**, 3704–3714.

Benjamin, D. C., and Teale, L. M. (1978). *J. Biol. Chem.* **253**, 8087–8092.

Bennett, W. S. and Steitz, T. A. (1978). *Proc. Natl. Acad. Sci. U.S.A.* **75**, 4848–4852.

Bennett, W. S., and Steitz, T. A. (1980a). *J. Mol. Biol.* **140**, 211–230.

Bennett, W. S., and Steitz, T. A. (1980b). *J. Mol. Biol.* **140**, 183–209.

Benson, S. W. (1960). "The Foundation of Chemical Kinetics." McGraw-Hill, New York.

Benz, F. W., and Roberts, G. C. K. (1973). *FEBS Lett.* **29**, 263–266.

Benz, F. W., and Roberts, G. C. K. (1975a). *J. Mol. Biol.* **91**, 345–365.

Benz, F. W., and Roberts, G. C. K. (1975b). *J. Mol. Biol.* **91**, 367–387.

Berendsen, H. J. C. (1972). *Proc. FEBS Meet.* **29**, 19–27.

Bergman, L. W., and Kuehl, W. M. (1979a). *J. Biol. Chem.* **254**, 5690–5694.

Bergman, L. W., and Kuehl, W. M. (1979b). *J. Biol. Chem.* **254**, 8869–8876.

Bergman, L. W., and Kuehl, W. M. (1979c). *J. Supramol. Struct.* **11**, 9–24.

Bergmann, J. E., and Lodish, H. F. (1979). *J. Biol. Chem.* **254**, 11927–11937.

Bergsma, J., Hol, W. G. J., Jansonius, J. N., Kalk, K. H., Poegman, J. H., and Smit, J. D. G. (1975). *J. Mol. Biol.* **98**, 637–643.

Bernal, J. D. (1958). *Discuss. Faraday Soc.* **25**, 7–18.

Bernhard, S. A., and Lau, S. J. (1972). *Cold Spring Harbor Symp. Quant. Biol.* **36**, 75–83.

Bernhard, S. A., and Rossi, J. L. (1968). "Structural Chemistry and Molecular Biology." Freeman, San Francisco, California.

Berzofsky, J. A., Curd, J. G., and Schechter, A. N. (1976). *Biochemistry* **15**, 2113–2121.

Beychok, S. (1968). *Annu. Rev. Biochem.* **37**, 437–462.

Biellmann, J. F., Samama, J. P., Bränden, C. I., and Eklund, H. (1979). *Eur. J. Biochem.* **102**, 107–110.
Biesecker, G., Harris, J. I., Thierry, J. C., Walker, J. E., and Wonacott, A. J. (1977). *Nature (London)* **266**, 328–333.
Bigelow, C. C., and Geschwind, I. I. (1961). *C. R. Trav. Lab. Carsberg* **32**, 89.
Bigelow, C. C., and Sonenberg, M. (1962). *Biochemistry* **1**, 197–204.
Bijvoet, J. M., Peerdeman, A. F., and van Bommel, A. J. (1951). *Nature (London)* **168**, 271–272.
Biltonen, R., Lumry, R., Madison, V., and Parker, H. (1965). *Proc. Natl. Acad. Sci. U.S.A.* **54**, 1412–1419.
Biltonen, R., Schwartz, A. T., and Watsö, I. (1971). *Biochemistry* **10**, 3417–3423.
Birk, Y., and Li, C. H. (1978). *Proc. Natl. Acad. Sci. U.S.A.* **75**, 2155–2159.
Birktoft, J. J., and Blow, D. M. (1972). *J. Mol. Biol.* **68**, 187–240.
Bishop, J., Leahy, J., and Schweet, R. (1960). *Proc. Natl. Acad. Sci. U.S.A.* **46**, 1030–1038.
Blake, C. C. F. (1977). *Nature (London)* **267**, 482–483.
Blake, C. C. F. (1980). *Nature (London)* **285**, 190–191.
Blake, C. C. F., Swan, I. D. A., Rerat, C., Berthou, J., Laurent, A., and Rerat, B. (1971). *J. Mol. Biol.* **61**, 217–224.
Blake, C. C. F., Evans, D. R., and Scopes, R. K. (1972). *Nature (London)* **235**, 195–198.
Blangy, D., Buc, H., and Monod, J. (1968). *J. Mol. Biol.* **31**, 13–35.
Blobel, G. (1980). *Proc. Natl. Acad. Sci. U.S.A.* **77**, 1496–1500.
Blobel, G., and Dobberstein, B. (1975a). *J. Cell Biol.* **67**, 835–851.
Blobel, G., and Dobberstein, B. (1975b). *J. Cell Biol.* **67**, 852–862.
Blobel, G., and Sabatini, D. D. (1971). *Biomembranes* **2**, 193–195.
Blout, E. R., and Asadourian, A. (1956). *J. Am. Chem. Soc.* **78**, 955–961.
Blow, D. M., and Steitz, T. A. (1970). *Annu. Rev. Biochem.* **39**, 63–97.
Blow, D. M., Irwin, M. J., and Nyborg, J. (1977). *Biochem. Biophys. Res. Commun.* **76**, 728–734.
Blum, A. D., Smallcombe, S. H., and Baldwin, R. L. (1978). *J. Mol. Biol.* **118**, 305–316.
Blundell, T. (1979). *TIBS* **4**, 51–54.
Bode, W. (1979). *J. Mol. Biol.* **127**, 357–374.
Bode, W., and Schwager, P. (1975). *J. Mol. Biol.* **98**, 693–717.
Bohnert, J. L., and Taniuchi, H. (1972). *J. Biol. Chem.* **247**, 4557–4560.
Bornmann, L., Hess, B., and Zimmermann-Telschow, H. (1974). *Proc. Natl. Acad. Sci. U.S.A* **71**, 1525–1529.
Bosshard, H. R., Koch, G. L. E., and Hartley, B. S. (1978). *J. Mol. Biol.* **119**, 337–389.
Bothwell, M. A., and Schachman, H. K. (1980). *J. Biol. Chem.* **255**, 1971–1977.
Boulter, D., Ramshaw, J. A. M., Thompson, E. W., Richardson, M., and Brown, R. H. (1972). *Proc. R. Soc. London, Ser. B* **181**, 441–455.
Bowers, W. F., Czubaroff, V. B., and Haschemeyer, R. H. (1970). *Biochemistry* **9**, 2620–2627.
Boyle, J., Robillard, G. T., and Kim, S. H. (1980). *J. Mol. Biol.* **139**, 601–625.
Bradbury, F. M., Brown, L., Downie, A. R., Eliott, A., Hanby, W. E., and McDonald, F. R. R. (1959) *Nature (London)* **183**, 1736–1737.
Bradbury, J., and King, N. L. R. (1971). *Aust. J. Chem.* **24**, 1703–1714.
Bradshaw, R. A., Kanarek, L., and Hill, R. L. (1967). *J. Biol. Chem.* **242**, 3789–3798.
Bragg, W. L., Kendrew, J. C., and Perutz, M. F. (1950). *Proc. R. Soc. London, Ser. A* **203**, 321–357.
Brahms, S., and Brahms, J. (1980). *J. Mol. Biol.* **138**, 149–178.
Brandts, J. F. (1964a). *J. Am. Chem. Soc.* **86**, 4291–4301.
Brandts, J. F. (1964b). *J. Am. Chem. Soc.* **86**, 4302–4314.
Brandts, J. F. (1969). *In* "Structure and Stability of Biological Macromolecules" S. N. (Timasheff and G. D. Fasman, eds.), pp. 213–290. Dekker, New York.

Brandts, J. F., and Hunt, L. (1967). *J. Am. Chem. Soc.* **89**, 4826–4838.

Brandts, J. F., and Lumry, R. (1963). *J. Phys. Chem.* **67**, 1484–1494.

Brandts, J. F., Halvorson, H. R., and Brennan, M. (1975). *Biochemistry* **14**, 4953–4963.

Brandts, J. F., Brennan, M., and Lin, L. N. (1977). *Proc. Natl. Acad. Sci. U.S.A.* **74**, 4178–4181.

Brant, D. A., and Flory, P. J. (1965a). *J. Am. Chem. Soc.* **87**, 663–664.

Brant, D. A., and Flory, P. J. (1965b). *J. Am. Chem. Soc.* **87**, 2791–2800.

Brant, D. A., Muller, W. G., and Flory, P. J. (1967). *J. Mol. Biol.* **23**, 47–65.

Brazhnikov, E. V., and Chirgadze, Y. N. (1978). *J. Mol. Biol.* **122**, 127–135.

Brown, J. C., and Horton, H. R. (1972). *Proc. Soc. Exp. Biol. Med.* **140**, 1451–1455.

Brown, J. C., Swaisgood, H. E., and Horton, H. R. (1972). *Biochem. Biophys. Res. Commun.* **48**, 1068–1073.

Brown, J. E., and Klee, W. A. (1969). *Biochemistry* **8**, 2876–2879.

Brown, J. E., and Klee, W. A. (1971). *Biochemistry* **10**, 470–476.

Brown, J. R., and Hartley, B. S. (1966). *Biochem. J.* **101**, 214–228.

Brown, K. G., Erfuth, S. C., Small, E. W., and Peticolas, W. L. (1972). *Proc. Natl. Acad. Sci. U.S.A.* **69**, 1467–1469.

Brown, L. R., de Marco, A., Wagner, G., and Wüthrich, K. (1976). *Eur. J. Biochem.* **62**, 103–107.

Brunner, H., and Sussner, H. (1972). *Biochim. Biophys. Acta* **271**, 16–22.

Brunner, H., Holz, M., and Gering, H. (1974). *Eur. J. Biochem.* **50**, 129–133.

Buehner, M., Ford, G. C., Moras, D., Olsen, K. W., and Rossmann, M. G. (1973). *Proc. Natl. Acad. Sci. U.S.A.* **70**, 3052–3054.

Bull, H. B., and Breese, K. (1973). *Arch. Biochem. Biophys.* **158**, 681–686.

Burgess, A. W., and Scheraga, H. A. (1975a). *J. Theor. Biol.* **53**, 403–420.

Burgess, A. W., and Scheraga, H. A. (1975b). *Proc. Natl. Acad. Sci. U.S.A.* **72**, 1221–1225.

Burgess, A. W., Ponnuswamy, P. K., and Scheraga, H. A. (1974). *Isr. J. Chem.* **12**, 239–286.

Burgess, A. W., Weinstein, L. I., Gable, D. G., and Scheraga, H. A. (1975). *Biochemistry* **14**, 197–200.

Burstein, Y., and Schechter, I. (1977). *Proc. Natl. Acad. Sci. U.S.A.* **74**, 716–720.

Burstein, Y., and Schechter, I. (1978). *Biochemistry* **17**, 2392–2400.

Campbell, I. D., Dobson, C. M., and Williams, R. J. P. (1975a). *Proc. R. Soc. London, Ser, B* **189**, 485–502.

Campbell, I. D., Dobson, C. M., and Williams, R. J. P. (1975b). *Proc. R. Soc. London, Ser. B* **189**, 503–509.

Campbell, I. D., Dobson, C. M., Moore, G. R., Perkins, S. J., and Williams, R. J. P. (1976). *FEBS Lett.* **70**, 96–100.

Campbell, J.W., Hogson, G. I., Watson, H. C., and Scopes, R. K. (1971). *J. Mol. Biol.* **61**, 257–259.

Campbell, J. W., Hogson, G. I., and Watson, H. C. (1972). *Nature* (*London*), *New Biol.* **240**, 137–139.

Canfield, R. E., and Anfinsen, C. B. (1963a). *Biochemistry* **2**, 1073–1078.

Canfield, R. E., and Anfinsen, C. B. (1963b). *In* "The Proteins" (H. Neurath, ed.), 2nd ed., Vol. 2, p. 311–378. Academic Press, New York.

Carlsson, U., Henderson, L. E., and Lindskoj, S. (1973). *Biochim. Biophys. Acta* **310**, 376–387.

Carlsson, U., Aasa, R., Henderson, L. R., Johnson, B. H., and Lindskoj, S. (1975). *Eur. J. Bio-Chem.* **52**, 25–36.

Carmichael, D. F., Morin, J. E., and Dixon, J. E. (1977). *J. Biol. Chem.* **252**, 7163–7167.

Carmichael, D. F., Keefe, M., Pace, M., and Dixon, J. E. (1979). *J. Biol. Chem.* **254**, 8386–8390.

Careri, G. (1974). *In* "Quantum Statistic Mechanisms in Natural Sciences" (B. Kursunoglu, S. L. Mintz, and S. M., Widmayer, eds.), pp. 15–35. Plenum, New York.

Careri, G., Fasella, P., and Gratton, E. (1975). *CRC Crit. Rev. Biochem.* **3**, 141–164.
Careri, G., Fasella, P., and Gratton, E. (1979). *Annu. Rev. Biophys. Bioeng.* **8**, 69–97.
Cave, A., Dobson, C. M., Parello, J., and Williams, R. J. P. (1976). *FEBS Lett.* **65**, 190–194.
Cecil, R., and McPhee, J. R. (1959). *Adv. Protein Chem.* **14**, 255–389.
Celada, F., and Strom, R. (1972). *Q. Rev. Biophys.* **5**, 395–425.
Cerf, R. (1973). *In.* "Dynamic Aspects of Conformation Changes in Biological Macromolecules" (C. Sadron, ed.), pp. 247–269. Reidel Publ. Dordrecht, Netherlands.
Cha, C. Y., and Scheraga, H. A. (1963). *J. Biol. Chem.* **238**, 2958–2964.
Chaffotte, A. F., Zakin, M. M., and Goldberg, M. E. (1980). *Biochem. Biophys. Res. Commun.* **92**, 381–388.
Chaiken, I. M., Freedman, M. H., Ligerla, J. R., and Cohen, J. (1973). *J. Biol. Chem.* **248**, 884–891.
Chan, S. J., Keim, P., and Steiner, D. F. (1976). *Proc. Natl. Acad. Sci. U.S.A.* **73**, 1964–1968.
Chan, W. W. C. (1970). *Biochem. Biophys. Res. Commun.* **41**, 1198–1204.
Chan, W. W. C., and Mawer, H. M. (1972). *Arch. Biochem. Biophys.* **149**, 136–145.
Chan, W. W. C., Mort, J. S., Chong, D. K. K., and Macdonald, P. D. M. (1973a) *J. Biol. Chem.* **248**, 2778–2784.
Chan, W. W. C., Schutt, H., and Brand, K. (1973b). *Eur. J. Biochem.* **40**, 533–541.
Chandrasekaran, R., Jardetzky, T. S., and Jardetzky, O. (1979). *FEBS Lett.* **101**, 11–14.
Chang, C. N., Model, P., and Blobel, G. (1979). *Proc. Natl. Acad. Sci. U.S.A.* **76**, 1251–1255.
Chang, L. T., Bauer, R. S., and Berliner, L. (1980). *J. Biol. Chem.* **255**, 5904–5906.
Chantrenne, H. (1961). "The Biosynthesis of Proteins," p. 122. Pergamon, New York.
Chavez, L. G., and Benjamin, D. C. (1978). *J. Biol. Chem.* **253**, 8081–8086.
Chavez, L. G., and Scheraga, H. A. (1977). *Biochemistry* **16**, 1849–1856.
Chavez, L. G., and Scheraga, H. A. (1979). *Biochemistry* **18**, 4386–4395.
Chavez, L. G., and Scheraga, H. A. (1980a) *Biochemistry* **19**, 996–1004.
Chavez, L. G., and Scheraga, H. A. (1980b). *Biochemistry* **19**, 1005–1012.
Chen, M. C., and Lord, R. C. (1976). *Biochemistry* **15**, 1889–1897.
Chen, M. C., Lord, R. C., and Mendelsohn, R. (1973). *Biochim. Biophys. Acta* **328**, 252–260.
Chen, M. C., Lord, R. C., and Mendelsohn, R. (1974). *J. Am. Chem. Soc.* **96**, 3038–3042.
Chen, R. F. (1967). *In* "Fluorescence" G. C. Guilbault, ed.), p. 443. Dekker, New York.
Chen, R. F., Edelhoch, H., and Steiner, R. F. (1969). *In* "Physical Principles and Techniques of Protein Chemistry" (S. J. Leach, ed.), Part A, p. 171. Academic Press, New York.
Chevallier, J., Yon, J., and Labouesse, J. (1969). *Biochim. Biophys. Acta* **181**, 73–81.
Chignell, D. A., Azhir, A., and Gratzer, W. (1972). *Eur, J. Biochem.* **26**, 37–42.
Chirckjian, J. G., Wright, H. T., and Fresco, J. R. (1972). *Proc. Natl. Acad. Sci. U.S.A.* **69**, 1638–1641.
Chothia, C. (1973). *J. Mol. Biol.* **75**, 295–302.
Chothia, C. (1974). *Nature (London)* **248**, 338–339.
Chothia, C. (1975). *Nature (London)* **254**, 304–308.
Chothia, C. (1976). *J. Mol. Biol.* **105**, 1–14.
Chothia, C., and Janin, J. (1975). *Nature (London)* **256**, 705–708.
Chothia, C., and Janin, J. (1978). *FEBS Meeting, Dresden, 12th* 117–126.
Chothia, C., and Lesk, A. M. (1980). *In* "Protein Folding" (R. Jaenicke, ed.), pp. 63–77. Elsevier/North-Holland, Amsterdam.
Chothia, C., Wodak, S., and Janin, J. (1976). *Proc. Natl. Acad. Sci. U.S.A.* **73**, 3793–3797.
Chothia, C., Levitt, M., and Richardson, D. (1977). *Proc. Natl. Acad. Sci. U.S.A.* **74**, 4130–4134.
Chothia, C., Levitt, M., and Richardson, D. C. (1981). *J. Mol. Biol.* **145**, 215–250.
Chou, P. Y., and Fasman, G. D. (1974a). *Biochemistry* **13**, 211–222.
Chou, P. Y., and Fasman, G. D. (1974b). *Biochemistry* **13**, 222–245.

Chou, P. Y., and Fasman, G. D. (1977). *J. Mol. Biol.* **115**, 135–175.

Chou, P. Y., and Fasman, G. D. (1978). *Annu. Rev. Biochem.* **47**, 251–276.

Chryssomalis, G. S., Torgerson, P. M., Drickamer, H. G., and Weber G. (1981). *Biochemistry* **20**, 3955–3959.

Citri, N. (1973). *Adv. Enzymol. Relat. Areas Mol. Biol.* **37**, 397–648.

Cohen, F. E., and Sternberg, M. J. E. (1980). *J. Mol. Biol.* **137**, 9–22.

Cohen, F. E., Richmond, T. J., and Richards, F. M. (1979). *J. Mol. Biol.* **132**, 275–288.

Cohen, F. E., Sternberg, M. J. E., and Taylor, W. (1980a). *In* "Protein Folding" (R. Jaenicke, ed.), pp. 131–148. Elsevier/North-Holland, Amsterdam.

Cohen, F. E., Sternberg, M. J. E., and Taylor, W. R. (1980b). *Nature (London)* **285**, 378–382.

Cohen, F. E., Sternberg, M. J. E., and Taylor, W. R. (1981). *J. Mol. Biol.* **148**, 253–272.

Cohlberg, J. A., Pigiet, V. P., and Schachman, H. K. (1972). *Biochemistry* **11**, 3396–3411.

Cone, J. C., Casamuno, C. L., Taniuchi, H., and Anfinsen, C. B. (1971). *J. Biol. Chem.* **246**, 3103–3131.

Cook, D. A. (1967). *J. Mol. Biol.* **29**, 167–171.

Cook, H., Schmid, F. X., and Baldwin, R. L. (1979). *Proc. Natl. Acad. Sci. U.S.A.* **76**, 6157–6161.

Cooke, J. P., Anfinsen, C. B., and Sela, M. (1963). *J. Biol. Chem.* **238**, 2034–2039.

Cooper, A. (1976). *Proc. Natl. Acad. Sci. U.S.A.* **73**, 2740–27411.

Corey, R. B., and Pauling, L. (1953). *Proc. R. Soc. London, Ser. B* **141**, 10.

Corey, R. B., and Pauling, L. (1956). *Proc. Int. Wool Text. Res. Conf., Sydney Geelong, Melbourne, 1955*, p. 249–267.

Coté, C., Solioz, M., and Schatz, G. (1979). *J. Biol. Chem.* **254**, 1437–1439.

Cotton, F. A., and Hazen, E. E. (1971). *In* "The Enzymes" (P. D. Boyer, ed.), Vol. IV, 3rd ed., pp. 153–204. Academic Press, New York.

Cournil, I. (1975). Thèse, Univ. de Paris-Sud.-Orsay.

Cournil, I., and Arrio-Dupont, M. (1975). *Biochem. Biophys. Res. Commun.* **64**, 1119–1125.

Cournil, I., Barba, J. M., Vergé, D., and Arrio-Dupont, M. (1975). *Eur. J. Biochem.* **250**. 8564–8568.

Craig, R. K., Brown, P. A., Harrison, O. S., McIlreavy, D., and Campbell, P. N. (1975). *Biochem. J.* **160**, 57–74.

Craik, C. S., Buchman, S. R., and Beychok, S. (1980). *Proc. Natl. Acad. Sci. U.S.A.* **77**, 1384–1388.

Crawford, J. L., Lipscomb, W. N., and Schellman, C. G. (1973). *Proc. Natl. Acad. Sci. U.S.A* **70**, 538–542.

Creighton, T. E. (1974a). *J. Mol. Biol.* **87**, 563–577.

Creighton, T. E. (1974b). *J. Mol. Biol.* **87**, 579–602.

Creighton, T. E. (1974c). *J. Mol. Biol.* **87**, 603–624.

Creighton, T. E. (1975a). *J. Mol. Biol.* **95**, 167–199.

Creighton, T. E. (1975b). *J. Mol. Biol.* **96**, 767–776.

Creighton, T. E. (1975c). *J. Mol. Biol.* **96**, 777–782.

Creighton, T. E. (1975d). *Nature (London)* **255**, 743–745.

Creighton, T. E. (1977a). *J. Mol. Biol.* **113**, 275–293.

Creighton, T. E. (1977b). *J. Mol. Biol.* **113**, 295–312.

Creighton, T. E. (1977c). *J. Mol. Biol.* **113**, 313–328.

Creighton, T. E. (1977d). *J. Mol. Biol.* **113**, 329–341.

Creighton, T. E. (1978a). *J. Mol. Biol.* **119**, 502–518.

Creighton, T. E. (1978b). *J. Mol. Biol.* **123**, 129–147.

Creighton, T. E. (1978c). *Prog. Biophys. Mol. Biol.* **33**, 231–297.

Creighton, T. E. (1979a). *J. Mol. Biol.* **129**, 235–264.

Creighton, T. E. (1979b). *J. Mol. Biol.* **129**, 411–431.

Creighton, T. E. (1980a). *J. Mol. Biol.* **137**, 61–80.
Creighton, T. E. (1980b). *FEBS Lett.* **118**, 283–288.
Creighton, T. E. (1980c). *J. Mol. Biol.* **144**, 521–550.
Creighton, T. E., and Pain, R. H. (1980). *J. Mol. Biol.* **137**, 431–436.
Creighton, T. E., Kalef, E., and Arnon, R. (1978), *J. Mol. Biol.* **123**, 129–147.
Crick, F. H. C. (1953). *Acta Crystallogr.* **6**, 689–697.
Crick, F. H. C. (1958). *Symp. Soc. Exp. Biol.* **13**, 138–163.
Crippen, G. M., and Kuntz, I. D. (1977). *J. Mol. Biol.* **66**, 47–61.
Crumpton, M. J. (1968). *Proc. Biochem. Soc.* **108**, 18P–19P.
Crumpton, M. J. (1974). *In* "The Antigens" (M. Sela, ed.), Vol. 2, pp. 1–78. Academic Press, New York.
Curd, J. G., Ludwig, D., and Schechter, A. N. (1976). *J. Biol. Chem.* **251**, 1283–1289.
d'Albis, A., and Béchet, J. J. (1967). *Biochim. Biophys. Acta* **140**, 435–458.
Darnall, D. W., and Klotz, I. M. (1972). *Arch. Biochem. Biophys.* **149**, 1–14.
Dautry-Varsat, A., and Cohen, G. (1977). *J. Biol. Chem.* **252**, 7685–7689.
Dautry-Varsat, A., and Garel, J. R. (1978). *Proc. Natl. Acad. Sci. U.S.A.* **75**, 5979–5982.
Dautry-Varsat, A., and Garel, J. R. (1981). *Biochemistry* **20**, 1396–1401.
Davies, D. R. (1965). *Prog. Biophys. Mol. Biol.* **15**, 189–222
Davies, D. R. (1967). *Annu. Rev. Biochem.* **36**, 321–364.
Davies, D. R., Padlan, A. E., and Segal, D. M. (1975). *Annu. Rev. Biochem.* **44**, 639–667.
Davis, B. D., and Tai, P. C. (1980). *Nature* (*London*) **283**, 433–438.
Dayhoff, M. O. (1969, 1972, 1973, 1976, 1978). "Atlas of Protein Sequence and Structure." Nat. Biomed, Res. Found., Silver Spring, Maryland.
Dayhoff, M. O., and Eck, R. V. (1967–1968). "Atlas of Protein Sequence and Structure." Nat. Biomed. Res. Found., Silver Spring, Maryland.
Deal, W. C. (1969). *Biochemistry* **8**, 2795–2805.
Deal, W. C., Rutter, I. W., and van Holde, K. E. (1963). *Biochemistry* **2**, 246–251.
Dean, J., and Schechter, A. N. (1979). *J. Biol. Chem.* **254**, 9185–9193.
Deisenhofer, J. O., and Steigemann, W. (1975). *Acta Crystallogr., Sect. B* **31**, 238–250.
de Lorenzo, F., Goldberger, R. F., Steers, E., Givol, D., and Anfinsen, C. B. (1966). *J. Biol. Chem.* **241**, 1562–1567
de Marco, A., Tzchesche, H., Wagner, G., and Wüthrich, K. (1977). *Biophys. Struct. Mech.* **3**, 303–315.
de Rosier, D. U., Oliver, R. M., and Reed, J. L. (1971). *Proc. Natl. Acad. Sci. U.S.A.* **68**, 1135–1137.
de Santis, P., Giglio, E., Liquori, A. M., and Ripamonti, A. (1965). *Nature* (*London*) **206**, 456–458.
Deschavanne, P. J., Viratelle, O. M., and Yon, J. M. (1978a). *J. Biol. Chem.* **253**, 833–837.
Deschavanne, P. J., Viratelle, O. M., and Yon, J. M. (1978b). *Proc. Natl. Acad. Sci. U.S.A.* **75**, 1892–1896.
Deslauriers, R., Leach, S. J., Maxfield, F. R., Minasian, E., McQuie, J. R., Mainwald, Y. C., Némethy, G., Pottle, M. S., Rae, J. D., Scheraga, H. A., Stimson, E. R., and van Nisken, J. W. (1979). *Proc. Natl. Acad. Sci. U.S.A.* **76**, 2512–2514.
Desmadril, M., and Yon, J. M. (1981). *Biochem. Biophys. Res. Commun.* **101**, 563–569.
Desmadril, M., and Yon, J. M. (1982). Submitted for publication.
Desmadril, M., Tempête-Gaillourdet, M., and Yon, J. M. (1982a). Submitted for publication.
Desmadril, M., Tempête-Gaillourdet, M., Amédée-Manesme, O., Crosetti, G., Mennecier, F., and Yon, J. M. (1982b). In preparation.
Devillers-Thierry, A., Kindt, T., Scheele, G., and Blobel, G. (1975). *Proc. Natl. Acad. Sci. U.S.A.* **772**, 5016–5050.

Dhondt, J. L., Dautrevaux, M., Biserte, G., and Farriaux, J. P. (1978). *Biochimie* **60**, 787–794.
Dickerson, R. E. (1964). *In* "The Proteins" (H. Neurath, ed.), Vol. 2, pp. 603–778. Academic Press, New York.
Dickerson, R. E. (1971). *J. Mol. Biol.* **1**, 26–45.
Dickerson, R. E., and Geis, I. (1969). "The Structure and Action of Proteins." Harper, New York.
Dickerson, R. E., and Timkovitch, R. (1975). *In* "The Enzymes" (P. D. Boyer, ed.), 3rd ed., Vol. 11, pp. 397–547. Academic Press, New York.
Dickerson, R. E., Takaro, T., Eisenberg, D., Kallai, O., Samson, L., Cooper, A., and Margoliash, E. (1971). *J. Biol. Chem.* **246**, 1511–1535.
Dintzis, H. M. (1961). *Proc. Natl. Acad. Sci. U.S.A.* **47**, 247–261.
Donovan, J. W. (1969a). *J. Biol. Chem.* **244**, 1961–1967.
Donovan, J. W. (1969b). *In* "Physical Principles and Techniques of Protein Chemistry" (S. J. Leach, ed.), Part A, pp. 101–170. Academic Press, New York.
Donovan, J. W. (1973). *In* "Enzyme Structure," Part C (C. H. W. Hirs and S. N. Timasheff, ed. Methods in Enzymology, Vol. 26, pp. 497–525. Academic Press, New York.
Doolittle, R. F. (1979). *In* "The Proteins" (H. Neurath and R. L. Hill, eds.), 3rd ed., Vol. 4, pp. 1–118. Academic Press, New York.
Doster, W., and Hess, B. (1981). *Biochemistry* **20**, 772–780.
Doty, P. (1959). *In* "Biophysical Science" (J. L. Oncley, ed.), p. 112. Wiley, New York.
Doty, P., and Yang, J. T. (1956). *J. Am. Chem. Soc.* **78**, 498–500.
Doty, P., Bradbury, J. H., and Holtzer, A. M. (1956). *J. Am. Chem. Soc.* **78**, 947–954.
Drenth, J., and Smith, J. D. G. (1971). *Biochem. Biophys. Res. Commun.* **45**, 1320–1322.
Drenth, J., Jansonius, J. M., Koekoek, R., and Wolthers, B. G. (1971). *Adv. Protein Chem.* **25**, 79–115.
Drenth, J., Low, B. W., Richardson, J. S., and Wright, C. S. (1980). *J. Biol. Chem.* **255**, 2652–2655.
Drew, H. R., and Dickerson, R. E. (1978). *J. Biol. Chem.* **253**, 8420–8427.
Dreyfus, M., Maigret, B., and Pullmann, A. (1970). *Theor. Chim. Acta* **17**, 109–119.
Drufton, M. J., and Hides, R. C. (1980). *TIBS* **5**, 53–56.
Dubs, A., Wagner, G., and Wüthrich, K. (1979). *Biochim. Biophys. Acta* **577**, 177–194.
Duguid, J. R., Steiner, D. F., and Chick, W. L. (1976). *Proc. Natl. Acad. Sci. U.S.A.* **73**, 3539–3543.
Dunfield, L. G., Burgess, A. W., and Scheraga, H. A. (1978). *J. Phys. Chem.* **82**, 2609–2616.
Dunhill, P. (1967). *Nature* (*London*) **215**, 621–622.
Dupont, M., and Yon, J. (1961). *J. Chem. Phys.* **58**. 683–689.
Dygert, M., Go, N., and Scheraga, H. A. (1975). *Macromolecules* **8**, 750–761.
Eagles, P. A. M., Iqbal, M., Johnson, L. N., Mosley, J., and Wilson, K. S. (1972). *J. Mol. Biol.* **71**, 803–806.
Eastlake, A., Sachs, D. H., Schechter, A. N., and Anfinsen, C. B. (1974). *Biochemistry* **13**, 1567–1571.
Eck, R. V., and Dayhoff, M. O. (1966). "Atlas of Protein Sequence and Structure." Nat. Biomed. Res. Found., Silver Spring, Maryland.
Edelman, G. M. (1970). *Biochemistry* **9**, 3197–3204.
Edelman, G. M., and Gally, G. A. (1968). *Brookhaven Symp. Biol.* **21**, 328.
Edelstein, S. J. (1980). *Biophys. Discuss. Meet. Proteins Nucleoproteins, Airlie, Va* pp. 323–332.
Edsall, J. T., and McKenzie, H. A. (1978). *Adv. Biophys.* **10**, 137–207.
Edsall, J. T., Flory, P. J., Kendrew, J. C., Liquori, A. M., Némethy, G., Ramachandran, G. N., and Scheraga, H. A. (IUPAC–IUB Commission of Biochemical Nomenclature)

(1966a). *J. Mol. Biol.* **15**, 399–407; *J. Biol. Chem.* **241**, 1004–1008 (1966b); *J. Mol. Biol.* **52**, 1–17 (1970); *Biochem. J.* **121**, 577–585 (1971).
Efimov, A. V. (1979). *J. Mol. Biol.* **134**, 23–40.
Egan, W., Shindo, H., and Cohen, J. S. (1977). *Annu. Rev. Biophys. Bioeng.* **6**, 383–417.
Eisenberg, M. A., and Schwert, G. W. (1951). *J. Gen. Physiol.* **34**, 583.
Eklund, H., and Branden, C. I. (1979). *J. Biol. Chem.* **254**, 3458–3461.
Eklund, H., Nordström, B., Zepperauer, E., Södesland, G., Ohlsen, I., Boiwe, T., Söderberg, B. O., Tapia, O., Branden, C. I., and Åkeson, A. (1976). *J. Mol. Biol.* **102**, 27–59.
Eldjarn, J., and Pihl, A. (1957). *J. Biol. Chem.* **225**, 499–510.
Elson, E. L. (1972a). *Biopolymers* **11**, 1499–1510.
Elson, E. L. (1972b). *J. Mol. Biol.* **63**, 469–475.
Elwell, M. (1976). Ph.D. Thesis, Univ. of Oregon, Eugene.
Elwell, M., and Schellman, J. (1975). *Biochim. Biophys. Acta* **386**, 309–323.
Elwell, M., and Schellman, J. (1977). *Biochim. Biophys. Acta* **494**, 367–383.
Engelhard, M., Rudolph, R., and Jaenicke, R. (1976). *Eur. J. Biochem.* **67**, 447–453.
Engelman, D. M., Henderson, R., McLachlan, A. D., and Wallace, B. A. (1980). *Proc. Natl. Acad. Sci. U.S.A.* **77**, 2023–2027.
England, P. J. (1980). *In* "The Enzymology of Post-Translational Modifications in Proteins" (R. B. Freedman and H. C. Hawkins, eds.), pp. 292–344. Academic Press, New York.
Englander, S. W. (1973). *Ann. N.Y. Acad. Sci.* **244**, 10–27.
Englander, S. W., and Englander, J. J. (1972). *In* "Enzyme Structure," Part C (C. H. W. Hirs and S. N. Timasheff, eds.), Methods in Enzymology, Vol. 26, pp. 406–424. Academic Press, New York.
Englander, S. W., and Mauel, C. (1972). *J. Biol. Chem.* **247**, 2387–2394.
Englander, S. W., and Rolfe, A. (1973). *J. Biol. Chem.* **248**, 4852–4861.
Englander, S. W., and Stanley, R. (1969). *J. Mol. Biol.* **45**, 277–295.
Englander, S. W., Downer, N. W., and Teitelbaum, H. (1972). *Annu. Rev. Biochem.* **41**, 903–924.
Epand, R. M., and Scheraga, H. A. (1968). *Biochemistry* **7**, 2864–2872.
Epp, O., Steigeman, W., Formanek, H., and Huber, R. (1971). *Eur. J. Biochem.* **20**, 432–437.
Epstein, C. J., and Anfinsen, C. B. (1962a). *J. Biol. Chem.* **237**, 2175–2179.
Epstein, C. J., and Anfinsen, C. B. (1962b). *J. Biol. Chem.* **237**, 3464–3467.
Epstein, C. J., and Goldberger, R. F. (1963). *J. Biol. Chem.* **238**, 1380–1383.
Epstein, C. J., Goldberger, R. F., Young, D. M., and Anfinsen, C. B. (1962). *Arch. Biochem. Biophys., Suppl.* No. 1, 223–231.
Epstein, C. J., Goldberger, R. F., and Anfinsen, C. B. (1963). *Cold Spring Harbor Symp. Quant. Biol.* **28**, 439–449.
Epstein, H. F., Schechter, A. N., Chen, R. F., and Anfinsen, C. B. (1971a). *J. Mol. Biol.* **60**, 499–508.
Epstein, H. F., Schechter, A. N., and Cohen, J. S. (1971b). *Proc. Natl. Acad. Sci. U.S.A.* **68**, 2042–2046.
Eyring, H., and Stearn, A. E. (1939). *Chem. Rev.* **24**, 253–529.
Faeder, E. J., and Hammes, G. G. (1971). *Biochemistry* **10**, 1041–1045.
Fasman, G. D., Chou, P. Y., and Adler, A. (1976). *Biophys. J.* **16**, 1201–1238.
Feldhammer, H., Bode, W., and Huber, R. (1977). *J. Mol. Biol.* **111**, 415–438.
Fern, E. B., and Garlik, P. J. (1976). *Biochem. J.* **156**, 189–192.
Finkelstein, A. V. (1977). *Biopolymers* **16**, 525–529.
Finkelstein, A. V., and Ptitsyn, O. B. (1976). *J. Mol. Biol.* **103**, 15–24.
Finkelstein, A. V., and Ptitsyn, O. B. (1977). *Biopolymers* **16**, 469–495.
Finney, J. L. (1975). *J. Mol. Biol.* **96**, 721–732.

Finney, J. L. (1978). *J. Mol. Biol.* **119**, 415–441.

Finney, J. L., Gellatly, B. J., Goldon, I. S., and Goodfellow, J. (1980). *Biophys. Discuss. Meet. Proteins Nucleoproteins, Airlie, Va.* pp. 131–142.

Fisher, W. R., Taniuchi, H., and Anfinsen, C. B. (1973). *J. Biol. Chem.* **248**, 3188–3195.

Fitch, W. M., and Margoliash, E. (1968). *Brookhaven Symp. Biol.* No. 21, 217–242.

Fletterick, R. J., Bates, D. J., and Steitz, T. A. (1975). *Proc. Natl. Acad. Sci. U.S.A.* **72**, 38–42.

Fletterick, R. J., Sygush, J., Semple, H., and Madsen, N. B. (1976). *J, Biol. Chem.* **251**, 6142–6146.

Flory, P. J. (1953). "Principles of Polymer Chemistry." Cornell Univ. Press, Ithaca, New York. Cited in Scheraga (1973b).

Flory, P. J. (1969). "Statistical Mechanics of Chain Molecules," pp. 249–255. Wiley (Interscience), New York.

Foster, T. (1960). *In* "The Plasma Proteins" (F. W. Putnam, ed.), p. 179. Academic Press, New York.

Franck, H. S., and Evans, M. J. (1945). *J. Chem. Phys.* **13**, 507–532.

Franck, H. S., and Wen, W. Y. (1957). *Discuss. Faraday Soc.* **24**, 133–140.

Frauenfelder, H., Petsko, G. A., and Tsernoglou, D. (1979). *Nature (London)* **280**, 558–563.

Freedman, R. (1979). *Nature (London)* **279**, 756–757.

Friedman, F. K., and Beychock, S. (1979). *Annu. Rev. Biochem.* **48**, 217–250.

Frost, A. A., and Pearson, R. G. (1953). "Kinetics and Mechanisms." Wiley, New York.

Fruchter, R. G., and Crestfield, A. M. (1965). *J. Biol. Chem.* **240**, 3875.

Fuchs, S., de Lorenzo, F., and Anfinsen, C. B. (1967). *J. Biol. Chem.* **242**, 398–402.

Fukushi, T., Imanishi, A., and Isemura, T. (1963). *Biochemistry* **63**, 409–416.

Furie, B., Schechter, A. N., Sachs, D. H., and Anfinsen, C. B. (1974). *Biochemistry* **13**, 1561.

Furie, B., Schechter, A. N., Sachs, B. H., and Anfinsen, C. B. (1975). *J. Mol. Biol.* **92**, 497–506.

Gaertner, F. H. (1978). *T.I.B.S.* **3**, 63–65.

Gagnon, J., Palmiter, R. D., and Walsh, K. A. (1978). *J. Biol. Chem.* **253**, 7464–7568.

Galat, A., Creighton, T. E., Lord, R. C., and Blout, E. R. (1981). *Biochemistry* **20**, 594–601.

Gally, J. A. (1973). *In* "Structure of Immunoglobulins in the Antigens" (M. Sela, ed.), Academic Press, New York. 162–298.

Garel, J. R. (1976). *Eur. J. Biochem.* **70**, 179–189.

Garel, J. R. (1978). *J. Mol. Biol.* **118**, 331–345.

Garel, J. R., and Baldwin, R. L. (1973). *Proc. Natl. Acad. Sci. U.S.A.* **70**, 3347–3351.

Garel, J. R., and Baldwin, R. L. (1975a). *J. Mol. Biol.* **94**, 611–620.

Garel, J. R., and Baldwin, R. L. (1975b). *J. Mol. Biol.* **94**, 621–632.

Garel, J. R., and Dautry-Varsat, A. (1980). *Proc. Natl. Acad. Sci. U.S.A.* **77**, 3379–3383.

Garel, J. R., and Labouesse, B. (1970). *J. Mol. Biol.* **47**, 41–56.

Garel, J. R., and Siffert, O. (1979). *Biochem. Biophys. Res. Commun.* **89**, 591–597.

Garel, J. R., Epely, S., and Labouesse, B. (1974). *Biochemistry* **13**, 3117–3123.

Garel, J. R., Nall, B. T., and Baldwin, R. L. (1976). *Proc. Natl. Acad. Sci. U.S.A.* **73**, 1853–1857.

Garnier, J., Gaye, P., Mercier, J. C., and Robson, B. (1980). *Biochimie* **62**, 231–239.

Gates, R. E. (1979). *J. Mol. Biol.* **127**, 345–351.

Gaye, P., Gautron, J. P., Mercier, J. C., and Haze, C. (1977). *Biochem. Biophys. Res. Commun.* **79**, 903–911.

Gekko, K., and Timasheff, S. N. (1981a). *Biochemistry* **20**, 4667–4676.

Gekko, K., and Timasheff, S. N. (1981b). *Biochemistry* **20**, 4677–4686.

Gelin, R. P., and Karplus, M. (1975). *Proc. Natl. Acad. Sci. U.S.A.* **72**, 2002–2006.

Gervais, M., Labeyrie, F., Risler, Y., and Vergnes, D. (1980). *Eur. J. Biochem.* **111**, 17–31.

Gertler, A., Walsh, K. A., and Neurath, H. (1974). *Biochemistry* **13**, 1302–1310.

Ghélis, C. (1971), Thèse, Univ. Paris-Sud. Orsay.
Ghélis, C. (1980). *Biophys. J.* **32**, 503–514.
Ghélis, C., and Hervé, G. (1978). *Int. Biophys. Congr. 6th, Kyoto* V-18.
Ghélis, C., and Yon, J. (1979). *C. R. Hebd. Seances Acad. Sci., Ser. D* **289**, 197–199.
Ghélis, C., and Yon, J. (1982). To be published.
Ghélis, C., and Zilber, E. (1982). To be published.
Ghélis, C., Garel, J. R., and Labouesse, J. (1970). *Biochemistry* **9**, 3902–3913.
Ghélis, C., Labouesse, J., and Labouesse, B. (1975). *Eur. J. Biochem.* **59**, 159–166.
Ghélis, C., Tempête,-Gaillourdet, M., and Yon, J. M. (1978). *Biochem. Biophys. Res. Commun.* **84**, 31–36.
Ghélis, C., Schechter, A. N., Tempête-Gaillourdet, M., Hatzfeld, A., and Yon, J. M. (1982). Unpublished results.
Ghose, R. C., and Englander, S. W. (1974). *J. Biol. Chem.* **249**, 7950–7955.
Gibbons, W. A., Némethy, G., Stern, A., and Craig, L. C. (1970). *Proc. Natl. Acad. Sci. U.S.A.* **67**, 239–246.
Gibson, K. D., and Scheraga, H. A. (1967). *Proc. Natl. Acad. Sci. U.S.A.* **58**, 420–427.
Gibson, K. D., and Scheraga, H. A. (1969). *Proc. Natl. Acad. Sci. U.S.A.* **63**, 242–245.
Gilbert, W. (1978). *Nature (London)* **271**, 501.
Ginsburg, A., and Carroll, W. R. (1965). *Biochemistry* **4**, 2159–2174.
Giudice, L. C., and Weintraub, B. D. (1979). *J. Biol. Chem.* **254**, 12679–12683.
Givol, D., Goldberger, R. F., and Anfinsen, C. B. (1964). *J. Biol. Chem.* **239**, 3114–3116.
Givol, D., de Lorenzo, F., Goldberger, R. F., and Anfinsen, C. B. (1965). *Proc. Natl. Acad. Sci. U.S.A.* **53**, 676–684.
Gō, N. (1976). *Adv. Biophys.* **9**, 65–113.
Gō, N., and Scheraga, H. A. (1969). *J. Chem. Phys.* **51**, 4751–4767.
Gō, N., and Scheraga, H. A. (1976). *Macromolecules* **9**, 535–542.
Gō, N., and Taketomi, H. (1978). *Proc. Natl. Acad. Sci. U.S.A.* **75**, 559–563.
Gō, N., and Taketomi, H. (1979a). *Int. J. Pept. Protein Res.* **13**, 235–252.
Gō, N., and Taketomi, H. (1979b). *Int. J. Pept. Protein Res.* **13**, 447–461.
Gō, N., Lewis, P. N., Gō, M., and Scheraga, H. A. (1971). *Macromolecules* **4**, 692–709.
Gō, N., Gō, M., and Scheraga, H. A. (1974). *Macromolecules* **7**, 137–139.
Gō, N., Abe, H., Mizuno, H., and Taketomi, H. (1980). *In* "Protein Folding" (R. Jaenicke, ed.), pp. 167–181. Elsevier/North-Holland, Amsterdam.
Goldberg, M. E. (1972). *In* "Dynamic Aspects of Conformation Changes in Biological Macromolecules" (C. Sadron, ed.), pp. 57–65. Reidel Publ. Dordrecht, Netherlands.
Goldberg, M. E., and Högberg-Raibaud, A. (1979). *J. Biol. Chem.* **254**, 7752–7757.
Goldberger, R. F., Epstein, C. J., and Anfinsen, C. B. (1963). *J. Biol. Chem.* **238**, 628–635.
Goldberger, R. F., Epstein, C. J., and Anfinsen, C. B. (1964). *J. Biol. Chem.* **239**, 1406–1410.
Goldman, B. M., and Blobel, G. (1978). *Proc. Natl. Acad. Sci. U.S.A.* **75**, 5066–5070.
Goldstein, A., and Brown, B. J. (1961). *Biochim. Biophys. Acta* **53**, 438–439.
Gordon, J. (1972). *Biochemistry* **11**, 1862–1870.
Goreki, M., and Zeelon, E. P. (1979). *J. Biol. Chem.* **254**, 525–529.
Graham, D. E., and Phillips, M. C. (1979). *J. Colloid Interface Sci.* **70**, 403–439.
Green, R. F., and Pace, C. N. (1974). *J. Biol. Chem.* **249**, 5388–5393.
Greenfield, N. J., and Fasman, G. D. (1969a). *Biochemistry* **8**, 4108–4116.
Greenfield, N. J., and Fasman, G. D. (1969a). *Biopolymers* **7**, 595–610.
Greenfield, N. J., Davidson, B., and Fasman, G. D. (1967). *Biochemistry* **6**, 1630–1637.
Groha, C., Bartholomes, P., and Jaenicke, R. (1978). *Eur. J. Biochem.* **92**, 437–441.
Guiard, B., and Lederer, F. (1979). *J. Mol. Biol.* **135**, 639–650.
Guidotti, G., Konigsberg, W., and Craig, L. C. (1963). *Proc. Natl. Acad. Sci. U.S.A.* **50**, 774–782.

Gurd, F. R. N., and Rothgeb, T. M. (1979). *Adv. Protein Chem.* **33**, 74–165.

Gutte, B. (1978). *Eur. J. Biochem.* **92**, 403–410.

Gutte, B., Lin, M. C., Caldi, D. G., and Merrifield, R. B. (1972). *J. Biol. Chem.* **247**, 4763–4767.

Guzzo, A. V. (1965). *Biophys. J.* **5**, 809–822.

Habener, J. F., Kemper, B., Potts, J. T., and Rich, A. (1975). *Biochem. Biophys. Res. Commun.* **67**, 1114–1121.

Habener, J. F., Potts, J. T., and Rich, A. (1976). *J. Biol. Chem.* **251**, 3893–3899.

Habener, J. F., Rosenblatt, M., Kemper, B., Kronenberg, H. M., Rich, R., and Potts, J. T. (1978). *Proc. Natl. Acad. Sci. U.S.A.* **75**, 2616–2620.

Haber, E., and Anfinsen, C. B. (1962). *J. Biol. Chem.* **237**, 1839–1844.

Haber, J. E., and Koshland, D. E. (1970). *J. Mol. Biol.* **50**, 617–639.

Hagerman, P. J. (1977). *Biopolymers* **16**, 731–747.

Hagerman, P. J., and Baldwin, R. L. (1976). *Biochemistry* 15, 1462–1473.

Hagerman, P. J., Schmid, F. X., and Baldwin, R. L. (1979). *Biochemistry* **18**, 293–297.

Hagler, A. T., and Hönig, B. (1978). *Proc. Natl. Acad. Sci. U.S.A.* **75**, 554–558.

Hagler, A. T., and Moult, J. (1978). *Nature* (*London*) **272**, 222–226.

Hagler, H. T., Scheraga, H. A., and Némethy, G. (1972). *J. Phys. Chem.* **76**, 3229–3243.

Hagler, A. T., Scheraga, H. A., and Némethy, G. (1973). *Ann. N.Y. Acad. Sci.* **204**, 51–78.

Hamaguchi, K., and Kurono, K. (1963). *J. Biochem.* (*Tokyo*) **54**, 111–122.

Hamilton, W., and Ibers, J. (1968). "Hydrogen Bonding in Solids." Benjamin, New York.

Hamlin, J., and Zabin, I. (1972). *Proc. Natl. Acad. Sci. U.S.A.* **69**, 412–416.

Hammerstedt, R. H. Möhler, H., Decker, K. A., and Wood, W. A. (1971). *J. Biol. Chem.* **246**, 2069–2074.

Hanson, A. W., Applebury, M. L., Coleman, J. E., and Wyckoff, H. (1970). *J. Biol. Chem.* **245**, 4975–4976.

Hanson, K. R. (1966). *J. Mol. Biol.* **22**, 405–409.

Hanson, K. R. (1968). *J. Mol. Biol.* **38**, 133–136.

Hantgan, R. R., and Taniuchi, H. (1977). *J. Biol. Chem.* **252**, 1367–1374.

Hantgan, R., and Taniuchi, H. (1978). *J. Biol. Chem.* **253**, 5373–5380.

Hantgan, R. R., Hammes, G. G., and Scheraga, H. A. (1974). *Biochemistry* **13**, 3421–3431.

Hardman, K. D., Wood, M. K., Schiffer, M., Edmundson, A. B., and Ainsworth, C. F. (1971). *Proc. Natl. Acad. Sci. U.S.A.* **68**, 1393–1397.

Hardy, P. M., Ridge, B., Rydon, H. N., and Dos Serraos, F. C. (1971). *J. Chem. Soc. C* pp. 1722–1731.

Harrington, W. F., and Sela, M. (1959). *Biochim. Biophys. Acta* **31**, 427–424.

Harrington, W. F., Josephs, R., and Segal, D. M. (1966). *Annu. Rev. Biochem.* **35**, 599–650.

Hartley, B. S. (1970). *Biochem. J.* **119**, 805–822.

Hartley, B. S., and Shotton, D. M. (1971). *In* "The Enzymes" (P. D. Boyer, ed.), 3rd ed., Vol. III, pp. 323–373. Academic Press, New York.

Hartley, B. S., Burleigh, B. D., Midwinter, G. G., Moore, C. H., Morris, H. R., Rigby, P. W. J., Smith, M. J., and Taylor, S. S. (1972). *Proc. FEBS Meet.* **29**, 151–176.

Hartsuck, J. A., and Lipscomb, W. N. (1971). *In* "The Enzymes" (P. D. Boyer, ed.), 3rd ed., Vol. III, pp. 1–56. Academic Press, New York.

Haugen, T. H., and Heath, E. C. (1979). *Proc. Natl. Acad. Sci. U.S.A.* **76**, 2689–2993.

Hawley, S. A., and Macleod, R. M. (1976). *J. Mol. Biol.* **103**, 655–657.

Hedgpeth, J., Clément, J. M., Marchal C., Perrin, D., and Hoffnung, M. (1980). *Proc. Natl. Acad. Sci. U.S.A.* **77**, 2621–2625.

Henderson, R., and Unwin, P. N. T. (1975). *Nature* (*London*) **257**, 28–31.

Henkens, R. W., and Turner, S. R. (1979). *J. Biol. Chem.* **254**, 8110–8112.

Hermann, T., Rudolph, R., and Jaenicke, R. (1979). *Nature* (*London*) **277**, 243–245.

Hermans, J., Jr. (1965). *Methods Biochem. Anal.* **13**, 81–111.
Hermans, J., Jr., and Acampora, G. (1967). *J. Am. Chem. Soc.* **89**, 1543–1552.
Herskovits, T. T. (1965). *J. Biol. Chem.* **240**, 628–638.
Herskovits, T. T. (1967). *In* "Enzyme Structure" (C. H. W. Hirs, ed.), Methods in Enzymology, Vol. 11, pp. 748–775. Academic Press, New York.
Herskovits, T. T., and Laskowski, M., Jr. (1962). *J. Biol. Chem.* **237**, 2481–2492.
Herskovits, T. T., and Laskowski, M., Jr. (1968). *J. Biol. Chem.* **243**, 2123–2129.
Herskovits, T. T., and Mescanti, L. (1965). *J. Biol. Chem.* **240**, 639–644.
Hess, G. P., McConn, J., Ku, E., and McConkey, G. L. (1970). *Philos. Trans. R. Soc. London, Ser. B*, **257**, 89–104.
Hetzel, R., Wüthrich, K., Deisenhofer, J., and Huber, R. (1976). *Biophys. Struct. Mech.* **2**, 159–180.
Hibbard, L. S., and Tulinsky, A. (1978). *Biochemistry* **17**, 5460–5468.
Higgins, W., Fairwell, T., and Miles, E. W. (1979). *Biochemistry* **18**, 4827–4835.
Hijazi, N. H., and Laidler, K. J. (1972). *J. Chem. Soc.* **68**, 1235–1242.
Hildebrandt, J. H. (1979a). *Proc. Natl. Acad. Sci. U.S.A.* **76**, 194.
Hildebrandt, J. H. (1979b). *Proc. Natl. Acad. Sci. U.S.A.* **76**, 4175–4176.
Hirs, C. H. W., Moore, S., and Stein, W. H. (1960). *J. Biol. Chem.* **235**, 633–647.
Hochman, J., Inbar, D., and Givol, D. (1973). *Biochemistry* **12**, 1130–1135.
Hochman, J., Gavish, M., Inbar, D., and Givol, D. (1976). *Biochemistry* **15**, 1706–2710.
Hodes, Z. I., Némethy, G., and Scheraga, H. A. (1979a). *Biopolymers* **18**, 1565–1610.
Hodes, Z. I., Némethy, G., and Scheraga, H. A. (1979b). *Biopolymers* **18**, 1611–1634.
Högberg-Raibaud, A., and Goldberg, M. E. (1977a). *Biochemistry* **16**, 4014–4020.
Högberg-Raibaud, A., and Goldberg, M. E. (1977b). *Proc. Natl. Acad. Sci. U.S.A.* **74**, 442–446.
Högberg-Raibaud, A., Raibaud, O., and Goldberg, M. E. (1975). *J. Biol. Chem.* **250**, 3352–3358.
Hoffmann, R., and Imamura, A. (1969). *Biopolymers* **7**, 207–213.
Holbrook, J. J., Liljas, A., Steindel, S. J., and Rossmann, M. G. (1975). *In* "The Enzymes" (P.D. Boyer, ed.), 3rd ed., Vol. 11, pp. 191–292. Academic Press, New York.
Holladay, L. A., Hammonds, R. G., Jr., and Puett, D. (1974). *Biochemistry* **13**, 1653–1661.
Holmgren, A. (1972). *FEBS Lett.* **24**, 351–354.
Hönig, B., Ray, A., and Levinthal, C. (1976). *Proc. Natl. Acad. Sci. U.S.A.* **73**, 1974–1978.
Hopfinger, A. J. (1971). *Macromolecules* **4**, 731–737.
Hopfinger, A. J. (1973). "Conformational Properties of Macromolecules." Academic Press, New York.
Howard, J. C., Ali, A., Scheraga, H. A., and Momany, F. A. (1975). *Macromolecules* **8**, 607–622.
Howarth, O. W. (1979). *Biochim. Biophys. Acta* **576**, 163–175.
Howlett, G. J., Blackburn, M. N., Compton, J. G., and Schachman, H. K. (1977). *Biochemistry* **16**, 5091–5099.
Hsu, L. S., and Neet, K. E. (1973). *Biochemistry* **12**, 586–595.
Hsu, L. S., and Neet, K. E. (1975). *J. Mol. Biol.* **97**, 351–367.
Huang, K. S., Bayley, H., Liao, M. J., London, E., and Khorana, H. G. (1981). *J. Biol. Chem.* **256**, 3802–3809.
Huber, R. (1979a). *Nature (London)* **280**, 538–539.
Huber, R. (1979b). T.I.B.S. **4**, 271–276.
Huber, R., and Steigemann, W. (1974). *FEBS Lett.* **48**, 235–236.
Huber, R., Epp, O., Steigemann, W., and Formanek, H. (1971). *Eur. J. Biochem.* **19**, 42–50.
Huber, R., Kukla, D., Rühlmann, A., and Steigemann, W. (1972). *Cold Spring Harbor Symp. Quant. Biol.* **36**, 141–150.
Huggins, M. L. (1943). *Chem. Rev.* **32**, 195–218.
Hull, W. H., and Sykes, B. D. (1975). *J. Mol. Biol.* **98**, 121–153.

Hurrell, J. G. R., Smith, J. A., and Leach, S. J. (1977). *Biochemistry* **16**, 175–285.
Hvidt, A., and Nielsen, S. O. (1966). *Adv. Protein Chem.* **21**, 287–386.
Hvidt, A., and Pedersen, E. J. (1974). *Eur. J. Biochem.* **48**, 333–338.
Ikai, A., and Tanford, C. (1971). *Nature* (*London*) **230**, 100–102.
Ikai, A., and Tanford, C. (1973). *J. Mol. Biol.* **73**, 145–154.
Ikai, A., Fish, W. W., and Tanford, C. (1973). *J. Mol. Biol.* **73**, 165–184.
Ikai, A., Tanaka, L., and Noda, H. (1979). *Arch. Biochem. Biophys.* **190**, 39–45.
Imahori, K., and Doty, P. (1957). Cited in Urnes and Doty (1961).
Imoto, T., and Rupley, M. A. (1973). *J. Mol. Biol.* **80**, 657–667.
Imoto, T., Johnson, I. N., North, A. C. T., Philipps, D. C., and Rupley, J. A. (1972). *In* "The Enzymes" (P. D. Boyer, ed.), 3rd ed., Vol. 7, pp. 665–868. Academic Press, New York.
Imoto, T., Fukuda, K. I., and Yagishita, K. (1974). *Biochem. Biophys. Acta* **336**, 264–269.
Imoto, T., Fukuda, K. I., and Yagishita, K. (1976). *J. Biochem.* (*Tokyo*) **80**, 1313–1318.
Inouye, M., and Halegoua, S. (1980). C.R.C. Crit. *Rev. Biochem.* 4, 339–371.
Inouye, M., Wang, S., Sekisawa, J., Halegoua, S., and Inouye, M. (1977). *Proc. Natl. Acad. Sci. U.S.A.* **74**, 1004–1008.
Irace, G., Balestrieri, K., Parlato, G., Servillo, L., and Colonna, G. (1981). *Biochemistry* **20**, 792–799.
Isemura, T., Takagi, T., Maeda, Y., and Imai, K. (1961). *Biochem. Biophys. Res. Commun.* **5**, 373–377.
Isemura, T., Takagi, T., Maeda, Y., and Yutani, K. (1963). *J. Biochem.* (*Tokyo*) **53**, 155–161.
Isenman, D. E., Lancet, D., and Pecht, I. (1979). *Biochemistry* **18**, 3327–3336.
Isogai, Y., Némethy, G., and Scheraga, H. A. (1977). *Proc. Natl. Acad. Sci. U.S.A.* **74**, 414–418.
Isogai, Y., Némethy, G., Rackowsky, S., Leach, S. J., and Scheraga, H. A. (1980). *Biopolymers* **19**, 1183–1210.
Jackson, W. M., and Brandts, J. F. (1970). *Biochemistry* **9**, 2294–2301.
Jaenicke, R. (1974). *Eur. J. Biochem.* **46**, 149–155.
Jaenicke, R. (1979). *Proc. FEBS Meet.* **52**, 187–198.
Jaenicke, R. (1981). *Ann. Rev. Biophys. Bioeng.* **10**, 1–67.
Jaenicke, R., and Rudolph, R. (1980). *In* "Protein Folding" (R. Jaenicke, ed.), pp. 525–548. Elsevier/North-Holland, Amsterdam.
Jaenicke, R., Rudolph, R., and Heider, I. (1979). *Biochemistry* **18**, 1217–1223.
Jaenicke, R., Vogel, W., and Rudolph, R. (1981). *Eur. J. Biochem.* **114**, 525–531.
Janin, J. (1971). *Cold Spring Harbor Symp. Quant. Biol.* **36**, 193–198.
Janin, J. (1976). *J. Mol. Biol.* **105**, 13–14.
Janin, J. (1979). *Bull Inst. Pasteur, Paris* **77**, 337–373.
Janin, J., and Chothia, C. (1978). *Proc. FEBS Meet.* **52**, 117–126.
Janin, J., Wodak, S., Levitt, M., and Maigret, B. (1978). *J. Mol. Biol.* **125**, 357–386.
Janofsky, C., and Crawford, P. (1972). *In* "The Enzymes" (P. D. Boyer, ed.), 3rd ed., Vol. 7, pp. 1–31. Academic Press, New York.
Jardetzky, O., Markley, J. L., Thielman, H., Arata, Y., and Williams, M. N. (1971). *Cold Spring Habor Symp. Quant. Biol.* **36**, 257–261.
Jaton, J. C., Klinman, N. R., Givol, D., and Sela, M. (1968). *Biochemistry* **7**, 4185–4195.
Jencks, W. P. (1975). *Adv. Enzymol. Relat. Areas Mol. Biol.* **43**, 219–402.
Jirgensons, B. (1967). *J. Biol. Chem.* **242**, 912–918.
Jirgensons, B. (1973). "Optical Activity of Proteins and Other Macromolecules." Springer-Verlag, Berlin and New York.
Jocelyn, P. C. (1972). "Biochemistry of the SH Group: The Occurrence, Chemical Properties, Metabolism and, Biological Function of Thiols and Disulphides." Academic Press, New York.

Johannes, K. J., and Hess, B. (1973). *J. Mol. Biol.* **76**, 181–205.

Johnson, E. R., Oh, K. J., and Wetlaufer, D. B. (1976). *J. Biol. Chem.* **251**, 3154–3157.

Johnson, E. R., Anderson, W. L., Wetlaufer, D. B., Lee, C. C., and Atassi, M. Z. (1978). *J. Biol. Chem.* **253**, 3408–3414.

Johnson, R. E., Adams, P., and Rupley, J. A. (1978). *Biochemistry* **17**, 1479–1484.

Jollès, J., Ibrahimi, I. M., Prayer, E. M., Schvengen, F., Jollès, P., and Wilson, A. C. (1979). *Biochemistry* **18**, 2744–2752.

Juillerat, M., Parr, G. R., and Taniuchi, H. (1980). *J. Biol. Chem.* **255**, 845–853.

Kabat, E. A. (1978). *Adv. Protein Chem.* **32**, 1–75.

Kabat, E. A., and Mayer, M. M. (1961). "Experimental Immunochemistry," 2nd ed., p. 120. Thomas, Springfield, Illinois.

Kabat, E. A., and Wu, T. T. (1973a). *Biopolymers* **12**, 751–774.

Kabat, E. A., and Wu, T. T. (1973b). *Proc. Natl. Acad. Sci. U.S.A.* **70**, 1473–1477.

Kanehisa, M. I., and Tsong, T. Y. (1978). *J. Mol. Biol.* **124**, 177–194.

Kanehisa, M. I., and Tsong, T. Y. (1979a). *Biopolymers* **18**, 1375–1388.

Kanehisa, M. I., and Tsong, T. Y. (1979b). *J. Mol. Biol.* **133**, 279–283.

Kannan, K. K., Liljas, A., Waara, I., Bergstén, P. G., Lövgren, S., Strauberg, B., Bengtsson, U., Carlbom, U., Fridborg, K., Järup, L., and Petef, M. (1972). *Cold Spring Harbor Symp. Quant. Biol.* **36**, 221–231.

Kannan, K. K., Nortrand, B., Fridberg, K., Lövgren, S., Ohlsson, A., and Peter, M. (1975). *Proc. Natl. Acad. Sci. U.S.A.* **72**, 51–55.

Kantrowitz, E., Pastra-Landis, S. C., and Lipcomb, W. N. (1980a). *TIBS* **5**, 124–128.

Kantrowitz, E., Pastra-Landis, S. C., and Lipcomb, W. N. (1980b). *TIBS* **5**, 150–153.

Kaplan, H., Stevenson, K. J., and Hartley, B. S. (1971). *Biochem. J.* **124**, 289–299.

Karplus, M., and McCammon, J. A. (1979). *Nature (London)* **277**, 578.

Karplus, M., and McCammon, J. A. (1981a). *CRC Crit. Rev.* **9**, 293–349.

Karplus, M., and McCammon, J. A. (1981b). *FEBS Lett.* **131**, 34–36.

Karplus, M., and Weaver, D. L. (1976). *Nature (London)* **260**, 404–406.

Karplus, S., Snyder, G. H., and Sykes, B. D. (1973). *Biochemistry* **12**, 1323–1329.

Kartha, G., Bello, J., and Harker, D. (1967). *Nature (London)* **213**, 862–865.

Kato, I., and Anfinsen, C. B. (1969). *J. Biol. Chem.* **244**, 1004–1007.

Katz, L., and Levinthal, C. (1972). *Annu. Rev. Biophys. Bioeng.* **1**, 465–504.

Kauzmann, W. (1954). *In* "The Mechanism of Enzyme Action". (W. D., McElroy and B. Glass, eds.), 70–120. Johns Hopkins Press, Baltimore, Maryland.

Kauzmann, W. (1959a). *Adv. Protein Chem.* **14**, 1–64.

Kauzmann W. (1959b). *In* "Sulfur in Proteins" (R. Benesh, P. D. Boyer, I. M. Klotz, W. R. Middlebrook, A. G. Szent-Györgii, and D. R. Schwartz, eds.), p. 70. Academic Press, New York.

Kauzmann, W., and Kuntz, I. D. (1974). *Adv. Protein Chem.* **28**, 239–345.

Kemper, B., Habener, J. F., Mulligen, J. F., Potts, J. T., and Rich, A. (1974). *Proc. Natl. Acad. Sci. U.S.A.* **71**, 3731–3735.

Kendrew, J. C. (1962). *Brookhaven Symp. Biol.* **15**, 216.

Kendrew, J. C., Dickerson, R. E., Stranberg, B. E., Hart, R. C., Davies, D. R., Philipps, D. C., and Shore, V. C. (1960). *Nature (London)* **185**, 422–427.

Kerbiriou, D., and Hervé, G. (1972). *J. Mol. Biol.* **64**, 379–392.

Khaled, M. A., Urry, D. W., and Okamote, K. (1976). *Biochem. Biophys. Res. Commun.* **72**, 162–169.

Khechinashvili, N. N., Privalov, P. L., and Tiktopulo, E. I. (1973). *FEBS Lett.* **30**, 57–60.

Kiho, Y., and Rich, A. (1964). *Proc. Natl. Acad. Sci. U.S.A.* **51**, 111–128.

Kim, P. S., and Baldwin, R. L. (1980). *Biochemistry* **19**, 6124–6129.

Kimura, M. (1979). Sci. Am. Nov. pp. 94–104.
Kirschner, K., Gallego, E., Schister, I., and Goodall, D. (1971). *J. Mol. Biol.* **58**, 29–50.
Kirschner, M. N., and Schachman, H. K. (1973). *Biochemistry* **12**, 2297–3004.
Kiselev, N. A., and Lerner, F. Y. (1971). *J. Mol. Biol.* **62**, 537–549.
Kitaigorodsky, A. I. (1961). *Tetrahedron* **14**, 230–236.
Kitaigorodsky, A. I. (1965). *Acta Crystallogr* **18**, 585–590.
Kita, N., Kuwajima, K., Nitta, K., and Sugai, S. (1976). *Biochim. Biophys. Acta* **427**, 350–358.
Kivirikko, K. I., and Myllylä, R. (1980). *In* "The Enzymology of Post-translational Modifications of Proteins" (R. B. Freeman and H. C. Hawkins. eds.), Vol. I, pp. 54–104, Academic Press, New York.
Klapper, M. H., and Klapper, I. Z. (1980). *Biophys. Discuss. Meet. Proteins Nucleoproteins, Airlie, Va.* pp. 25–26.
Klee, W. A. (1967). *Biochemistry* **6**, 3736–3742.
Klee, W. A. (1968). *Biochemistry* **7**, 2731–2736.
Klotz, I. M. (1958). *Science* **128**, 815–822.
Klotz, I. M. (1960). *Brookhaven Symp. Biol.* **13**, 25.
Klotz, I. M., Langerman, N. R., and Darnall, D. W. (1970). *Annu. Rev. Biochem.* **39**, 25–62.
Klotz, I. M., Darnall, D. W., and Langerman, N. R. (1975). *In* "The Proteins" (H. Neurath, R. L., Hilland, and C. L. Boeder, eds.), 3rd ed., pp. 293–411. Academic Press, New York.
Klug, A. (1967). *Symp. Int. Soc. Cell Biol.* **6**, 1–18.
Knapp, J. A., and Pace, C. N. (1974). *Biochemistry* **13**, 1289–1294.
Koenig, J. L. (1972), *J. Polym. Sci., Part D* **6**, 60–177.
Koenig, J. L., and Frushour, B. G. (1972). *Biopolymers* **11**, 2505–2520.
Konishi, Y., and Scheraga, H. A. (1980a). *Biochemistry* **19**, 1308–1316.
Konishi, Y., and Scheraga, H. A. (1980b). *Biochemistry* **19**, 1316–1322.
Kosen, P. A., Creighton, T. E., and Blout, E. R. (1980). *Biochemistry* **19**, 4936–4944.
Koshland, D. E. (1970). *In* "The Enzymes" (P. D. Boyer, ed.), 3rd ed., Vol. I, pp. 341–396. Academic Press, New York.
Koshland, D. E., and Neet, K. E. (1968). *Annu. Rev. Biochem.* **37**, 359–410.
Koshland, D. E., Némethy, G., and Filmer, D. (1966). *Biochemistry* **5**, 365–385.
Kotelchuck, D., and Scheraga, H. A. (1968). *Proc. Natl. Acad. Sci. U.S.A.* **61**, 1163–1170.
Kotelchuck, D., and Scheraga, H. A. (1969). *Proc. Natl. Acad. Sci. U.S.A.* **62**, 14–21.
Kraut, J. (1965). *Annu. Rev. Biochem.* **34**, 247–265.
Kraut, J., Robertus, J. D., Birktoft, J. J., Alden, R. A., Wilcox, P. E., and Powers, J. C. (1972). *Cold Spring Harbor Symp. Quant. Biol.* **36**, 117–123.
Krebs, H., Rudolph, R., and Jaenicke, R. (1979). *Eur. J. Biochem.* **100**, 359–364.
Kreibich, G., Freienstein, C. M., Pereyra, B. N., Ulrech, B. L., and Sabatini, D. D. (1978a). *J. Cell Biol.* **77**, 488–506.
Kreibich, G., Ulrich, B. L., and Sabatini, D. D. (1978b). *J. Cell Biol.* **77**, 464–487.
Kress, L. F., and Laskowski, M., Jr. (1967). *J. Biol. Chem.* **242**, 4925–4929.
Kretsinger, R. H., and Nickolls, C. E. (1973). *J. Biol. Chem.* **248**, 3313–3326.
Krieger, D. T., and Liotta, A. S. (1979). *Science* **205**, 366–372.
Krimm, S., and Venkatachalam, C. M. (1971). *Proc. Natl. Acad. Sci. U.S.A.* **68**, 2468–2471.
Kronvall, G., Seal, U. S., Finshed, T., and Williams, R. D. (1970). *J. Immunol.* **104**, 140–147.
Kunitz, M. (1948). *J. Gen. Physiol.* **32**, 241–263.
Kuntz, I. D. (1972). *J. Am. Chem. Soc.* **94**, 4009–4012.
Kuntz, I. D., Crippen, G. M., Kollman, P. A., and Kimelman, D. (1976). *J. Mol. Biol.* **106**, 983–994.
Kuwajima, K. (1977). *J. Mol. Biol.* **114**, 241–258.

Kuwajima, K., Nitta, K., Yoneyama, M., and Sugai, S. (1976). *J. Mol. Biol.* **106**, 359–373.
Kuwajima, K., Ogawa, Y., and Sugai, S. (1981). *J. Biochem.* **89**, 759–770.
Labhardt, A. M., and Baldwin, R. L. (1979a). *J. Mol. Biol.* **135**, 231–244.
Labhardt, A. M., and Baldwin, R. L. (1979b). *J. Mol. Biol.* **135**, 245–254.
Lagerkvist, U., Rymo, L., Lindquist, O., and Andersen, O. (1972). *J. Biol. Chem.* **247**, 3897–3899.
Lakowicz, J. R., and Weber, G. (1973). *Biochemistry* **12**, 4171–4179.
Lakshminarayan, A. V., Sasisekharan, A. V., and Ramachandran, G. N. (1967). *In* "Conformations of Biopolymers" (G. N. Ramachandran, ed.), pp. 61–82. Academic Press, New York.
Langone, J. J. (1978). *J. Immunol. Methods* **24**, 269–285.
Langone, J. J., and Levine, L. (1978). *Anal. Biochem.* **95**, 472–478.
Langone, J. J., Boyle, M. D., and Borson, T. (1979). *Anal. Biochem.* **93**, 207–215.
Lapanje, S., and Rupley, J. A. (1973). *Biochemistry* **12**, 2370–2372.
Laskowski, M., and Scheraga, H. A. (1954). *J. Am. Chem. Soc.* **76**, 6305–6319.
Lau, H. K., Rosenberg, J. S., Beeler, D. L., and Rosenberg, R. D. (1979). *J. Biol. Chem.* **254**, 8751–8761.
Laurell, C. B. (1966). *Anal. Biochem.* **15**, 45–52.
Laurent, M., Seydoux, F. J., and Dessen, P. (1979). *J. Biol. Chem.* **254**, 7515–7520.
Lavialle, F., de Foresta, B., Vacher, M., Nicot, C., and Alfsen, A. (1979). *Eur. J. Biochem.* **95**, 561–567.
Leberman, R. (1971). *J. Mol. Biol.* **55**, 23–30.
Lee, B., and Richards, F. M. (1971). *J. Mol. Biol.* **55**, 379–400.
Lehrer, G. M., and Barker, R. (1971). *Biochemistry* **10**, 1705–1714.
Lehrer, S. S., and Fasman, G. (1967). *Biochem. Biophys. Res. Commun.* **29**, 767–772.
Lenormant, H., and Blout, E. R. (1953). *Nature (London)* **172**, 770–771.
Lenstra, J. A. (1977). *Biochim. Biophys. Acta* **491**, 333–338.
Lenstra, J. A., Hofsteenge, J., and Beintema, J. J. (1977). *J. Mol. Biol.* **109**, 185–193.
Lentz, B. R., Hagler, A. T, and Scheraga, H. A. (1974). *J. Phys. Chem.* **78**, 1531.
Lesk, A. M., and Chothia, C. (1980). *J. Mol. Biol.* **136**, 225–270.
Lessard, J. L., Carlton, D., Rei, D. C., and Akeson, R. (1979). *Anal. Biochem.* **94**, 140–149.
Levine, M., Muirhead, H., Stammers, D. K., and Stuart, D. I. (1978). *Nature (London)* **271**, 626–630.
Levinthal, C. (1966). Sci. Am. **214**, 642–652.
Levinthal, C. (1968) *J. Chim. Phys.* **65**, 44–45.
Levinthal, C. (1969) Mössbauer Spectroscopy in Biological Systems Meeting, Univ. of Illinois at Urbana-Champaign, Urbana, cited by Wetlaufer.
Levitt, M. (1976). *J. Mol. Biol.* **104**, 59–107.
Levitt, M., and Chothia, C. (1976). *Nature (London)* **261**, 552–558.
Levitt, M., and Greer, J. (1977). *J. Mol. Biol.* **114**, 181–293.
Levitt, M., and Warshel, A. (1975). *Nature (London)* **253**, 694–698.
Lewis, P. N., and Scheraga, H. A. (1971a). *Arch. Biochem. Biophys.* **144**, 576–583.
Lewis, P. N., and Scheraga, H. A. (1971b). *Arch. Biochem. Biophys.* **144**, 584–588.
Lewis, P. N., Gō, N., and Kotelchuck, D. (1970). *Proc. Natl. Acad. Sci. U.S.A.* **65**, 810–815.
Lewis, P. N., Momany, F. A., and Scheraga, H. A. (1971). *Proc. Natl. Acad. Sci. U.S.A.* **68**, 2293–2297.
Lewis, P. N., Momany, F. A., and Scheraga, H. A. (1973a). *Biochim. Biophys. Acta* **303**, 211–229.
Lewis, P. N., Momany, F. A., and Scheraga, H. A. (1973b). *Isr. J. Chem.* **11**, 121–152.
Li, C. H., Blake, J., and Hagashida, T. (1978). *Biochem. Biophys. Res. Commun.* **82**, 2117–2122.
Lie, G. C., and Clementi, E. (1975). *J. Chem. Phys.* **62**, 2195–2199

Lie, G. C., Clementi, E., and Yoshimine, M. (1976). *J. Chem. Phys.* **64**, 2314.
Liem, R. K. M., Calhoun, D. B., Englander, J. J., and Englander, S. W. (1968). *J. Biol. Chem.* **255**, 10687–10694.
Lifson, S., and Roig, A. (1961). *J. Chem. Phys.* **34**, 1963–1974
Lifson, S., and Sander, C. (1979). *Nature* (*London*) **282**, 109–111.
Lifson, S., and Sander, C. (1980a). *J. Mol. Biol.* **139**, 627–639.
Lifson, S., and Sander C. (1980b). *In* "Protein Folding" (R. Jaenicke, ed.), pp. 289–316. Elsevier/North-Holland, Amsterdam.
Light, A., and Odorzynski, T. W. (1979). *J. Biol. Chem.* **254**, 9162–9166.
Light, A., and Sinha, N. K. (1976). *Biochem. Biophys. Res. Commun.* **68**, 1188–1193.
Light, A., Taniuchi, H., and Chen, R. F. (1974). *J. Biol. Chem.* **249**, 2285–2293.
Liljas, A., and Rossmann, M. G. (1974). *Annu. Rev. Biochem.* **43**, 475–507.
Lim, V. I. (1974a). *J. Mol. Biol.* **88**, 857–872.
Lim, V. I. (1974b). *J. Mol. Biol.* **88**, 873–894.
Lim, V. I. (1978). *FEBS Lett.* **89**, 10–14.
Lim, V. I., and Efimov, A. V. (1977). *FEBS Lett.* **78**, 279–283.
Lin, L. N., and Brandts, J. F. (1978). *Biochemistry* **17**, 4102–4110.
Lin, L. N., and Brandts, J. F. (1979). *Biochemistry* **18**, 43–47.
Lin, M. C. (1970). *J. Biol. Chem.* **245**, 6726–6731.
Lin, M. C., Gutte, B., Moore, S., and Merrifield, R. B. (1970). *J. Biol. Chem.* **245**, 5169–5170.
Linderstrøm-Lang, K. U. (1942). *C. R. Trav. Lab. Carlsberg, Ser. Chim.* **24**, 1–48.
Linderstrøm-Lang, K. U. (1950). *Cold Spring Harbor Symp. Quant. Biol.* **14**, 117–126.
Linderstrøm-Lang, K. U. (1952). "Proteins and Enzymes," Lane Medical Lectures, No. VI, p. 58. Stanford Univ. Press, Stanford, California.
Linderstrøm-Lang, K. U. (1955). *Chem. Soc., Spec. Publ.* No. 2, 1–20.
Linderstrøm-Lang, K. U., and Schellman, J. A. (1959). *In* "The Enzymes" (P. D. Boyer, H. Lardy, and K. Myrbäck, eds.), 2nd ed., Vol. 1, pp. 444–510. Academic Press, New York.
Lingappa, V. R., Katz, F. N., Lodish, H. F., and Blobel, G. (1978a). *J. Biol. Chem.* **253**, 8667–8670.
Lingappa, V. R., Shield, D., Woo, S. L. C., and Blobel, G. (1978b). *J. Cell Biol.* **79**, 567–572.
Lingappa, V. R., Lingappa, J. R., and Blobel, G. (1979) Nature (*London*) **281**, 117–121.
Liotta, A. S., Loudes, C., McKelvy, J. F., and Krieger, D. T. (1980). *Proc. Natl. Acad. Sci. U.S.A.* **77**, 1880–1884
Lippert, J. L., Gorczyca, L. E., and Keikeljohn, G. (1975). *Biochim. Biophys. Acta* **382**, 51–57.
Lippincott, E. R., and Schroeder, R. (1955). *J. Chem. Phys.* **23**, 1099–1106.
Liquori, A. M. (1966). *Princ. Biomol. Organ. Ciba Found. Symp. 1965* p.40.
Liquori, A. M. (1969). *Nobel Symp.* **11**, 101–121.
Lomedico, P. T., and Sauenders, G. F. (1976). *Nucleic Acids Res.* **3**, 381–391.
Lomedico, P. T., Chan, S. J., Steiner, D. F., and Sauenders, G. F. (1977). *J. Biol. Chem.* **252**, 7971–7978.
London, J., Skrzynia, C., and Goldberg, M. E. (1974). *Eur. J. Biochem.* **47**, 409–415.
Lonsdale-Eccles, J. D., Kerr, M. A., Walsh, K. A., and Neurath, H. (1979). *FEBS Lett.* **100**, 157–160.
Lord, R. C., and Mendelsohn, C. R., (1972). *J. Am. Chem. Soc.* **94**, 2133–2135.
Louvard, D., Maroux, S., and Desnuelle, P. (1975). *Biochim. Biophys. Acta* **389**, 389–400.
Louvard, D., Vannier, C., Maroux, S., Pagès, J. M., and Lazdunski, C. (1976). *Anal. Biochem.* **76**, 83–94.
Low, B. W., Lovell, F. M., and Rudko, A. D. (1968). *Proc. Natl. Acad. Sci. U.S.A.* **60**, 1519–1526.
Lowe, D., and Halliman, T. (1973). *Biochem. J.* **136**, 825–828.

Lumry, R., and Eyring, H. (1954). *J. Phys. Chem.* **58**, 110–120.
Lumry, R., Biltonen, R., and Brandts, J. F. (1966). *Biopolymers* **4**, 917–944.
Lynen, F. (1972). *Biochem. Soc. Symp.* **35**, 5–26.
Matthei, J. H., and Nirenberg, N. W. (1961). *Proc. Natl. Acad. Sci. U.S.A.* **47**, 1580–1588.
McCammon, J. A., and Karplus, M. (1977). *Nature (London)* **268**, 765–776.
McCammon, J. A., Gelin, B. R., Karplus, M., and Wolynes, P. G. (1976). *Nature (London)* **262**, 325–326.
McCammon, J. A., Gelin, B. R., and Karplus, M. (1977). *Nature (London)* **267**, 585–590.
Maccechini, M. L., Rudin, Y., Blobel, G., and Schatz, C. G. (1979). *Proc. Natl. Acad. Sci. U.S.A.* **76**, 343–347.
McCoy, L. F., Rowe, E. S., and Wong, K. P. (1980). *Biochemistry* **19**, 4738–4743.
McCoy, L. F., and Wong, K. P. (1981). *Biochemistry* **20**, 3062–3067.
McDonald, C. C., Phillips, W. D., and Glickson, J. (1971). *J. Am. Chem. Soc.* **93**, 235–246.
McLachlan, A. D. (1972a). *J. Mol. Biol.* **64**, 417–437.
McLachlan, A. D. (1972b). *Nature (London), New Biol.* **240**, 83–85.
McLachlan, A. D. (1979a), *J. Mol. Biol.* **128**, 49–79.
McLachlan, A. D. (1979b). *J. Mol. Biol.* **133**, 557–563.
McLachlan, A. D. (1980a). *Nature (London)* **285**, 267–268.
McLachlan, A. D. (1980b). In "Protein Folding" (R. Jaenicke, ed.), pp. 79–99. Elsevier/North-Holland, Amsterdam.
McLachlan, A. D., Bloomer, A. C., and Butler, P. J. G. (1980). *J. Mol. Biol.* **136**, 203–224.
McLendon, G., and Smith, M. (1978). *J. Biol. Chem.* **253**, 4004–4008.
McPhie, P. (1980). *J. Biol. Chem.* **255**, 4048–4052.
Magat, M., and Reinish, L. *Coll. Solvay Bruxelles, 1972*
Maigret, B. (1971). Thèse de Spécialité, Univ. Paris-VI.
Maigret, B., Pullman, B., and Dreyfus, M. (1970). *J. Theor. Biol.* **26**, 321–333.
Mains, R. E., and Eipper, B. A. (1979). *J. Biol. Chem.* **254**, 7885–7894.
Malin, E. L., and Englander, S. W. (1980). *J. Biol. Chem.* **255**, 10695–10701.
Maklin, L. I., and Rich, A. (1967). *J. Mol. Biol.* **26**, 329–346.
Mancini, G., Carbonara, A. D., and Heremans, J. F. (1965). *Immunochemistry* **2**, 235–254.
Margoliash, E. (1971). *J. Biol. Chem* **246**, 1511–1535.
Margoliash, E. (1972). *Harvey Lect.* **66**, 177–247.
Margoliash, E., and Shejter, A. (1966). *Adv. Protein Chem.* **21**, 113–286.
Margoliash, E., and Smith, E. L. (1965). *In* "Evolving Genes and Proteins" (V. Bryson and H. J. Vogel, eds.), pp. 221–242. Academic Press, New York.
Marinetti, T. D., Snyder, G. H., and Sykes, B. D. (1976). *Biochemistry* **15**, 4600–4608.
Markley, J. L. (1975a). *Biochemistry* **14**, 3546–3554.
Markley, J. L. (1975b). *Biochemistry* **14**, 3554–3561.
Markley, J. L., and Finkenstadt, W. R. (1975). *Biochemistry* **14**, 3562–3566.
Marks, N. (1979). *In* "Peptides in Neurobiology" (H. Gainer, ed.), pp. 221–258. Plenum, New York.
Marx, J. L. (1980). *Science* **207**, 164–167.
Masson, A., and Wüthrich, K. (1973). *FEBS Lett.* **31**, 114–118.
Matheson, R. R., and Scheraga, H. A. (1978). *Macromolecules* **11**, 819–929.
Matheson, R. R., and Scheraga, H. A. (1979a). *Biochemistry* **18**, 2438–2445.
Matheson, R. R., and Scheraga, H. A. (1979b). *Biochemistry* **18**, 2446–2450.
Matheson, R. R., Van Wart, H. E., Burgess, A. W., Weinstein, L. I., and Scheraga, H. A. (1977a). *Biochemistry* **16**, 396–403.
Matheson, R. R., Dugas, H., and Scheraga, H. A. (1977b). *Biophys. Biochem. Res. Comm.* **74**, 869–876.
Matthews, B. W. (1972). *Macromolecules* **5**, 818–819.

Matthews, B. W. (1975). *Biochim. Biophys. Acta.* **405**, 442–451.

Matthews, B. W. (1977). *In* "The Proteins" (H. Neurath and R. L. Hill, eds.), 3rd ed., Vol. 3, pp. 403–590. Academic Press, New York.

Matthews, B. W., and Bernhard, S. A. (1973). *Annu. Rev. Biophys. Bioeng.* **2**, 257–317.

Matthews, B. W., Weaver, L. H., and Kester, W. R. (1974). *J. Biol. Chem.* **249**, 8030–8044.

Matthews, C. R., and Crisanti, M. M. (1981). *Biochemistry* **20**, 784–792.

Matthews, C. R., Crisanti, M. M., Gerner, G. L., Velicelebi, G., and Sturtevant, J. M. (1980). *Biochemistry* **19**, 1290–1293.

Matthews, F. S., Argos, P., and Levine, M. (1972). *Cold Spring Harbor Symp. Quant. Biol.* **36**, 387–395.

Matthyssens, G. E., and Kanarek, L. (1974). *Eur. J. Biochem.* **43**, 363–369.

Matthyssens, G. E., Simons, G., and Kanarek, L. (1972). *Eur. J. Biochem.* **26**, 449–454.

Matthyssens, G. E., Giebens, G., and Kanarek, L. (1974). *J. Biochem.* **43**, 353–362.

Maxfield, F. R., and Scheraga, H. A. (1976). *Biochemistry* **15**, 5138–5153.

Menez, A., Bonnet, F., Gushlbauer, W., and Fromageot, P. (1980). *Biochemistry* **19**, 4166–4172.

Mercier, J. C., Haze, G., Gaye, P., and Hue, D. (1978). *Biochem. Biophys. Res. Commun.* **82**, 1236–1245.

Mevel-Ninio, M., Risler, Y., and Labeyrie, F. (1977). *Eur. J. Biochem.* **73**, 131–140.

Milstein, C., Brownlee, G. C., Harrison, T. M., and Matthews, M. B. (1972). *Nature* (*London*), *New Biol.* **239**, 117–120.

Mirsky, A. E., and Anson, M. L. (1935). *J. Gen. Physiol.* **18**, 307–323.

Mirsky, A. E., and Pauling, L. (1936). *Proc. Natl. Acad. Sci. U.S.A.* **22**, 439–447.

Mizushima, S., and Shimanouchi, T. (1961). *Adv. Enzymol. Relat. Subj. Biochem.* **23**, 1–27.

Molday, R. S., Englander, S. W., and Kallen, R. G. (1972). *Biochemistry* **11**, 150–158.

Momany, F. A., Vanderkooi, G., and Scheraga, H. A. (1968). *Proc. Natl. Acad. Sci. U.S.A.* **61**, 429–436.

Momany, F. A., McGuire, R. F., Yan, J. F., and Scheraga, H. A. (1971). *J. Phys. Chem.* **75**, 2286–2297.

Momany, F. A., Carruthers, L. M., McGuire, R. F., and Scheraga, H. A. (1974a) *J. Phys. Chem.* **78**, 1595–1620.

Momany, F. A., Carruthers, L. M., and Scheraga, H. A. (1974b). *J. Phys. Chem.* **78**, 1621–1630.

Momany, F. A., McGuire, R. F., Burgess, A. W., and Scheraga, H. A. (1975). *J. Phys. Chem.* **79**, 2361–2381.

Monod, J., Wyman, J., and Changeux, J. P. (1965). *J. Mol. Biol.* **12**, 88–118.

Moreno, F., Fowler, A., Hall, M., Silhavy, T. J., Zabin, I., and Schwartz, M. (1980). *Nature* (*London*) 286, 356–359.

Morgan, R. S., Miller, S. L., and McAdon, J. M. (1979). *J. Mol. Biol.* **127**, 31–39.

Morino, Y., and Snell, E. E. (1967). *J. Biol. Chem.* **242**, 5591–5601.

Mozhaev, V. V., and Martinek, K. (1981). *Eur. J. Biochem.* **115**, 143–147.

Muller K., Kratky, O., Röschlan, P., and Hess, B. (1972), *Hoppe Seyler's Z. Physiol. Chem.* **353**, 803–809.

Myer, Y. P., Pande, A., and Saturno, A. F. (1981). *J. Biol. Chem.* **256**, 1576–1581.

Nagano, K. (1973). *J. Mol. Biol.* **75**, 401–420.

Nagano, K. (1974). *J. Mol. Biol.* **84**, 337–372.

Nagano, K. (1977a). *J. Mol. Biol.* **109**, 235–250.

Nagano, K. (1977b). *J. Mol. Biol.* **109**, 251–274.

Nagano, K. (1980). *J. Mol. Biol.* **138**, 797–832.

Nakanishi, M., and Tsuboi, M. (1974). *J. Mol. Biol.* **83**, 379–391.

Nakanishi, M., Tsuboi, M., and Ikegami, A. (1972). *J. Mol. Biol.* **70**, 351–361.

Nakanishi, M., Tsuboi, M., and Ikegami, A. (1973). *J. Mol. Biol.* **75**, 673–682.

Nakanishi, M., Inoue, A., Kita, T., Wakamura, M., Chang, A. C. Y., Cohen, S. N., and Numa, S. (1979). *Nature* (*London*) **278**, 423–427.
Nall, B. T., and Baldwin, R. L. (1977). *Biochemistry* **16**, 3572–3576.
Nall, B. T., Garel, J. R., and Baldwin, R. L. (1978). *J. Mol. Biol.* **118**, 317–330.
Naugthon, M. A., and Dintzis, H. M. (1962). *Proc. Natl. Acad. Sci. U.S.A.* **48**, 1822–1830.
Nelson, C. A., and Hummel, J. P. (1962). *J. Biol. Chem.* **237**, 1567–1574.
Nemenoff, R. A., Snirr, J., and Scheraga, H. A. (1978a). *J. Phys. Chem.* **82**, 2504–2512.
Nemenoff, R. A., Snirr, J., and Scheraga, H. A. (1978b). *J. Phys. Chem.* **82**, 2513–2526.
Némethy, G. (1967). *Angew. Chem., Int. Ed. Engl.* **6**, 195–206.
Némethy, G. (1968). *In* "Low Temperature Biology and Foodstuffs" (J. Hawthorne, ed.), pp. 1–21. Pergamon, Oxford.
Némethy, G. (1969). *Ann. N.Y. Acad. Sci.* **155**, 492–506.
Némethy, G. (1972). *Comment.—Pontif. Accad. Sci.* **3**(51), 1–24.
Némethy, G. (1974). *PAABS* (*Pan-Am. Assoc. Biochem. Soc.*) *Rev.* **3**, 51–56.
Némethy, G., and Printz, M. P. (1972). *Macromolecules* **5**, 755–758.
Némethy, G., and Scheraga, H. A. (1962a). *J. Chem. Phys.* **36**, 3382–3400.
Némethy, G., and Scheraga, H. A. (1962b). *J. Chem. Phys.* **36**, 3401–3417.
Némethy, G., and Scheraga, H. A. (1962c). *J. Phys. Chem.* **66**, 1773–1782.
Némethy, G., and Scheraga, H. A. (1965). *Biopolymers* **3**, 155–184.
Némethy, G., and Scheraga, H. A. (1977). *Q. Rev. Biophys.* **10**, 3, 239–352.
Némethy, G., and Scheraga, H. A. (1979). *Proc. Natl. Acad. Sci. U.S.A.* **76**, 6050–6054.
Némethy, G., and Scheraga, H. A. (1980a). *Biochem. Biophys. Res. Commun.* **95**, 320–327.
Némethy, G., Steinberg, I. Z., and Scheraga, H. A. (1963). *Biopolymers* **1**, 43–60.
Némethy, G., Leach, S. J., and Scheraga, H. A. (1966). *J. Phys. Chem.* **70**, 998–1004.
Némethy, G., Phillips, D. C., Leach, S. J., and Scheraga, H. A. (1967). *Nature* (*London*) **214**, 363–365.
Némethy, G., Hodes, Z. I., and Scheraga, H. A. (1978), *Proc. Natl. Acad. Sci. U.S.A.* **75**, 5760–5764.
Neurath, H., and Bull, H. B. (1938). *Chem. Rev.* **23**, 391.
Neurath, H., and Walsh, K. A. (1976). *Proc. Natl. Acad. Sci. U.S.A.* **73**, 3825–3832.
Neurath, H., and Walsh, K. A. (1977). *Proc. FEBS Meet.* **47**, 1–14.
Neurath, H., Cooper, G. R., and Erickson, J. O. (1942). *J. Chem. Phys.* **46**, 203.
Nielsen, S. O. (1960). *Biochim. Biophys. Acta* **37**, 146–147.
Nishikawa, K., and Ooi, T. (1974). *J. Theor. Biol.* **43**, 351–374.
Nishikawa, K., and Scheraga, H. A. (1976). *Macromolecules* **2**, 395–407.
Nishikawa, K., Ooi, T., Isogai, Y., and Saito, N. (1972). *J. Phys. Soc. Jpn.* **32**, 1331–1337.
Nishikawa, K., Momany, F. A., and Scheraga, H. A. (1974). *Macromolecules* **7**, 797–806.
Nitta, K., and Sugai, S. (1972). *Biopolymers* **11**, 1893–1901.
Nitta, K., Kitta, N., Kuwajima, K., and Sugai, S. (1977a). *Biochim. Biophys. Acta* **490**, 200–208.
Nitta, K., Segawa, T., Kuwajima, K., and Sugai, S. (1977b). *Biopolymers* **16**, 703–706.
Nojima, H., Ikai, A., Oshima, T., and Noda, H. (1977). *J. Mol. Biol.* **116**, 429–442.
Nojima, H., Hon-Nami, K., Oshima, T., and Noda, H. (1978). *J. Mol. Biol.* **122**, 33–42.
Northrop, J. H. (1932). *J. Gen. Physiol.* **16**, 323–337, 339–348.
Northrup, S. H., Pear, M. R., McCammon, J. A., and Karplus, M. (1980). *Nature* (*London*) **286**, 304–305.
Nozaka, N., Kuwajima, K., Nitta, K., and Sugai, S. (1978). *Biochemistry* **17**, 3753–3758.
Nozaki, Y., and Tanford, C. (1963). *J. Biol. Chem.* **238**, 4074–4081.
Nozaki, Y., and Tanford, C. (1965). *J. Biol. Chem.* **240**, 3568–3573.
Nozaki, Y., and Tanford, C. (1970). *J. Biol. Chem.* **245**, 1648–1652.
Nozaki, Y., and Tanford, C. (1971). *J. Biol. Chem.* **246**, 2211–2217.

Odani, S., and Ikenaka, T. (1973). *J. Biochem.* (*Tokyo*) **74**, 857–860.

Odorzynski, T. W., and Light, A. (1979). *J. Biol. Chem.* **254**, 4291–4295.

Ogasahara, K., and Hamagushi, K. (1967). *J. Biochem.* (*Tokyo*) **61**, 189–210.

Ohlsson, I., Nordström, B., and Branden, C. I. (1974). *J. Mol. Biol.* **89**, 339–354.

Okuda, T., and Sugai, S. (1977). *J. Biochem.* (*Tokyo*) **81**, 1051–1056.

Ooi, T., and Nishikawa, K. (1973). *In* "Conformation of Biological Molecules and Polymers" (B. D. Bergman and B. Pullman, eds.), pp. 173–187. Academic Press, New York.

Ooi, T., and Scheraga, H. A. (1964). *Biochemistry* **3**, 648–652.

Ooi, T., Rupley, J. A., and Scheraga, H. A. (1963). *Biochemistry* **2**, 432–437.

Oppenheimer, H. L., Labouesse, B., and Hess, G. P. (1966). *J. Biol. Chem.* **242**, 2720–2730.

Orsini, G., and Goldberg, M. E. (1978). *J. Biol. Chem.* **253**, 3453–3458.

Orsini, G., Skrzynia, C., and Goldberg, M. E. (1975). *Eur. J. Biochem.* **59**, 433–440.

Ottesen, M. (1971). *Methods Biochem. Anal.* **20**, 135–168.

Ovchinnikov, Y. A., Abdulaev, N. G., Feigina, M. Y., Kiselev, A. V., and Lobanov, N. A. (1979). *FEBS Lett.* **100**, 219–224.

Overman, L. E., and O'Connor, E. M. (1976). *J. Am. Chem. Soc.* **98**, 771–775.

Owicki, J. C., and Scheraga, H. A. (1977a). *Chem. Phys. Lett.* **47**, 600.

Owicki, J. C., and Scheraga, H. A. (1977b). *J. Am. Chem. Soc.* **99**, 7403–7412.

Owicki, J. C., and Scheraga, H. A. (1977c). *J. Am. Chem. Soc.* **99**, 7413–7418.

Owicki, J. C., and Scheraga, H. A. (1978). *J. Phys. Chem.* **82**, 1257–1264.

Owicki, J. C., Lentz, B. R., Hagler, A. T., and Scheraga, H. A. (1975). *J. Phys. Chem.* **79**, 2352–2367.

Oxender, D. L., Anderson, J. L., Daniels, C. J., Landick, R., Gunsalus, R. P., Zurawsky, G., and Yanofsky, C. (1980). *Proc. Natl. Acad. Sci. U.S.A.* **77**, 2005–2009.

Pabo, C. O., Suer, R. T., Sturtevant, J. M., and Ptaschne, M. (1979). *Proc. Natl. Acad. Sci. U.S.A.* **76**, 1608–1612.

Pace, C. N. (1975). *CRC Crit. Rev. Biochem.* **3**, 1–43.

Pace, C. N., and McGrath, T. (1980). *J. Biol. Chem.* **255**, 3862–3865.

Pace, C. N., and Tanford, C. (1968). *Biochemistry* **7**, 198–208.

Painter, P. C., and Koenig, J. (1976). *Biopolymers* **15**, 241–255.

Palade, G. (1975). *Science* **189**, 347–358.

Paladini, A. A., and Weber, G. (1981). *Biochemistry* **20**, 2587–2593.

Palmiter, R. D., Gagnon, J., Ericson, L. H., and Walsh, K. A. (1977a). *J. Biol. Chem.* **252**, 6386–6393.

Palmiter, R. D., Thibodeau, S. N., Gagnon, J., and Walsh, K. A. (1977b). *Proc. FEBS Meet.* 89–101.

Palmiter, R. D., Gagnon, J., and Walsh, K. A. (1978). *Proc. Natl. Acad. Sci. U.S.A.* **75**, 94–98.

Pandin, M., Padlan, E., di Bello, C., and Chaiken, I. M. (1976). *Proc. Natl. Acad. Sci. U.S.A.* **73**, 1844–1847.

Parr, G. R., and Hammes, G. G. (1975). *Biochemistry* **14**, 1600–1605.

Parr, G. R., and Hammes, G. G. (1976). *Biochemistry* **15**, 857–862.

Parr, G. R., and Taniuchi, H. (1979). *J. Biol. Chem.* **254**, 4836–4842.

Parr, G., and Taniuchi, H. (1980a). *J. Biol. Chem.* **255**, 2616–2623.

Parr, G., and Taniuchi, H. (1980b). *J. Biol. Chem.* **255**, 8914–8918.

Parr, G., and Taniuchi, H. (1981). *J. Biol. Chem.* **256**, 125–132.

Parr, G. R., Hantgan, R. R., and Taniuchi, H. (1978). *J. Biol. Chem.* **253**, 5381–5388.

Pauling, L. (1940). *J. Am. Chem. Soc.* **62**, 2643–2657.

Pauling, L., and Corey, R. M. (1951a). *Proc. Natl. Acad. Sci. U.S.A.* **37**, 235–240.

Pauling, L., and Corey, R. B. (1951b). *Proc. Natl. Acad. Sci. U.S.A.* **37**, 241–250.

Pauling, L., and Corey, R. B. (1951c). *Proc. Natl. Acad. Sci. U.S.A.* **37**, 251–256.
Pauling, L., and Corey, R. B. (1951d). *Proc. Natl. Acad. Sci. U.S.A.* **37**, 256–261.
Pauling, L., and Corey, R. B. (1951e). *Proc. Natl. Acad. Sci. U.S.A.* **37**, 729–740.
Pauling, L., and Corey, R. B. (1952). *Proc. Natl. Acad. Sci. U.S.A.* **38**, 86–93.
Pauling, L., and Corey, R. B. (1953a). *Proc. Natl. Acad. Sci. U.S.A.* **39**, 247–252.
Pauling, L., and Corey, R. B. (1953b). *Proc. Natl. Acad. Sci. U.S.A.* **39**, 253–256.
Pauling, L., Corey, R. B., and Branson, H. R. (1951). *Proc. Natl. Acad. Sci. U.S.A.* **37**, 205.
Pauling, L., Corey, R. B., and Yakel, H. L. (1952). *Nature (London)* **169**, 920.
Pederson, D. M., and Foster, J. F. (1969). *Biochemistry* **8**, 2357–2363.
Penhoet, E., Kochman, M., Valentine, R., and Rutter, W. J. (1967). *Biochemistry* **6**, 2940–2949.
Perahia, D. (1971). Thèse de Spécialité, Univ. Paris-VI.
Perkins, S. J., and Wüthrich, K. (1978). *Biochem. Biophys. Acta* **536**, 406–420.
Perkins, S. J., and Wüthrich, K. (1980). *J. Mol. Biol.* **138**, 43–64.
Perutz, M. F. (1962). *Nature (London)* **194**, 914–917.
Perutz, M. F. (1978). *Science* **201**, 1187–1191.
Perutz, M. F., and Raidt, H. (1975). *Nature (London)* **255**, 256–259.
Perutz, M. F., and Ten Eyck, L. F. (1972). *Cold Spring Harbor Symp. Quant. Biol.* **36**, 295–310.
Perutz, M. F., Muirhead, H., Cox, J. M., and Goaman, L. C. G. (1968a). *Nature (London)* **219**, 131–139.
Perutz, M. F., Muirhead, H., Cox, J. M., Goaman, L. C. G., Matthews, F. S., McGandy E. L., and Webb, L. E. (1968b). *Nature (London)* **219**, 29–32.
Peticolas, W. L. (1975). *Biochimie* **57**, 417–428.
Pfeil, W., and Privalov, R. L. (1976). *Biophys. Chem.* **4**, 23–33.
Pfeil, W., and Privalov, R. L. (1976). *Biophys. Chem.* **4**, 33–40.
Pfeil, W., and Privalov, R. L. (1976). *Biophys. Chem.* **4**, 41–50.
Phelps, C. F. (1980). *In* "The Enzymology of Post-translational Modifications of Proteins," (R. B. Freedman and H. C. Hawkins, ed.), Vol I, pp. 105–155. Academic Press, New York.
Phillips, D. C. (1966). *Sci. Am.* **215**, 78–90.
Phillips, D. C. (1967). *Proc. Natl. Acad. Sci. U.S.A.* **57**, 484–495.
Phillips, D. C., Sternberg, M. J. E., Thornton, J. M., and Wilson, I. A. (1978). *J. Mol. Biol.* **119**, 329–351.
Pickover, C. A., McKay, D. B., Engelman, D. M., and Steitz, T. A. (1979). *J. Biol. Chem.* **254**, 11323–11329.
Pimentel, G. C., and McClellan, A. L. (1960). "The Hydrogen." Freeman, San Francisco, California.
Pincus, M. R., and Scheraga, H. A. (1977). *J. Phys. Chem.* **81**, 1579–1583.
Pohl, F. M. (1968a). *Eur. J. Biochem.* **4**, 373–377.
Pohl, F. M. (1968b). *Eur. J. Biochem.* **7**, 146–152.
Pohl, F. M. (1969). *FEBS Lett.* **3**, 60–64.
Pohl, F. M. (1972a). *Angew. Chem., Int. Ed. Engl.* **10**, 894–906.
Pohl, F. M. (1972b). *Angew. Chem., Int. Ed. Engl.* **11**, 394–906.
Pohl, F. M. (1976). *FEBS Lett.* **65**, 293–296.
Poland, D., and Scheraga, H. A. (1965). *Biopolymers* **3**, 401–419.
Poljak, R. J., Amzel, L. M., Chen, B. L., Phizackerley, R. P., and Saul, F. (1973). *Proc. Natl. Acad. Sci. U.S.A.* **71**, 3440–3444.
Pompon, D., Guiard B., and Lederer, F. (1980). *Eur. J. Biochem.* **110**, 565–570.
Ponnuswamy, P. K., Warme, P. K., and Scheraga, H. A. (1973). *Proc. Natl. Acad. Sci. U.S.A.* **70**, 830–833.
Pople, J. A. (1951). *Proc. R. Soc. London, Ser. A* **205**, 163–178.

Poplis, V. A., Strickland, D. K., and Castellino, E. S. (1981). *Biochemistry* **20**, 15–21.
Popov, E. M., Dashevskii, V. G., Lipkind, G. M., and Arkhipova, S. F. (1968a). *Mol. Biol. (Moscow)* **2**, 491–497.
Popov, E. M., Lipkind, G. M., Dashevskii, V. G., and Arkhipova, S. F. (1968b). *Mol. Biol. (Moscow)* **2**, 498–504.
Porter, R. R. (1959). *Biochem. J.* **73**, 119–127.
Porubcan, R. S., Watters, K. L., and McFarland, J. J. (1978). *Arch. Biochem. Biophys.* **186**, 255–264.
Potts, J. T., Young, D. M., Anfinsen, C. B., and Sandoval, A. (1964). *J. Biol. Chem.* **239**, 3781–3786.
Printz, A. P., Nemethy, G., and Bleich, H. (1972). *Nature (London), New Biol.* **237**, 135–140.
Privalov, P. L. (1963). *Biofizika* **8**, 308–316.
Privalov, P. L. (1974). *FEBS Lett.* **40**, S140–S143.
Privalov, P. L., and Khechinashvili, N. N. (1974). *J. Mol. Biol.* **86**, 665–684.
Privalov, P. L., Khechinashvili, N. N., and Atanosov, B. P. (1971). *Biopolymers* **10**, 1865–1890.
Privalov, P. L., Tiktopulo, E. I., and Khechinashvili, N. N. (1973). *Int. J. Pept. Protein Res.* **5**, 229–237.
Prothero, J. W. (1966). *Biophys. J.* **6**, 367–370.
Provencher, S. W., and Glöckner, J. (1981). *Biochemistry* **20**, 33–37.
Ptitsyn, O. B. (1969). *J. Mol. Biol.* **42**, 501–510.
Ptitsyn, O. B. (1973). *Dokl. Akad. Nauk SSSR* **210**, 1213–1215.
Ptitsyn, O. B. (1978). *FEBS Lett.* **93**, 1–4.
Ptitsyn, O. B., and Finkelstein, A. V. (1970). *Biofizika* **15**, 757.
Ptitsyn, O. B., and Finkelstein, A. V. (1978). *Proc. FEBS Meet.* **52**, 105–115.
Ptitsyn, O. B., and Rashin, A. A. (1973). *Dokl. Akad. Nauk SSSR* **213**, 473–475.
Ptitsyn, O. B., and Rashin, A. A. (1975). *Biophys. Chem.* **3**, 1–20.
Ptitsyn, O. B., Lim, V. I., and Finkelstein, A. V. (1972), *Proc. FEBS Meet.* **25**, 421–429.
Ptitsyn, O. B., Finkelstein, A. V., and Falk (Bendzko), P. (1979). *FEBS Lett.* **101**, 1–5.
Puett, D. (1973). *J. Biol. Chem.* **248**, 4623–4634.
Puett, D., Friebele, E., and Hammonds, R. G. (1973). *Biochim. Biophys. Acta* **328**, 261–277.
Pullman, A., and Pullman, B. (1974). *Q. Rev. Biophys.* **7**, 505–566.
Pullman, B., and Pullman, A. (1974). *Adv. Protein Chem.* **28**, 347–526.
Quigley, G. J., and Rich, A. (1976). *Science* **194**, 795–806.
Quiocho, F. A., and Lipscomb, W. N. (1971). *Adv. Protein Chem.* **25**, 1–78.
Rackovsky, S., and Scheraga, H. A. (1978). *Macromolecules* **11**, 1–8.
Raghavendra, K., and Sasisekharan, V. (1979). *Int. J. Pept. Protein Res.* **14**, 326–338.
Rahman, A., and Stillinger, F. H. (1971). *J. Chem. Phys.* **55**, 3336–3359.
Rahman, A., and Stillinger, F. H. (1973). *J. Am. Chem. Soc.* **95**, 7943–7948.
Rahman, A., Stillinger, F. H., and Lemberg, H. I. (1975). *J. Chem. Phys.* **63**, 5223–5230.
Raibaud, O. (1977). Thèse d'Etat, Univ. Paris-VII.
Raibaud, O., and Goldberg, M. E. (1973). *J. Biol. Chem.* **248**, 3451–3455.
Raibaud, O., and Goldberg, M. E. (1976a). *J. Biol. Chem.* **251**, 2814–2819.
Raibaud, O., and Goldberg, M. E. (1976b). *J. Biol. Chem.* **251**, 2820–2824.
Raibaud, O., and Goldberg, M. E. (1977). *Eur. J. Biochem.* **73**, 591–599.
Ralston, G. B. (1974). *C. R. Trav. Lab. Carlsberg* **39**, 299–413.
Ramachandran, G. N. (1973). *Jerusalem Symp. Quantum Chem. Biochem.* **5**, 1–11.
Ramachandran, G. N. (1974). *In* "Peptides, Polypeptides and Proteins" (E. R. Blout, F. A. Bovey, M. Goodman, and N. Lotan, eds.) pp. 14–34. Wiley (Interscience) New York.
Ramachandran, G. N., and Mitra, A. K. (1976). *J. Mol. Biol.* **107**, 85–92.
Ramachandran, G. N., and Sasisekharan, V. (1968). *Adv. Protein Chem.* **23**, 283–438.

Ramachandran, G. N., Ramakrishnan, C., and Sasisekharan, V. (1963a). *In* "Aspects of Protein Structure" (G. N. Ramachandran, ed.), pp. 121–135. Academic Press, New York.
Ramachandran, G. N., Ramakrishnan, C., and Sasisekharan, V. (1963b). *J. Mol. Biol.* **7**, 95–99.
Ramachandran, G. N., Venkatachalan, C. M., and Krimm, S. (1966). *Biophys. J.* **6**, 849–872.
Ramakrishnan, C. (1964). *Proc.–Indian Acad. Sci., Sect. A* **59**, 327–343.
Rao, S. T., and Rossmann, M. G. (1973). *J. Mol. Biol.* **76**, 241–256.
Rashin, A. A., and Yudman, B. H. (1979). *FEBS Lett.* **101**, 6–10.
Record, M. T., Jr., Anderson, C., and Lohman, T. M. (1978). *Q. Rev. Biophys.* **11**, 103–178.
Redfield, R. R., and Anfinsen, C. B. (1956). *J. Biol. Chem.* **221**, 385.
Reichlin, M. (1975). *Adv. Immunol.* **20**, 71–123.
Reichlin, M., Bucci, E., Antonini, E., Wyman, J., and Rossi-Fanelli, A. (1964). *J. Mol. Biol.* **9**, 785–788.
Remington, S. J., and Matthews, B. W. (1978). *Proc. Natl. Acad. Sci. U.S.A.* **75**, 2180–2184.
Remington, S. J., and Matthews, B. W. (1980). *J. Mol. Biol.* **140**, 77–99.
Remington, S. J., Anderson, W. F., Owen, J., Ten Eyck, L. F., Graizer, C. T., and Matthews, B. W. (1978). *J. Mol. Biol.* **118**, 81–98.
Ricard, J., Meunier, J. C., and Buc, J. (1974). *Eur. J. Biochem.* **49**, 195–208.
Richards, F. M. (1958). *Proc. Natl. Acad. Sci. U.S.A.* **44**, 162–168.
Richards, F. M. (1974). *J. Mol. Biol.* **82**, 1–14.
Richards, F. M. (1977). *Annu. Rev. Biophys. Bioeng.* **6**, 151–176.
Richards, F. M. (1979). *Carlsberg Res. Commun.* **44**, 47–63.
Richards, F. M., and Vithayathil, P. J. (1959). *J. Biol. Chem.* **234**, 1459–1465.
Richards, F. M., Richmond, T. J., Sternberg, M. J. E., and Cohen, F. E. (1980). *In* "Protein Folding" (R. Jaenicke, ed.), pp. 117–129. Elsevier/North-Holland, Amsterdam.
Richardson, J. S. (1976). *Proc. Natl. Acad. Sci. U.S.A.* **73**, 2619–2623.
Richardson, J. S. (1977). *Nature (London)* **268**, 495–500.
Richardson, J. S. (1979). *Biochem. Biophys. Res. Commun.* **90**, 285–290.
Richardson, J. S. (1980). *In* "Protein Folding" (R. Jaenicke, ed.), pp. 326–338. Elsevier North-Holland, Amsterdam.
Richardson, J. S. (1981). *Adv. Protein Chem.* **34**, 167–339.
Richardson, J. S., Richardson, D. C., Thomas, K. A., Silverton, E. W., and Davies, R. (1976). *J. Mol. Biol.* **102**, 221–235.
Richardson, J. S., Getzoff, E. D., and Richardson, D. C. (1978). *Proc. Natl. Acad. Sci. U.S.A.* **75**, 2574–2578.
Richarz, R., and Wüthrich, K. (1977). *FEBS Lett.* **79**, 64–68.
Richarz, R., and Wüthrich, K. (1978). *Biochemistry* **17**, 2263–2269.
Richarz, R., Sehr, P., Wagner, G., and Wüthrich, K. (1979). *J. Mol. Biol.* **130**, 19–30.
Richarz, R., Nagayama, K., and Wüthrich, K. (1980). *Biochemistry* **19**, 5189–5196.
Richmond, T. J., and Richards, F. M. (1978). *J. Mol. Biol.* **119**, 537–555.
Riddiford, L. M. (1966). *J. Biol. Chem.* **241**, 2792–2802
Ristow, S. S., and Wetlaufer, D. B. (1973). *Biochem. Biophys. Res. Commun.* **50**, 544–550.
Roberts, J. L., and Herbert, E. (1977a). *Proc. Natl. Acad. Sci. U.S.A.* **74**, 4826–4830.
Roberts, J. L., and Herbert, E. (1977b). *Proc. Natl. Acad. Sci. U.S.A.* **74**, 5300–5304.
Roberts, J. L., Phillips, M., Rosa, P. A., and Herbert, E. (1978). *Biochemistry* **17**, 3609–3618.
Robertus, J. D., Alden, R. A., Birktoft, J. J., Kraut, J., Powers, J. C., and Wilcox, P. E. (1972). *Biochemistry* **11**, 2439–2449.
Robinson, D. R., and Jencks, W. P. (1965). *J. Am. Chem. Soc.* **87**, 2462–2470.
Robson, B. (1975). *Nature (London)* **254**, 386–388.
Robson, B. (1977). *Nature (London)* **267**, 577–578.
Robson, B., and Pain, R. H. (1971). *J. Mol. Biol.* **58**, 237–259.

Robson, B., and Pain, R. H. (1973). *Jerusalem Symp. Quantum Chem. Biochem.* **5**, 161–172.
Robson, B., and Pain, R. H. (1974a). *Biochem. J.* **141**, 883–897.
Robson, B., and Pain, R. H. (1974b). *Biochem. J.* **141**, 897–904.
Robson, B., and Suzuki, E. (1976). *J. Mol. Biol.* **107**, 327–356.
Rodiguin, N. M., and Rodiguina, E. N. (1964). "Consecutive Chemical Reactions." Van Nostrand, Princeton, New Jersey.
Rosa, J. J., and Richards, F. M. (1979). *J. Mol. Biol.* **133**, 399–416.
Rose, G. D. (1978). *Nature (London)* **272**, 586–590.
Rose, G. D. (1979). *J. Mol. Biol.* **134**, 447–470.
Rose, G. D., and Wetlaufer, D. B. (1977). *Nature (London)* **268**, 769–770.
Roseman, D., and Jencks, W. P. (1975). *J. Am. Chem. Soc.* **97**, 631–640.
Ross, J. B. A., Rousslang, K. W., and Kwiram, A. L. (1980). *Biochemistry* **19**, 376–383.
Rossi, G. L., and Bernhard, S. A. (1971). *J. Mol. Biol.* **55**, 215–230.
Rossmann, M. G., and Argos, P. (1975). *J. Biol. Chem.* **250**, 7525–7532.
Rossmann, M. G., and Argos, P. (1976). *J. Mol. Biol.* **105**, 75–95.
Rossmann, M. G., and Argos, P. (1977). *J. Mol. Biol.* **109**, 99–129.
Rossmann, M. G., and Argos, P. (1981). *Annu. Rev. Biochem.* **50**, 497–532.
Rossmann, M. G., and Liljas, A. (1974a). *J. Mol. Biol.* **85**, 177–181.
Rossmann, M. G., Jeffrey, B. A., Main, P., and Warren, S. (1967). *Proc. Natl. Acad. Sci. U.S.A.* **57**, 515–524.
Rossmann, M. G., Adams, M. J., Buchner, M., Ford, G. G., Hackert, M. L., Lentz, P. J., McPherson, A., Schevitz, R. W., and Smiley, I. E. (1972). *Cold Spring Harbor Symp. Quant. Biol.* **36**, 179–191.
Rossmann, M. G., Moras, D., and Olsen, K. W. (1974). *Nature (London)* **250**, 194–199.
Rossmann, M. G., Liljas, A., Brändén, C. I., and Banaszak, L. J. (1975). *In* "The Enzymes" (P. D. Boyer, ed.), 3rd ed., Vol. 11, pp. 61–102. Academic Press, New York.
Rothman, J. E., and Lennard, J. (1977). *Science* **195**, 743–753.
Rothman, J. E., and Lodish, H. F. (1977). *Nature (London)* **269**, 775–780.
Rousslang, K. W., Ross, J. B. A., Duranleau, D. A., and Kwiren, A. L. (1978). *Biochemistry* **17**, 1087–1092.
Rousslang, K. W., Thomasson, J. M., Ross, J. B. A., and Kwiren, A. L. (1979). *Biochemistry* **18**, 2296–2300.
Rowe, E. S., and Tanford, C. (1973). *Biochemistry* **12**, 4822–4827.
Rudolph, R., and Jaenicke, R. (1976). *Eur. J. Biochem.* **63**, 409–417.
Rudolph, R., Engelhard, M., and Jaenicke, R. (1976). *Eur. J. Biochem.* **67**, 455–462.
Rudolph, R., Heider, I., and Jaenicke, R. (1977a). *Biochemistry* **16**, 3527–3531.
Rudolph, R., Heider, I., and Jaenicke, R. (1977b). *Eur. J. Biochem.* **81**, 563–570.
Rudolph, R., Heider, I., Westhof, E., and Jaenicke, R. (1977c). *Biochemistry* **16**, 3384–3390.
Rudolph, R., Westhof, E., and Jaenicke, R. (1977d). *FEBS Lett.* **73**, 204–206.
Rudolph, R., Zettlmeissl, G., and Jaenicke, R. (1979). *Biochemistry* **18**, 5572–5575.
Rupley, J. A., and Scheraga, H. A. (1963). *Biochemistry* **2**, 421–431.
Sachs, D. H., Schechter, A. N., Eastlake, A., and Anfinsen, C. B. (1972a). *Biochemistry* **11**, 4268–4273.
Sachs, D. H., Schechter, A. N., Eastlake, A., and Anfinsen, C. B. (1972b). *J. Immunol.* **109**, 1300–1310.
Sachs, D. H., Schechter, A. N., Eastlake, A., and Anfinsen, C. B. (1972c). *Proc. Natl. Acad. Sci. U.S.A.* **69**, 3790–3794.
Sachs, D. H., Schechter, A. N., Eastlake, A., and Anfinsen, C. B. (1974). *Nature (London)* **251**, 242–244.
Salahuddin, A., and Tanford, C. (1970). *Biochemistry* **9**, 1342–1347.

Salemme, F. R. (1981). *J. Mol. Biol.* **146**, 143–156.

Salemme, F. R., and Weatherford, D. W. (1981a). *J. Mol. Biol.* **146**, 101–117.

Salemme, F. R., and Weatherford, D. W. (1981b). *J. Mol. Biol.* **146**, 119–141.

Sanger, F. (1956). *In* "Currents in Biochemical Research" (D. E. Green, ed.), p. 434. Wiley (Interscience), New York.

Sarthy, A., Fowler, A., Zabin, I., and Beckwith, J. (1979). *J. Bacteriol.* **139**, 932–939.

Sasaki, D. M., Martin, P. D., Dosher, M. S., and Tsernoglou, D. (1979). *J. Mol. Biol.* **135**, 301–304.

Sasisekharan, V. (1962). *In* "Collagen" (N. Ramanathan, ed.), *Proc. Symp. Madras NZ9.30* Wiley (Interscience), New York.

Savithi, H. S., and Light, A. (1980). *Biochem. Biophys. Res. Commun.* **94**, 360–365.

Sawyer, L., Shotton, D. M., Campbell, J. W., Wendell, R. C., Muirhead, H., Watson, H. C., Diamond, R. R., and Ladner, R. C. (1978). *J. Mol. Biol.* **118**, 137–208.

Saxena, P., and Wetlaufer, D. B. (1970). *Biochemistry* **9**, 5015–5022.

Saxena, V. P., and Wetlaufer, D. B. (1971). *Proc. Natl. Acad. Sci. U.S.A.* **68**, 969–972.

Schade, B. C., Rudolph, R., Lüdeman, H. D., and Jaenicke, R. (1981). *Biophys. Chem.* **11**, 257–263.

Schaffer, S. W., Ahmed, A. K., and Wetlaufer, D. B. (1975). *J. Biol. Chem.* **250**, 8483–8486.

Schechter, A. N., Chen, R. F., and Anfinsen, C. B. (1970). *Science* **167**, 886–887.

Schechter, Y., Patchornik, A., and Bernstein, Y. (1973). *Biochemistry* **12**, 3407–3413.

Scheele, G., Dobberstein, B., and Blobel, G. (1978). *Eur. J. Biochem.* **82**, 593–599.

Schellman, J. A. (1955a). *C. R. Trav. Lab. Carlsberg., Ser. Chim.* **29**, 230.

Schellman, J. A. (1955b). *J. Phys. Chem.* **64**, 1917.

Schellman, J. A. (1958). *C. R. Trav. Lab. Carlsberg, Ser. Chim.* **30**, 450.

Schellman, J. A. (1975). *Biopolymers* **14**, 999–1018.

Schellman, J. A. (1978). *Biopolymers* **17**, 1305–1322.

Schellman, J. A., and Schellman, C. (1964). *In* "The Proteins" (H. Neurath, ed.), 2nd ed., Vol. 2, Chap. 2. Academic Press, New York.

Scheraga, H. A. (1961). "Protein Structure." Academic Press, New York.

Scheraga, H. A. (1963). *In* "The Proteins" (H. Neurath, ed.), 2nd ed., Vol. 1, pp. 478–594. Academic Press, New York.

Scheraga, H. A. (1968). *Adv. Phys. Org. Chem.* **6**, 103–184.

Scheraga, H. A. (1969). *Nobel Symp.* No. 11, 43–78.

Scheraga, H. A. (1971). *Chem. Rev.* **71**, 195–217.

Scheraga, H. A. (1973a). *Jerusalem Symp. Quantum Chem. Biochem.* **5**, 51–68.

Scheraga, H. A. (1973b). *In* "Current Topics in Biochemistry" (C. B. Anfinsen and A. N. Schechter, eds.), pp. 1–42. Academic Press, New York.

Scheraga, H. A. (1974). *In* "Peptides, Polypeptides and Proteins" (E. R. Blout, F. A. Bovey, M. Goodman, and N. Lotan eds.), pp. 49–70. Wiley, New York.

Scheraga, H. A. (1977). *Ann. N. Y. Acad. Sci.* **303**, 2–9.

Scheraga, H. A. (1979). *Acc. Chem. Res.* **12**, 7–14.

Scheraga, H. A. (1980). *In* "Protein Folding" (R. Jaenicke ed.), pp. 261–288. Elsevier/North-Holland, Amsterdam.

Scheraga, H. A., Némethy, G., and Steinberg, I. Z. (1962). *J. Biol. Chem.* **237**, 2506–2508.

Scheraga, H. A., Scott, A., Vanderkooi, G., Leach, S. J., Gibson, K. D., Ooi, T., and Némethy, G. (1967). *In* "Conformation of Biopolymers" (G. N. Ramachandran, ed.), Vol. 1, p. 47. Academic Press, New York.

Scheule, R. K., Van Wart, H. E., Vallee, B. L., and Scheraga, H. A. (1980). *Biochemistry* **19**, 759–766.

Schiffer, M., and Edmundson, A. B. (1967). *Biophys. J.* **7**, 121–135.

Schiffer, M., Girling, R. L., Ely, K. R., and Edmundson, A. B. (1973). *Biochemistry* **12**, 4620–4631.

Schmid, F. X. (1980). *In* "Protein Folding" (R. Jaenicke, ed.), pp. 387–400. Elsevier/North-Holland, Amsterdam.

Schmid, F. X. (1981). *Eur. J. Biochem.* **114**, 105–109.

Schmid, F. X., and Baldwin, R. L. (1978a). *Proc. Natl. Acad. Sci. U.S.A.* **75**, 4766–4768.

Schmid, F. X., and Baldwin, R. L. (1978b). *Proc. FEBS Meet.* **52**, 173–185.

Schmid, F. X., and Baldwin, R. L. (1979a). *J. Mol. Biol.* **133**, 285–287.

Schmid, F. X., and Baldwin, R. L. (1979b). *J. Mol. Biol.* **135**, 199–215.

Schmid, F. X., and Blaschek, H. (1981). *Eur. J. Biochem.* **114**, 111–117.

Schmid, G., Lüdeman, H. D., and Jaenicke, R. (1975). *Biophys. Chem.* **3**, 90–98.

Schmid, G., Lüdeman, H. D., and Jaenicke, R. (1978). *Eur. J. Biochem.* **86**, 219–226.

Schmidt, M. F. G., and Schlesinger, M. J. (1979). *Cell* **17**, 813–819.

Schoenborn, D. P. (1971). *In* "Probes of Structure and Function of Macromolecules and Membranes" (B. Chance, T. Yonetani, and A. S. Mildvan, eds.), Vol. 2, pp. 181. Academic Press, New York.

Schoenborn, D. P., Watson, H. C., and Kendrew, J. C. (1965). *Nature* (*London*) **207**, 28–30.

Schreier, A. A., and Baldwin, R. L. (1976). *J. Mol. Biol.* **105**, 409–426.

Schreier, A. A., and Baldwin, R. L. (1977). *Biochemistry* **16**, 4203–4209.

Schroeder, R., and Lippincott, E. R. (1957). *J. Phys. Chem.* **61**, 921.

Schulz, G. E. (1977). *Angew. Chem.* (*Int. Ed. Engl.*) **16**, 23–32.

Schulz, G. E. (1980). *J. Mol. Biol.* **138**, 335–347.

Schulz, G. E., and Schirmer, R. H. (1974). *Nature* (*London*) **250**, 142–144.

Schulz, G. E., and Schirmer, R. H. (1979). "Principles of Protein Structure." Springer-Verlag, Berlin and New York.

Schulz, G. E., Barry, C. D., Friedman, J., Chou, P. Y., Fasman, G. D., Finkelstein, D. V., Lim, V. I., Ptitsyn, O., Kabat, E. A., Wu, T. T., Levitt, M., Robson, B., and Nagano, K. (1974). *Nature* (*London*) **250**, 140–142.

Scott, R. A., and Scheraga, H. A. (1963). *J. Am. Chem. Soc.* **85**, 3866–3873.

Scott, R. A., and Scheraga, H. A. (1966). *J. Chem. Phys.* **45**, 2091–2101.

Seeburg, P. H., Shine, J., Martial, J. A., Baxter, J. D., and Goodman, H. B. (1977). *Nature* (*London*) **270**, 486–494.

Sela, M., and Lifson, M. (1959). *Biochim. Biophys. Acta* **36**, 471–478.

Sela, M., White, F. H., and Anfinsen, C. B. (1957). *Science* **125**, 691–692.

Sela, M., Schechter, B., Schechter, I., and Forek, F. (1967). *Cold Spring Harbor Symp. Quant. Biol.* **32**, 537–545.

Shen, L. L., and Hermans, J. (1972). *Biochemistry* **11**, 1836–1841.

Shoham, M., and Steitz, T. A. (1980). *J. Mol. Biol.* **140**, 1–14.

Shore, G. C., and Tata, J. R. (1977). *Biochim. Biophys. Acta* **472**, 197–236.

Shrake, A., and Rupley, J. A. (1973). *J. Mol. Biol.* **79**, 351–372.

Siekevitz, P., and Palade, G. E. (1960). *J. Biophys. Biochem. Cytol.* **7**, 619–630.

Sielecki, A. R., Hendrickson, W. A., Broughton, C. G., Delbaere, L. T. J., Brager, G. D., and James, M. N. G. (1979). *J. Mol. Biol.* **134**, 781–804.

Silhavy, T. J., Shuman, H. A., Beckwith, J., and Schwartz, M. (1977). *Proc. Natl. Acad. Sci. U.S.A.* **74**, 5411–5415.

Silverman, D. N., Kotelchuck, D., Taylor, G. T., and Scheraga, H. A. (1972). *Arch. Biochem. Biophys.* **150**, 757–766.

Singer, J. J. (1962). *Adv. Protein Chem.* **17**, 1–68.

Singhal, R. P., and Atassi, M. Z. (1970). *Biochemistry* **9**, 4252–4259.

Sinha, N. K., and Light, A. (1975). *J. Biol. Chem.* **250**, 8624–8629.

Skrzynia, C., London, J., and Goldberg, M. E. (1974). *J. Biol. Chem.* **249**, 2325–2326.
Slaby, I., and Holmgren, A. (1975). *J. Biol. Chem.* **250**, 1340–1347.
Slaby, I., and Holmgren, A. (1979). *Biochemistry* **18**, 5584–5599.
Snirr, J., Nemenoff, R. A., and Scheraga, H. A. (1978a). *J. Phys. Chem.* **82**, 2497–2503.
Snirr, J., Nemenoff, R. A., and Scheraga, H. A. (1978b). *J. Phys. Chem.* **82**, 2527–2530.
Snyder, G. H., Rowan, R., III, Karplus, S., and Sykes, B. D. (1975). *Biochemistry* **14**, 3765–3777.
Snyder, G. H., Rowan, R., III, and Sykes, B. D. (1976). *Biochemistry* **15**, 2275–2283.
Snyder, J. P., and Carlsen, L. (1977). *J. Am. Chem. Soc.* **99**, 2931–2942.
Srinivasan, R., Balasubramanian, R., and Rajan, S. S. (1975). *J. Mol. Biol.* **98**, 739–747.
Stark, G. R. (1977). *TIBS* **4**, 64–66.
Stark, G. R., Stein, W. H., and Moore, S. (1960). *J. Biol. Chem.* **235**, 3177–3181.
Staros, J. V., and Richards, F. M. (1974). *Biochemistry* **13**, 2720–2726.
States, D. J., Dobson, C. M., Karplus, M., and Creighton, T. E. (1980). *Nature* (*London*) **286**, 630–632.
Stein, E. A., Junge, J. M., and Fisher, E. H. (1960). *J. Biol. Chem.* **235**, 371–378.
Steiner, D. F., and Clark, J. L. (1968). *Proc. Natl. Acad. Sci. U.S.A.* **60**, 622–629.
Steiner, R. F., de Lorenzo, F., and Anfinsen, C. B. (1965). *J. Biol. Chem.* **240**, 4648–4651.
Steinhardt, J., and Zaiser, E. M. (1951). *J. Am. Chem. Soc.* **73**, 5568–5572.
Steitz, T. A. (1971). *J. Mol. Biol.* **61**, 695–700.
Steitz, T. A., Fletterick, R. T., and Hwang, K. J. (1973). *J. Mol. Biol.* **78**, 551–561.
Steitz, T. A., Fletterick, R. T., Anderson, W. F., and Anderson, C. M. (1976). *J. Mol. Biol.* **104**, 197–222.
Stellwagen, E. (1977). *Acc. Chem. Res.* **10**, 92–98.
Stellwagen, E. (1979). *J. Mol. Biol.* **135**, 217–229.
Stellwagen, E., and Schachman, H. K. (1962). *Biochemistry* **1**, 1056–1069.
Stellwagen, E., and Wilgus, H. (1978). *Nature* (*London*) **275**, 342–343.
Stellwagen, E., Carr, R., Thompson, S. I., and Woody, M. (1975). *Nature* (*London*) **257**, 716–719.
Sternberg, M. J. E., and Thornton, J. M. (1976). *J. Mol. Biol.* **105**, 367–382.
Sternberg, M. J. E., and Thornton, J. M. (1977a). *J. Mol. Biol.* **110**, 269–283.
Sternberg, M. J. E., and Thornton, J. M. (1977b). *J. Mol. Biol.* **110**, 285–296
Sternberg, M. J. E., and Thornton, J. M. (1978). *Nature* (*London*) **271**, 15–20.
Sternberg, M. J. E., Grace, D. E. P., and Phillips, D. C. (1979). *J. Mol. Biol.* **130**, 231–253.
Stewart, W. E., Mandelkern, L., and Glick, R. E. (1967). *Biochemistry* **6**, 143–150, 150–153.
Stillinger, F. H., and Rahman, A. (1972). *J. Chem. Phys.* **57**, 1281–1292.
Stillinger, F. H., and Rahman, A. (1974a). *J. Chem. Phys.* **60**, 1545–1557.
Stillinger, F. H., and Rahman, A. (1974b). *J. Chem. Phys.* **61**, 4973–4980.
Stillinger, F. H. (1980). *Science* **209**, 451–457.
Strauss, A. W., Bennett, C. D., Donohue, A. M., Rodkey, J. A., and Alberts, A. W. (1977). *J. Biol. Chem.* **252**, 6846–6855.
Subramanian, E., Swan, I. D. A., Liu, M., Davies, D. R., Jenkins, J. A., Tickle, I. J., and Blundell, T. L. (1977). *Proc. Natl. Acad. Sci. U.S.A.* **74**, 556–559.
Suchanek, G., Kindas-Mugge, J., Kreil, G., and Schreier, M. H. (1975). *Eur. J. Biochem.* **60**, 309–315.
Suchanek, G., Kreil, G., and Hermodson, M. A. (1978). *Proc. Natl. Acad. Sci. U.S.A.* **75**, 701–704.
Sugai, S., Yashiro, H., and Nitta, K. (1973). *Biochim. Biophys. Acta* **328**, 35–41.
Sugimoto, K., Sugisaki, H., Okamoto, T., and Takanami, M. (1977). *J. Mol. Biol.* **111**, 487–507.
Summers, M. R., and McPhie, P. (1972). *Biochem. Biophys. Res. Commun.* **47**, 831–837.
Sund, H., and Weber, K. (1966). *Ange. Chem., Int. Ed. Engl.* **5**, 231–245.
Sundaram, T. K., Wright, I. P., and Wilkinson, A. E. (1980). *Biochemistry* **19**, 2017–2022.

Sutcliffe, J. G. (1979). *Proc. Natl. Acad. Sci. U.S.A.* **75**, 3737–3741.
Sutties, J. W. (1980). *In* "The Enzymology of Post-translational Modifications of Proteins" (R. B. Freedman and H. C. Hawkins, eds.), pp. 214–258. Academic Press, New York.
Sykes, G. H., Rowan, R., III, Karplus, S., and Sykes, B. D. (1975). *Biochemistry* **14**, 3765–3777.
Takagi, T., and Isemura, T. (1963). *Biochem. Biophys. Res. Commun.* **13**, 353–359.
Takagi, T., and Isemura, T. (1964). *J. Biochem.* (*Tokyo*) **56**, 344–350.
Takahashi, S., Kontani, T., Yoneda, M., and Ooi, T. (1977). *J. Biochem.* (*Tokyo*) **82**, 1127–1133.
Takahashi, S., Ihara, S., and Ooi, T. (1978). *Nature* (*London*) **276**, 735–736.
Takano, T., Swanson, R., Kallai, O. B., and Dickerson, R. E. (1972). *Cold Spring Harbor Symp. Quant. Biol.* **36**, 397–404.
Takesada, H., Nakanishi, M., and Tsuboi, M. (1973). *J. Mol. Biol.* **77**, 605–614.
Taketomi, H., Ueda, Y., and Gō, N. (1975). *Int. J. Pept. Protein Res.* **7**, 445–459.
Tallan, H. H., and Stein, W. H. (1953). *J. Biol. Chem.* **200**, 507–514.
Tanaka, S., and Scheraga, H. A. (1975). *Proc. Natl. Acad. Sci. U.S.A.* **72**, 3802–3806.
Tanaka, S., and Scheraga, H. A. (1976a). *Macromolecules* **9**, 159–167.
Tanaka, S., and Scheraga, H. A. (1976b). *Macromolecules* **9**, 168–182.
Tanaka, S., and Scheraga, H. A. (1976c). *Macromolecules* **9**, 812–833.
Tanaka, S., and Scheraga, H. A. (1976d). *Macromolecules* **9**, 945–950.
Tanaka, S., and Scheraga, H. A. (1977). *Macromolecules* **10**, 291–304.
Tanford, C. (1958). *In* Symposium on "Protein Structure" ed. by (A. Neuberger, ed.), p. 35. Methuen, London.
Tanford, C. (1961). *J. Am. Chem. Soc.* **83**, 1628–1633.
Tanford, C. (1962a). *J. Am. Chem. Soc.* **84**, 4240–4247.
Tanford, C. (1962b). *Adv. Protein Chem.* **17**, 69–165.
Tanford, C. (1964). *J. Am. Chem. Soc.* **86**, 2050–2059.
Tanford, C. (1968). *Adv. Protein Chem.* **23**, 121–282.
Tanford, C. (1970). *Adv. Protein Chem.* **24**, 1–95.
Tanford, C. (1972). *Ciba Found. Symp.* **7**, 125–146.
Tanford, C. (1979). *Proc. Natl. Acad. Sci. U.S.A.* **76**, 4175–4176.
Tanford, C., and Aune, K. C. (1970). *Biochemistry* **9**, 206–211.
Tanford, C., and Taggart, V. (1961). *J. Am. Chem. Soc.* **83**, 1634–1638.
Tanford, C., De, P. K., and Taggart, V. G. (1960). *J. Am. Chem. Soc.* **82**, 6028–6034.
Tanford, C., Kawahara, K., and Lapanje, S. (1966a). *J. Biol. Chem.* **241**, 1921–1923.
Tanford, C., Pain, R. H., and Otchin, N. S. (1966b). *J. Mol. Biol.* **15**, 489–504.
Tanford, C., Kawahara, K., Lapanje, S., Hooker, T. M., Zarlengo, M. H., Salahuddin, A., Aune, K. C., and Takagi, T. (1967). *J. Am. Chem. Soc.* **89**, 5023–5029.
Tanford, C., Aune, K. C., and Ikai, A. (1973). *J. Mol. Biol.* **73**, 185–197.
Tang, J., James, M. N. G., Nsu, I., Jenkins, J. A., and Blundell, T. L. (1978). *Nature* (*London*) **271**, 618–621.
Taniuchi, H. (1970). *J. Biol. Chem.* **245**, 5459–5468.
Taniuchi, H., and Anfinsen, C. B. (1968). *J. Biol. Chem.* **243**, 4778–4786.
Taniuchi, H., and Anfinsen, C. B. (1969). *J. Biol. Chem.* **244**, 3864–3875.
Taniuchi, H., and Anfinsen, C. B. (1971). *J. Biol. Chem.* **246**, 2291–2301.
Taniuchi, H., and Bohnert, J. L. (1973). *Fed. Proc., Fed Am Soc Exp. Biol.* **32**, 458.
Taniuchi, H., and Bohnert, J. L. (1975). *J. Biol. Chem.* **250**, 2388–2394.
Taniuchi, H., Moraveck, L., and Anfinsen, C. B. (1969). *J. Biol. Chem.* **244**, 4600–4606.
Taniuchi, H., Parker, D. S., and Bohnert, J. L. (1977). *J. Biol. Chem.* **252**, 125–140.
Tata, J. R. (1978). *Proc. FEBS Meet.* **53**, 11–20.
Teale, J. M., and Benjamin, D. C. (1976a). *J. Biol. Chem.* **251**, 4603–4608.
Teale, J. M., and Benjamin, D. C. (1976b). *J. Biol. Chem.* **251**, 4609–4615.

Teale, J. M., and Benjamin, D. C. (1977). *J. Biol. Chem.* **252**, 4521–4526.

Teipel, J. W. (1972). *Biochemistry* **11**, 4100–4107.

Teipel, J. W., and Koshland, D. E. (1971a). *Biochemistry* **10**, 792–798.

Teipel, J. W., and Koshland, D. E. (1971b). *Biochemistry* **10**, 798–805.

Thibodeau, S. N., Lee, D. C., and Palmiter, R. D. (1978). *J. Biol. Chem.* **253**, 3771–3774.

Thiery, C., Nabedryk-Viala, E., Menez, A., Fromageot, P., and Thiery, J. M. (1980). *Biochem. Biophys. Res. Commun.* **93**, 889–897.

Thomas, K. A., and Schechter, A. N. (1980). *In* "Biological Regulation and Development" (R. F. Goldberger, ed.), Vol. 2, pp. 33–99. Plenum, New York.

Thompson, S. T., and Stellwagen, E. (1976). *Proc. Natl. Acad. Sci. U.S.A.* **73**, 361–365.

Thompson, S. T., Carr, K. H., and Stellwagen, E. (1975). *Proc. Natl. Acad. Sci. U.S.A.* **72**, 669–672.

Thornton, J. M. (1981), *J. Mol. Biol.* **151**, 261–287.

Tietze, F. (1969). *Anal. Biochem.* **27**, 502–522.

Tiktopulo, E. I., and Privalov, P. L. (1974). *Biophys. Chem.* **1**, 349–357.

Tiktopulo, E. I., and Privalov, P. L. (1978). *FEBS Lett.* **91**, 57–58.

Timchenko, A. A., Ptitsyn, O. B., Troitsky, A. V., and Denesyuk, A. I. (1978). *FEBS Lett.* **88**, 109–113.

Tobin, M. C. (1972). *In* "Enzyme Structure," Part B (C. H. W. Hirs and S. N. Timasheff, eds.), Methods in Enzymology, Vol. 25, pp. 473–497. Academic Press, New York.

Tsong, T. Y. (1973). *Biochemistry* **12**, 2209–2214.

Tsong, T. Y. (1974). *J. Biol. Chem.* **249**, 1988–1990.

Tsong, T. Y. (1975). *Biochemistry* **14**, 1542–1547.

Tsong, T. Y., and Baldwin, R. L. (1972a). *J. Mol. Biol.* **69**, 145–148.

Tsong, T. Y., and Baldwin, R. L. (1972b). *J. Mol. Biol.* **69**, 149–153.

Tsong, T. Y., Hearn, R. P., Wrathall, D. P., and Sturtevant, J. M. (1970). *Biochemistry* **2**, 2666–2667.

Tsong, T. Y., Baldwin, R. L., and Elson, E. (1971). *Proc. Natl. Acad. Sci. U.S.A.* **68**, 2712–2715.

Tsong, T. Y., Baldwin, R. L., and Mc Phie, P. (1972a) *J. Mol. Biol.* **63**, 453–475.

Tsong, T. Y., Baldwin, R. L., and Elson, E. L. (1972b) *Proc. Natl. Acad. Sci. U.S.A.* **69**, 1809–1812.

Tsong, T. Y., Baldwin, R. L., McPhie, P., and Elson, E. L. (1972c) *J. Mol. Biol.* **63**, 453–475.

Ueda, Y., Taketomi, H., and Gō, N. (1978). *Biopolymers* **17**, 1531–1548.

Uenoyama, K., and Ono, T. (1972). *Biochem. Biophys. Acta* **281**, 124–129.

Ullmann, A., and Monod, J. (1969). *Biochem. Biophys. Res. Commun.* **35**, 35–42.

Urnes, P., and Doty, P. (1961). *Adv. Protein Chem.* **16**, 401–544.

Urry, D. W., Mitchell, L. W., Ohnishi, T., and Long, M. N. (1975). *J. Mol. Biol.* **96**, 101–117.

Van Wart, H. E., and Scheraga, H. A. (1977). *Proc. Natl. Acad. Sci. U.S.A.* **74**, 13–17.

Van Wart, H. E., and Scheraga, H. A. (1978). *In* "Enzyme Structure," Part F (S. N. Timasheff and C. H. W. Hirs, eds.), Methods in Enzymology, Vol. 48, pp. 67–149. Academic Press, New York.

Velicelebi, G., and Sturtevant, J. M. (1979). *Biochemistry* **18**, 1180–1185.

Venetianer, P., and Straub, F. B. (1963). *Biochim, Biophys. Acta* **67**, 166–168.

Venetianer, P., and Straub, F. B. (1964). *Biochim. Biophys. Acta* **89**, 189–190.

Venkatachalam, C. M. (1968). *Biopolymers* **6**, 1425–1436.

Venkatachalam, C. M., and Ramachandran, G. N. (1967). *In* "Conformations in Biopolymers" (G. N. Ramachandran, ed.), pp. 83–108. Academic Press, New York.

Veron, M., Falcoz-Kelly, F., and Cohen, G. N. (1972). *Eur. J. Biochem.* **28**, 520–527.

Vimard, C., Orsini, G., and Goldberg, M. E. (1975). *Eur. J. Biochem.* **51**, 521–527.

Viratelle, O. M., and Yon, J. M. (1973). *Eur. J. Biochem.* **33**, 110–116.

Wagner, G. (1980). *FEBS Lett.* **112**, 280–284.
von Heijne, G., and Blomberg, C. (1977). *J. Mol. Biol.* **117**, 821–824.
von Heijne, G., and Blomberg, C. (1978). *Biopolymers* **17**, 2033–2037.
von Hippel, P. H., and Schleich, T. (1969). *In* "Structure and Stability of Biological Molecules" (S. N. Timasheff and G. D. Fasman, eds.), Vol. 2, pp. 417–574. Academic Press, New York.
von Hippel, P. H., and Wong, K. Y. (1964). *Nature (London)* **145**, 577–580.
von Hippel, P. H., and Wong, K. Y. (1965). *J. Biol. Chem.* **240**, 3909–3923.
Wagner, G., and Wüthrich, K. (1978a). *Nature (London)* **275**, 247–248.
Wagner, G., and Wüthrich, K. (1978b). *Trends Biochem. Sci.* **3**, 227–230.
Wagner, G., and Wüthrich, K. (1979a). *J. Mol. Biol.* **130**, 31–37.
Wagner, G., and Wüthrich, K. (1979b). *J. Mol. Biol.* **134**, 75–94.
Wagner, G., de Marco, A., and Wüthrich, K. (1975). *J. Magn. Res.* **20**, 565–569.
Wagner, G., de Marco, A., and Wüthrich, K. (1976). *Biophys. Struct. Mech.* **2**, 139–158.
Wagner, G., Wüthrich, K., and Tschesche, H. (1978a). *Eur. J. Biochem.* **89**, 376–377.
Wagner, G., Wüthrich, K., and Tschesche, H. (1978b). *Eur. J. Biochem.* **86**, 67–76.
Wagner, G., Gilboa, J., and Wüthrich, K. (1979a). *Eur. J. Biochem.* **95**, 249–253.
Wagner, G., Tschesche, H., and Wüthrich, K. (1979b). *Eur. J. Biochem.* **95**, 239–248.
Walker, J. E. (1978). *Proc. FEBS Meet.* **52**, 211–225.
Wallevick, K. (1973). *J. Biol. Chem.* **248**, 2650–2655.
Warme, P. K., Momany, F. A., Rumballs, S. V., Tuttle, R. W., and Scheraga, H. A. (1974). *Biochemistry* **13**, 768–782.
Watson, H. C. (1968). *Prog. Stereochem.* **4**, 299–333.
Watson, H. C., and Banaszak, L. J. (1964). *Nature (London)* **204**, 918–920.
Weatherford, D. W., and Salemme, F. R. (1979). *Proc. Natl. Acad. Sci. U.S.A.* **76**, 19–23.
Weber, G. (1965). *In* "Molecular Biophysics" (B. Pullman and W. Weissbluth, eds.), 369. Academic Press, New York.
Weber, G. (1973a). *Biochemistry* **12**, 4161–4170.
Weber, G. (1973b). *Biochemistry* **12**, 4171–4179.
Weber, G. (1975). *Adv. Protein Chem.* **29**, 1–83.
Weber, G. (1979). *Symp. New Horizons Biochem., Nagoya, Jpn.* "The effects of high-pressure upon proteins and other biomolecules" (personal communication)
Weber, G., and Teale, F. W. J. (1965). *In* "The Proteins" (H. Neurath, ed.), 2nd ed., Vol. 3, pp. 445–521. Academic Press, New York.
Weber, P. C., Bartsch, R. G., Cusanovich, M. A., Hamlin, R. C., Howard, A., Jordan, S. R., Kamen, M. D., Meyer, T. E., Weatherford, D. W., Xvong, N. H., and Salemme, F. R. (1980). *Nature (London)* **286**, 302–304.
Wendell, P. L., Bugart, T. N., and Watson, H. C. (1972). *Nature* New Biol. *(London)* **240**, 134–136.
Westmoreland, D. C., and Matthews, C. R. (1973). *Proc. Natl. Acad. Sci. U.S.A.* **70**, 914–918.
Wetlaufer, D. B. (1961). *Nature (London)* **190**, 1113.
Wetlaufer, D. B. (1962). *Adv. Protein Chem.* **17**, 303–390.
Wetlaufer, D. B. (1973). *Proc. Natl. Acad. Sci. U.S.A.* **70**, 697–701.
Wetlaufer, D. B. (1981). *Adv. Prot. Chem.* **34**, 61–92.
Wetlaufer, D. B., and Ristow, S. (1973). *Annu. Rev. Biochem.* **42**, 135–158.
Wetlaufer, D. B., Malik, S. K., Stetter, L., and Coffin, R. L. (1964). *J. Am. Chem. Soc.* **86**, 508–514.
White, F. H. (1961). *J. Biol. Chem.* **236**, 1353–1360.
White, F. H. (1976). *Biochemistry* **15**, 2906–2912.
Wichman, A., Svenson, A., and Anderson, L. O. (1977). *Eur. J. Biochem.* **79**, 339–344.

Wiley, D. C., and Lipscomb, W. N. (1968). *Nature (London)* **218**, 1119–1121.
Williams, R. J. P. (1978). *Proc. R. Soc. London, Ser. B* **200**, 353.
Williams, R. W., Cutera, T., Dunker, A. K., and Peticolas, W. L. (1980). *FEBS Lett.* **115**, 306–308.
Williams, R. W., and Dunker, A. K. (1981). *J. Mol. Biol.* **152**, 783–813.
Witt, J. J., and Roskoski, R., Jr. (1980). *Biochemistry* **19**, 143–148.
Wodak, S., and Janin, J. (1978). *J. Mol. Biol.* **124**, 323–342.
Wodak, S., and Janin, J. (1980). *Proc. Natl. Acad. Sci. U.S.A.* **77**, 1736–1740.
Wold, F. (1981). *Ann. Rev. Biochem.* **50**, 783–814.
Wong, K. P., and Hamlin, L. M. (1975). *Arch. Biochem. Biophys.* **170**, 12–22.
Wong, K. P., and Tanford, C. (1970). *Fed. Proc., Fed. Am. Soc. Exp. Biol.* **29**, 335.
Wong, K. P., and Tanford, C. (1973). *J. Biol. Chem.* **248**, 8519–8523.
Wong, K. P., Allen, S. R., and Hamlin, L. M. (1972). *Fed. Proc., Fed. Am. Soc. Exp. Biol.* **312**, 923.
Woodbury, R. G., and Neurath, H. (1980). *FEBS Lett.* **114**, 189–196.
Woodward, C. K., and Hilton, B. D. (1979). *Annu. Rev. Biophys. Bioeng.* **8**, 99–127.
Woodward, C. K., and Rosenberg, A. (1970). *Proc. Natl. Acad. Sci. U.S.A.* **66**, 1067–1074.
Woodward, C. K., and Rosenberg, A. (1971). *J. Biol. Chem.* **246**, 4114–4121.
Wooten, J. B., and Cohen, J. S. (1979). *Biochemistry* **18**, 4188–4191.
Wright, H. T. (1973). *J. Mol. Biol.* **79**, 1–11.
Wu, H. (1931). *Chin. J. Physiol.* **5**, 221.
Wu, T. T., and Kabat, E. A. (1971). *Proc. Natl. Acad. Sci. U.S.A.* **68**, 1501–1506.
Wu, T. T., and Kabat, E. A. (1973). *J. Mol. Biol.* **75**, 13–31.
Wüthrich, K., and Baumann, R. (1976). *Org. Magn. Reson.* **8**, 532–535.
Wüthrich, K., and Wagner, G. (1975). *FEBS Lett.* **50**, 265–268.
Wüthrich, K., and Wagner, G. (1978). *TIBS* **3**, 227–230.
Wüthrich, K., and Wagner, G. (1979). *J. Mol. Biol.* **130**, 1–18.
Wüthrich, K., Grathwohl, C., and Schwyzer, R. (1974). *In* "Peptides, Polypeptides and Proteins" (E. R. Blout, F. A. Bovey, M. Goodman, and N. Lotar, eds.), p. 300. Wiley, New York
Wüthrich, K., Wagner, G., Richarz, R., and Perkins, S. J. (1978). *Biochemistry* **17**, 2253–2262.
Wüthrich, K., Wagner, G., Richarz, R., and Braun, W. (1980). *Biophys. J.* **52**, 549–558.
Wyckoff, H. W., Hardman, K. D., Allewell, N. M., Tadaski, I., Johnson, L. N., and Richards, F. M. (1967). *J. Biol. Chem.* **242**, 3984–3988.
Wyckoff, H. W., Tsernoglou, D., Hanson, A. W., Knox, J. R., Lee, B., and Richards, F. M. (1970). *J. Biol. Chem.* **245**, 305–328.
Yagamata, S., and Snell, E. E. (1979). *Biochemistry* **18**, 2964–2967.
Yang, J. T., and Doty, P. (1957). *J. Am. Chem. Soc.* **79**, 761–775.
Yazgan, A., and Henkens, R. W. (1972). *Biochemistry* **11**, 1314–1318.
Yon, J. (1955a). *J. Chim. Phys. Phys.-Chim. Biol.* **52**, 413–438.
Yon, J. (1955b). *J. Chim. Phys. Phys.-Chim. Biol.* **52**, 452–416.
Yon, J. (1958a). *Biochim. Biophys. Acta* **27**, 111–121.
Yon, J. (1958b). *Bull. Soc. Chim. Biol.* **40**, 379–396.
Yon, J. (1969). "Structure et Dynamique Conformationnelle des Protéines." Hermann, Paris.
Yon, J. (1976). *Biochimie* **58**, 61–69.
Yon, J. (1978). *Biochimie* **60**, 581–591.
Yoshida, A., and Tobita, T. (1960). *Biochim. Biophys. Acta* **37**, 513–520.
Young, D. M., and Potts, J. T. (1963). *J. Biol. Chem.* **238**, 1995–2002.
Young, N. S., Curd, J. G., Eastlake, A., Furie, B., and Schechter, A. N. (1975). *Proc. Natl. Acad. Sci. U.S.A.* **72**, 4759–4763.

Young, N. S., Eastlake, A., and Schechter, A. N. (1976). *J. Biol. Chem.* **251**, 6431–6438.
Yu, I. J., Lippert, J. L., and Peticolas, W. L. (1973). *Biopolymers* **12**, 2161.
Yu, K. T., and Liu, C. S. (1972). *J. Am. Chem. Soc.* **94**, 5127–5128.
Yutani, K., Takaji, T., and Isemura, T. (1965). *J. Biochem.* (*Tokyo*) **57**, 590–597.
Yutani, K., Yutani, A., Imanishi, A., and Isemura, T. (1968). *J. Biochem.* (*Tokyo*) **64**, 449–455.
Zakin, M. M., Garel, J. R., Dautry-Varsat, A., Cohen, G. N., and Boulot, G. (1978), *Biochemistry* **17**, 4318–4323.
Zana, R. (1975). *Biopolymers* **14**, 2425–2428.
Zeppezauer, M., Soderberg, B. O., and Bränden C. I. (1967). *Acta Chem. Scand.* **21**, 1099–1101.
Zerner, B., and Bender, M. L. (1964). *J. Am. Chem. Soc.* **86**, 3669–3674.
Zetina, C. R., and Goldberg, M. E. (1980a). *J. Biol. Chem.* **255**, 4381–4385.
Zetina, C. R., and Goldberg, M. E. (1980b). *J. Mol. Biol.* **137**, 401–414.
Zettlmeissl, G., Rudolph, R., and Jaenicke, R. (1979a). *Biochemistry* **18**, 5567–5571.
Zettmeissl, G., Rudolph, R., and Jaenicke, R. (1979b). *Eur. J. Biochem.* **100**, 593–598.
Zilber, E. (1979). Thèse de Spécialité, Univ. Paris-Sud, Orsay.
Zimm, B. H., and Bragg, J. K. (1959). *J. Chem. Phys.* **31**, 526.
Zimmerman, S. S., and Scheraga, H. A. (1976). *Macromolecules* **9**, 408–416.
Zimmerman, S. S., and Scheraga, H. A. (1977a). *Proc. Natl. Acad. Sci. U.S.A.* **74**, 4126–4129.
Zimmerman, S. S., and Scheraga, H. A. (1977b). *Biopolymers* **16**, 811–843.
Zimmerman, S. S., and Scheraga, H. A. (1978a). *Biopolymers* **17**, 1849–1869.
Zimmerman, S. S., and Scheraga, H. A. (1978b). *Biopolymers* **17**, 1871–1884.
Zimmerman, S. S., and Scheraga, H. A. (1978c). *Biopolymers* **17**, 1885–1890.
Zimmerman, S. S., Pottle, M. S., Némethy, G., and Scheraga, H. A. (1977). *Macromolecules* **10**, 1–9.
Zipser, D. (1963). *J. Mol. Biol.* **7**, 739–751.
Zipser, D., and Perrin, D. (1963). *Cold Spring Harbor Symp. Quant. Biol.* **28**, 533–537.

Index

Molecular Biology

An International Series of Monographs and Textbooks

Editors

HAROLD A. SCHERAGA. Protein Structure. 1961

STUART A. RICE AND MITSURU NAGASAWA. Polyelectrolyte Solutions: A Theoretical Introduction, *with a contribution by Herbert Morawetz*. 1961

SIDNEY UDENFRIEND. Fluorescence Assay in Biology and Medicine. Volume I–1962. Volume II–1969

J. HERBERT TAYLOR (Editor). Molecular Genetics. Part I–1963. Part II–1967. Part III–Chromosome Structure–1979

ARTHUR VEIS. The Macromolecular Chemistry of Gelatin. 1964

M. JOLY. A Physico-chemical Approach to the Denaturation of Proteins. 1965

SYDNEY J. LEACH (Editor). Physical Principles and Techniques of Protein Chemistry. Part A–1969. Part B–1970. Part C–1973

KENDRIC C. SMITH AND PHILIP C. HANAWALT. Molecular Photobiology: Inactivation and Recovery. 1969

RONALD BENTLEY. Molecular Asymmetry in Biology. Volume I–1969. Volume II–1970

JACINTO STEINHARDT AND JACQUELINE A. REYNOLDS. Multiple Equilibria in Protein. 1969

DOUGLAS POLAND AND HAROLD A. SCHERAGA. Theory of Helix-Coil Transitions in Biopolymers. 1970

JOHN R. CANN. Interacting Macromolecules: The Theory and Practice of Their Electrophoresis, Ultracentrifugation, and Chromatography. 1970

WALTER W. WAINIO. The Mammalian Mitochondrial Respiratory Chain. 1970

LAWRENCE I. ROTHFIELD (Editor). Structure and Function of Biological Membranes. 1971

ALAN G. WALTON AND JOHN BLACKWELL. Biopolymers. 1973

WALTER LOVENBERG (Editor). Iron-Sulfur Proteins. Volume I, Biological Properties–1973. Volume II, Molecular Properties–1973. Volume III, Structure and Metabolic Mechanisms–1977

A. J. HOPFINGER. Conformational Properties of Macromolecules. 1973

R. D. B. FRASER AND T. P. MACRAE. Conformation in Fibrous Proteins. 1973

OSAMU HAYAISHI (Editor). Molecular Mechanisms of Oxygen Activation. 1974

FUMIO OOSAWA AND SHO ASAKURA. Thermodynamics of the Polymerization of Protein. 1975

LAWRENCE J. BERLINER (Editor). Spin Labeling: Theory and Applications. Volume I, 1976. Volume II, 1978

T. BLUNDELL AND L. JOHNSON. Protein Crystallography. 1976

HERBERT WEISSBACH AND SIDNEY PESTKA (Editors). Molecular Mechanisms of Protein Biosynthesis. 1977

TERRANCE LEIGHTON AND WILLIAM F. LOOMIS, JR. (Editors). The Molecular Genetics of Development: An Introduction to Recent Research on Experimental Systems. 1980

ROBERT B. FREEDMAN AND HILARY C. HAWKINS (Editors). The Enzymology of Post-Translational Modification of Proteins, Volume 1. 1980

WAI YIU CHEUNG (Editor). Calcium and Cell Function, Volume I: Calmodulin. 1980. Volume II. 1982

OLEG JARDETZKY and G. C. K. ROBERTS. NMR in Molecular Biology. 1981

DAVID A. DUBNAU (Editor). The Molecular Biology of the Bacilli, Volume I: *Bacillus subtilis.* 1982

GORDON G. HAMMES. Enzyme Catalysis and Regulation. 1982

GUNTER KAHL and JOSEF S. SCHELL (Editors). Molecular Biology of Plant Tumors. 1982

P. R. CAREY. Biochemical Applications of Raman and Resonance Raman Spectroscopies. 1982

OSAMU HAYAISHI and KUNIHIRO UEDA (Editors). ADP-Ribosylation Reactions. 1982

G. O. Aspinall. The Polysaccharides, Volume 1. 1982

Charis Ghelis and Jeannine Yon. Protein Folding. 1982

In preparation

Wai Yiu Cheung (Editor). Calcium and Cell Function, Volume III. 1983

Alfred Stracher (Editor). Muscle and Non-Muscle Motility, Volume 1. 1982